MILLIMETER WAVE TECHNOLOGY IN WIRELESS PAN, LAN, AND MAN

WIRELESS NETWORKS AND MOBILE COMMUNICATIONS

Dr. Yan Zhang, Series Editor
Simula Research Laboratory, Norway
E-mail: yanzhang@ieee.org

Unlicensed Mobile Access Technology: Protocols, Architectures, Security, Standards and Applications
Yan Zhang, Laurence T. Yang and Jianhua Ma
ISBN: 1-4200-5537-2

Wireless Quality-of-Service: Techniques, Standards and Applications
Maode Ma, Mieso K. Denko and Yan Zhang
ISBN: 1-4200-5130-X

Broadband Mobile Multimedia: Techniques and Applications
Yan Zhang, Shiwen Mao, Laurence T. Yang and Thomas M Chen
ISBN: 1-4200-5184-9

The Internet of Things: From RFID to the Next-Generation Pervasive Networked Systems
Lu Yan, Yan Zhang, Laurence T. Yang and Huansheng Ning
ISBN: 1-4200-5281-0

Millimeter Wave Technology in Wireless PAN, LAN, and MAN
Shao-Qiu Xiao, Ming-Tuo Zhou and Yan Zhang
ISBN: 0-8493-8227-0

Security in Wireless Mesh Networks
Yan Zhang, Jun Zheng and Honglin Hu
ISBN: 0-8493-8250-5

Resource, Mobility and Security Management in Wireless Networks and Mobile Communications
Yan Zhang, Honglin Hu, and Masayuki Fujise
ISBN: 0-8493-8036-7

Wireless Mesh Networking: Architectures, Protocols and Standards
Yan Zhang, Jijun Luo and Honglin Hu
ISBN: 0-8493-7399-9

Mobile WIMAX: Toward Broadband Wireless Metropolitan Area Networks
Yan Zhang and Hsiao-Hwa Chen
ISBN: 0-8493-2624-9

Distributed Antenna Systems: Open Architecture for Future Wireless Communications
Honglin Hu, Yan Zhang and Jijun Luo
ISBN: 1-4200-4288-2

AUERBACH PUBLICATIONS
www.auerbach-publications.com
To Order Call: 1-800-272-7737 • Fax: 1-800-374-3401
E-mail: orders@crcpress.com

MILLIMETER WAVE TECHNOLOGY IN WIRELESS PAN, LAN, AND MAN

Edited by
Shao-Qiu Xiao
Ming-Tuo Zhou
Yan Zhang

CRC Press is an imprint of the
Taylor & Francis Group, an **informa** business
AN AUERBACH BOOK

CRC Press
Taylor & Francis Group
6000 Broken Sound Parkway NW, Suite 300
Boca Raton, FL 33487-2742

First issued in paperback 2019

ISBN-13: 978-0-8493-8227-7 (hbk)
ISBN-13: 978-0-367-38727-3 (pbk)

Library of Congress Cataloging-in-Publication Data

Xiao, Shao-Qiu.
Millimeter wave technology in wireless Pan, Lan, and Man / Shao-Qiu Xiao, Ming-Tuo Zhou, and Yan Zhang.
p. cm.
Includes bibliographical references and index.
ISBN-13: 978-0-8493-8227-7 (alk. paper) 1. Broadband communication systems. 2. Wireless communication systems. 3. Millimeter waves. I. Zhou, Ming-Tuo. II. Zhang, Yan, 1977- III. Title.

TK5103.4.X53 2007
004.6'8--dc22 2007011241

Visit the Taylor & Francis Web site at
http://www.taylorandfrancis.com

and the CRC Press Web site at
http://www.crcpress.com

Contents

List of Contributors

Dan An
Millimeter-wave INnovation Technology Research Center
Department of Electronic Engineering
Dongguk University
Seoul, Korea

Pantel D. M. Arapoglou
National Technical University of Athens
Athens, Greece

T. Chen
CREATE-NET International Research Center
Trento, Italy

I. Chlamtac
CREATE-NET International Research Center
Trento, Italy

Panayotis G. Cottis
National Technical University of Athens
Athens, Greece

X. James Dong
University of California
Berkeley, California

James P. K. Gilb
SiBEAM
Fremont, California

Joerg Habetha
Philips Research Laboratories
Aachen, Germany

Sung-Chan Kim
Millimeter-wave INnovation Technology Research Center
Department of Electronic Engineering
Dongguk University
Seoul, Korea

Andrey Kobyakov
Corning Inc.
Corning, New York

J. Laskar
Georgia Institute of Technology
Atlanta, Georgia

Bok-Hyung Lee
Millimeter-wave INnovation Technology Research Center
Department of Electronic Engineering
Dongguk University
Seoul, Korea

J.-H. Lee
Georgia Institute of Technology
Atlanta, Georgia

Sheung L. Li
SiBEAM
Fremont, California

Huilai Liu
University of Electronics Science and Technology of China
Chengdu, China

Jim Misener
University of California
Berkeley, California

John Mitchell
University College London
London, U.K.

Athanasios D. Panagopoulos
National Technical University of Athens
Athens, Greece

Javier del Pardo
Philips Semiconductors
Sophia Antipolis, France

S. Pinel
Georgia Institute of Technology
Atlanta, Georgia

Emma Regentova
University of Nevada
Las Vegas, Nevada

Jin-Koo Rhee
Millimeter-wave INnovation Technology Research Center
Department of Electronic Engineering
Dongguk University
Seoul, Korea

Michael Sauer
Corning Inc.
Corning, New York

M. M. Tentzeris
Georgia Institute of Technology
Atlanta, Georgia

Pravin Varaiya
University of California
Berkeley, California

Bing-Zhong Wang
University of Electronics Science and Technology of China
Chengdu, China

Jianpeng Wang
University of Electronics, Science and Technology of China
Chengdu, China

H. Woesner
CREATE-NET International Research Center
Trento, Italy

Shao-Qiu Xiao
University of Electronics, Science and Technology of China
Chengdu, China

Felix Yanovsky
The National Aviation University
Kiev, Ukraine

Jijun Yao
National University of Singapore
Singapore

Wenbing Zhang
University of California
Berkeley, California

Yan Zhang
Simula Research Laboratory
Lysaker, Norway

Jun Zheng
Queens College
City University of New York
New York, New York

Ming-Tuo Zhou
National Institute of Information
and Communications Technology
Singapore

Chapter 1

Millimeter-Wave Monolithic Integrated Circuit for Wireless LAN

Jin-Koo Rhee, Dan An, Sung-Chan Kim, and Bok-Hyung Lee

Contents

1.1 Introduction

The increase in high-performance personal computers (PCs) and multimedia equipment in offices and homes requires high-speed and broadband wireless data transmission. This requirement makes wireless indoor communication systems such as wireless local area networks (WLAN) and personal area networks (PAN) because of their portable convenience. For such short-range indoor broadband WLAN and PAN systems, the millimeter-wave band offers significant advantages in supplying enough bandwidth for the transmission of various multimedia content. In particular, there has been an increasing requirement for the development of the V-band WLAN for commercial applications. The frequency of 60 GHz is very useful for short-distance wireless communications due to the strong absorption characteristic by oxygen in the atmosphere. Therefore, frequency efficiency is improved compared to other frequency bands. In the last few decades, many research groups in the world have developed millimeter-wave LAN systems. For example, Communications Research Laboratory (CRL, now NiCT) took up the project of developing indoor WLAN systems using millimeter waves in 1992 [1]. The final goal of the millimeter-wave WLAN systems is to provide point-to-multipoint access with transmission rates higher than 100 Mbps for the connectivity of broadband integrated services digital networks (B-ISDNs) or conventional fast Ethernet. For these millimeter-wave WLAN applications, we have to solve some problems. First of all, we have to reduce the size and cost of the systems. The millimeter-wave systems are generally fabricated using HIC (hybrid integrated circuit) technology, causing a large system. MMIC (monolithic millimeter-wave integrated circuit) technology is regarded as an alternative to HIC technology due to its ability to integrate active with passive elements on a single semiconductor substrate [2–6]. The MMIC has advantages, such as small size, high reliability, high productivity, and low cost due to using semiconductor technologies compared to the conventional HIC. The main objective of this chapter is to discuss the MMIC technology and its applications for millimeter-wave wireless LAN. First, millimeter-wave WLAN will be introduced. Then, the modeling of active and passive devices will be described. The design and fabrication technologies of millimeter-wave circuits are presented. Finally, millimeter-wave monolithic circuits for WLAN applications are explained.

1.2 Millimeter-Wave Wireless Local Area Network

Recently, a broadband and high-speed indoor network for office and home environments has been required. Additionally, microwave frequency bands have been saturated and there is growing necessity to exploit new frequency bands that have not yet been utilized for commercial applications. For this reason, utilization of the millimeter-wave band has been recommended, and much research has been devoted to developing millimeter-wave wireless LAN. Advantages of millimeter-wave communication are very wide frequency band, high-speed transmission, and radiated power limitation for unlicensed use. Therefore, millimeter-wave WLAN can be utilized in short-range communication and indoor networks. In particular, a 60-GHz band is very useful for wireless short-distance communications due to strong absorption by oxygen. Thus, compared to other frequency bands, frequency efficiency is improved. Figure 1.1 shows the atmospheric absorption versus frequency [7]. Features to be noted are:

1. Good coexistence between millimeter-wave system and 802.11a/b/g & Bluetooth due to large frequency difference
2. Higher speed transmission, more than 1 Gbps
3. Exploitation of antenna directivity
4. Simple modulation/demodulation
5. Simple signal processing

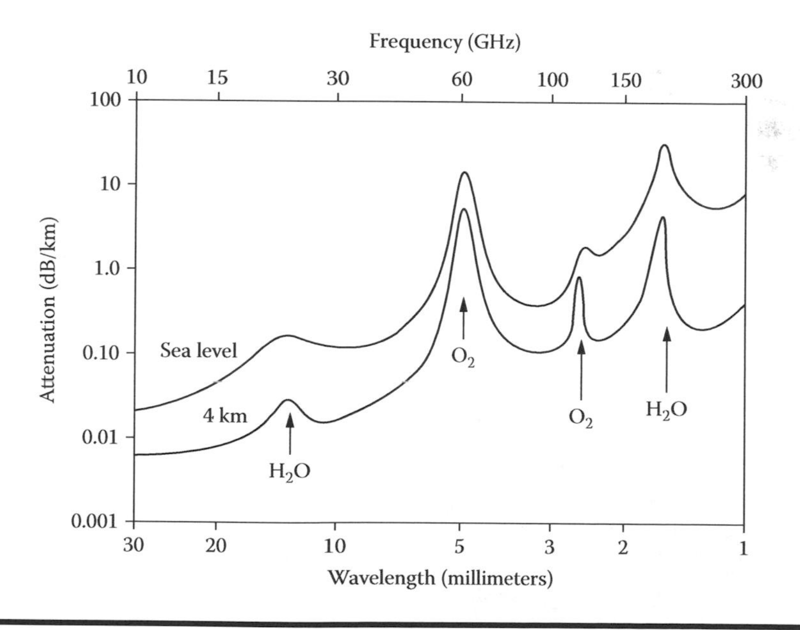

Figure 1.1 Average atmospheric absorption of frequency.

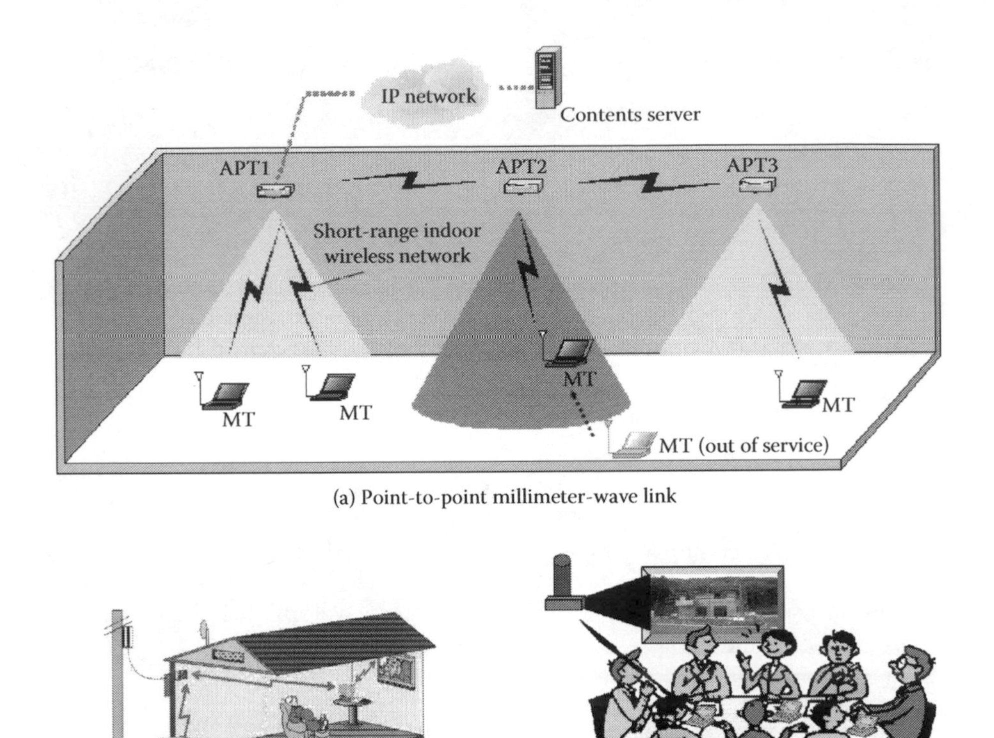

(a) Point-to-point millimeter-wave link

(b) Home utilization

(c) Office utilization

Figure 1.2 Utilizations of millimeter-wave wireless LAN.

Millimeter-wave WLAN is possible for wireless networks such as multimedia equipment, home appliances, videosignal transmissions, and personal computers. Utilization of millimeter-wave WLAN is explained in Figure 1.2. A millimeter-wave circuit and system have been developed using a waveguide module, hybrid integrated circuit method, resulting in large size, high cost, and low productivity. Use of these systems in wireless LAN has many problems due to small size and low cost in wireless LAN. To overcome these problems, high-speed devices such as high electron mobility transistors (HEMTs) and MMTCs need to be added to the millimeter-wave wireless LAN. Figure 1.3 shows normal millimeter-wave WLAN and millimeter-wave circuit components. Millimeter-wave components are usually composed of a low-noise amplifier (LNA), a power amplifier, an oscillator, and an up/down mixer. Also, passive components such as a filter, an antenna, a coupler, and a circulator are required. These components can be varied with system architecture. Millimeter-wave circuits must be

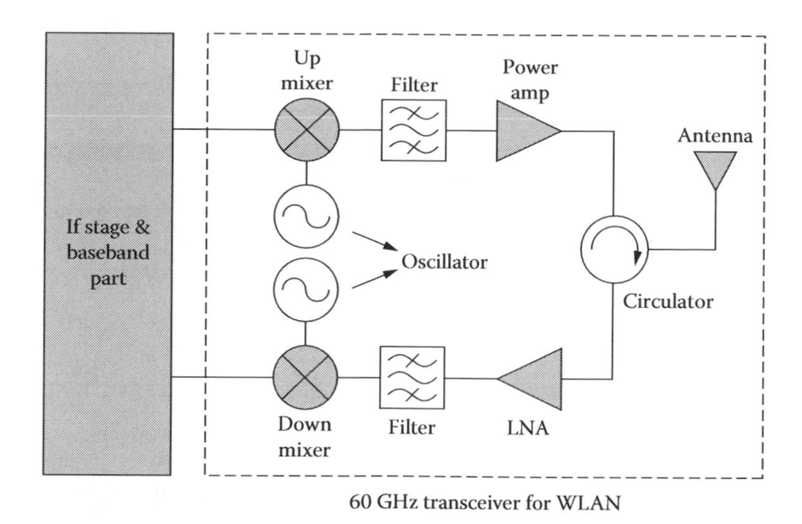

Figure 1.3 Millimeter-wave WLAN and millimeter-wave circuit components (IF, intermediate frequency).

realized high-speed operating characteristics, high linearity, small size, and low cost. For this reason, many technologies such as device modeling, circuit design, circuit fabrication, and measurement technology need to be developed for MMICs.

1.3 Millimeter-Wave Monolithic Integrated Circuit Technology

1.3.1 Millimeter-Wave Device and Modeling

1.3.1.1 Millimeter-Wave HEMT Technology

To fabricate the MMIC devices for the 60-GHz wireless LAN system, the development of active devices that have a high frequency, low noise, and high power performance is an essential theme. Since introduced in 1981 [8], the AlGaAs/GaAs HEMT has been widely used for the microwave region hybrid and monolithic circuits. However, the frequency performances of the conventional AlGaAs/GaAs HEMT devices cannot satisfy the millimeter-wave region (30–300 GHz) MMIC applications. In order to obtain the high-frequency performance, a Pseudomorphic HEMT (PHEMT), which has a relatively low energy band gap characteristic for higher conduction band offset, has been developed. The PHEMT epitaxial structure is shown in Figure 1.4. The epitaxial structure of the device consists of the following layers: an 500 nm GaAs buffer layer, an 18.5/1.5 nm AlGaAs/GaAs × 10 super-lattice buffer layer, a silicon planar doped layer ($1 \times 10^{12}/\text{cm}^2$), a 6-nm

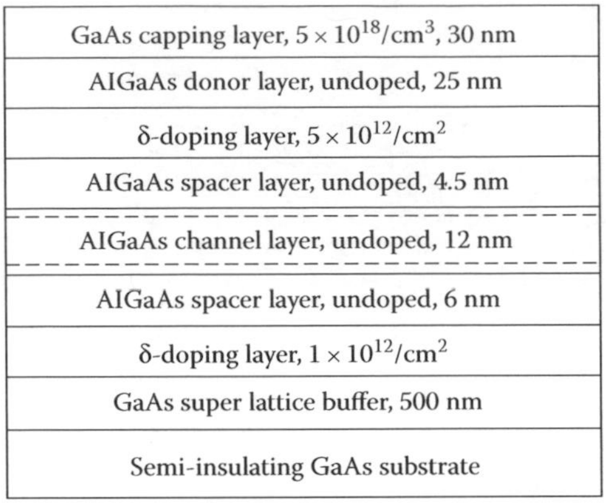

GaAs capping layer, $5 \times 10^{18}/cm^3$, 30 nm
AlGaAs donor layer, undoped, 25 nm
δ-doping layer, $5 \times 10^{12}/cm^2$
AlGaAs spacer layer, undoped, 4.5 nm
AlGaAs channel layer, undoped, 12 nm
AlGaAs spacer layer, undoped, 6 nm
δ-doping layer, $1 \times 10^{12}/cm^2$
GaAs super lattice buffer, 500 nm
Semi-insulating GaAs substrate

Figure 1.4 The epitaxial structure for PHEMT fabrication. (Copyright 2004. With permission from Elsevier.)

AlGaAs lower spacer layer, a 12-nm InGaAs channel layer, a 4-nm AlGaAs upper spacer layer, a silicon planar doped layer ($5 \times 10^{12}/cm^2$), a 25-nm AlGaAs donor layer, and a 30-nm GaAs cap layer [9–10]. In this chapter, the DC characteristics of the 70 μm × 2 PHEMTs were measured by an HP 4156A DC parameter analyzer. The obtained DC performances show a knee voltage (Vk) of 0.6 V, a pinch-off voltage (Vp) of –1.5 V, a drain-source saturation current (Idss) density of 384 mA/mm and a maximum extrinsic transconductance of 367.9 mS/mm, as shown in Figure 1.5. Radio frequency (RF) characteristics of the PHEMTs were examined by an HP 8510C vector network analyzer. The measurement of S-parameters was performed in a frequency range of 1–50 GHz. For this RF measurement, the drain and gate bias conditions of 2 V and –0.6 V were used. We obtained a current gain cut-off frequency (ft) of 113 GHz, a maximum frequency of oscillation (fmax) of 180 GHz, and a measured S21 gain of 3.9 dB at 50 GHz, as shown in Figure 1.6. HEMTs on InP substrates have demonstrated superior microwave and low-noise performances compared to PHEMTs on GaAs substrates. The excellent device performances of the InP-based HEMTs operating in the millimeter-wave region is mostly based on the InGaAs/InAlAs/InP material system. However, compared to GaAs-based wafers, InP-based wafers have some critical drawbacks, such as the mechanical fragility of the wafers and the higher material cost. Moreover, InP-based HEMTs are not quite proper for large-scale production because the backside etching rate for the InP material is much slower. In recent decades, active research has been done on GaAs-based metamorphic HEMTs (MHEMTs) to address the needs for both high microwave performance and low device cost [11–13]. The use of metamorphic buffers on GaAs substrates was introduced to accommodate the lattice mismatch between the substrate and the active layers, as

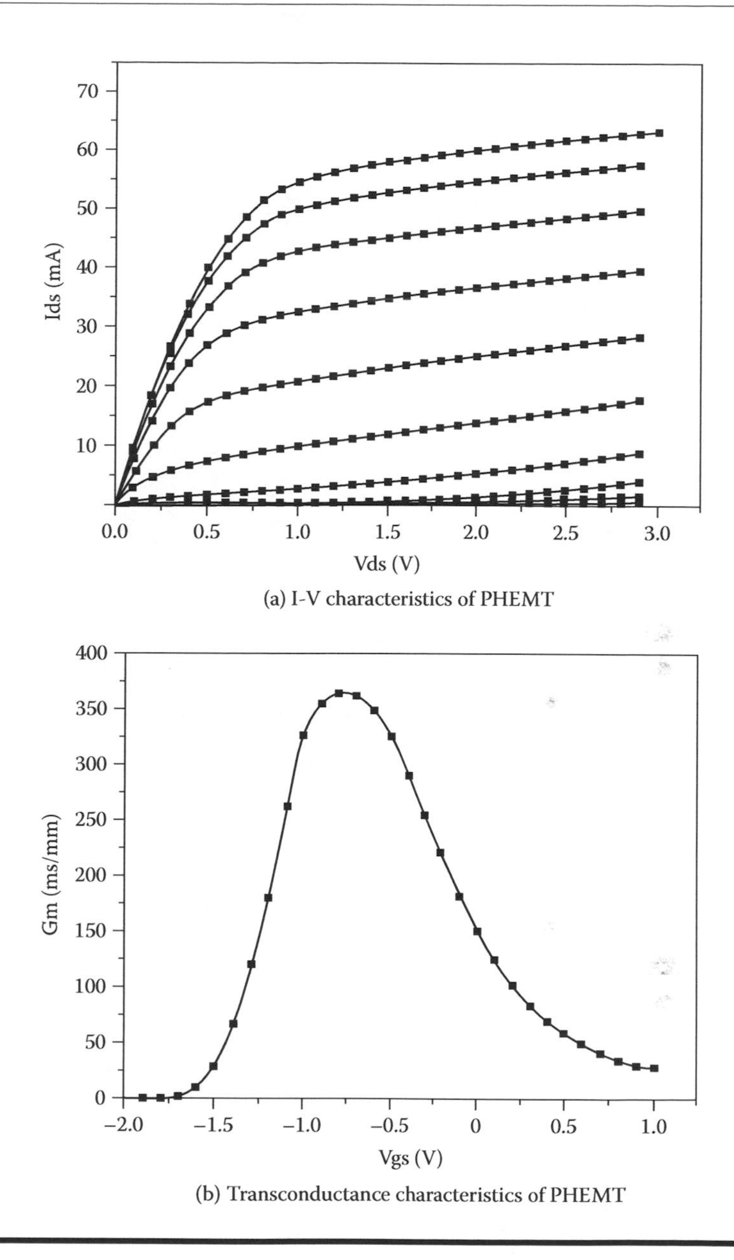

Figure 1.5 DC characteristics of GaAs pseudomorphic HEMT (gate length: 0.1 μm, total gate width: 140 μm). (Copyright 2004. With permission from Elsevier.)

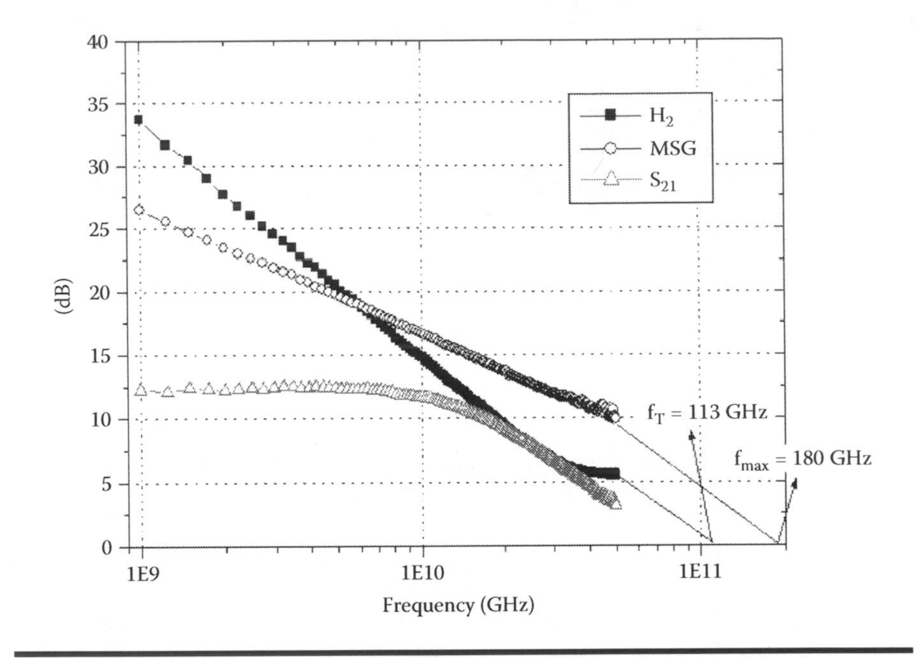

Figure 1.6 The RF characteristics of pseudomorphic HEMT (gate length: 0.1 μm, total gate width: 140 μm). (Copyright 2004. With permission from Elsevier.)

well as to avoid the InP substrates. By using the metamorphic buffers, unstrained InGaAs/InAlAs hetero-structures could be grown over a wide range of indium (In) contents for the InGaAs channels, thereby exhibiting device performances comparable to those of InP-based HEMTs. In Figure 1.7, the device active layers of MHEMT are grown on a strain-relaxed, compositionally graded, metamorphic buffer layer. The buffer layer provides the ability to adjust the lattice constant to any indium content channel desired, and therefore allows the device designer an additional degree of freedom to optimize the transistors for high frequency gain, power, linearity, and low noise and to trap dislocations and prevent them from propagating into the device channel. The DC and transfer characteristics of the MHEMT were measured using an HP 4156A semiconductor parameter analyzer. As shown in Figure 1.8, a pinch-off voltage of –1.5 V and a drain saturation current of 96 mA were measured at a gate voltage (Vgs) of 0 V. The fabricated MHEMT also showed a maximum transconductance of 760 mS/mm at a Vgs of –0.3 V and a drain voltage (Vds) of 1.8 V. The S-parameters of the MHEMTs were measured using an ME7808A Vector Network Analyzer in a frequency range from 0.04 to 110 GHz. In Figure 1.9, we showed the measured S21, h21, and maximum stable gain (MSG) of the MHEMT. The measured S21, ft, and fmax of the MHEMT were 6 dB (at 110 GHz), 195 GHz, and 391 GHz, respectively.

InGaAs capping layer, $6 \times 10^{18}/cm^3$, 15 nm
InAIAs donor layer, undoped, 15 nm
δ-doping layer, $4.5 \times 10^{12}/cm^2$
InAIAs spacer layer, undoped, 3 nm
InGaAs channel layer, undoped, 23 nm
InAIAs spacer layer, undoped, 4 nm
δ-doping layer, $1.3 \times 10^{12}/cm^2$
InAIAs buffer layer, undoped, 400 nm
$In_xAI_{1-x}As$ (x = 0~0.5) Metamorphic buffer, undoped, 1000 nm
Semi-insulating GaAs substrate

Figure 1.7 The epitaxial structure for MHEMT fabrication.

1.3.1.2 Millimeter-Wave Active Device Modeling

A small-signal equivalent model is widely used to analyze the characteristics of gains and noises of active devices. These models provide a vital connection between measured S-parameters and the electrical operating characteristics of the device. A more precise small-signal model provides some of the most important information in designing the devices and circuits as well. The components in the small-signal equivalent circuit are composed of a lumped element approximation to some aspect of the device physics. The small-signal equivalent model consists of both extrinsic and intrinsic elements. In general, the operating characteristics of devices may be mainly decided by the intrinsic elements such as Cgs, Cgd, Cds, gm, gds, and Ri. The values of intrinsic elements may be varied by the values of extrinsic elements. Therefore, through the precise extraction of the values of extrinsic elements, a more precise small-signal equivalent model may be formed. The small-signal parameter extraction of the HEMT is very useful for the device modeling and analysis in the design of millimeter-wave circuits. The small-signal equivalent circuit of an HEMT is shown in Figure 1.10. Basically, this equivalent circuit can be divided into two parts: (1) the intrinsic elements gm, Rds, Cgs, Cgd, Cds, Ri, and which are functions of the biasing conditions; and (2) the extrinsic elements Lg, Ld, Ls, Rg, Rd, Rs, Cpg, and Cpd, which are independent of the biasing conditions.

The extraction method of small-signal parameters has usually used the Dambrine method [14]. Figure 1.11 and Table 1.1 show the procedure of small-signal extraction and the extracted small-signal parameters of normal GaAs HEMTs, respectively. A small-signal model depicts linear characteristics of the device, while a large-signal model expresses nonlinear

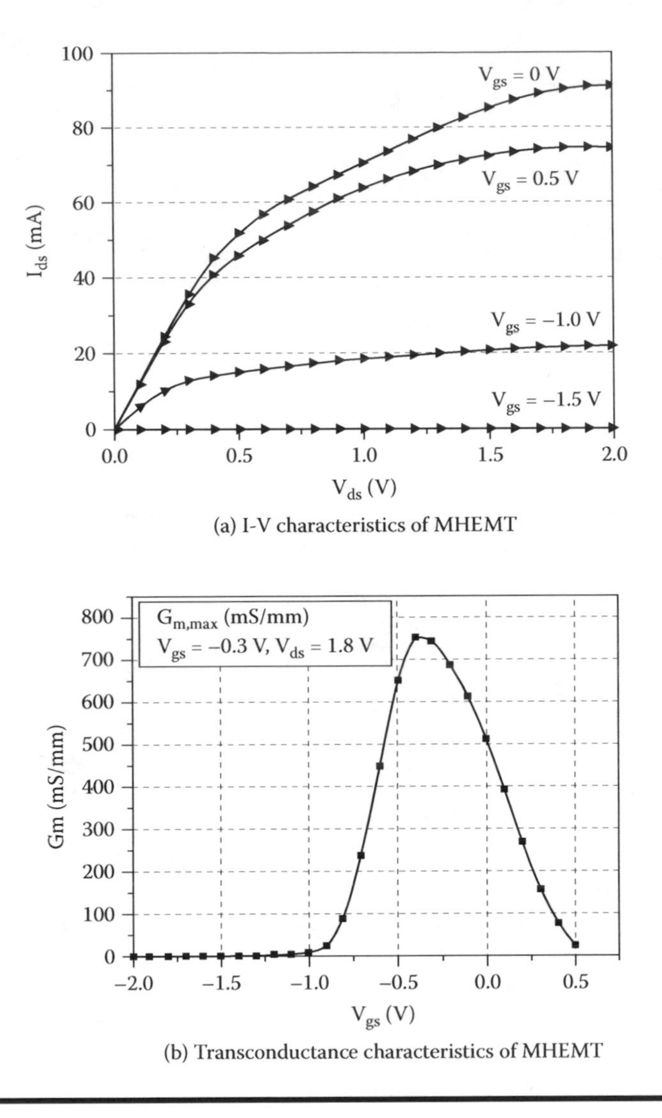

Figure 1.8 DC characteristics of GaAs metamorphic HEMT (gate length: 0.1 μm, total gate width: 140 μm).

characteristics of the device. Analytical large-signal models approximate the nonlinear properties of an active device using a unique set of analytical equations. These nonlinear characteristics can be related to elements of the large-signal equivalent circuit. The large-signal parameters and the equivalent circuit are shown in Table 1.2 and Figure 1.12 respectively [15]. The large-signal model of the device can be used to analyze the performance of the nonlinear components, such as a power amplifier, an oscillator, a mixer, etc. Several studies have developed a large-signal model that

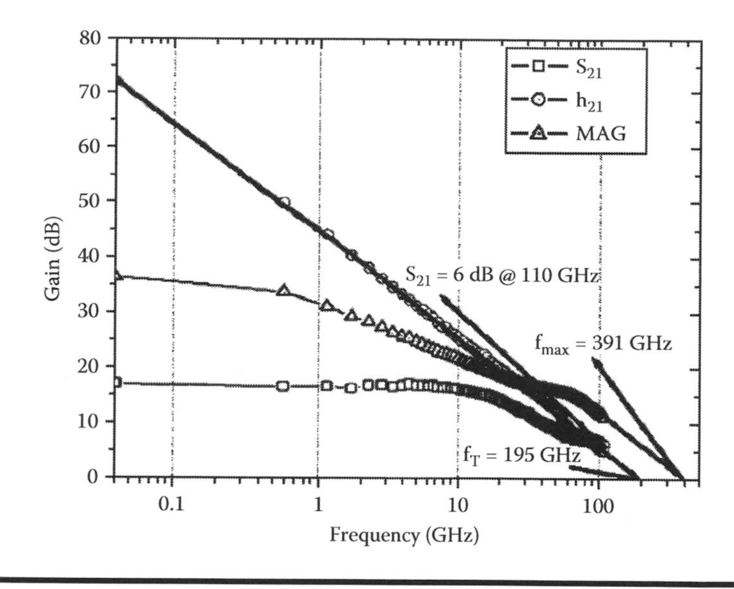

Figure 1.9 The RF characteristics of metamorphic HEMT (gate length: 0.1 μm, total gate width: 140 μm).

describes the nonlinear characteristics of a device. These models include the Curtice model, Statz model, TriQuint Own Model (TOM), Root, etc. The large-signal modeling method is generally following procedures and is shown in Figure 1.13: (1) device measurement (DC, RF characteristics), (2) extraction of parameter set using analytical equations, (3) a simulation

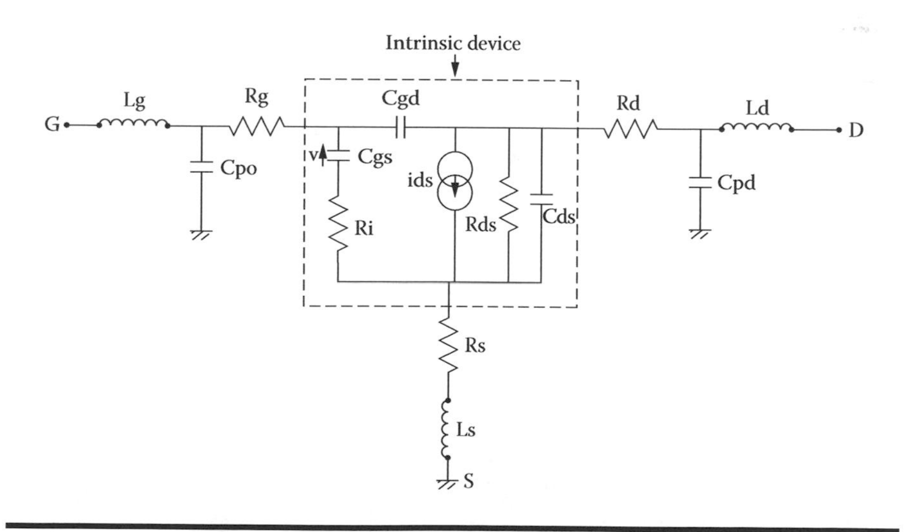

Figure 1.10 A small-signal equivalent circuit of HEMT.

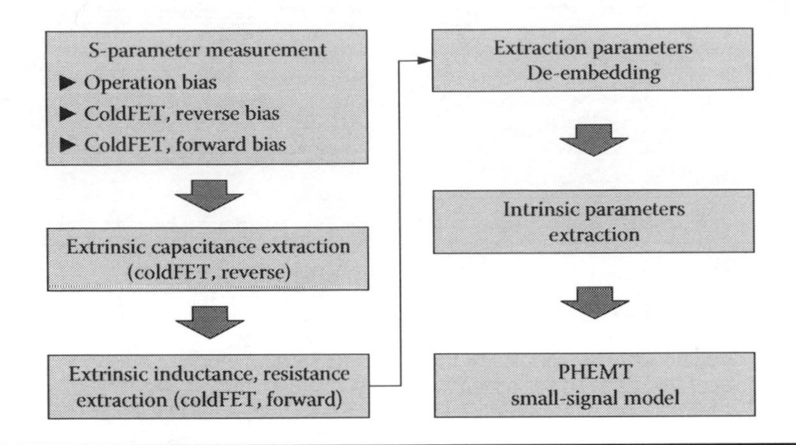

Figure 1.11 The procedure of small-signal modeling.

Table 1.1 The Extracted Small-Signal Parameters of GaAs, PHEMTs, and MHEMTs (gate length: 0.1 μm; total gate width: 140 μm)

Extrinsic Parameter	*Value*	*Intrinsic Parameter*	*Value*
(a) GaAs Pseudomorphic HEMT			
R_g [Ω]	1.980	C_{gs} [pF]	0.178
R_d [Ω]	6.900	C_{gd} [pF]	0.010
R_s [Ω]	3.120	C_{ds} [pF]	0.012
L_g [nH]	0.085	R_{ds} [Ω]	782.2
L_d [nH]	0.140	G_m [mS]	58.59
L_s [nH]	0.011	τ [psec]	1.190
C_{pg} [pF[	0.060	R_i [Ω]	2.310
C_{pd} [pF]	0.040		
(b) GaAs Metamorphic HEMT			
R_g [Ω]	2.560	C_{gs} [pF]	0.062
R_d [Ω]	6.730	C_{gd} [pF]	0.005
R_s [Ω]	2.600	C_{ds} [pF]	0.003
L_g [nH]	0.210	R_{ds} [Ω]	1143
L_d [nH]	0.094	G_m [mS]	83.73
L_s [nH]	0.006	τ [psec]	1.030
C_{pg} [pF[	0.054	R_i [Ω]	1.720
C_{pd} [pF]	0.052		

Table 1.2 The Large-Signal Parameters of GaAs PHEMTs and MHEMTs (gate length: 0.1 μm; total gate width: 140 μm; EEHEMT1 model)

Parameter	*Description*	*PHEMT (Unit)*	*MHEMT (Unit)*
Vto	Zero-bias threshold parameter	−1.679 (V)	−1.237 (V)
Vgo	Gate-source voltage where transconductance is maximum	−634.2 (mV)	−687.0 (mV)
Gm, max	Peak transconductance parameter	46.83 (mS)	76.21 (mS)
Vsat	Drain-source current saturation parameter	973.0 (mV)	193.2.0 (mV)
Kappa	Output conductance parameter	0.013 (1/V)	0.025 (1/V)
C11o	Maximum input capacitance for Vds = Vdso and Vdso > Deltds	129.2 (fF)	72.4(fF)
C11th	Minimum input capacitance for Vds = Vdso	184.9 (aF)	67.57 (aF)
Vinfl	Inflection point in C11-Vgs characteristic	−1.120 (V)	−1.958 (V)
Deltgs	C11th to C11o transition voltage	11.831 (V)	12.071 (V)
Deltds	Linear region to saturation region transition parameter	1.103 (V)	0.160 (V)
Lambda	C11-Vds characteristics slope parameter	0.348 (1/V)	0.696 (1/V)
C12sat	Input transcapacitance for Vgs = Vinfl and Vds > Deltds	20.46 (fF)	5.934 (fF)
Cgdsat	Gate-drain capacitance for Vds >Deltds	15.40 (fF)	21.77(fF)
Cdso	Drain-source inter-electrode capacitance	15.20 (fF)	20.48 (aF)
Rdb	Dispersion source output impedance	1,000 (Gohms)	1.300 (Gohms)
Cds	Dispersion source capacitance	160.0 (fF)	183.0 (fF)
Gdbm	Additional d—b branch conductance at Vds = Vdsm	51.31 (uS)	50.87 (uS)
Gmmaxac	Peak transconductance parameter (AC)	42.57 (mS)	63.81 (mS)
Vtoac	Zero-bias threshold parameter (AC)	−1.703 (V)	−2.376 (V)
Gammaac	Vds-dependent threshold parameter (AC)	35.14 (mS)	43.63 (mS)
Kappaac	Output conductance parameter (AC)	65.13u (1/V)	67.25u (1/V)
Peffac	Channel-to-backside self-heating parameter (AC)	15.75 (W)	29.56 (W)
Vco	Voltage where transconductance compression begins	−660.2 (mV)	−1.071 (mV)
Mu	Parameter that adds Vds dependence to transconductance compression	0.0024 (dimensionless)	0.0043 (dimensionless)
Vba	Transconductance compression "tail-off" parameter	1.001 (V)	1.331 (V)
Vbc	Transconductance roll-off to tail-off transition voltage	701.4 (mV)	849.8 (mV)
Deltgm	Slope of transconductance compression characteristic	267.2 (mS/V)	51.11 (mS/V)
Alpha	Transconductance saturation-to-compression transition parameter	2.094 (V)	1.033 (V)

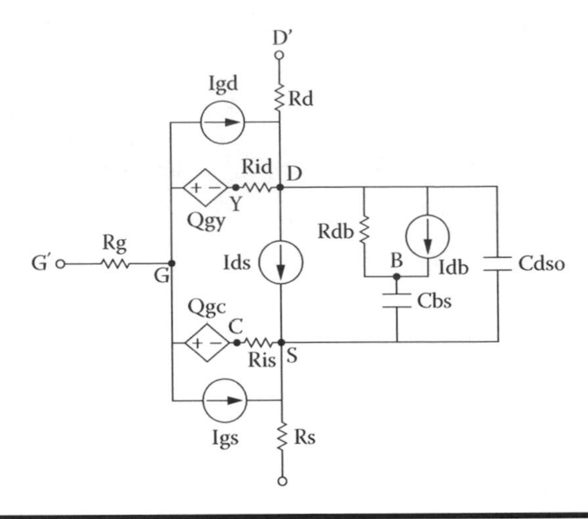

Figure 1.12 The equivalent circuit of an HEMT large-signal model (EEHEMT1 model).

using extracted parameter, (4) a comparison of measured characteristics and simulated characteristics, (5) an optimization of nonlinear. Figure 1.14 shows the comparison of measured and simulated results of a large-signal model (0.1 GaAs MHEMT) in frequency range from 1 to 110 GHz at the gate voltage of –1.0 V and the drain voltage of 1.8 V.

1.3.1.3 Millimeter-Wave Passive Device Modeling

In MMIC design, passive components are used for impedance matching, DC biasing, phase-shifting, and many other functions. These elements include not only the distributed transmission lines such as a coplanar waveguide

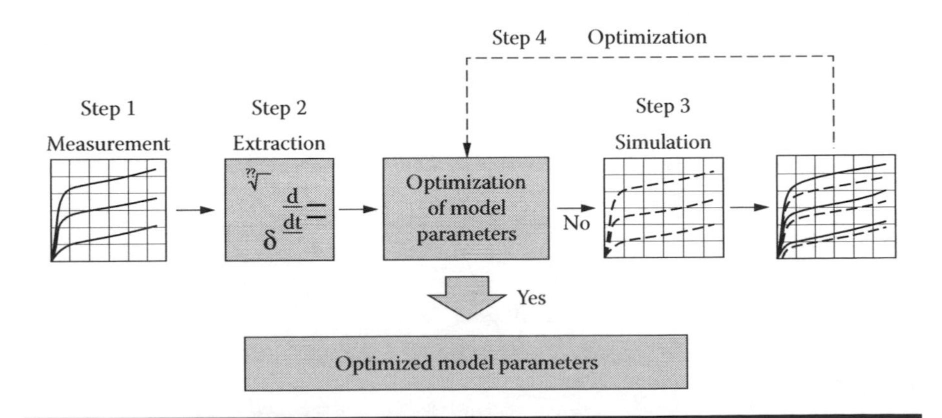

Figure 1.13 The procedure of nonlinear large-signal modeling.

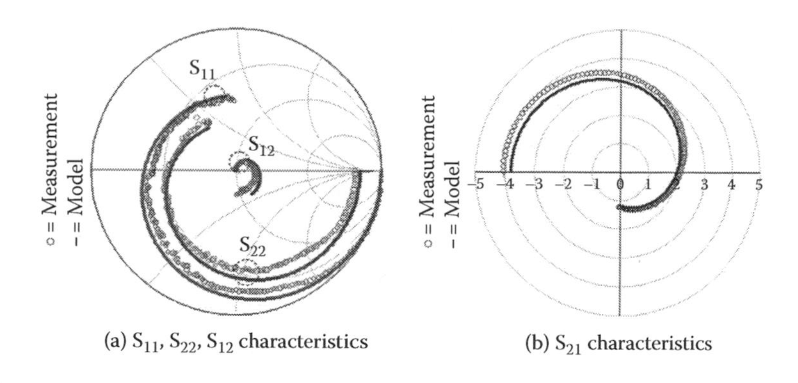

Figure 1.14 The modeling result of GaAs MHEMT (gate length: 0.1 μm; total gate width: 140 μm, gate voltage: –1.0 V; drain voltage: 1.8 V; —: model; o: measurement; frequency range: 1 ~ 110 GHz).

(CPW) and a microstrip line but also the lumped capacitors and resistors. In millimeter-wave range, the distributed transmission lines have been mainly used because of low resonance frequency of lumped elements. In this chapter, the CPW structure was employed, because it has many advantages over the microstrip line structure in millimeter wave. It is well known that CPW-based MMIC processes may be cheaper than microstrip-based processes with holes and have high yield because backside processes are not needed [16–20]. Furthermore, CPW structure increases the packing density of the circuit and reduces the substrate dispersion characteristics for millimeter-wave operation. Figure 1.15 shows the CPW structure on GaAs substrate. The CPW models include curves, T junctions, and cross junctions as well as common elements, as shown in Figure 1.16. For simulation of the designed CPW, we used commercial software such as LineCalc of ADS from Agilent Incorporated. Figure 1.17 explains the extraction technology of passive device modeling. The passive libraries were finally completed after optimization by comparing the measured S-parameters with the momentum-simulated S-parameters. Metal–insulator–metal (MIM) capacitors and the

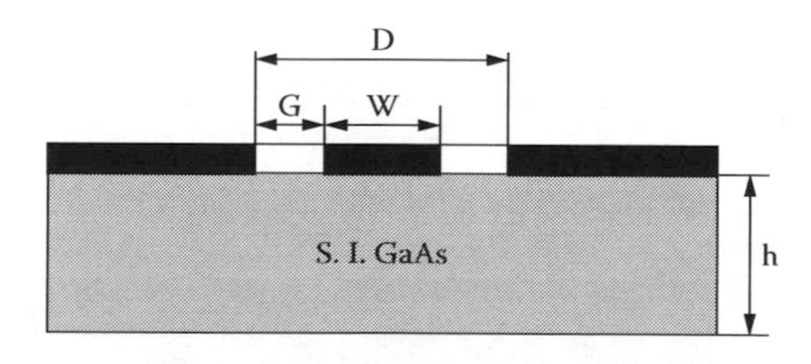

Figure 1.15 The coplanar waveguide (CPW) structure.

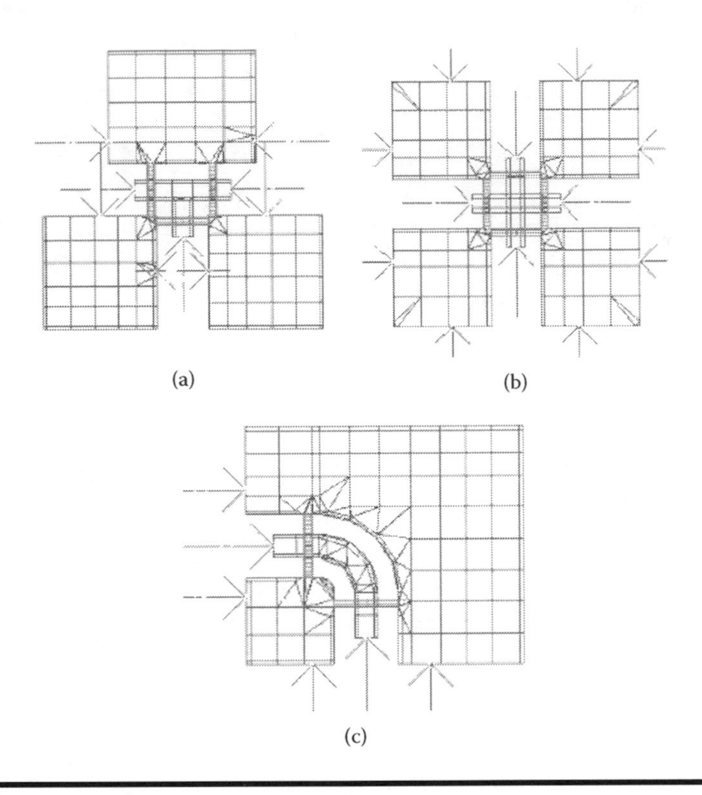

Figure 1.16 The CPW discontinuity patterns for EM analysis: (a) CPW "*Tee*," (b) CPW cross, and (c) CPW curve.

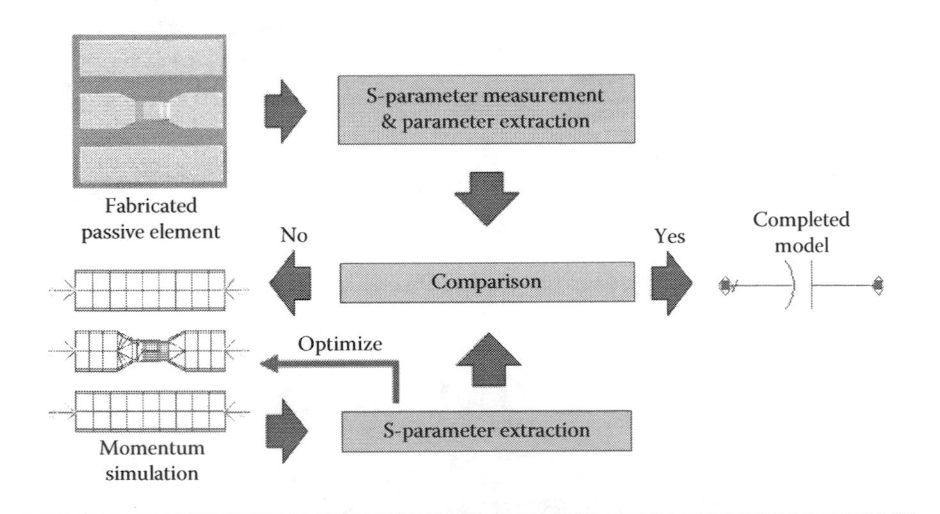

Figure 1.17 The extraction technology of the passive model.

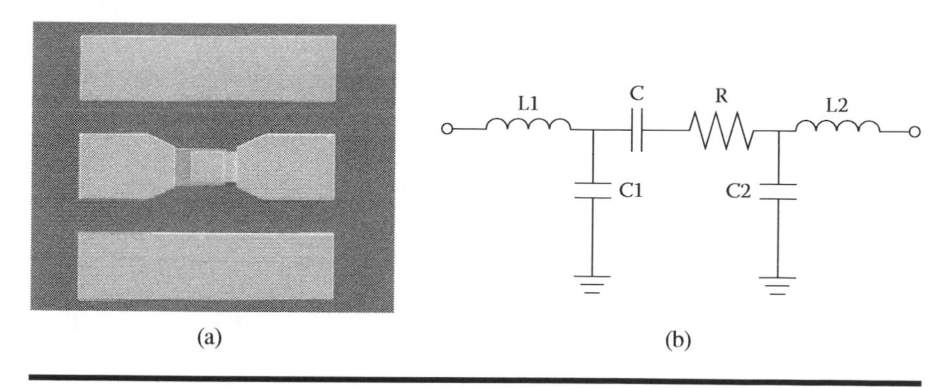

Figure 1.18 The fabricated metal–insulator–metal (MIM) capacitor: (a) the micrograph of the MIM capacitor (capacitor size: 45 μm × 45μm), and (b) the equivalent circuit of the MIM capacitor.

Ti thin-film resistors were fabricated and modeled. Capacitors are used in MMICs for blocking and bypass purposes. Due to the high dielectric constant (r = 5.5–7.5) and breakdown field (> 106 V/cm), the Si3N4 is mostly preferred as a dielectric layer for MIM capacitors [21]. An Si3N4 film was deposited using the PECVD (plasma-enhanced chemical vapor deposition) system for the MIM capacitor. The connection between the top capacitor plate and adjacent metallization is an airbridge connection in order to avoid problems caused by edges and slopes. The SEM photography of the fabricated MIM capacitor is shown in Figure 1.18(a), and the applied lumped equivalent circuit is shown in Figure 1.18(b). Resistors are used in MMICs for several purposes including feedback, isolation, terminations, and voltage dividers (or self-biasing) in a bias network. Ti thin-film resistors allow precise control of resistance due to their small sheet resistances and have a large current capacity per unit width compared to resistors composed of the active GaAs material [22]. The SEM photography of the fabricated thin-film resistor is shown in Figure 1.19a, and the applied lumped equivalent circuit is shown in Figure 1.19b.

1.3.2 Design Technology of Millimeter-Wave Monolithic Circuits

In the field of a millimeter-wave monolithic circuit, more detailed design technologies are needed than for a microwave monolithic circuit. These needs can be met only by more accurate circuit simulation both for chip yield and electrical performance due to high frequency. Although an electromagnetic (EM) field analysis was used in millimeter wave, the efficient design method is required because of reduction of simulation time and stable design. The efficient design technology of a millimeter-wave monolithic

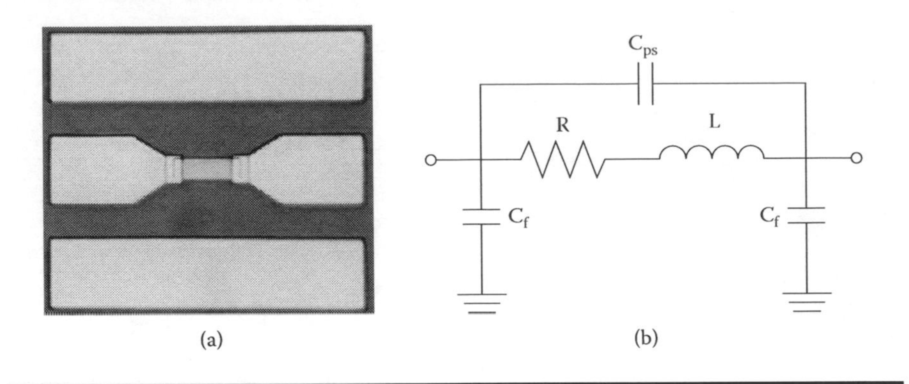

Figure 1.19 The fabricated Ti thin-film resistor: (a) the micrograph of the Ti thin-film resistor (length of 60 and width of 30) and (b) the equivalent circuit.

circuit is as follows: (1) perform modeling of active and passive devices, (2) determine the circuit topology, (3) perform initial design using circuit simulation based on CAD, (4) perform the preliminary circuit layout, (5) analyze using EM simulation of critical areas, (6) modify the circuit design and layout, (7) perform the resimulation for the modified circuit, (8) optimization through repeating phases (5–7). Figure 1.20 shows the design flow of a millimeter-wave monolithic circuit. Although design of a millimeter-wave monolithic circuit is similar to a microwave circuit design in phases 1–5, EM simulation and layout optimization are required due to the parasitic effect for accurate design. However, EM simulation requires a very long time and significant computing capacity, if the total MMIC pattern is

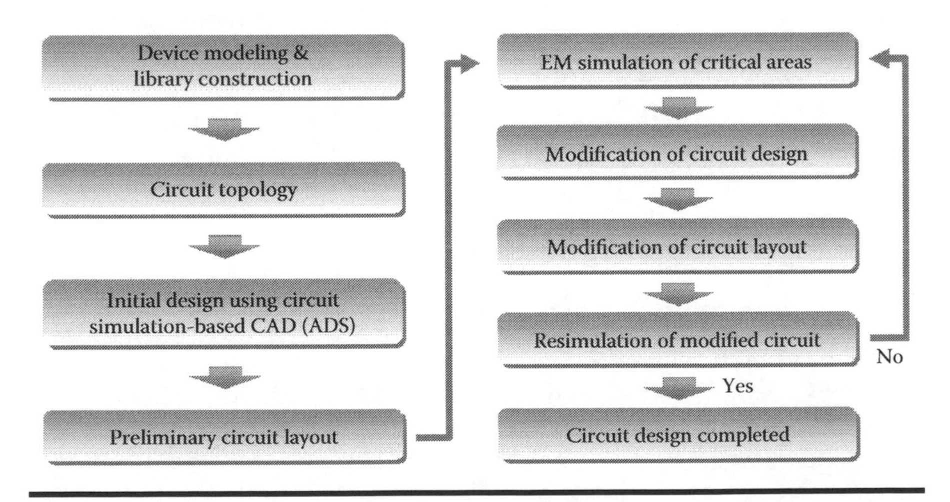

Figure 1.20 The design procedure of a millimeter-wave monolithic integrated circuit.

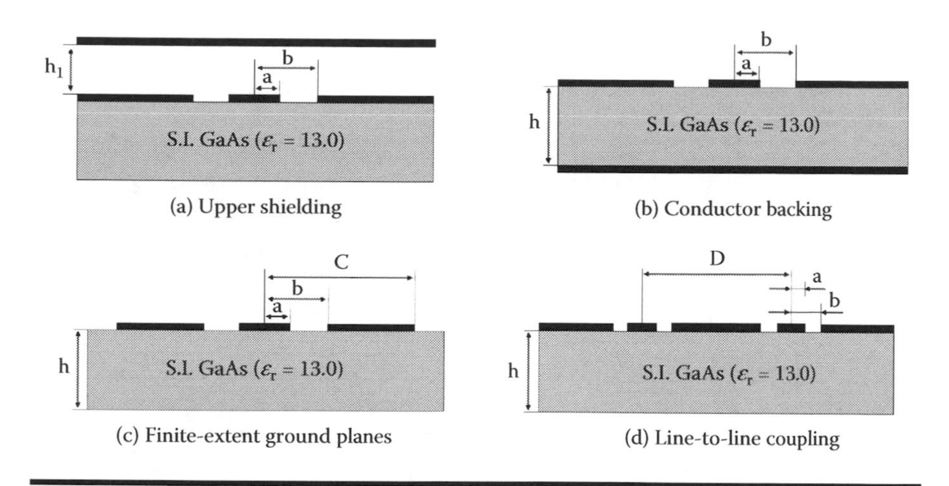

Figure 1.21 Parasitic effects of a CPW transmission line. (Taken from [23] IEEE 1987.)

simulated. Also, it is not easy for a designer to revise the circuit pattern through EM simulation of the total pattern. For design technology to solve this problem, analysis of a basic parasitic effect can be presented [23]. CPW offers several advantages over conventional microstrips for millimeter-wave monolithic circuit applications on GaAs substrates. But, CPW has parasitic effects such as upper shielding, conductor backing, lateral ground plane truncation, and line-to-line coupling. These parasitic effects change MMIC characteristics for various pattern shapes. Figure 1.21 explains the parasitic effects of a CPW transmission line. In millimeter-wave circuit design, some analysis concerning CPW lines on GaAs substrates can be used for the aim of practical design criteria. The effect of upper shielding basically results in a reduction of the line impedance. This is clearly seen in Figure 1.22, where the constant Zol curves are given in the plane a/b-h_1/b, while the substrate thickness was set to h/b = 1. Although the minimum cover height needed to avoid significant impedance lowering depends on the line impedance itself, as a conservative estimate the cover height should be at least hl = 4b. As expected, conductor-backed coplanar lines are slightly less sensitive than upper shielding coplanar waveguides for the same characteristic impedance. However, also for this case, the design criterion $h_1 > 3b$ is a conservative estimate. If the minimum substrate thickness needs to be independent of line impedance, then a rough estimate for a 50 line suggests $h/b > 3$ as a reasonable value (error with respect to $h/b \rightarrow \infty$ less than 2 percent). Figure 1.23 shows constant-impedance conductor-backed coplanar waveguide as a function of the shape ratio a/b. Finite ground-plane width leads to a slight increase of the line impedance with respect to the ideal case ($c \rightarrow \infty$, or $b/c \rightarrow 0$) as shown in Figure 1.24 (h/b = 1, C/b ranging from 1 to 3.33). In case

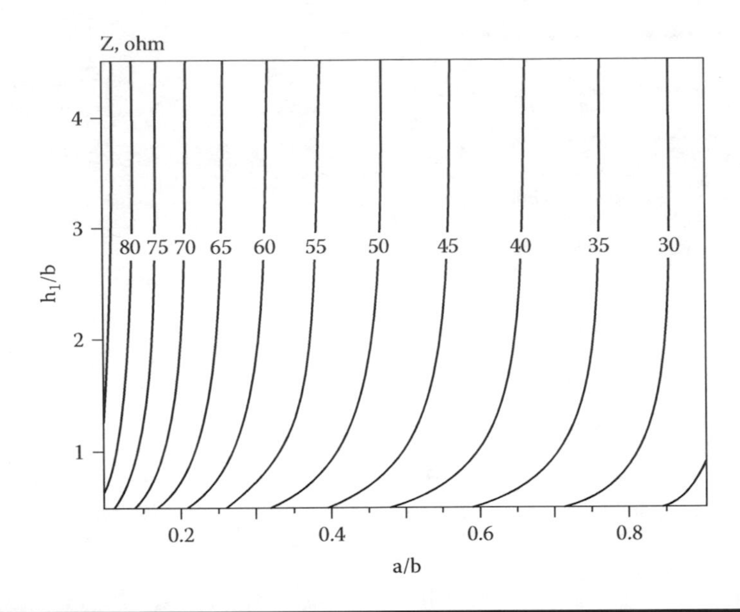

Figure 1.22 Constant impedance with upper shielding coplanar waveguide as a function of the shape ratio, a/b and the cover height, h_l/b, with substrate thickness h/b = 1 and GaAs substrate permittivity $\varepsilon_r = 13$. (Taken from [23] IEEE 1987.)

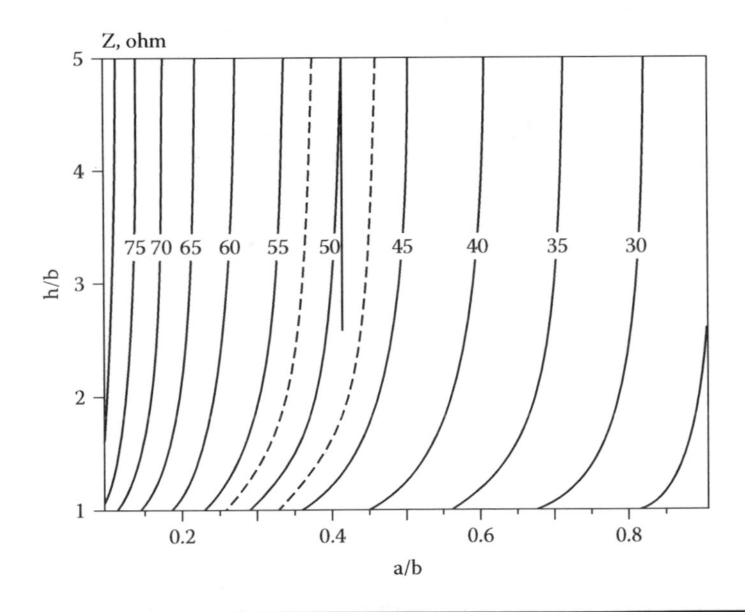

Figure 1.23 Constant-impedance conductor-backed coplanar waveguide without upper shielding as a function of the shape ratio, a/b, and the substrate thickness, h/b, with GaAs substrate permittivity $\varepsilon_r = 13$. (Taken from [23] IEEE 1987.)

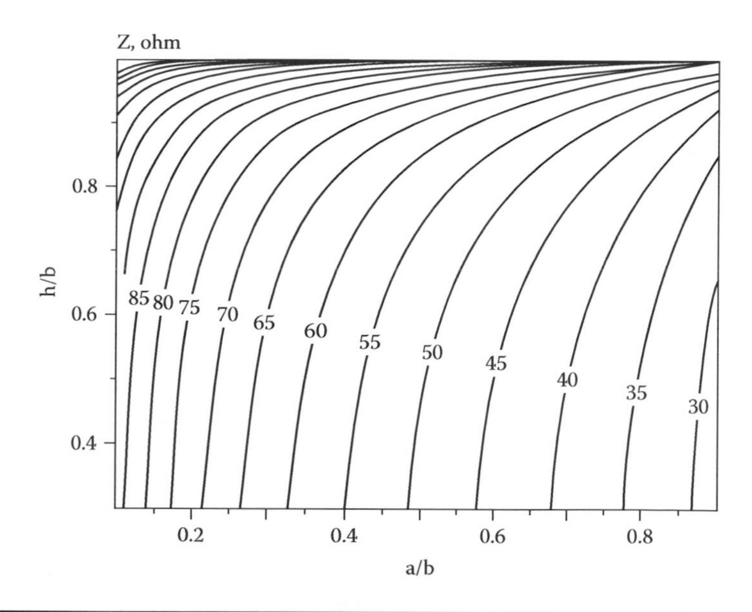

Figure 1.24 Constant-impedance coplanar waveguide with finite ground planes as a function of the shape ratio, a/b, and the inverse of the ground plane width, b/C, with substrate thickness h/b = 1 and GaAs substrate permittivity $\varepsilon_r = 13$. (Taken from [23] IEEE 1987.)

of very narrow lateral ground planes, the impedance is largely increased. As a conservative estimate, one should have C/b = 4 at least to ensure that the variation is negligible. As a last point, let us consider line-to-line coupling. A chart for evaluating this effect is shown in Figure 1.25. It can be seen that coupling weakly depends on the shape ratio a/b, whereas, as is obvious, it is strongly influenced by the line spacing D. From Figure 1.25, the minimum D needed to ensure coupling less than a given value can be obtained. As a conservative estimate of maximum coupling allowed, one can assume, for instance, the value of 40 dB, thereby requiring line spacing to be at least D/b = 7. Based on analysis of the CPW parasitic effect, we can expect a critical area with an electrical characteristic that is very different compared to the result of circuit-based simulation (ideal case). Figure 1.26 shows the preliminary layout of millimeter-wave amplifier and critical areas. Figure 1.27 is an EM simulation pattern for critical areas. The design example is a W-band MMIC amplifier employed in a two-stage structure with 70 μm × 2 MHEMTs. The matching circuit was designed using CPW transmission lines. The open stubs and transmission lines were used for matching circuits of input/output and the interstage. Also, the W-band MMIC amplifier was designed as a low Q-factor matching circuit structure to improve the broadband characteristic. The bias line was designed using $\lambda/4$ (at 100 GHz) short stub [24]. The initial designed W-band amplifier was

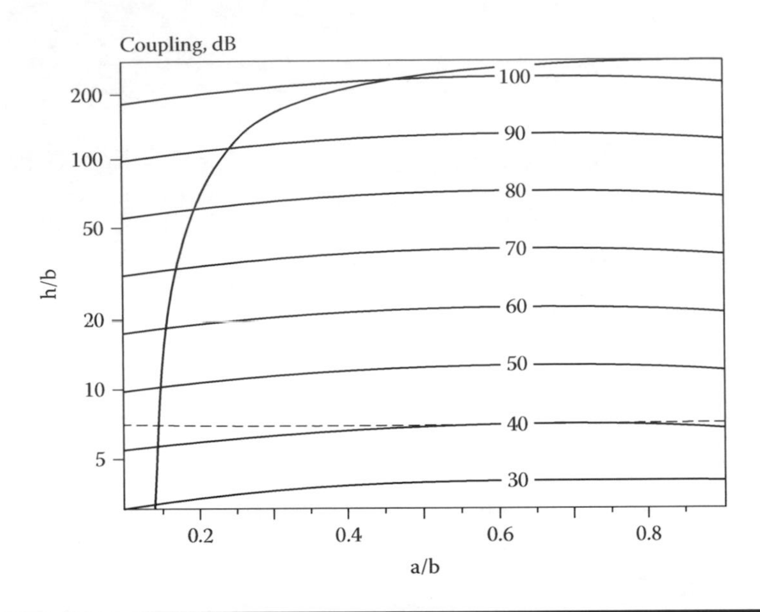

Figure 1.25 Constant-coupling curves for parallel coplanar lines on infinitely thick GaAs substrate ($\varepsilon_r = 13$) as a function of the shape ratio, a/b, and the normalized distance, D/b (log scale). (Taken from [23] IEEE 1987.)

optimized by the EM simulator (ADS Momentum™). When EM simulation was performed, only the critical area was optimized without total pattern simulation. Therefore, reduction of simulation time and efficient design are possible. Figure 1.28 shows comparison data of circuit-based simulation and EM simulation for the pattern in Figure 1.27. From analysis results, an

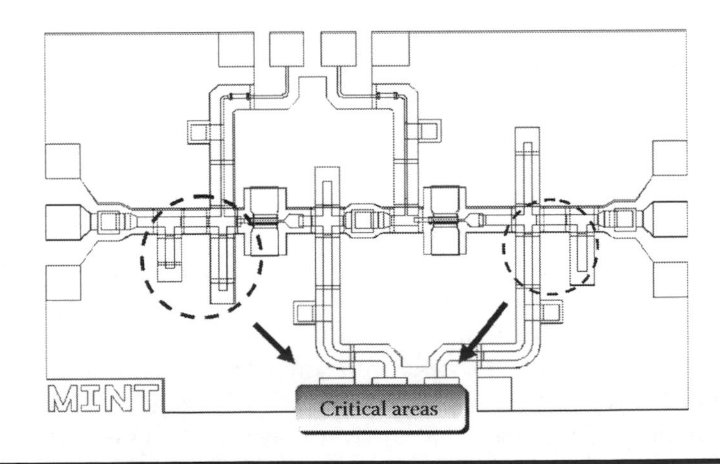

Figure 1.26 The preliminary layout of a W-band amplifier and critical areas.

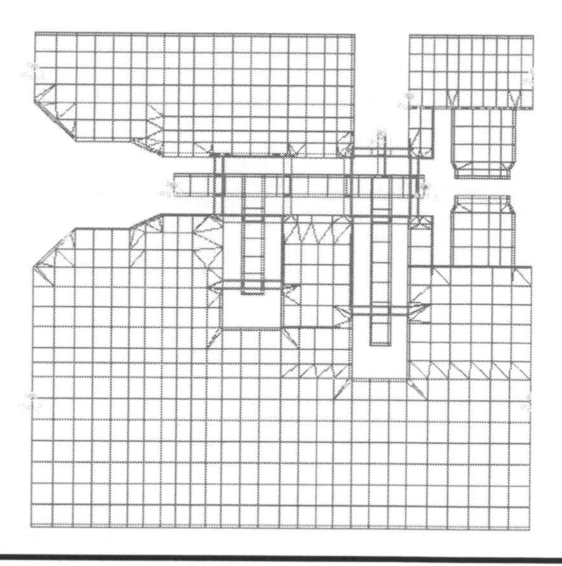

Figure 1.27 EM simulation pattern for critical areas.

apparent difference between circuit-based simulation and EM simulation of critical areas was obtained. The EM simulation exhibited higher impedance characteristics than circuit-based simulation (ideal case) due to parasitic effect (finite ground, line-to-line coupling). After the critical area was optimized, the completed amplifier circuit is shown in Figure 1.29. From the measured results, W-band MMIC amplifiers exhibit a broadband characteristic from the S21 gains of 11 ± 2 dB in a W-band frequency range of

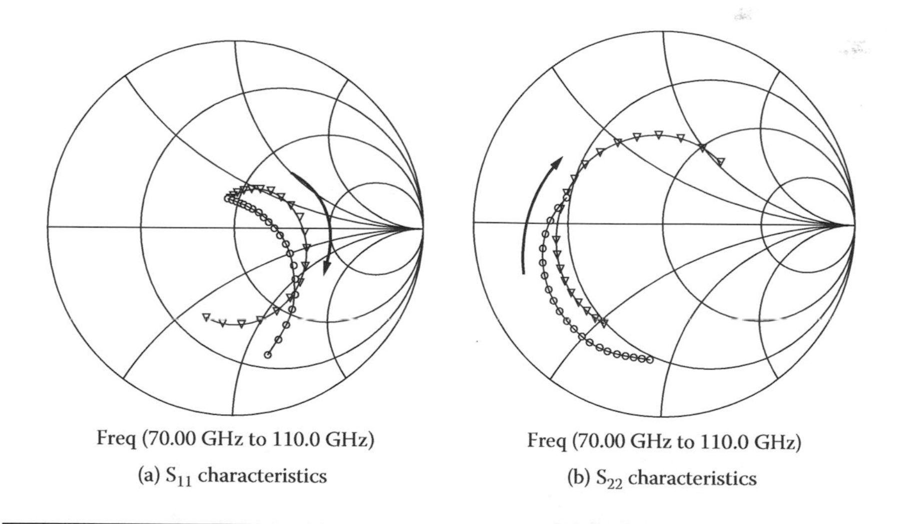

Figure 1.28 Comparison data of circuit-based simulation and EM simulation for critical areas (○: circuit-based simulation, ▽: EM simulation).

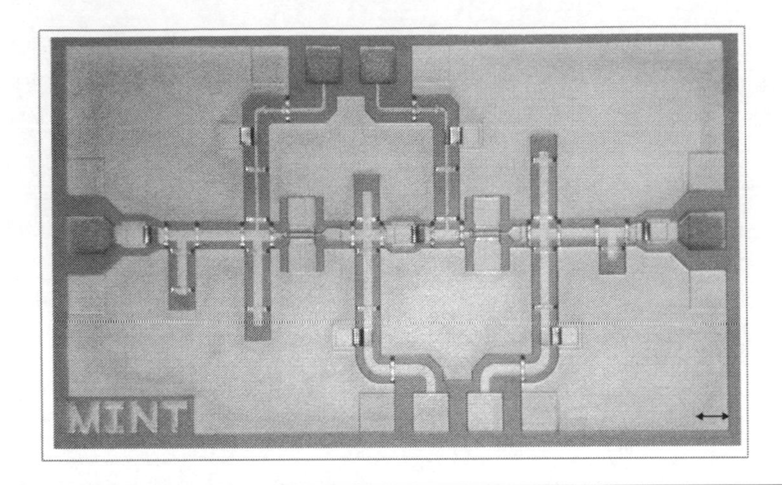

Figure 1.29 The fabricated MMIC W-band amplifier circuit (Millimeter-wave INnovation Technology Research Center (MINT) size: 1.8 × 1.0 mm^2. (Taken from [24] IEEE 2004.)

70–100 GHz with good return losses. Also, a good agreement was obtained with the S-parameter measurements over the entire frequency range from 70 to 100 GHz compared to the simulated data. As shown in Figure 1.30, this result demonstrates that the presented design technology is a reasonable and efficient method in millimeter-wave ranges.

1.3.3 Millimeter-Wave Monolithic Integrated Circuit for WLAN

1.3.3.1 60-GHz Band MMIC Amplifier and Oscillator

Amplifiers are the basic building block of a wireless LAN system and mainly perform an amplification of weak signal. Amplifiers can be classified as low noise, drive, power, and linear. Also, HEMT and HBT devices have been generally adopted for MMIC amplifiers of millimeter-wave WLAN and so on. Figure 1.31 shows the performance of the reported 60-GHz MMIC low-noise amplifier and power amplifier for millimeter-wave WLAN [25–37]. A low-noise amplifier shows an S21 gain of 14–22.8 dB and a noise figure of 2.2–5.8 dB in a frequency range of 58–62 GHz, respectively. In the case of the MMIC power amplifier, an S21 gain of 7.5–13.8 dB and output power of 23–26.8 dBm were reported. A V-band MMIC low noise amplifier chip for millimeter-wave WLAN is shown in Figure 1.32. The MMIC low-noise amplifier is a three-stage circuit structure and GaAs PHEMTs are used. Circuit performances show an S21 gain of a typical 20 dB and a noise figure of 4.2 dB in the 55–65 GHz frequency range. The measured results of

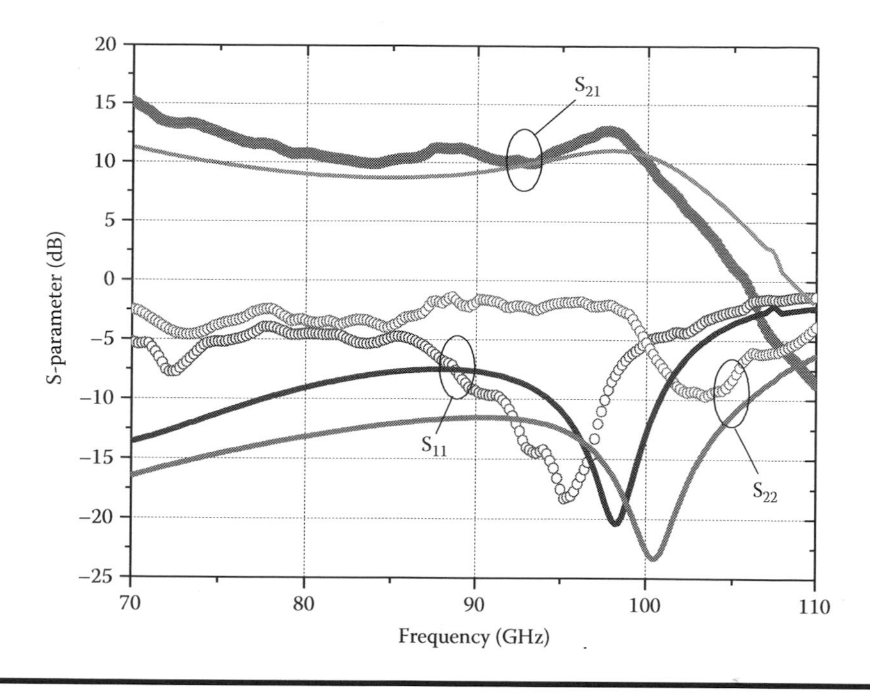

Figure 1.30 The simulated and measured results of the MMIC W-band amplifier circuit (o: measured data; —: simulated data). (Taken from [24] IEEE 2004.)

an MMIC low-noise amplifier are shown in Figure 1.33. This MMIC circuit can be used for the amplification block of a receiver in millimeter-wave WLAN. Oscillators are an essential block for millimeter-wave WLAN. Oscillators generate a millimeter-wave signal source using negative resistance. An ideal oscillator produces a pure sinusoidal carrier with fixed amplitude, frequency, and phase. Practical oscillators, however, generate carrier waveforms with parameters (oscillation frequency, output power) that may vary in time due to temperature changes and component characteristics. This phenomenon appears as phase and amplitude fluctuations at the oscillator output and will be of main concern in a wireless LAN system. Phase noise generation by the local oscillator at the receiver can significantly affect the performance of a wireless system. Figure 1.34 shows performance of the reported 60-GHz MMIC oscillator for millimeter-wave WLAN [38–41]. Oscillators show phase noise of −80 to −104 dBc/Hz and output power of 2.5–11.1 dBm in a frequency range of 56–62.5 GHz, respectively. Figure 1.35 is a photograph of the 60-GHz MMIC voltage-controlled oscillator (VCO), which was designed using GaAs PHEMTs. The VCO structure adopted an injection-locked VCO structure. CPW transmission lines and intrinsic gate capacitance of the HEMT are used for resonance at 60 GHz. To adjust the output frequency, a varactor diode of 280 width HEMT is used. A CPW

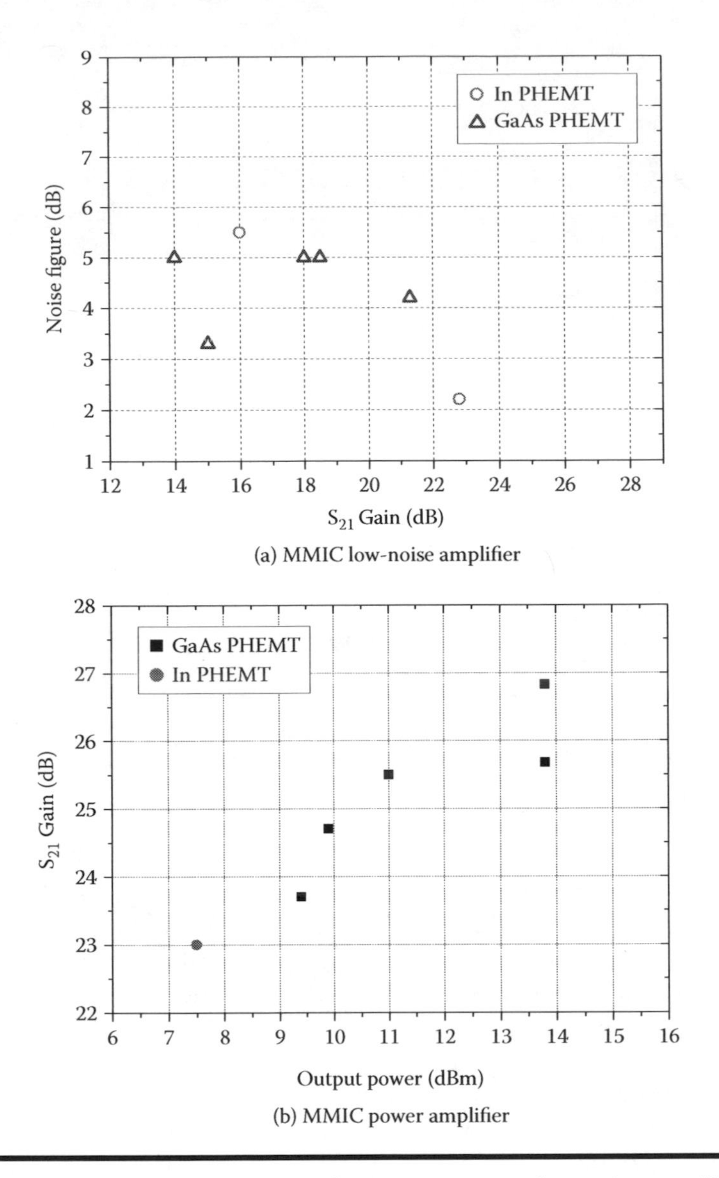

Figure 1.31 Performance of reported 60-GHz MMIC low-noise amplifier and power amplifier for millimeter-wave WLAN [25–37].

coupler is designed to inject the reference signal. The output phase of VCO is locked by the injected signal through the coupler. The major purpose of the buffer amplifier is to ensure proper matching of the oscillator output and isolation to the output port of VCO. Circuit performances show output power of typically 2.5 dBm and phase noise of 83 dBc/Hz (at 1 MHz offset) at 60 GHz oscillation frequency.

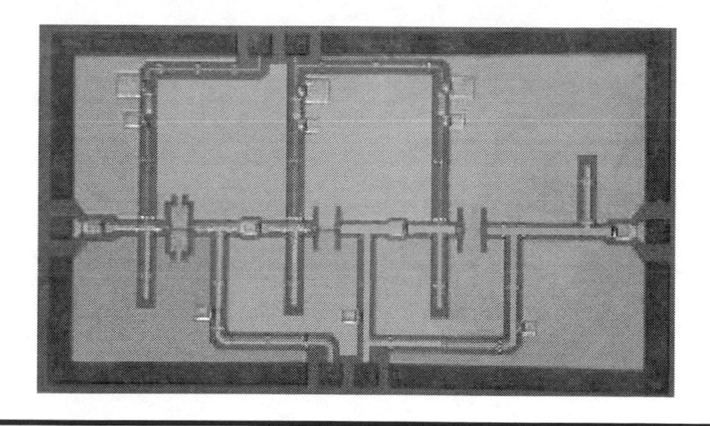

Figure 1.32 V-band MMIC low-noise amplifier for millimeter-wave wireless LAN (Millimeter-wave INnovation Technology Research Center (MINT); size: 2.6 × 1.5 μm^2) [25].

1.3.3.2 60-GHz Band MMIC Mixer

A mixer, or frequency converter, has the prime function of converting a signal from one frequency to another with minimum loss of the signal and minimum noise performance degradation. A mixer is one of the fundamental blocks of a wireless LAN system. A mixer faithfully preserves the amplitude and phase properties of the RF signal at the input. Therefore, signals can be translated into frequency without affecting their modulation properties. An ideal mixer multiplies the input signal by the sinusoidal signal generated by a local oscillator. This results in a mixed product that consists of higher and lower frequency components. Devices that exhibit

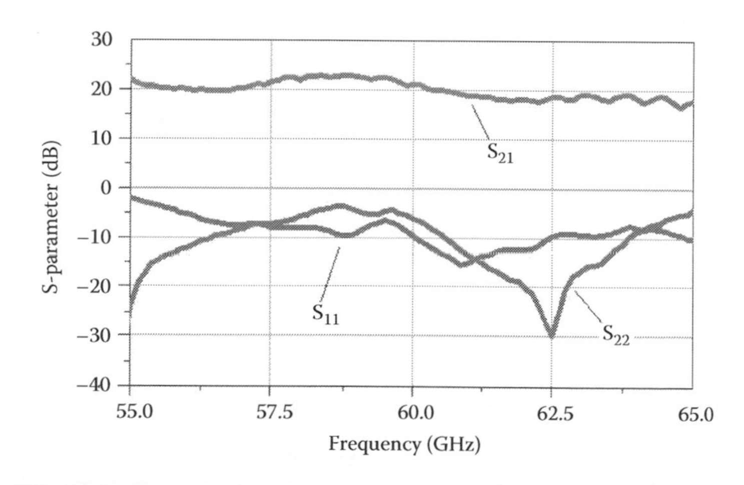

Figure 1.33 The measured results of a V-band MMIC low-noise amplifier [25].

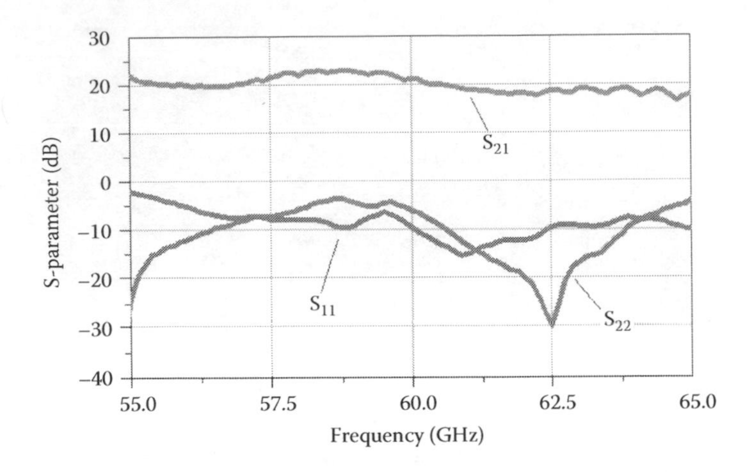

Figure 1.34 Performance of reported 60-GHz MMIC oscillator for millimeter-wave WLAN [38–41].

nonlinear or rectifying characteristics are good candidates for designing mixers. Diodes and HEMT devices are commonly used in the design of millimeter-wave mixers, because of their rectifying and nonlinear characteristics. Schottky barrier diodes have broad bandwidth and are low cost. Besides, diodes do not need DC bias to operate and have fast switching capability. On the other hand, HEMT or HBT devices have lower noise, better frequency response, and increased power handling ability. Also, HEMT and HBT devices are amenable to monolithic circuit integration. The

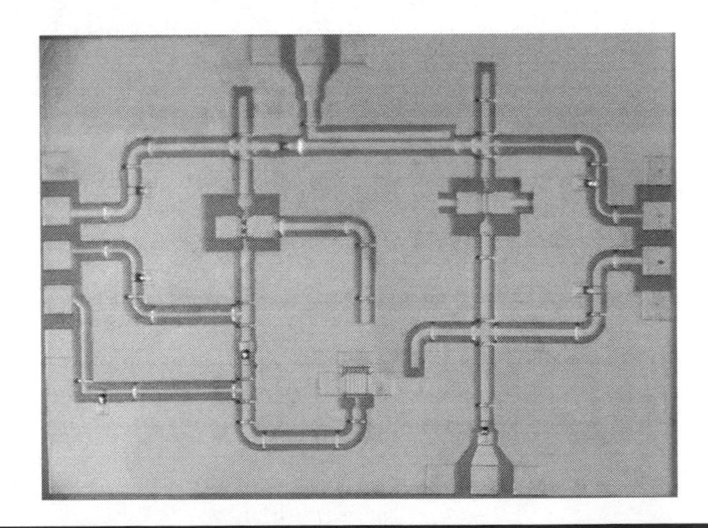

Figure 1.35 The 60-GHz MMIC voltage control oscillator (Millimeter-wave INnovation Technology Research Center (MINT); size: 2.2 × 1.6 μm^2).

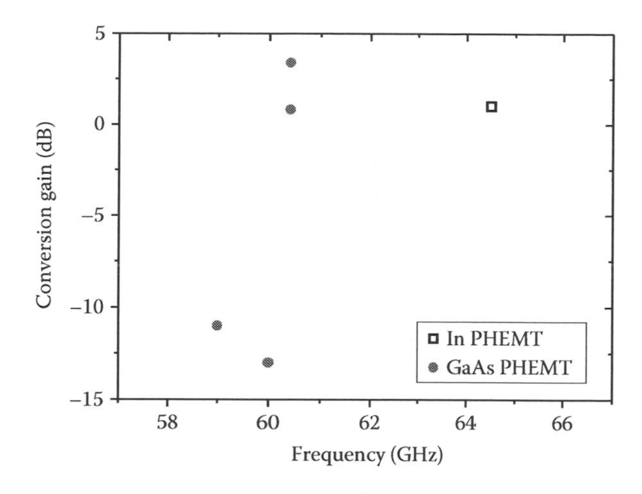

Figure 1.36 Performance of the reported 60-GHz MMIC mixer for millimeter-wave WLAN [42–46].

distortion caused by the inherent nonlinearities of the diodes is reduced in HEMT mixers. Figure 1.36 shows the performance of the reported 60-GHz MMIC mixer for a wireless LAN [42–46]. Conversion gains exhibit −13 to 3.4 dB in the frequency range of 59–64.5 GHz. Circuit structures have been studied in a single-ended diode mixer, balanced HEMT mixer, and balanced resistive mixer. A 60-GHz MMIC mixer chip for a millimeter-wave wireless LAN is shown in Figure 1.37. A circuit structure is a double-balanced

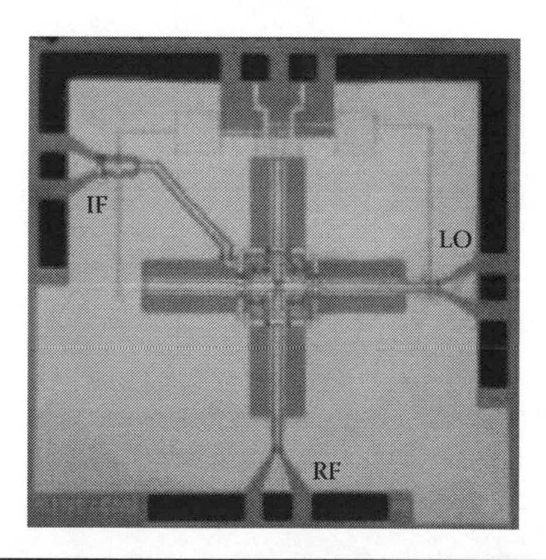

Figure 1.37 A 60-GHz MMIC mixer chip for millimeter-wave wireless LAN (Millimeter-wave INnovation Technology Research Center (MINT); size: 1.5 × 1.5 mm^2. (Taken from [45] IEEE 2005.)

star mixer and GaAs PHEMT is used. The novel star mixer is composed of gate-drain (GD)-connected PHEMT diodes where the gate-source junctions are reverse biased to pinch off. Due to the reverse bias, the mixing occurs mainly by the DS conductance rather than by the gate-source junction diode. Consequently, the conversion loss does not suffer degradation due to the hetero-junction diode. Besides, the DS conductance is not as highly nonlinear as the diode point of the structure; the star mixer has a simpler topology, whereas the resistive ring mixer requires three separate baluns. More wideband operation is possible with the removal of the IF (intermediate frequency) balun [45]. Circuit performances show conversion loss of a typical 13 dB and isolation of a 35 dB at all ports. This circuit can be used for up/down frequency converters in the millimeter-wave WLAN. Furthermore, a subharmonic mixer has been researched in millimeter-wave ranges because a stable local oscillator (LO) signal is indispensable to the mixers. However, it is practically difficult to fabricate a stable oscillator operating at 60 GHz or above. Thus, continuous research and development efforts on the subharmonic mixers have been made because they can utilize lower LO frequencies than the conventional mixers [47]. This approach allows the use of a local oscillator of a relatively low frequency because an LO frequency is located at some integer fraction (1/n) of the fundamental LO frequency. For this reason, the subharmonic mixers with antiparallel diode structure were evaluated at millimeter-wave frequencies [48]. Figure 1.38 shows the designed circuit schematic of the 60-GHz MMIC subharmonic mixer. In this

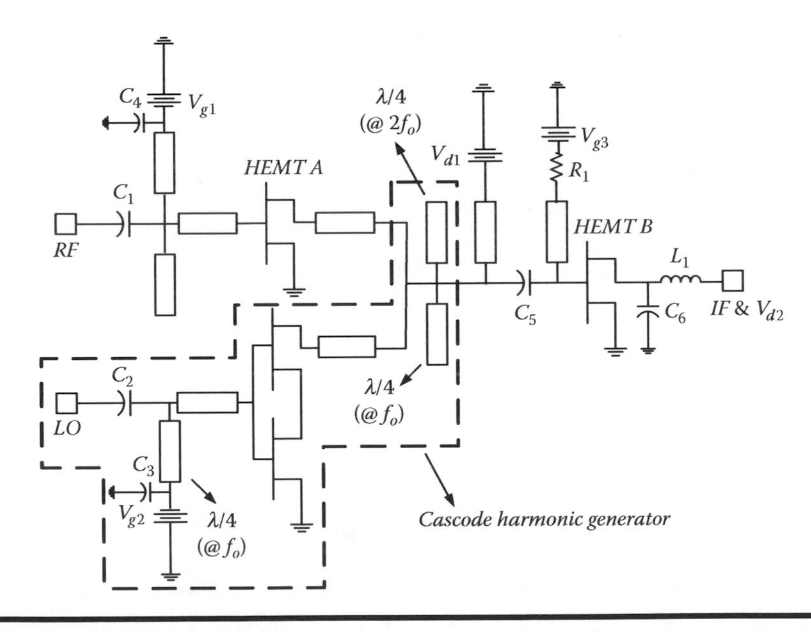

Figure 1.38 Circuit schematic of the 60-GHz MMIC subharmonic mixer. (Taken from [43] IEEE 2003.)

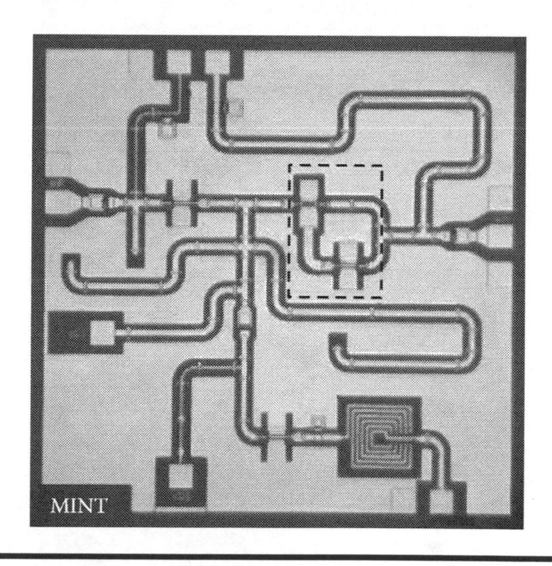

Figure 1.39 Micrograph of the fabricated MMIC subharmonic mixer (Millimeter-wave INnovation Technology Research Center (MINT); size: 1.9 × 1.8 mm^2; 2003 IEEE International Microwave Symposium; the cascode pair is shown in the dashed lines). (Taken from [43] IEEE 2003.)

work, a cascode harmonic generator is proposed to improve the conversion gain of the quadruple subharmonic mixer. The circuit of the subharmonic mixer was designed using the architecture of a gate mixer. An IF stage device (HEMT B) mixes the generated fourth harmonic signal and the RF signal. An RF stage device (HEMT A) not only amplifies the RF signal but also improves the LO–RF isolation due to the isolation characteristics of reverse direction. Matching circuits for the RF and the LO ports were designed using the CPW transmission lines. Figure 1.39 is a micrograph of the fabricated subharmonic mixer [43]. The measured results of the subharmonic mixer demonstrated that the conversion gain is 3.4 dB, which is a good conversion gain at the LO power of 13 dBm, as shown in Figure 1.40. Also, the conversion gain versus the LO input power was measured. The conversion gain is saturated at an LO input power level higher than 13 dBm, as shown in Figure 1.40. As shown in Figure 1.41, the fabricated subharmonic mixers show a good LO-to-IF isolation of −53.6 dB and LO-to-RF isolation of −46.2 dB at 14.5 GHz, respectively. The measurement results exhibited a high degree of isolation characteristics. The MMIC subharmonic mixer circuit provides high performance through the proposed cascode harmonic generator, and the cost of a wireless LAN system can be reduced because of utilization of a lower LO frequency than a conventional mixer. In addition, Figure 1.42 shows the circuit schematic and planar view micrograph of the fabricated 94-GHz MMIC resistive mixer chip using metamorphic

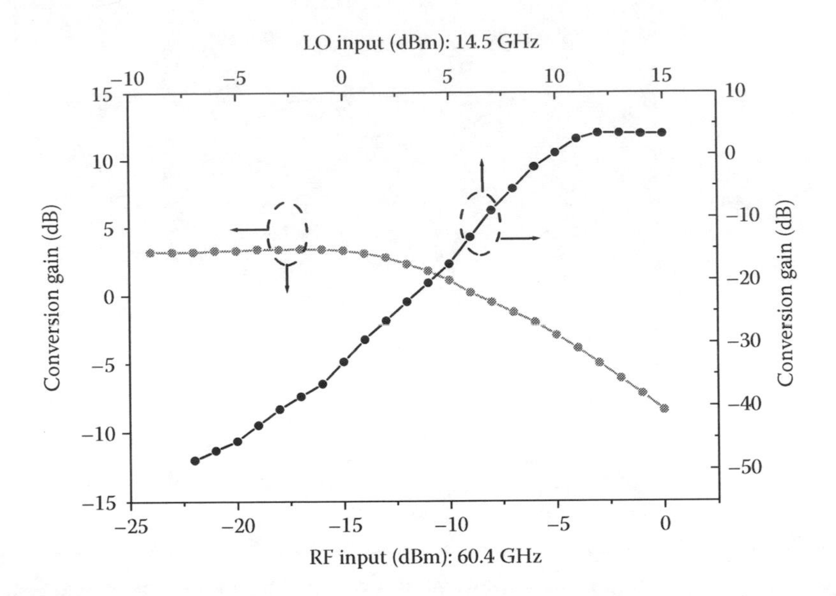

Figure 1.40 Conversion gain vs. RF input and LO input (RF frequency: 60.4 GHz; LO frequency: 14.5 GHz; IF frequency: 2.4 GHz). (Taken from [43] IEEE 2003.)

HEMTs [49]. Resistive mixers are widely used due to good conversion loss, low distortion, and no drain bias. Also, a frequency of 94 GHz has been actively researched as millimeter image sensor and FMCW (frequency modulated continuous wave) radar applications. However, the development of

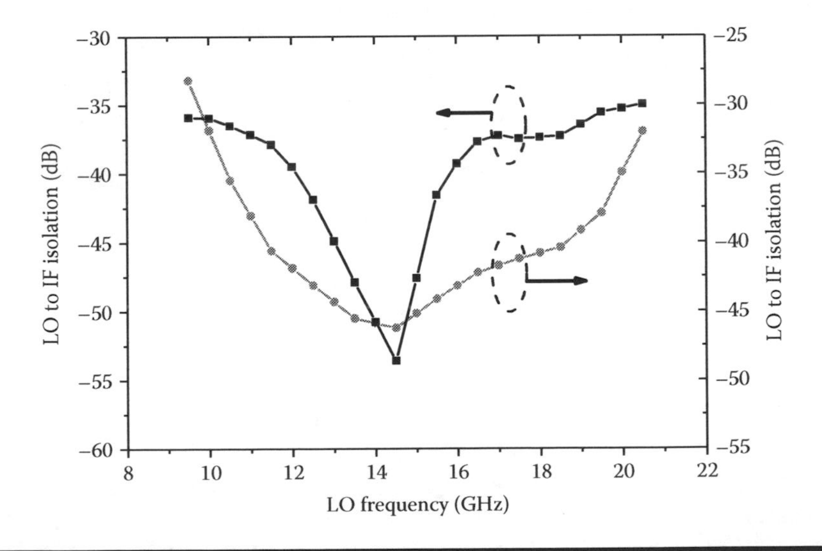

Figure 1.41 The measured results of isolation characteristics. (Taken from [43] IEEE 2003.)

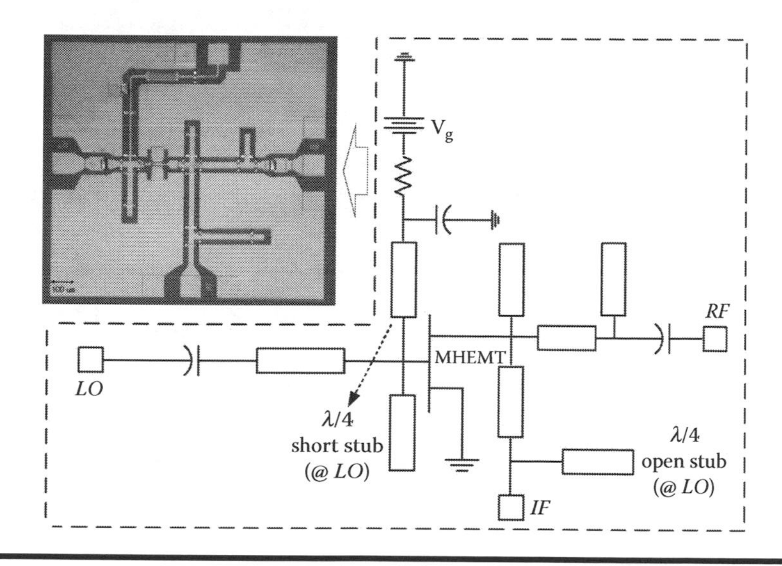

Figure 1.42 The circuit schematic and planar view micrograph of the fabricated 94-GHz MMIC resistive mixer (MINT), size: 1.2 × 1.1 μm^2. (Taken from [49] IEEE 2003.)

high-performance resistive mixers is difficult because of the limitations of the active device. Recently, GaAs-based metamorphic HEMTs (MHEMTs) adopted the low-cost GaAs substrates using the metamorphic buffers and higher indium mole fraction for the InGaAs channels than that of PHEMTs. Therefore, the MHEMTs materialize the same structure of InGaAs/InAlAs hetero-junction with the InP-based HEMTs and exhibit comparable RF performance for millimeter-wave applications [50]. The 94-GHz resistive mixer in this work was designed using MHEMTs for high performance. LO and RF matching circuits of the resistive mixer were implemented using coplanar waveguide transmission lines. A gate bias circuit was designed using a quarter-wavelength (4 at 94 GHz) short stub and a Ti thin-film resistor. A structure of a quarter-wavelength open stub was added at the IF stage for suppressing the LO signal at the IF port. The designed MMIC mixer was fabricated using the metamorphic HEMT-based MMIC process. Conversion loss of the mixer was measured with an applied RF signal of 94.075 GHz and an LO signal at 94.240 GHz. Conversion loss versus RF frequencies were obtained at an RF power of −20 dBm and LO power of 7 dBm. Figure 1.43 shows the simulated and measured conversion loss as a function of LO input power and RF frequencies. As shown in the plot, the resistive mixer exhibited a very low conversion loss of 8.2 dB at an LO power of 7 dBm. Compared to previously reported W-band (75–110 GHz) MMIC resistive mixers based on PHEMTs [51] or InP-based HEMTs [52], the MHEMT-based MMIC resistive mixer presented in this work has shown superior conversion loss.

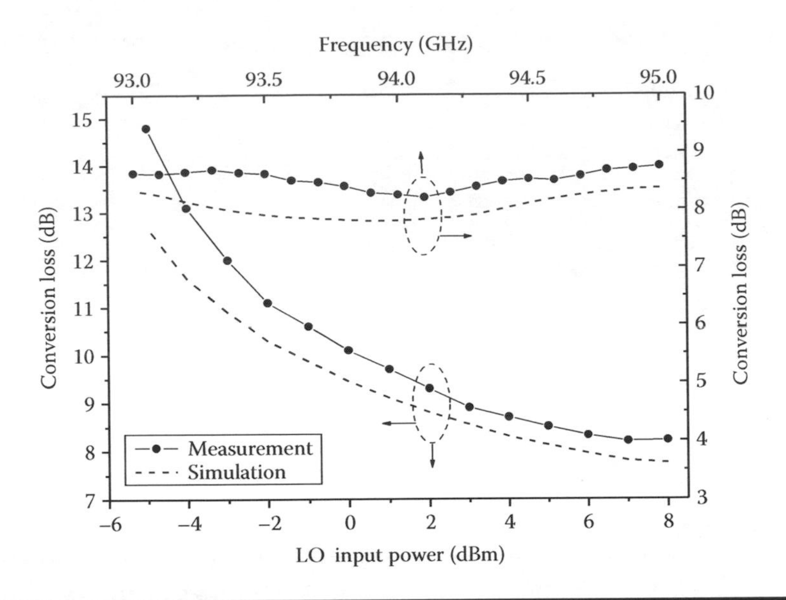

Figure 1.43 Conversion loss versus LO input power (at an LO frequency of 94.240 GHz, an RF frequency of 94.075 GHz, and an RF input power of −20 dBm) and RF frequencies (at an RF input power of −20 dBm and an LO input power of 7 dBm) measured for the MMIC resistive mixer. (Taken from [49] IEEE 2005.)

1.3.4 Fabrication Process of Millimeter-Wave Monolithic Integrated Circuits

There were few differences between the fabrication process for PHEMT-based MMICs and MHEMT-based MMICs, such as etching solution and metal thickness for ohmic contacts. However, a basic factor of fabrication is the same. In this chapter, we shall limit discussion of MHEMT-based MMICs. Fabrication processes for the MMIC were performed in the following sequence. First, mesa etching was done by the phosphoric acid-based etching solution. AuGe/Ni/Au for ohmic contacts were evaporated and annealed. After Ti metal evaporating for thin-film resistor, a 100-nm T-gate was patterned through a triple-layer resist pattern using electron beam lithography technology. Then, we performed a gate recess etching to control current of the devices. After the gate metal formation of Ti/Au, we performed the first-level metallization. And then, we deposited Si3N4 film for the passivation of MHEMTs and interlayer dielectric of the MIM (metal–insulator–metal) capacitor. After the RIE (reactive ion etching) process for opening the contact window, we finally performed the airbridge (second-level metallization) in order to connect between isolated electrodes.

1.3.4.1 Mesa Isolation Process

Isolation serves a number of purposes. In active devices, it restricts the current flow to the desired path. Separate devices are electrically isolated from each other. Isolation reduces parasitic capacitances and resistances. Isolation also provides a sufficiently insulating surface for construction of capacitors and transmission lines. Three principal methods are used to achieve isolation: mesa etching, ion bombardment, and selective implantation [22]. Mesa etching is the simplest method of providing the isolation and for this reason has been widely used, especially in research and development environments. The shape of the mesa edge is important because metal patterns must generally step over in this edge. If the edge profile is too steep or undercut, then it can cause problems for metal crossovers. In order to solve this problem, we performed the development and etching process three times in turns. As a result, we could get the steps shape of the mesa edge, which is a suitable profile for the metallization, as shown in Figure 1.44. Mesa etching was done by H3PO4:H2O2:H2O = 1:1:60. The etching rate of this etchant was 14.4/s, and 2300 of epilayer was removed.

1.3.4.2 Ohmic Contact Process

The purpose of an ohmic contact on a semiconductor is to allow electrical current to flow into or out of the semiconductor. The contact should have linear I-V characteristics, be stable over time and temperature, and contribute as little resistance as possible. The most common approach to fabricating ohmic contacts on a GaAs substrate is to apply an appropriate metallization to the wafer in the desired pattern, and then alloy the metal

Figure 1.44 SEM photography of the mesa etching profile.

into the GaAs substrate. Candidate species for this function are Si, Ge, Sn, and Te for n-type GaAs and Zn, Cd, Be, and Mg for p-type GaAs [22]. Gold-germanium (AuGe) contact is common on h-GaAs substrate; therefore, discussion for ohmic contacts will be focused on AuGe. AuGe is usually applied with an overlay of another metal, such as nickel (Ni) or silver (Ag); nickel is the more common choice. The germanium is used for doping the GaAs during alloy. The nickel acts as a wetting agent and prevents the AuGe metal from balling up during the alloy [22]. This balling up is likely to occur if AuGe alone is used and results in contacts exhibiting poor contact resistance, poor morphology, and irregular edges. It also aids in the diffusion of the Ge into the GaAs [22]. Applying a thick gold overcoating before alloy is usually desirable. Such an overcoat has several potential advantages. Alloyed AuGe/Ni metallization has poor sheet resistance, and so it usually requires that an overcoat of gold be added in a subsequent step to enhance sheet conductivity. An overlaying Au layer alleviates this problem and also enhances the ability to make accurate measurements using probe for ohmic contact formation. The extra gold results in improved surface morphology. For nano-gate fabrication using e-beam lithography, e-beam alignment marks are usually formed as part of the ohmic contact pattern. The extra gold layer over the AuGe/Ni usually results in markers having improved contrast and edge definition [22]. Therefore this work investigated an AuGe/Ni/Au ohmic contact, for the above reasons. For ohmic contact formation, an image reversal photoresist of AZ5214 was used to achieve a well-defined overhang profile for the liftoff process, as subscribed in the previous section. The metal system was thermally evaporated and annealed using RTA (rapid thermal annealing). AuGe/Ni/Au of 1440/280/1600 metal system yielded the lowest ohmic contact resistance at the alloy temperature of 300° for 60 s. Contact-specific resistivity was 1×10^{-6} Ω.cm^2.

1.3.4.3 Gate Formation Process

It is difficult to define the short gate footprint by the single e-beam exposure method because the emitted electrons are scattered when they pass through the thick resistance. Therefore, we used the double exposure method. In this process, the wide gate head is first exposed and developed before the second exposure of the gate footprint. When forming the gate footprint, we have only 1/5 bottom layers thickness of whole resistance thickness. Therefore, when we perform the second exposure for the gate footprint, the forward scattering effect of electrons decreases further. Therefore, we can obtain a small line width and improve the uniformity after development. Furthermore, this tri-layer e-beam resistance structure was used to ensure good liftoff of the strips. In order to achieve a gate length of 100 nm, we produced a small resistance thickness of 150 nm, which will define the T-gate foot, by using a mixing liquid of PMMA950K and MCB (monochlorobenzene) (1.5:1). The thickness of the applied e-beam resistance stack, (PMMA950k:MCB=1.5:1)/P(MMA-MAA)/PMMA950k, is 1500/7000/3000, respectively. MCB, IPA-diluted methanol (1:1), and IPA-diluted MIBK (methyl isobutyl ketone) (1:3) were used as a developer for each resistance layer in turn. After performing an experiment to observe the variation of the gate length caused by the change of the development time and the exposure dose, we determined optimum fabrication conditions for a development time of 100 seconds and a dose of 900 C/cm^2 for the 100-nm gate length. After gate patterning, we performed a gate recess using a pH-adjusted solution of succinic acid and H2O2 to selectively etch the heavily doped InGaAs ohmic layer over the InAlAs Schottky layer. The etch selectivity of the InGaAs ohmic layer over the InAlAs Schottky layer was obtained over 100 [53, 54]. A part of the InAlAs Schottky layer was etched using a phosphoric acid solution for reducing the gate-channel distance. Reducing the gate-channel distance can effectively increase fT characteristics because transconductance is improved due to an increase of average electron velocity under the gate control. An SEM photograph of a 100-nm T-gate after gate metal (Ti/Au = 500/4000) evaporation, liftoff, and passivation with a Si3N4 film is shown in Figure 1.45.

1.3.4.4 Si_3N_4 Passivation Process

Dielectric films are used in GaAs processing for environmental encapsulation, capacitor dielectrics, and crossover insulator [22]. The primary purpose

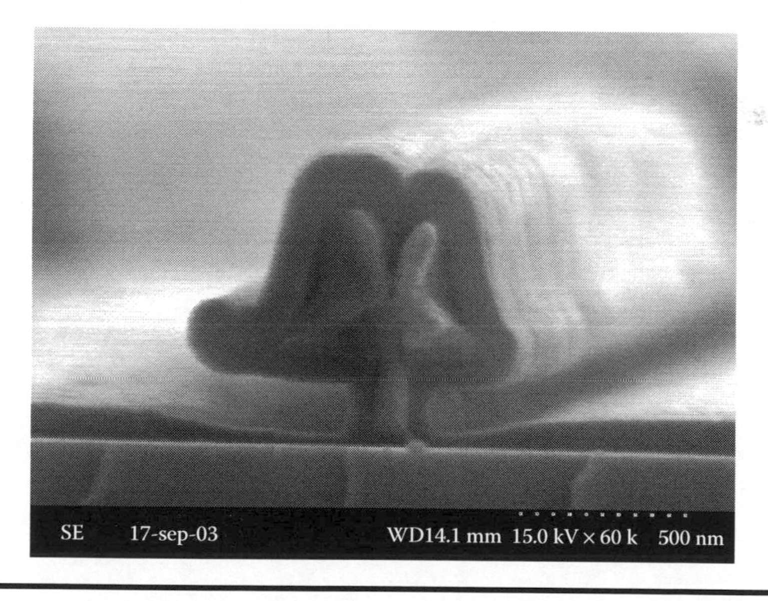

Figure 1.45 SEM photograph showing a 100-nm T-gate after liftoff and Si_3N_4 passivation.

of dielectric formation is simply protective encapsulation. This protects critical portions of the wafer surface from environmental contamination and mechanical damage. The encapsulating dielectric is usually also used for the MIM capacitors. The encapsulating dielectric is usually silicon dioxide (SiO2) or silicon nitride (Si3N4). These two materials are easily applied to the wafer either by sputtering or by PECVD. Silicon dioxide films have a lower dielectric constant than silicon nitride films, thus making them preferable for a crossover dielectric due to lower crossover capacitance. On the other hand, the higher dielectric constant of silicon nitride makes it preferable for use in capacitor formation. Besides, silicon nitride is less permeable to ions than silicon dioxide and therefore makes the superior encapsulation [22]. In this chapter, the PECVD system was used to deposit the silicon nitride films. The films with different characteristics were deposited by changing the flow ratio of silane (SiH4) to ammonia (NH_3) in order to obtain good quality films due to low H2 concentration in films. After these experiments, we determined optimum fabrication conditions for the good quality silicon nitride film that can be applied to the passivation process and dielectric layer of the MIM capacitor [55]. Refractive index and dielectric constant were 1.98 and 6.9, respectively. The resistivity of 6×10^{12} Ω.cm at a field of 2 MV/cm and breakdown strength of 7.2 MV/cm at a current of 1 were obtained.

1.3.4.5 Airbridge Process

Airbridges are used extensively in MMIC applications for interconnections. They may be used to interconnect sources of HEMTs, to cross over a lower level of metallization, to connect the top plate of an MIM capacitor to adjacent metallization, and to compensate for defective ground potential due to connection of the ground plates separated by signal lines of a CPW

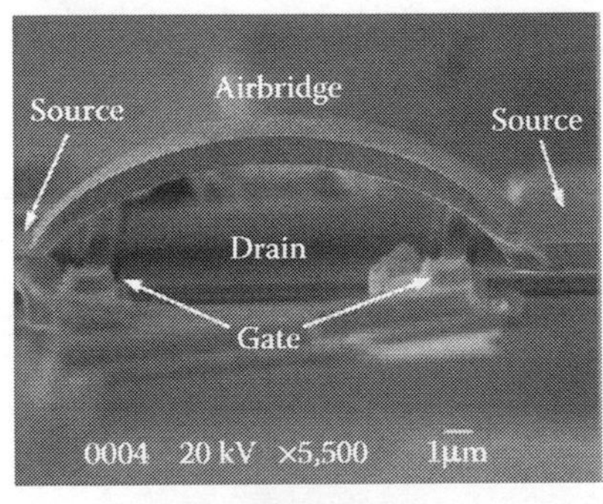

(a) HEMT airbridge

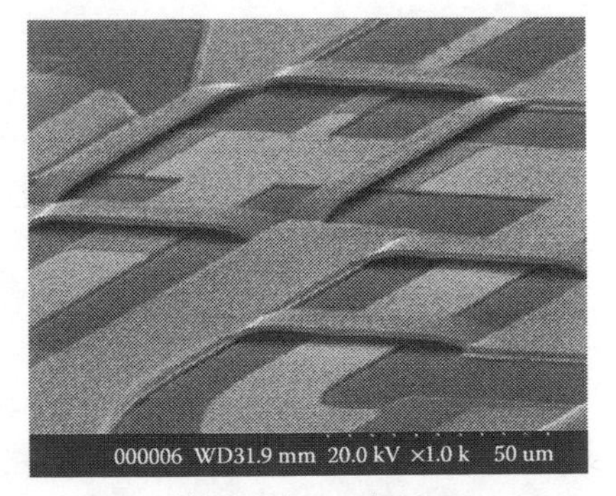

(b) CPW line airbridge

Figure 1.46 SEM photograph of the fabricated airbridge.

structure. Airbridges have several advantages over the dielectric crossover, typically used in digital circuits. These include low parasitic capacitance, immunity to edge profile problems, and the ability to carry substantial current [22]. We evaporated the Ti/Au (200/10,000) of the airbridge [56] in order to connect between isolated electrodes and/or grounds, as shown in Figure 1.46.

1.4 WLAN Applications of Millimeter-Wave Monolithic Circuits

A millimeter-wave RF front-end system for WLAN has been developed using metal waveguides, HIC methods, and waveguide filters, resulting in a large size, high cost, and low productivity. Use of these systems in wireless LAN is difficult because WLAN requires small size and low cost for commercial applications. The MMIC technology that integrates active and passive elements is one of the key solutions of these problems, because a multichip module (MCM) using MMICs has low cost, small size, and high productivity. For the system architecture of a millimeter-wave wireless LAN, there are a wide variety of communication architectures, such as a heterodyne and a super-heterodyne scheme. Also, they are mainly affected by integration technology, cost, and power consumption. A super-heterodyne needs two mixers in the transmitter and receiver. Therefore, system architectures are very complicated and expensive due to several components. A heterodyne scheme also requires a local oscillator to pump an up- and down-mixer. Because a stable LO signal is practically difficult operating at 60 GHz or above, the system cost increases due to an expensive LO component. To reduce system cost, a V-band wireless transceiver system was presented by using a DSB self-heterodyne structure with no filter and LO in the receiver blocks [57, 58]. For the high-productivity and low-cost solution, MMIC chip sets for the transceiver system were also used. The block diagram of the V-band self-heterodyne transceiver is shown Figure 1.47. Besides MMIC modules, a switch for the TDD (time-division duplexing), a horn antenna, and a circulator were used for constructing the self-heterodyne transceiver. V-band MMIC modules were developed to demonstrate the self-heterodyne transceiver. The MMIC chips were designed using CPWs and GaAs PHEMT technologies. To interface the MMIC chips with the component modules for the self-heterodyne transceiver, CPW-to-waveguide fin-line transitions of WR-15 type were designed and fabricated [59]. The DSB self-heterodyne transceiver consists of several components, including a low-noise amplifier module, a medium-power amplifier module, an up/down-mixer module, an LO source (only transmitter), a circulator, and an antenna. In Figure 1.48, photographs of the

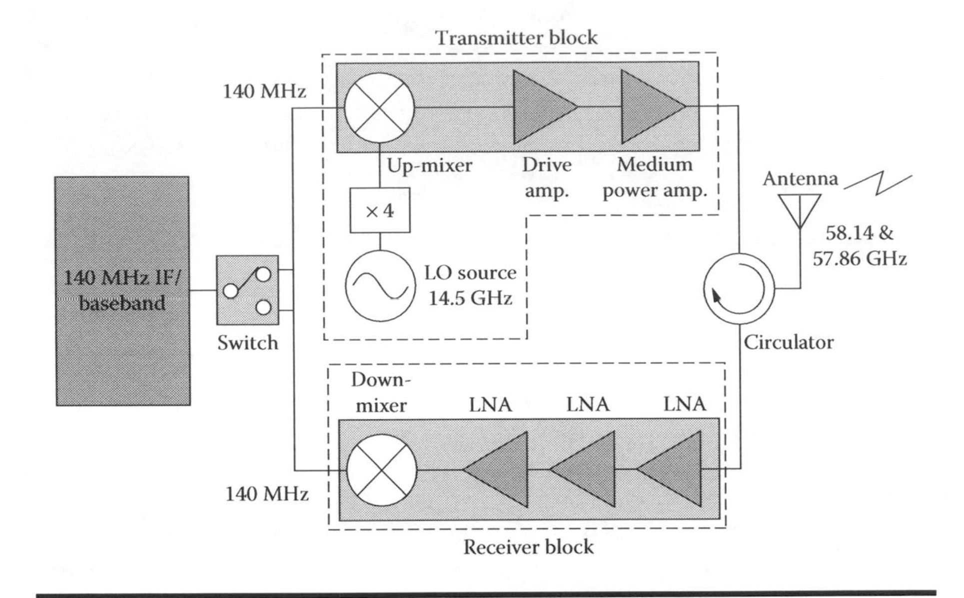

Figure 1.47 The block diagram of the V-band self-heterodyne transceiver.

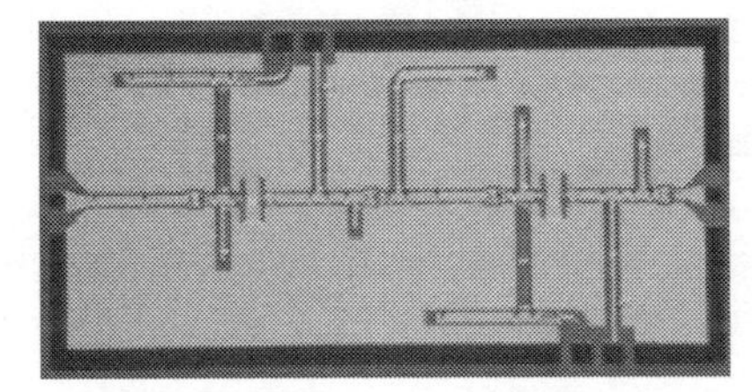

(a) V-band MMIC medium power amplifier

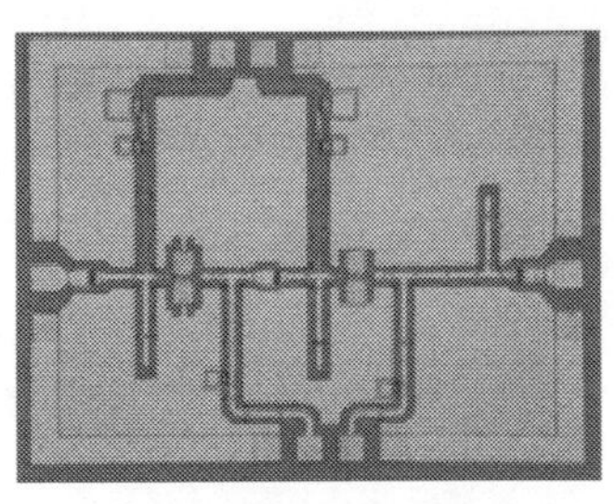

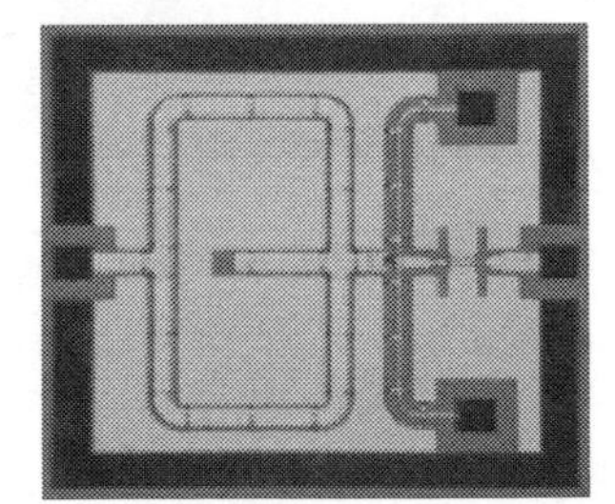

(b) V-band MMIC low noise amplifier (c) V-band MMIC down-mixer

Figure 1.48 V-band MMIC chipsets for millimeter-wave self-heterodyne WLAN transceiver (Millimeter-wave INnovation Technology Research Center [MINT]).

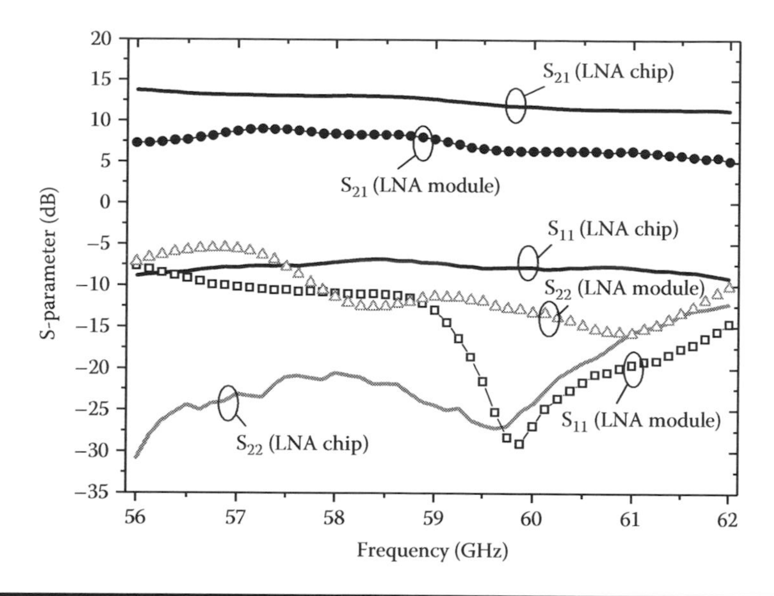

Figure 1.49 The measured S-parameters of the V-band MMIC LNA chip and module.

MMIC chip sets are shown. In the circuit construction of the two-stage LNAs, gate bias lines were designed to have short stubs of a quarter wavelength (4) with the thin-film resistors added to improve the circuit stability. The series feedback was also used to enhance the noise characteristics with a broadband matching. The MMIC medium-power amplifier was optimized for the class-A operation to obtain good linearity characteristics, and a conjugate matching method was used for achieving a good power gain [60]. The MMIC down-mixer adopted a square-law detector structure using LO leakages of the transmitter. The LO and RF were inputted in a gate port, while the IF output was located in the drain port. The V-band MMIC chips were fabricated using a 0.1 GaAs PHEMT MMIC process. S_{21} gains of the V-band MMIC LNA chip and the module were 13.1 dB and 8.3 dB at 58 GHz, respectively. The V-band LNA chip and the module showed noise figures of 3.6 and 5.6 dB, respectively, at 58 GHz. The measured S-parameters are illustrated in Figure 1.49. The MMIC medium-power amplifier (MPA) chip showed an S_{21} gain of 11.04 dB at 58 GHz, and the MPA module showed an S_{21} gain of 6.85 dB at 58 GHz after packaging. As shown in Figure 1.50, the output powers (P1dB) of the MPA MMIC chip and the module were 6.9 and 5.4 dBm, respectively, at 58 GHz. From the V-band MMIC down-mixer, the conversion gains of down-mixer chip and module were 15.2 dB and 19.7 dB at an LO power of 0 dBm, respectively. Figure 1.51 illustrates the conversion losses versus the LO input power of the MMIC down-mixer chip

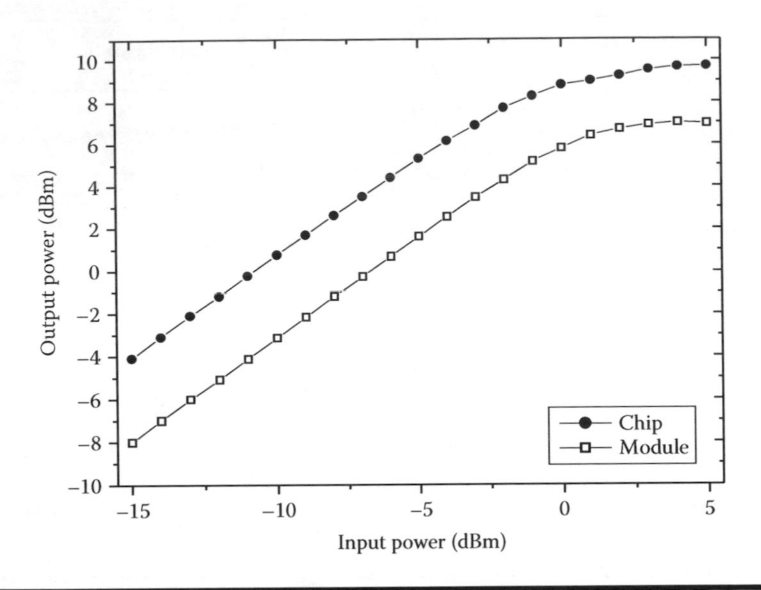

Figure 1.50 Output powers of the V-band MMIC MPA chip and the module.

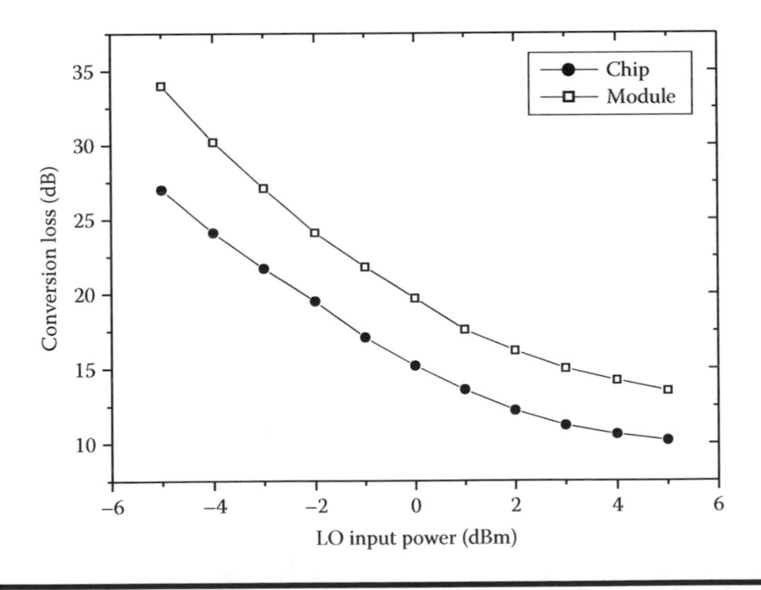

Figure 1.51 Conversion losses versus LO input power of the V-band MMIC down-mixer chip and the module.

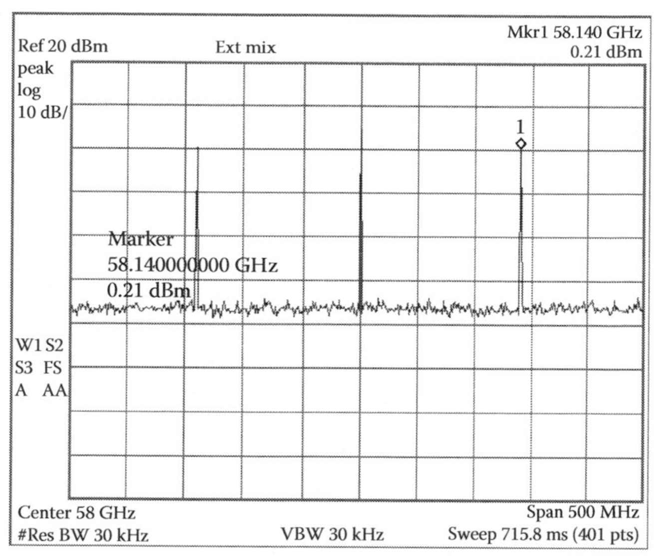

(a) Output spectrum of the V-band transmitter
(If input frequency = 140 MHz, If input power = 0 dB m)

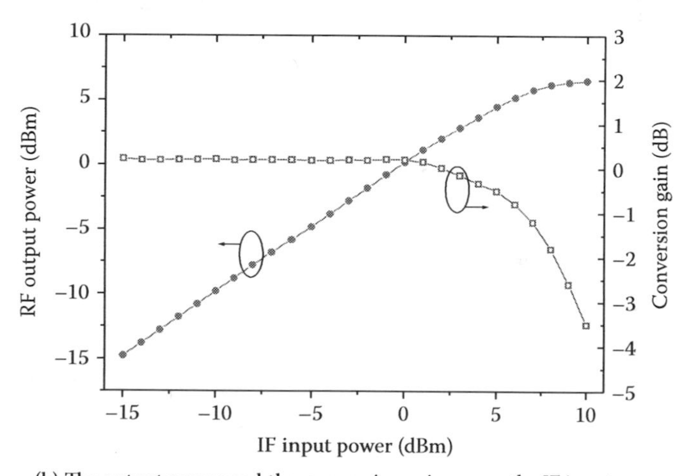

(b) The output power and the conversion gain versus the IF input power

Figure 1.52 The output power and the conversion gain of the V-band transmitter.

and the module. From measured results of the self-heterodyne transceiver, we obtained a conversion gain of 0.21 dB and an output power (1 dB compression point) of 5.2 dBm from the transmitter. Shown in Figure 1.52 are the output power and the conversion gain of the V-band transmitter. From the receiver part, we obtained a conversion gain of 2.1 dB and an output power of −18.6 dBm. Figure 1.53 shows the output powers and the

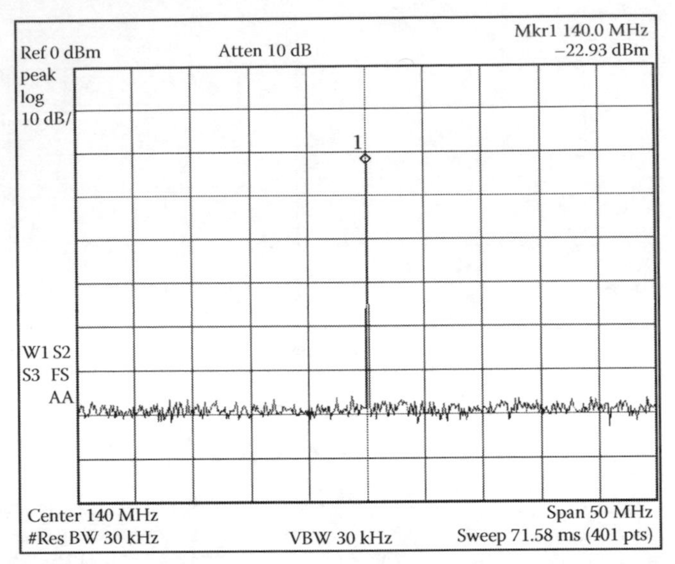

(a) Output spectrum of the V-band receiver (RF input frequency = 58.14 and 57.86 GHz, RF input power = −25 dBm, LO frequency = 58 GHz, LO input power = −22 dBm)

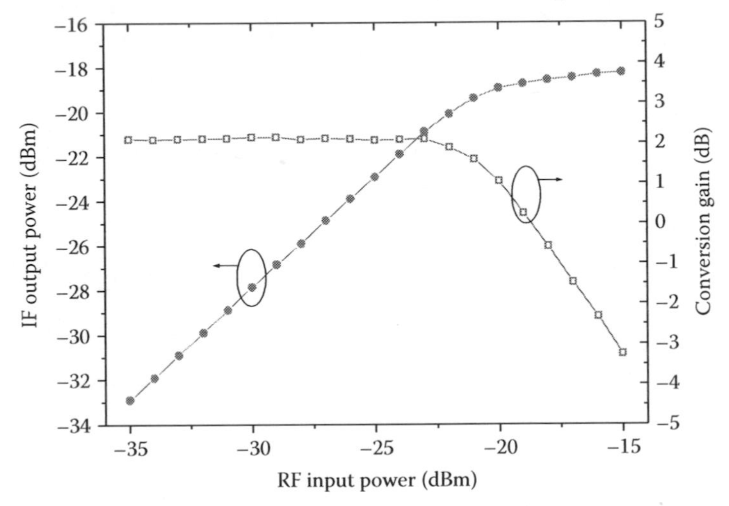

(b) Output powers and the conversion gain versus the RF input power

Figure 1.53 The output powers and the conversion gain of the V-band receiver.

conversion gain of the V-band receiver at an LO input power of −22 dBm. Performance of the V-band self-heterodyne WLAN transceiver is summarized in Table 1.3. From the transmitter, an output power of 5.2 dBm and a conversion gain of 0.2 dB were obtained, while a conversion gain of 2.1 dB was measured from the receiver. Figure 1.54 shows the fabricated V-band

Table 1.3 Performance of the V-Band Self-Heterodyne WLAN Transceiver

Item	*Transmitter*	*Receiver*
Input frequency	140 MHz	RF: 58.14 & 57.86 GHz LO: 58 GHz (From antenna)
Output frequency	RF: 58.14 & 57.86 GHz LO:58 GHz (To antenna)	140 MHz
Amplification block gain	14.5 dB (at 58 GHz)	22.6 dB (at 58 GHz)
Mixer conversion loss	14.3 dB (at LO = 15 dBm)	19.7 dB (at LO = 0 dBm)
Conversion gain	0.2 dB (at LO = 15 dBm)	2.1 dB (at LO = −22 dBm)
Output power (P_{1dB})	5.2 dBm	−18.9 dBm

self-heterodyne WLAN transceiver. From the link tests of the transceiver, the V-band self-heterodyne WLAN transceiver demonstrated a successful data transfer, showing a BER (bit error rate) smaller than 10^{-6}. To reduce self-heterodyne transceiver size, carrier type modules were proposed and developed [61]. In this system, the transmitter sends out the local carrier as well as the modulated RF signal (double-side-band signal), and the received signal is demodulated by a MMIC self-mixer (i.e., a square-law-type detector). The overall implementation cost of the transceiver can be reduced significantly because the transmitter can utilize a low-cost LO source with relatively poor phase noise performance, and the receiver does not need an LO. In addition, the transmitter does not need high-Q filters. The transmitter design is as follows: the 140-MHz IF signal is amplified by an

Figure 1.54 The fabricated V-band self-heterodyne WLAN transceiver using MMIC. (Millimeter-wave INnovation Technology Research Center [MINT] [58].)

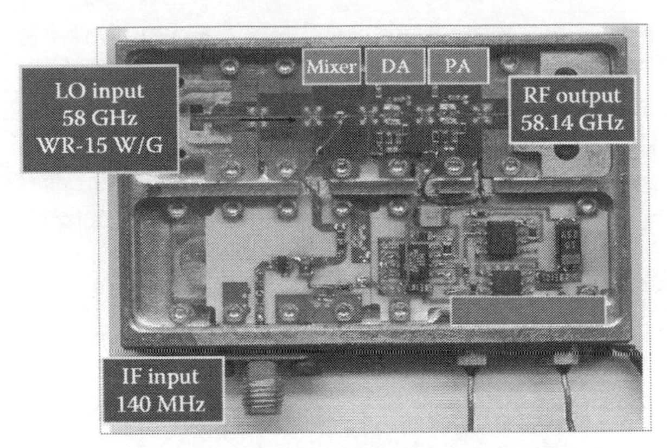

(a) V-band transmitter carrier module

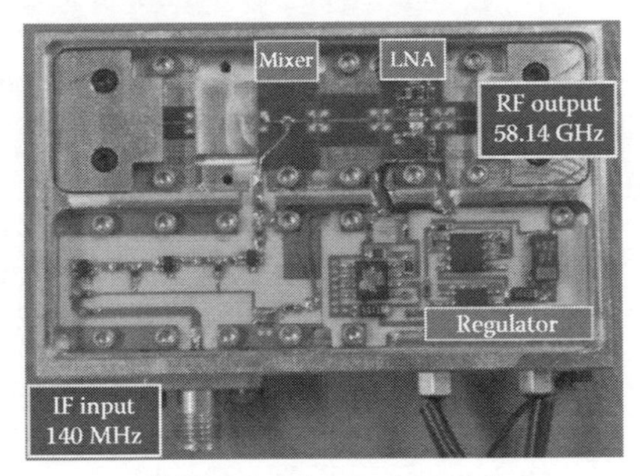

(b) V-band receiver carrier module

Figure 1.55 The implemented V-band self-heterodyne carrier module (Millimeter-wave INnovation Technology Research Center (MINT); size: 69 × 44 × 13 μm^3; 2005 European Microwave Conference) [61].

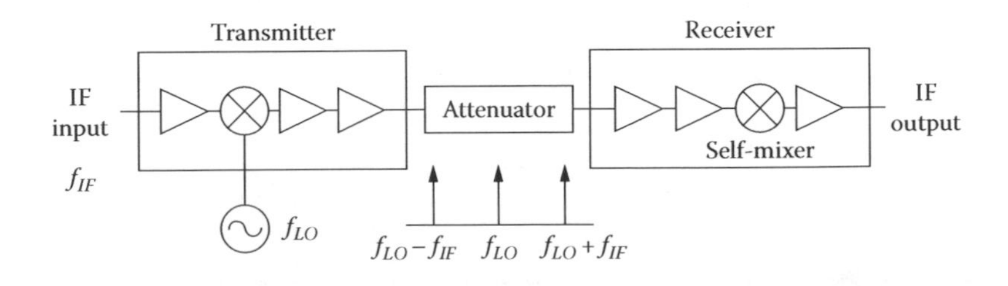

Figure 1.56 Test setup for the V-band self-heterodyne carrier transceiver.

amplifier (Sirenza SGA-4363), which has the power gain of 14 dB. Then, the IF signal passes through an attenuator. This attenuator is used for adjusting the IF power level with the LO leakage level. The IF signal is mixed in the MMIC mixer (Velocium MDB169) with an LO signal. The LO-to-RF leakage (25 dB) is used intentionally to transmit the LO power together with the RF signals. Then, the DSB RF signals and the leaked LO signal are amplified by two MMIC MPAs (Velocium ABH209), which provide sufficient linearity. The receiver design is as follows: the received RF signals and the local carrier are amplified by two MMIC low-noise amplifiers (Velocium ALH382-0). Then, the RF signals with the local carrier are down-converted by the self-mixer. The down-converted IF signal is amplified by two IF amplifiers (Sirenza SGA-4563) for a sufficient gain of 50 dB. For the substrate of the RF front-end, 5-mil Duroid 5880 is used. In order to test the DC and RF properties of each circuit component, a unit circuit block is defined. The unit circuit block contains one major circuit component (such as an MMIC, an MMIC, a mixer, etc.), transmission lines, and bias circuitry. In order to install MMIC components and to attach the Duroid substrate, a carrier, which is made of molybdenum of the size 298 × 600-mil (7.56 × 15.25 mm) and the thickness of 15 mil, is used. The completed transmitter and receiver modules are shown in Figure 1.55. The size of both modules is 69 × 44 × 13 mm^3. After characterizing the DC and RF properties of each unit circuit block, the transmitter/receiver modules are constructed as a cascade connection of these unit circuit blocks. Since the operation frequency is very high (i.e., V-band), the separation distance between these circuit blocks should be minimized, preferably less than 4 mil, in order to minimize the performance degradation. For low-loss interconnections and measurement facilitation, microstrip-to-CPW transitions are used at each end of the unit circuit blocks. To test the module performance, the output of the transmitter and the input of the receiver are connected through a fixed attenuator, which is used to simulate the transmission channel loss, as shown in Figure 1.56. Figure 1.57 shows the output power variation as the input power of the transmitter is increased. From the graph, we obtain the power gain of 13 dB and the output P1dB of +8 dBm. Because two RF signals and a local carrier are amplified, the amplifier is saturated at a lower power level than the P1dB in the data sheet. Figure 1.58 shows the output signal power variation as the RF-to-local-carrier power ratio of the transmitter changes. The receiver IF output power depends on the power distribution between the transmitted local carrier and the RF signals, and the maximum IF output power can occur when the power level of the LO and RF signals are equal [58]. Also, the phase noise characteristics of the IF input signal of the transmitter and the IF output signal of the receiver are obtained. As shown in Table 1.4, the phase noise characteristics of the IF output signal are degraded by less than 3 dB compared with that of the IF input signal over the entire range of the frequency offset. Therefore, the

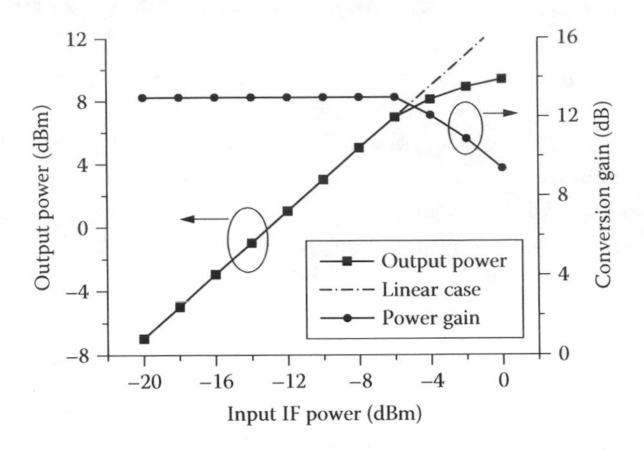

Figure 1.57 The output power variation for the input power of the V-band transmitter.

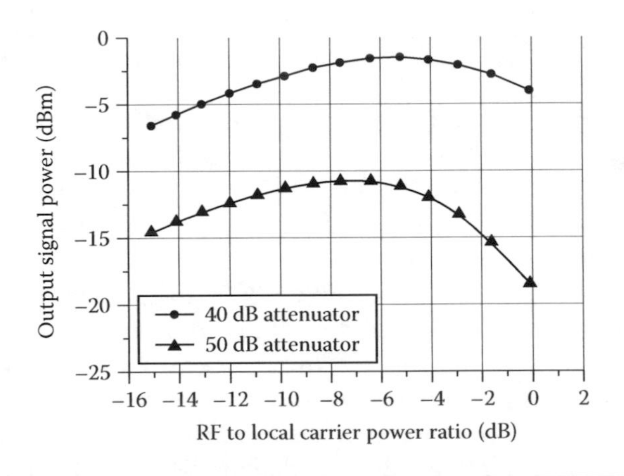

Figure 1.58 IF output power variation as the RF-to-local-carrier power ratio of the transmitter changes [61].

Table 1.4 Comparison of the Input IF and Output IF Phase Noise [unit: dBc/Hz]

Offset freq. [KHz]	*1*	*10*	*100*
Input IF	−8.5	−92.5	−112.8
Output IF	−80.2	−92.1	−110.0

phase noise degradation by the self-heterodyne receiver could be negligible even if the LO of the transmitter has poor phase noise characteristics.

1.5 Conclusion

This chapter was concerned mainly about the design and the process technology of the millimeter-wave monolithic integrated circuit for WLAN applications. The device modeling technologies of millimeter-wave devices (active and passive) were described. HEMT devices (pseudomorphic HEMT and metamorphic HEMT) were mainly discussed in this chapter. Also, the design technology of a monolithic integrated circuit was presented in millimeter-wave frequency. In the millimeter-wave MMIC circuit design, analysis concerning CPW parasitic effects can be used for the practical MMIC design criteria. The presented design technology has the advantage of simulation time reduction and efficient EM simulation. In addition, various millimeter-wave monolithic integrated circuits were described. Results of V-band MMICs for millimeter-wave WLAN were mainly introduced. MMIC process technologies such as 0.1-gate formation, dielectric passivation, and airbridge were explained. Finally, this chapter described WLAN applications of the millimeter-wave monolithic integrated circuits. At the system architecture for millimeter-wave wireless LAN, a low-cost V-band DSB self-heterodyne transceiver with no filter and an LO in the receiver blocks was presented. For the high-productivity and low-cost solution, MMIC chip sets and self-heterodyne architecture were used, including the miniature carrier-type transceiver. Furthermore, continuous research of high-performance millimeter-wave circuits needs to improve the characteristics of millimeter-wave WLAN. A low-cost MCM using MMIC is one of the best candidates for millimeter-wave wireless LAN systems. Finally, the development of the fully integrated one-chip MMIC transceiver including amplifier, mixer, filters, oscillators, and planar antennas is an aggressive part of further study in advanced millimeter-wave WLAN applications.

References

[1] Yukio Takimoto, Recent Activities on Millimeter-Wave Indoor LAN System Development in Japan, in Proceedings of IEEE Microwave System Conference 1995, pp. 7–10.

[2] Sung-Chan Kim, Dan An, Dong-Hoon Shin, and Jin-Koo Rhee, High Performance Quadruple Sub-Harmonic Mixer for Millimeter-Wave Circuits, *Microelectrics Journal*, vol. 34, no. 11, pp. 1093–1098, Nov. 2003.

[3] Won-Young Uhm, Woo-Suk Sul, Hyo-Jong Han, Sung-Chan Kim, Han-Sin Lee, Dan An, Sam-Dong Kim, Dong-Hoon Shin, Hyung-Moo Park, and Jin-Koo Rhee, A High Performance V-band Monolithic Quadruple

Sub-Harmonic Mixer, in Proceedings of IEEE Int. Microwave Symp. 2003, Philadelphia, USA, June 2003, pp. 1319–1322.

[4] Dan An, Bok Hyung Lee, Yeon Sik Chae, Hyun Chang Park, Hyung Moo Park, and Jin Koo Rhee, Low LO Power V-band CPW Down-Converter Using a GaAs PHEMT, *J. Korean Phys. Soc.*, vol. 41, no. 6, pp. 1013–1016, Dec. 2002.

[5] Sung-Chan Kim, Dan An, Byeong Ok Lim, Tae Jong Baek, Dong Hoon Shin, and Jin Koo Rhee, High-Performance 94-GHz Single Balanced Mixer Using 70-nm MHEMTs and Surface Micromachined Technology, *IEEE Elec. Device Lett.*, vol. 27, no. 1, pp. 28–30, Jan. 2006.

[6] Sung-Chan Kim, Dan An, Woo-Suk Sul, Han-Shin Lee, Hyo-Jong Han, Won-Young Uhm, Hyung-Moo Park, Sam-Dong Kim, Dong-Hoon Shin, and Jin-Koo Rhee, High Conversion Gain Cascode Quadruple Subharmonic Mixer for Millimeter-Wave Applications, *Current Applied Physics* (CAP), vol. 5, issue 3, pp. 231–236, March 2005.

[7] Federal Communications Commission (FCC), Millimeter-Wave Propagation: Spectrum Management Implications, FCC document, 1997.

[8] T. Mimura, S. Hiyamizu, and S. Hikosak, Enhancement Mode High Electron Mobility Transistors for Logic Applications, *Japanese J. Appl. Phys.*, p. L317, 1981.

[9] B-H Lee, S-C Kim, M-K Lee, W-S Sul, B-O Lim, W-Y Uhm, and J-K Rhee, Q-band High Conversion Gain Active Sub-Harmonic Mixer, *Current Applied Physics*, vol. 4, pp. 69–73, Feb. 2004. Reprinted from Publication title, Copyright (2004), with permission from Elsevier.

[10] Seong-Dae Lee, Sung-Chan Kim, Bok-Hyoung Lee, Woo-Suk Sul, Byeong-Ok Lim, Dan An, Youg-Soon Yoon, Sam-Dong Kim, Dong-Hoon Shin, Jin-Koo Rhee, Design and Fabrication of the 0.1-Shaped Gate PHEMTs for Millimeter-Waves, *J. Korea Electromagnetic Eng. Soc.*, vol. 1, no. 1, pp. 73–78, May 2001.

[11] B-H Lee, S-D Kim, and J-K Rhee, Small Signal Analysis of High fmax 0.1-μm Off-set-Shaped Gate InGaAs/InAlAs/GaAs Metamorphic HEMTs, *Japanese J. Appl. Phys.*, vol. 43, pp. 1914–1918, 2004.

[12] B-H Lee, S-D Kim, and J-K Rhee, High Maximum Frequency of Oscillation of 0.1-μm Off-set-shaped gate InGaAs/InAlAs/GaAs Metamorphic HEMTs, *Journal of Korea Physics Society*, vol. 42, pp. 427–430, 2003.

[13] M-S Son, B-H Lee, M-R Kim, S-D Kim, and J-K Rhee, Simulation of the DC and Millimeter-Wave Characteristics of 0.1-μm Offset T-shaped Gate InxGa1-xAs/ In0.52Al0.48As/GaAs MHEMTs, *Journal of Korea Physics Society*, vol. 44, pp. 408–417, 2004.

[14] Gilles Dambrine, Alain Cappy, Frederic Heliodore, and Edouard Playez, A New Method for Determining the FET Small-Signal Equivalent Circuit, *IEEE Trans. Microwave Theory Tech.*, vol. 36, no. 7, pp. 1151–1159, July 1988.

[15] Agilent Technologies, High-Frequency Model Tutorials. Agilent IC-CAP 5.3, 2000.

[16] Sang Jin Lee, Du Hyun Ko, Jin Man Jin, Dan An, Mun Kyo Lee, Chang Shik Cho, Bok Hyung Lee, Yong Hyun Baek, Yeon Sik Chae, Hyung Moo Park, and Jin Koo Rhee, The Low Conversion Loss and LO Power V-Band MIMIC Up-Mixer, in Proceedings of AWAD 2004, July 2004, pp. 39–42.

[17] Dan An, Bok Hyung Lee, Yeon Sik Chae, Hyung Moo Park, and Jin Koo Rhee, Low LO Power V-Band CPW Mixer Using GaAs PHEMT, in 32nd European Microwave Conference, Milan, Italy, Sep. 2002, pp. 773–776.

[18] Sang-Jin Lee, Dan An, Mun-Kyo Lee, Jin-Man Jin, Du-Hyun Ko, Chang-Sik Cho, Seong-Dae Lee, Tae-Jong Baek, Seong-Chan Kim, Hyung-Moo Park, Jin-Koo Rhee, Design and Fabrication of the Low Conversion Loss Self Oscillating Mixer for Q-band, in International Technical Conference on Circuits/Systems, *Computers and Communications*, pp. 371–372, July 2005.

[19] Woo Suk Sul, Dan An, Sung Dae Lee, Bok Hyoung Lee, Sung Chan Kim, Byeong Ok Lim, Dong Hoon Shin, Sam Dong Kim, Do Hyun Kim, and Jin Koo Rhee, Design and Fabrication of V-Band CPW Power Amplifier using GaAs PHEMT, in Proceedings of the Int. Conference on Electronics Information and Communications (ICEIC 2002), July 2002, pp. 324–327.

[20] Tae-Sin Kang, Seong-Dae Lee, Bok-Hyoung Lee, Sam-Dong Kim, Hyun-Chang Park, Hyung-Moo Park, and Jin-Koo Rhee, Design and Fabrication of a Low-Noise Amplifier for the V-Band, *J. Korean Phys. Soc.*, vol. 41, no. 4, pp. 533–538, Oct. 2002.

[21] Ravender Goyal, *Monolithic Microwave Integrated Circuits: Technology & Design*. Norwood, MA: Artech House, 1989.

[22] Ralph Williams, *Modern GaAs Processing Methods*. Norwood, MA: Artech House, 1990.

[23] Giovanni Ghione, and Carlo U. Naldi, Coplanar Waveguides for MMIC Applications: Effect of Upper Shielding, Conductor Backing, Finite-Extent Ground Planes, and Line-to-Line Coupling, *IEEE Trans. on Microwave Theory and Tech.*, vol. MTT-35, no. 3, pp. 260–267, March 1987.

[24] Bok-Hyung Lee, Dan An, Mun-Kyo Lee, Byeong-Ok Lim, Sam-Dong Kim, and Jin-Koo Rhee, Two-Stage Broadband High-Gain W-Band Amplifier Using 0.1 Metamorphic HEMT Technology, *IEEE Elec. Device Lett.*, vol. 25, no. 12, pp. 766–768, Dec. 2004.

[25] Chang-Shik Cho, Dan An, Seong-Dae Lee, Jin-Man Jin, Tae-Jong Baek, Seok-Gyu Choi, Sam-Dong Kim, and Jin-Koo Rhee, Design and Fabrication of V-Band Low Noise Amplifiers Using GaAs PHEMTs, 2004 International SoC Design Conference, pp. 172–175, Oct. 2004.

[26] Kohei Fujii, Miroslaw Adamski, Piero Bianco, Daniel Gunyan, John Hall, Ron Kishimura, Camille Lesko, Matthias Schefer, Steve Hessel, Henrik Morkner, and Antoni Niedzwiecki, A 60GHz MMIC Chipset for 1-Gbit Wireless Links, 2002 IEEE MIT-S Digest, pp. 1725–1728.

[27] Kenjiro Nishikawa, Belinda Piernas, Kenji Kamogawa, Tadao Nakagawa, and Katsuhiko Araki, Compact LNA and VCO 3-D MMICs Using Commercial GaAs PHEMT Technology for V-Band Single-Chip TRX MMIC, 2002 IEEE MIT-S Digest, pp. 1717–1721.

[28] A. Fujihara, E. Mizuki, H. Miyamoto, Y. Makino, K. Yamanoguchi, and N. Samoto, High Performance 60-GHz Coplanar MMIC LNA Using InP Hetero-Junction FETs with AlAs/InAs Super-Lattice Layer, 2000 IEEE MIT-S Digest, pp. 21–24.

[29] J. M. Tanskanen, P. Kangaslahti, H. Ahtola, P. Jukkala, T. Karttaavi, M. Lahdes, J. Varis, and J. Tuovinen, Cryogenic Indium-Phosphide HEMT

Low-Noise Amplifiers at V-Band, *IEEE Trans. on Microwave Theory and Tech.*, vol. 18, no. 7, pp. 1283–1286, July 2000.

[30] Yutaka Mimino, Kannichi Nakamura, Yuichi Hasegawa, Yoshio Aoki, Shigeru Kuroda, and Tsuneo Tokumitsu, A 60 GHz Millimeter-Wave MMIC Chipset for Broadband Wireless Access System Front-End, 2002 IEEE MIT-S Digest, pp. 1721–1724.

[31] C. A. Zelley, A. R. Barnes, D. C. Bannister, and R. W. Ashcroft, A 60 GHz integrated Sub-Harmonic Receiver MMIC, 2003 IEEE GaAs Digest, pp. 175–178.

[32] Wendell M. T. Kong, Sujane C. Wang, Pane-Chane Chao, Der-Wei Tu, Kuichul Hwang, O. S. A. Tang, Shih-Ming Liu, Pin Ho, Kirby Nichols, and John Heaton, Very High Efficiency V-Band Power InP HEMT MMICs, *IEEE Elec. Device Lett.*, vol. 21, no. 11, pp. 521–523, Nov. 2000.

[33] R. Lai, M. Nishmito, Y. Hwang, M. Biedenbender, B. Kasody, C. Geiger, Y. C. Chen, and G. Zell, A High Efficiency 0.15 μm 2-mil Thick InGaAs/AlGaAs/GaAs V-Band Power HEMT MMIC, 1996 IEEE GaAs IC Symposium, pp. 225–227.

[34] Seng-Woon Chen, Phillip M. Smith, Shih-Ming J. Liu, William F. Kopp, and Thomas J. Rogers, A 60-GHz High Efficiency Monolithic Power Amplifier Using 0.1-μm PHEMTs, *IEEE Microwave and Guided Wave Lett.*, vol. 5, no. 6, pp. 201–203, June 1995.

[35] R. E. Kasody, G. S. Dow, A. K. Sharma, M. V. Aust, D. Yamauchi, R. Lai, M. Biedenbender, K. L. Tan, and B. R. Allen, A High Efficiency V-Band Monolithic HEMT Power Amplifier, *IEEE Microwave and Guided Wave Lett.*, vol. 4, no. 9, pp. 303–304, Sep. 1994.

[36] Arvind K. Sharma, Gerald P. Onak, Richard Lai, and Kin L. Tan, A V-Band High-Efficiency Pseudomorphic HEMT Monolithic Power Amplifier, *IEEE Trans. on Microwave Theory and Tech.*, vol. 42, no. 12, pp. 2603–2609, Dec. 1994.

[37] O. S. Andy Tang, K. H. George Duh, S. M. Joseph Liu, Phillip M. Smith, William F. Kopp, Thomas J. Rogers, and David J. Pritchard, Design of High-Power, High-Efficiency 60-GHz MMICs Using an Improved Nonlinear PHEMT Model, *IEEE Journal of Solid-State Circuits*, vol. 32, no. 9, pp. 1326–1333, Sep. 1997.

[38] Huei Wang, Kwo-Wei Chang, Dennis Chung-Wen Lo, Liem T. Tran, John C. Cowles, Thomas R. Block, Gee Sam Dow, Aaron Oki, Dwight C. Streit, and Barry R. Allen, A 62-GHz Monolithic InP-Based HBT VCO, *IEEE Microwave and Guided Wave Lett.*, vol. 5, no. 1, pp. 388–390, Nov. 1995.

[39] C. Lee, T. Yao, A. Mangan, K. Yau, M. A. Copeland, and S. P. Voinigescu, SiGe BiCMOS 65-GHz BPSK Transmitter and 30 to 122 GHz LC- Varactor VCOs with up to 212003, *IEEE CSIC Digest*, pp. 179–182.

[40] Takuo Kashiwa, Takao Ishida, Takayuki Katoh, Hitoshi Kurusu, Hiroyuki Hoshi, and Yasuo Mitsui, V-Band High-Power Low Phase-Noise Monolithic Oscillators and Investigation of Low Phase-Noise Performance at High Drain Bias, *IEEE Trans. on Microwave Theory and Tech.*, vol. 46, no. 10, pp. 1559–1565, Oct. 1998.

[41] Keiichi Qhata, Takashi Inoue, Masahiro Funabashi, Akihiko Inoue, Yukio Takimoto, Toshihide Kuwabara, Satoru Shinozaki, Kenichi Mamhashi,

Kenichi Hosaya, and Hiroaki Nagai, Sixty-GHz-Band Ultra-Miniature Monolithic T/R Modules for Multimedia Wireless Communication Systems, *IEEE Trans. on Microwave Theory and Tech.*, vol. 44, no. 2, pp. 2354–2360, Dec. 1996.

[42] Won-Young Uhm, Woo-Suk Sul, Hyo-Jong Han, Sung-Chan Kim, Han-Sin Lee, Dan An, Sam-Dong Kim, Dong-Hoon Shin, Hyung-Moo Park, and Jin-Koo Rhee, A High Performance V-Band Monolithic Quadruple Sub-Harmonic Mixer, in Proceedings of IEEE Int. Microwave Symp. 2003, Philadelphia, USA, June 2003, pp. 1319–1322.

[43] Dan An, Sung-Chan Kim, Woo-Suk Sul, Hyo-Jong Han, Han Shin Lee, Won Young Uhm, Hyung-Moo Park, Sam-Dong Kim, Dong-Hoon Shin, and Jin-Koo Rhee, High Conversion Gain V-Band Quadruple Subharmonic Mixer Using Cascode Structure, in IEEE MTT-S Int. Microwave Symp. Dig., Philadelphia, USA, June 2003, pp. 911–914.

[44] A. Orzati, F. Robin, H. Meier, H. Benedikter, and W. Bechtold, A V-Band Up-Converting InP HEMT Active Mixer with Low LO-Power Requirements, *IEEE Microwave and Wireless Components Lett.*, vol. 13, no. 6, pp. 202–204, June 2003.

[45] Kyung-Whan Yeom and Du-Hyun Ko, A Novel 60-GHz Monolithic Star Mixer Using Gate-Drain-Connected pHEMT Diodes, *IEEE Trans. on Microwave Theory and Tech.*, vol. 53, no. 7, pp. 2435–2440, July 2005.

[46] C. A. Zelley, A. R. Barnes, D. C. Bannister, and R. W. Ashcroft, A 60 GHz Integrated Sub-Harmonic Receiver MMIC, 2003 IEEE GaAs Digest, pp. 175–178.

[47] Michael J. Roberts, Stavros Iezekiel, and Christopher M. Snowdan, A W-Band Self-Oscillating Subharmonic MMIC Mixer, *IEEE Trans. Microwave Theory and Tech.*, vol. 46, no. 12, pp. 2104–2108, Dec. 1998.

[48] Michael W. Chapman and Sanjay Raman, A 60 GHz Uniplanar MMIC 4 × Subharmonic Mixer, in IEEE MTT-S Digest, 2001, pp. 95–98.

[49] Dan An, Bok-Hyung Lee, Byeong-Ok Lim, Mun-Kyo Lee, Sung-Chan Kim, Jung-Hun Oh, Sam-Dong Kim, Hyung-Moo Park, Dong-Hoon Shin, and Jin-Koo Rhee, High Switching Performance 0.1- Metamorphic HEMTs for Low Conversion Loss 94 GHz Resistive Mixers, *IEEE Elec. Device Lett.*, vol. 26, no. 10, pp. 707–709, Oct. 2005.

[50] C. S. Whelan et al., Low Noise In0.32(AlGa)0.68As/In0.43Ga0.57As Metamorphic HEMT on GaAs Substrate with 850 mW/mm Output Power Density, *IEEE Electron Device Lett.*, vol. 21, no. 1, pp. 5–8, Jan. 2000.

[51] K. W. Chang, E. W. Lin, H. Wang, K. L. Tan, and W. H. Ku, A W-band Monolithic, Singly Balanced Resistive Mixer with Low Conversion Loss, *IEEE Microwave Guided Wave Lett.*, vol. 4, no. 9, pp. 301–302, Sep. 1994.

[52] R. S. Virk, L. Tran, M. Matloubian, M. Le, M. G. Case, and C. Ngo, A Comparison of W-Band MMIC Mixers using InP HEMT Technology, in IEEE MTT-S Digest, 1997, pp. 435–438.

[53] S. C. Kim, B. O. Lim, H. S. Lee, D. H. Shin, S. K. Kim, H. C. Hark, and J. K. Rhee, Sub-100nm T-gate fabrication using a positive resist ZEP520/P(MMA-MAA) PMMA trilayer by double exposore at 50 KV e-beam lithography, *Material Science in Semiconductor Processing* no. 7, pp. 7–11, 2004. Reprinted from Publication title. Copyright (2004), with permission from Elsevier.

[54] H. S. Kim, B. O. Lim, S. C. Kim, S. D. Lee, D. H. Shin, and J. K. Rhee, Study of the Fabrication of PHEMTs for a 0.1 Scale Gate Using Electron Beam Lithography: Structure, Fabrication, and Characteristics, *Microelectronics Eng.*, vol. 63, pp. 417–431, 2002.

[55] J. W. Shin, Y. S. Yoon, S. D. Lee, H. C. Park, and J. K. Rhee, Effect of He Gas on Hydrogen Content and Passivation of GaAs PHEMT with SiN Film, in Proc. Asia-Pacific Workshop on Fundamental and Application of Advanced Semiconductor Devices, 2000, pp. 121–124.

[56] I. H. Lee, S. D. Lee, and J. K. Rhee, Studies on Air-Bridge Processes for mm-Wave MMIC's Applications, *J. Korean Phys. Soc.*, vol. 35, no. 94, pp. S1043–S1046, 1999.

[57] Y. Shoji, K. Hamaguchi, and H. Ogawa, A Low-Cost and Stable Millimeter-Wave Transmission System Using a Transmission-Filter-Less Double-Side-Band Millimeter-Wave Self-Heterodyne Transmission Technique, *IEICE Trans. Commun.*, vol. E86-B, no. 6, pp. 1884–1892, June 2003.

[58] Dan An, Mun-Kyo Lee, Sang-Jin Lee, Du-Hyun Ko, Jin-Jin Man, Sung-Chan Kim, Sam-Dong Kim, Hyun-Chang Park, Hyung-Moo Park, and Jin-Koo Rhee, V-Band Self-heterodyne Wireless Transceiver using MMIC Modules, *Journal of Semiconductor Technology and Science*, vol. 5, no. 3, pp. 69–79, Sep. 2005.

[59] D. H. Ko, J. Y. Moon, D. An, M. K. Lee, S. J. Lee, J. M. Jin, Y. S. Chae, S. W. Yun, S. D. Kim, H. M. Park, and J. K. Rhee, V-Band Waveguide-to-Coplanar Waveguide Transition for 60 GHz Wireless LAN Application, in 34th European Microwave Conference, Oct. 2004, pp. 641-644.

[60] Du Hyun Ko, Jin Man Jin, Sang Jin Lee, Dan An, Mun Kyo Lee, Seong Dae Lee, Byong Chul Jun, Min Han, Yeon Sik Chae, Hyung Moo Park, and Jin Koo Rhee, V-Band CPW Medium Power Amplifier for 60 GHz Wireless LAN Application, in 2004 AWAD conference, July 2004, pp. 49–52.

[61] Tae-Gyu Kim, Dong-Sik Woo, Kang Wook Kim, Implementation of Low-Cost 60 GHz Self-Heterodyne Transceiver Modules for WPAN Applications, in 35th European Microwave Conference, Paris, France, Oct. 2005, pp. 741–744.

Chapter 2

Package Technology for Millimeter-Wave Circuits and Systems

J.-H. Lee, S. Pinel, J. Laskar, and M. M. Tentzeris

Contents

2.1 Integrated Passives

2.2 Introduction

Millimeter-wave electronics for commercial applications, such as short-range broadband wireless communications and automotive collision avoidance radars, require low manufacturing costs, excellent performance, and a high level of integration. Especially, the 60-GHz band is of much interest because this is the band in which massive amounts of spectrum space (7 GHz) have been allocated worldwide for dense wireless local communications [1]. There are a number of multimedia applications for short-range communications, such as high-speed Internet access, video streaming, content downloads, and wireless data bus for cable replacement [2]. Such emerging applications with data rates greater than 2 Gb/s require real estate efficiency, low-cost manufacturing, and excellent performance achieved by a high level of integration of embedded functions [1]. The multilayer low-temperature co-fired ceramic (LTCC) system-on-package (SOP) approach is very well suited for these requirements because it offers a great potential for passives' integration and enables microwave devices to be fabricated with high reliability while maintaining the low cost. LTCC has been widely used as a packaging material because of its process maturity/stability and its relatively high dielectric constant that enables a significant reduction in the module/function dimensions. As an alternative, liquid crystal polymer (LCP) is an organic material that offers a unique low-cost all-in-one solution for high-frequency designs due to its ability to act as both a high-performance flexible substrate ($\epsilon_r = 2.9$–3.0, $\tan\delta = 0.002 - 0.004$) and a near-hermetic package for multilayer modules. These characteristics make LCP very appealing for many applications, and it can be viewed as a prime technology for enabling system-on-package radio frequency (RF) and millimeter-wave designs.

In the first section of this chapter, we introduce the three-dimensional (3D), multilayer module concept using SOP technologies. In the second section, we address the development of various passive components using SOP technology as the candidates for 3D integration of compact, low-cost wireless front-end systems used in RF and up to V-band frequency ranges. The concept of the modern soft surface (SS) is applied to improve the radiation pattern of patch antennas. The single-mode, slotted-patch filter with a transverse cut on each side has been exploited at the operating frequencies of 58–60 GHz. Vertically stacked, 3D, low-loss cavity bandpass filters (BPFs) are developed for receiver (Rx) and transmitter (Tx) channels to realize a fully integrated, compact filter solution. In the third section, the fully integrated cavity filters and the dual-polarized antenna that covers Rx and Tx channels are proposed, employing the presented designs of the filters and air cavities. The stringent demand of high isolation between two channels induces the advanced design of a duplexer and an antenna as a fully integrated function for the V-band front-end module.

2.3 3D Integrated Module Concept

Figure 2.1a illustrates the proposed 3D, multilayer module concept [3]. Two stacked, SOP multilayer substrates are used, and board-to-board, vertical transition is ensured by μBGA balls. Standard alignment equipment is used

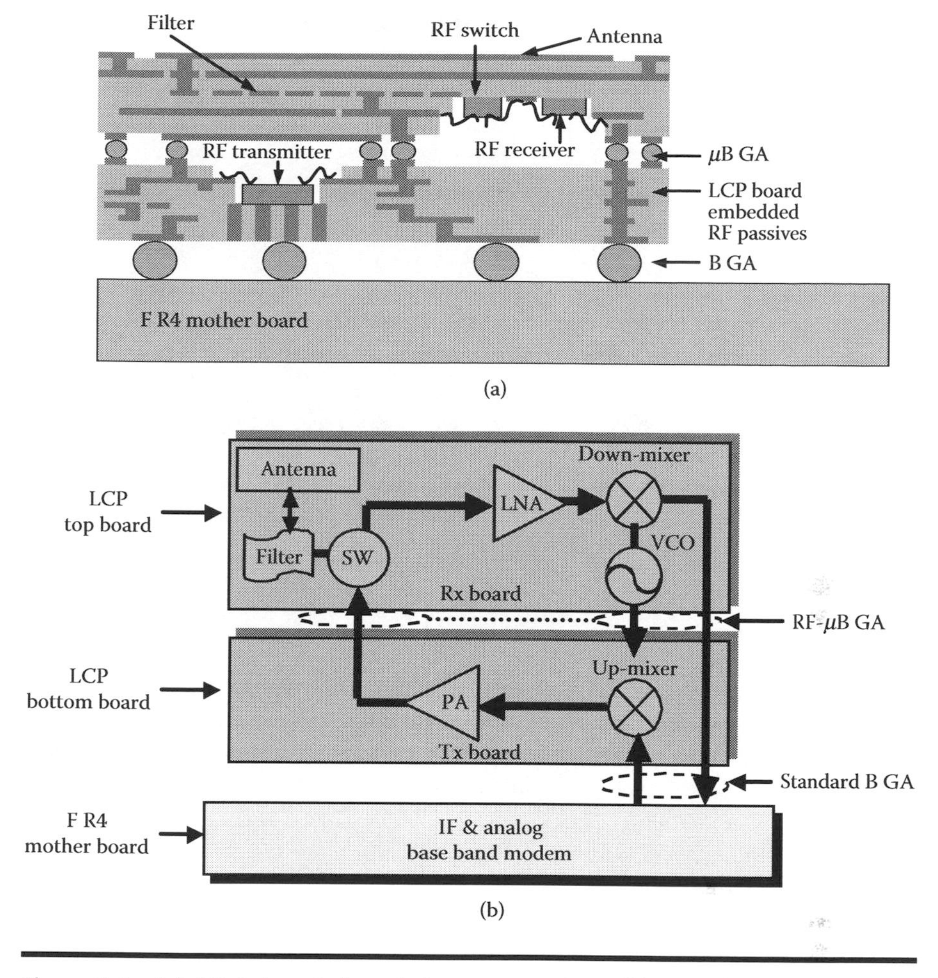

Figure 2.1 (a) 3D integrated module concept view, (b) Rx and Tx board block diagram.

to stack the board and, thus, provide a compact, high-performance, and low-cost assembly process. Multi-stepped cavities into the SOP boards provide spacing for the embedded RF active devices (RF switch, RF receiver, and RF transmitter) chipset and thus, lead to significant volume reduction by minimizing the gap between the boards. Active devices can be flip-chipped as well as wire-bonded. Cavities also provide an excellent opportunity for the easy integration of RF micro-electro-mechanical systems (MEMS) devices, such as MEMS switches or tuners. Passive components, off-chip matching networks, embedded filters, and antennas are implemented directly into the SOP boards using multilayer technologies [4–5]. Standard BGA balls ensure the effective broadband interconnection of this high-density module with motherboards such as FR4 board. The top and bottom substrates are dedicated respectively to the receiver and transmitter building blocks of the RF front-end module. Figure 2.1b shows the RF block diagram of each board. The receiver board includes the antenna, BPF, active switch, and RF receiver chipset (LNA, low-noise amplifier; VCO, voltage-controlled oscillator; and down-conversion mixer). The transmitter board includes the RF transmitter chipset (up-conversion mixer and power amplifier) and off-chip matching networks. Ground planes and vertical via walls are used to address the isolation issues between the transmitter and the receiver functional blocks. Arrays of vertical vias are added into the transmitter board to achieve better thermal management.

2.3.1 Patch Antenna Using Soft Surface Structure

The radiation performance of patch antennas on a large-size substrate can be significantly degraded by the diffraction of surface waves at the edge of the substrate. Most of the modern techniques for surface-wave suppression are related to periodic structures, such as photonic bandgap (PBG) or electromagnetic bandgap (EBG) geometries [6–8]. However, those techniques require a considerable area to form a complete bandgap structure. In addition, it is usually difficult for most printed-circuit technologies to realize such a perforated structure. We apply the concept of the modern SS to improve the radiation pattern of patch antennas [9]. A single square ring in the form of shorted quarter-wavelength metal strips is employed to suppress the outward propagating surface wave radiated by the patch in the center, thus alleviating the diffraction at the edge of the substrate. Since only a single ring of metal strips is involved, the formed SS structure is compact and easily integrated with 3D modules.

For the sake of simplicity, we consider a probe-fed square-patch antenna on a square grounded substrate with thickness H and a dielectric constant ϵ_r. The patch antenna is surrounded by an ideal compact SS structure, which consists of a square ring of metal strip short-circuited to the ground plane by a metal wall along the outer edge of the ring, as shown in Figure 2.2. The substrate area has a size of $L \times L$, much larger than the size of the square patch ($L_p \times L_p$). The inner length of the SS ring (denoted by L_s) was found to be approximately one wavelength plus L_p. The width of the metal strip (W_s) is approximately equal to a quarter of the guided wavelength.

To demonstrate the feasibility of this technology on the implementation of the SS, we first simulated a benchmarking prototype that was constructed to replace the shorting wall with a ring of vias. The utilized LTCC material had a dielectric constant of 5.4. The whole module consists of a total of 11 LTCC layers (layer thickness = 100 μm) and 12 metal layers (layer thickness = 10 μm). The diameter of each via was specified by the fabrication process to be 100 μm, and the distance between the centers of two adjacent vias

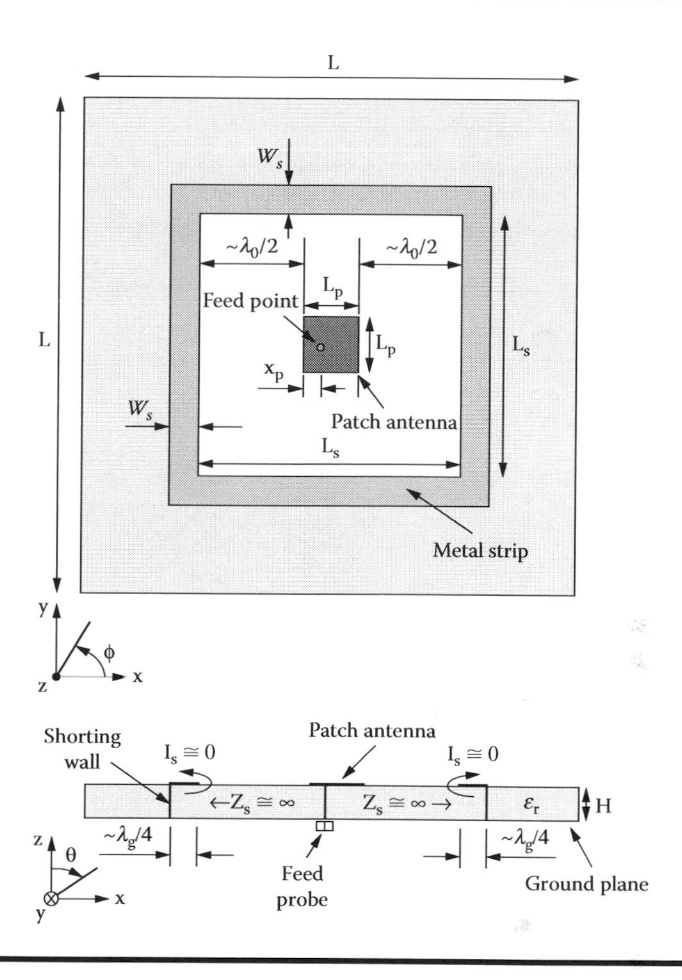

Figure 2.2 Patch antenna surrounded by an ideal compact soft surface structure, which consists of a ring of metal strip and a ring of shorting wall. (I_s = the surface current on the top surface of the soft surface ring; Z_s = the impedance looking into the shorted metal strip.)

was 500 μm. To support the vias, a metal pad is required on each metal layer. To simplify the simulation, all pads on each metal layer are connected by a metal strip with a width of 600 μm. Simulation showed that the width of the pad metal strips has little effect on the performance of the SS structure as long as it is less than the width of the metal strips for the SS ring (W_S). The size of the LTCC board was 30 mm × 30 mm. The operating frequency was set within the Ku band (the design frequency f_o = 16.5 GHz). The optimized values for L_S and W_S were, respectively, 22.2 mm and 1.4 mm, which led to a total via number of 200 (51 vias on each side of the square ring). Including the width (300 μm) of the pad metal strip, the total metal strip width for the SS ring was found to be 1.7 mm.

Since the substrate was electrically thick at $f_o = 16.5$ GHz ($>0.1\ \lambda_g$), a stacked configuration was adopted for the patch antenna to improve its input impedance performance. By adjusting the distance between the stacked square patches, a broadband characteristic for the return loss can be achieved [10]. For the present case, the upper and lower patches (with the same size, 3.4 mm × 3.4 mm) were respectively printed on the first LTCC layer and the seventh layer from the top, leaving a distance between the two patches of six LTCC layers. The lower patch was connected by a via hole to a 50-Ω microstrip feed line that is on the bottom surface of the LTCC substrate. The ground plane was embedded between the second and third LTCC layers from the bottom. Figure 2.3 shows the layout and

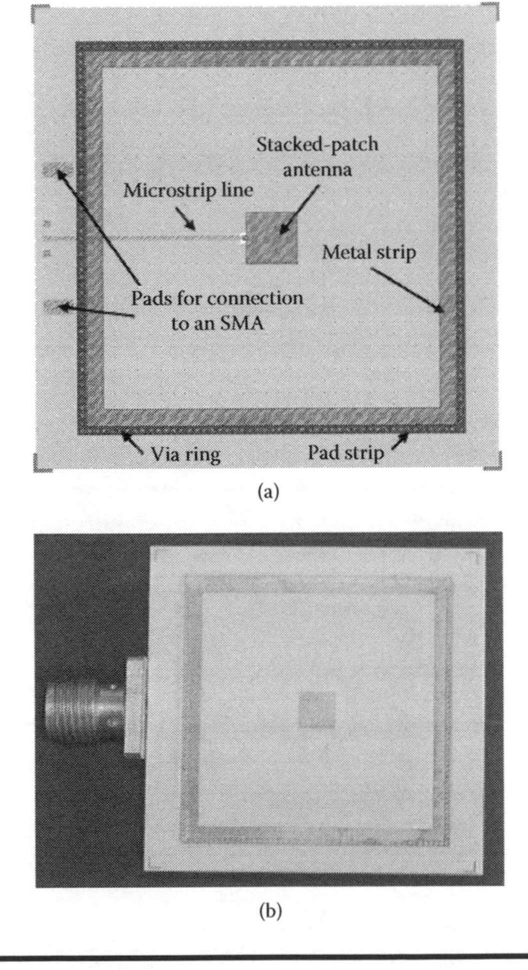

Figure 2.3 (a) Layout and (b) prototype of a stacked-patch antenna surrounded by a compact SS structure implemented on LTCC technology.

a prototype of the stacked-patch antenna surrounded by a compact SS structure implemented on the LTCC technology. The inner conductor of an SMA (semi-miniaturized type-A) connector was connected to the microstrip feed line while its outer conductor was soldered on the bottom of the LTCC board to a pair of pads that were shorted to the ground through via metallization. Note that the microstrip feed line was printed on the bottom of the LTCC substrate to avoid its interference with the SS ring and to alleviate the contribution of its spurious radiation to the radiation pattern at broadside. For comparison, the same stacked-patch antenna on the LTCC substrate but without the SS ring was also built.

The simulated and measured results for the return loss are shown in Figure 2.4, and good agreement is observed. Note that because the impedance performance of the stacked-patch antenna is dominated by the coupling between the lower and upper patches, the return loss for the stacked-patch antenna seems more sensitive to the SS structure than that for the previous, thinner single-patch antenna. The measured return loss is close to –10 dB over the frequency range 15.8–17.4 GHz (about 9% in bandwidth). The slight discrepancy between the measured and simulated results is mainly due to the fabrication issues (such as the variation of dielectric constant and the deviation of via positions) and the effect of the transition between the microstrip line and the SMA connector. It is also noted that there is a frequency shift of about 0.3 GHz (about 1.5 % up). This may be caused by the LTCC material, which may have a real dielectric constant a little bit lower than the overestimated design value. Note that it is normal for practical dielectric substrates to have a dielectric constant with 2% deviation.

The radiation patterns measured in the E- and H-planes are compared with simulated results in Figure 2.5. The radiation patterns compared here are at the frequency of 17 GHz, where the maximum gain of the patch antenna with the SS was observed. From Figure 2.5, we can see good agreement for the co-polarized components. It is confirmed that the radiation at broadside is enhanced and the backside level is reduced. Also, the beamwidth in the E-plane is significantly reduced by the soft surface. It is noted that the measured cross-polarized component has a higher level and more ripples than the simulation result. This is because the simulated radiation patterns were plotted in two ideal principal planes; i.e., $\phi = 0°$ and $\phi = 90°$ planes. From simulation, we found that the maximum cross-polarization may happen in the plane $\phi = 45°$ or $\phi = 135°$. During measurement, a slight deviation from the ideal planes can cause a considerable variation for the cross-polarized component because the spatial variation of the cross-polarization is quick and irregular. Also, a slight polarization mismatch or objects near the antenna (such as the connector or the connection cable) may contribute considerably to the high cross-polarization.

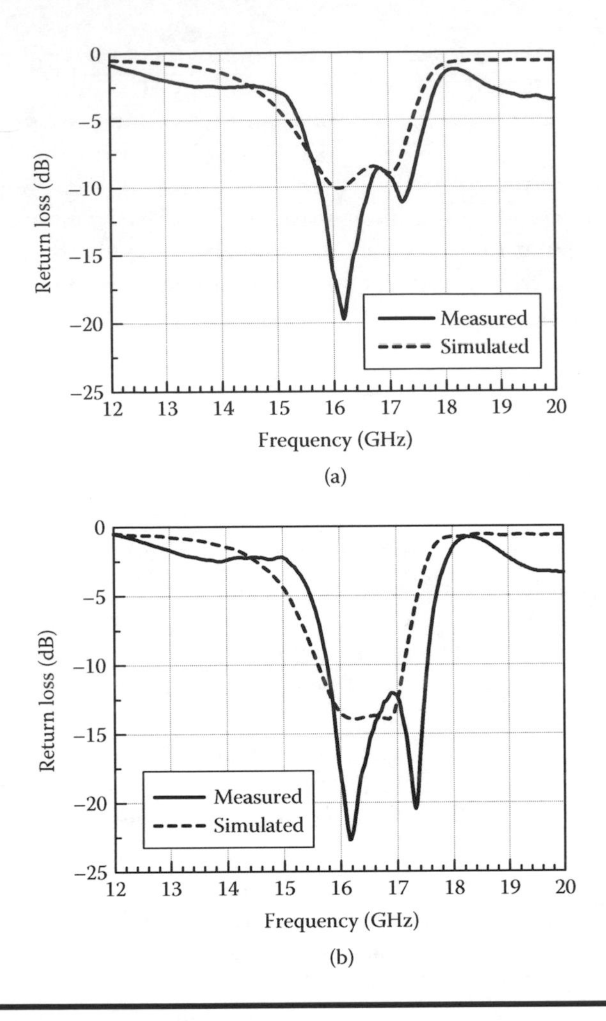

Figure 2.4 Comparison of return loss between simulated and measured results for the stacked-patch antennas (a) with and (b) without the soft surface implemented on LTCC technology.

In addition, the maximum gain measured for the patch with the SS is near 9 dBi, about 3 dB higher than the maximum gain and 7 dB higher than the gain at broadside for the antenna without the SS.

2.3.2 Miniaturized Single-Mode Patch Resonator

Integrating filters on-package in LTCC multilayer technology is a very attractive option for RF front ends up to millimeter-wave frequency range because it can miniaturize the size by vertical deployment of filter elements, reduce the number of components, and reduce assembly costs by

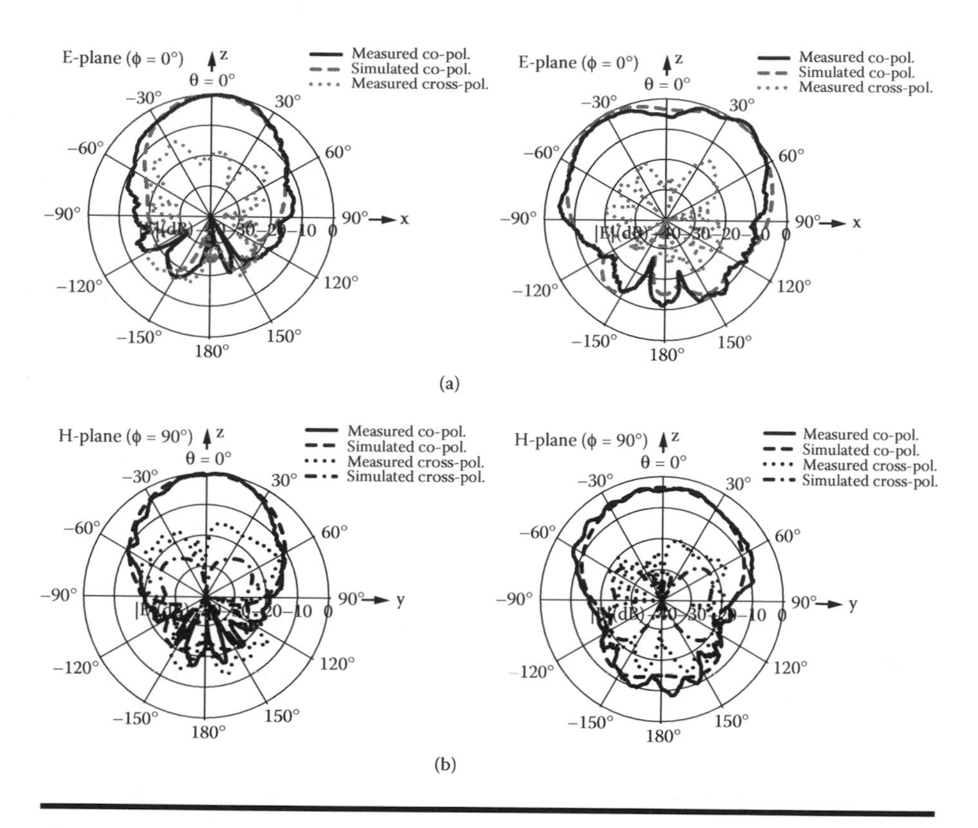

Figure 2.5 Comparison between simulated and measured radiation patterns for the stacked-patch antennas with and without the soft surface implemented on LTCC technology ($f_o = 17$ GHz). (a) *E*-plane ($\phi = 0°$) (b) *H*-plane ($\phi = 90°$).

eliminating the demand for discrete filters. In millimeter-wave frequencies, the BPFs are commonly realized using slotted-patch resonators because of their miniaturized size, making them an excellent compromise between size, power handling, and easy-to-design layout [11]. In this section, the design of a single-pole slotted-patch filter [12] is presented for the operating frequency band of 58–60 GHz. All designs have been simulated using the method of moments (MoM)-based, 2.5-D, full-wave solver IE3D.

Figure 2.6 shows a top-view comparison between a basic, half-wavelength ($\lambda/2$) square-patch resonator ($L \times L = 0.996 \times 0.996$ mm) [13] and the new configuration ($L \times L = 0.616$ mm $\times$ 0.616 mm) (Figure 2.6b), capable of providing good tradeoffs between miniaturization and power handling. A 5 side view of the 60-GHz patch resonator is shown in Figure 2.7. In the design of the $\lambda/2$ square patch, the planar single-mode patch is located at metal 3 (M3 in Figure 2.7) and uses the end-gap capacitive coupling between the feed lines and the resonator itself in order to achieve 3% of 3-dB bandwidth and <3 dB of insertion loss around the

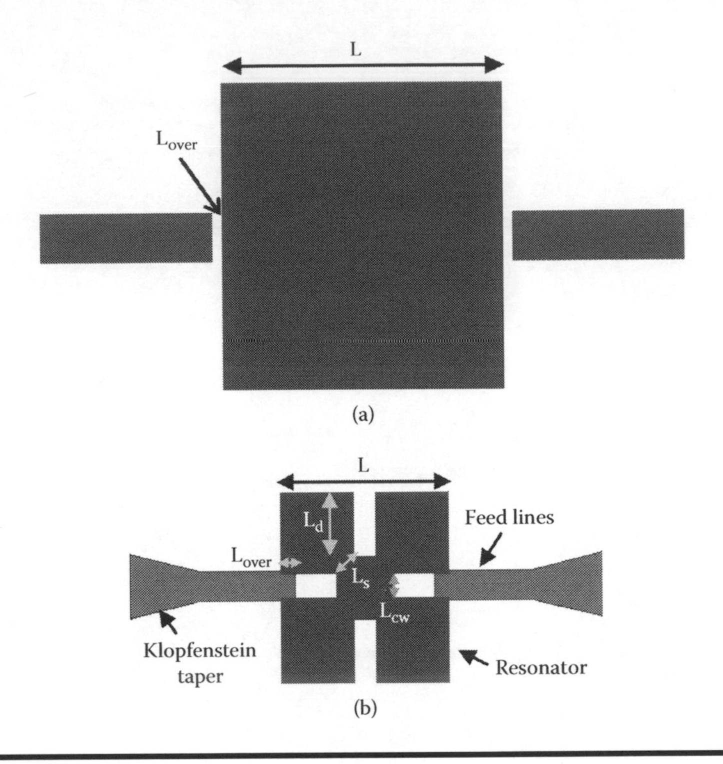

Figure 2.6 Top view of (a) a conventional λ/2 square-patch and (b) a miniaturized-patch resonator.

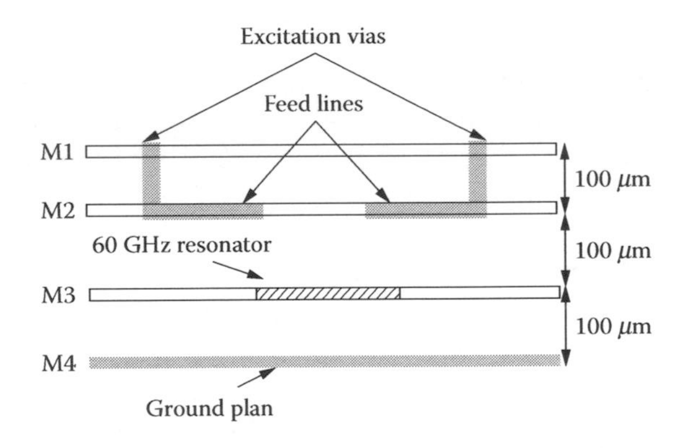

Figure 2.7 Side view of the 60-GHz slotted-patch resonator.

center frequency of 60 GHz. However, the required coupling capacitances to obtain design specifications could not be achieved because of the LTCC design rule limitations.

In order to maximize the coupling strength, while minimizing the effects of the fabrication, the proposed novel structure takes advantage of the vertical deployment of filter elements by placing the feed lines and the resonator into different vertical metal layers as shown in Figure 2.6. This transition also introduces a 7.6% frequency downshift because of the additional capacitive coupling effect as compared to the basic $\lambda/2$ square-patch resonator (Figure 2.6a) directly attached by feed lines. Transverse cuts have been added on each side of the patch in order to achieve significant miniaturization of the patch by adding additional inductance.

Figure 2.8 shows the simulated response for the center frequency and the insertion loss as the length of cuts (L_{cl} in Figure 2.6b) increases while the fixed width of cuts ($L_{cw} = L/8$ in Figure 2.6b) is determined by the fabrication tolerance. It can be observed that the operating frequency range shifts further downward (about 33%) as the length of cut (L_{cl} in Figure 2.6b) increases by approximately 379 μm. Additional miniaturization is limited by the minimum distance (L_s in Figure 2.6b) between the corners of adjacent orthogonal cuts. Meanwhile, as the operating frequency decreases, the shunt conductance in the equivalent circuit of the single patch also decreases because its value is reciprocal to the exponential function of the operating frequency [13]. This fact additionally causes the reduction of radiation loss since it is proportionally related to the conductance in the

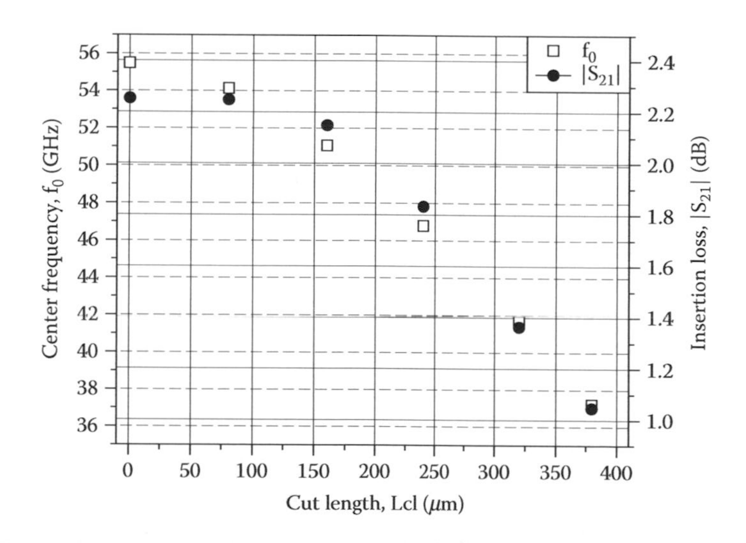

Figure 2.8 Side view of the 60-GHz slotted-patch resonator.

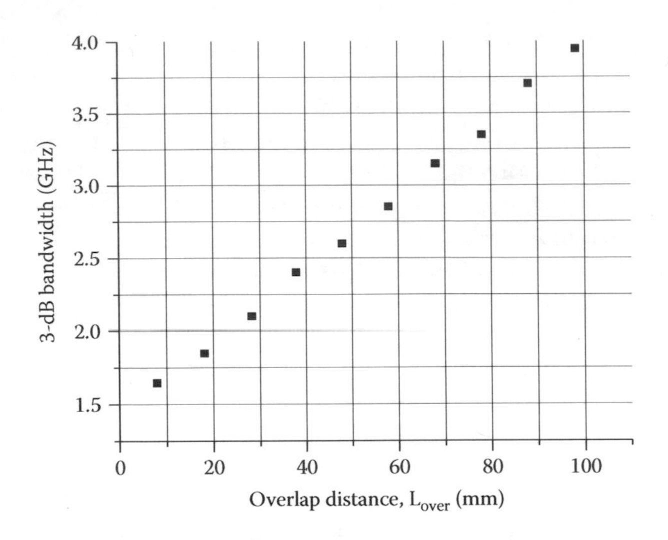

Figure 2.9 Side view of the 60-GHz slotted-patch resonator.

absence of conductor loss [14]. Therefore, insertion loss at resonance is improved from 2.27 dB to 1.06 dB by an increase of L_{cl} in Figure 2.6b.

The patch size is reduced significantly from 0.996 mm to 0.616 mm. The modification of bandwidth due to the patch's miniaturization can be compensated by adjusting the overlap distance (L_{over}). Figure 2.9 shows the simulated response for a 3-dB bandwidth as L_{over} increases. It is observed that the 3-dB bandwidth increases as L_{over} increases due to a stronger coupling effect, and then L_{over} is determined to be 18 μm, corresponding to a 1.85-GHz 3-dB bandwidth.

The proposed embedded microstrip line filters are excited through vias connecting the coplanar waveguide (CPW) signal pads on the top metal layer (M1 in Figure 2.7), reducing radiation loss compared to microstrip lines on the top (surface) layer. As shown in Figure 2.6b, Klopfenstein impedance tapers are used to connect the 50 Ω feeding line and the via pad on the second metal layer (M2 in Figure 2.7). The overlap ($L_{over} \approx L/31$) and transverse cuts ($L_{cw} \approx L/8$, $L_{cl} \approx L/3.26$) have been finally determined to achieve desired filter characteristics with the aid of IE3D. The filters with CPW pads were fabricated in LTCC ($\epsilon_r = 5.4$, tan$\delta = 0.0015$) with a dielectric layer thickness of 100 μm and metal thickness of 9 μm. The overall size is 4.018 mm × 1.140 mm × 0.3 mm, including the CPW measurement pads. As shown in Figure 2.10, the experimental and the simulated results agree very well. It can be easily observed that the insertion loss is <2.3 dB, the return loss is >25.3 dB over the pass band, and the 3-dB bandwidth is about 1 GHz. The center frequency shift from 59.85 GHz to 59.3 GHz can be attributed to the fabrication accuracy (vertical

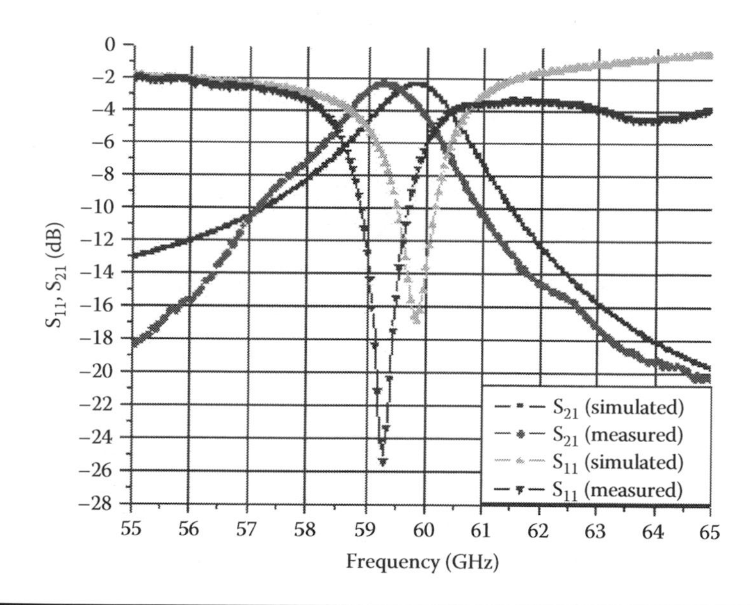

Figure 2.10 Measured and simulated S-parameters of the 60-GHz slotted-patch resonator.

coupling overlap affected by the alignment between layers, layer thickness tolerance). This was the first fabrication iteration, the differences could be corrected in the second and third iterations.

2.3.3 Rectangular Cavity Resonator

The cavity resonator (Figure 2.11), which is the most fundamental component of the cavity filter, is built based on the conventional rectangular cavity resonator approach [14]. The cavity resonator shown in Figure 2.11 consists of one LTCC cavity, two microstrip lines for input and output, and two vertically coupling slots etched on the ground planes of the cavity. The resonant frequency of the fundamental TE_{101} mode can be determined by [14]

$$f_{res} = \frac{c}{2\pi\sqrt{\epsilon_r}}\sqrt{\left(\frac{m\pi}{L}\right)^2 + \left(\frac{n\pi}{H}\right)^2 + \left(\frac{l\pi}{W}\right)^2} \tag{2.1}$$

where f_{res} is the resonant frequency, c the speed of light, r the dielectric constant, L the length of cavity, W the width of cavity, H the height of cavity, and $m = l = 1$, $n = 0$ the indexes for TE_{101} mode. The resonant frequency at 60.25 GHz establishes the initial dimensions of the cavity resonator enclosed by perfectly conducting walls. For the purpose of

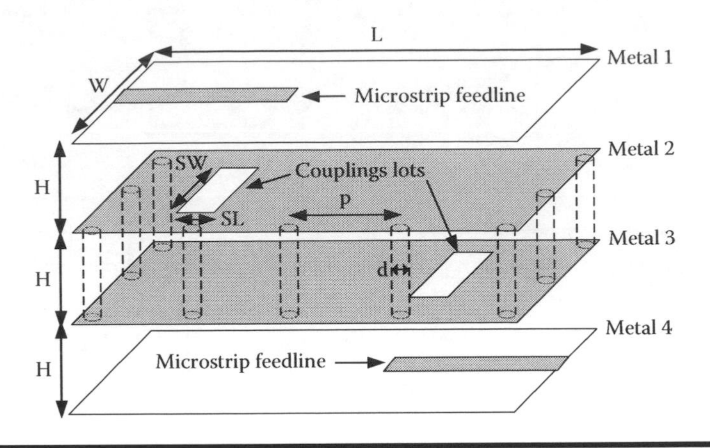

Figure 2.11 3D overview of the LTCC cavity resonator employing slot excitation with microstrip feed lines.

compactness, the height (H) is determined to be 0.1 mm (one substrate layer). And the vertical conducting walls are replaced by double rows of via posts that are sufficient to suppress the field leakage and enhance the quality factor (Q) [15]. In addition, the size and spacing of via posts are properly chosen to prevent electromagnetic field leakage and achieve stop-band characteristic at the desired resonant frequency according to the guidelines specified by Hill et al. [15]. In our work, the minimum value (390 μm $= p$ in Figure 2.11) of center-to-center vias spacing and the minimum value (130 μm $= d$ in Figure 2.11) of via diameter of the LTCC design rules are used. The final dimensions of the via-based cavity are determined using the tuning analysis of an HFSS full-wave simulator (L = 1.95 mm, $W = 1.275$ mm, $H = 0.1$ mm).

With the cavity size determined, microstrip lines are utilized as the feeding structures to excite the cavity via coupling slots that couple energy magnetically from the microstrip lines into the cavity. For a preliminary testing of the vertical intercoupling of the 3-pole cavity BPF, the input and output feed lines are placed on metal 1 and metal 4, respectively, as shown in Figure 2.11. The coupling coefficient can be controlled by the location and size of the coupling slots etched on metal 2 and metal 3 in Figure 2.11. The coupling slots are located a quarter of the cavity length from the sides, and the slot length (SL in Figure 2.11) is varied with the fixed slot width ($SW \approx \lambda_g/4$ at 60.25 GHz) to achieve the desired frequency response [15]. To accurately estimate the unloaded quality factor (Q_u), the weakly coupled cavity resonator [16] with a relatively small value of the slot length (SL in Figure 2.11) is implemented in the HFSS simulator. The unloaded quality factor (Q_u) can be extracted from the external quality factor (Q_{ext}) and the loaded quality factor (Q_l) using Equations (2.2)–(2.4) [16].

$$Q_l = \frac{f_{res}}{\Delta f} \tag{2.2}$$

$$Q_{ext} = 10^{-S_{21}(dB)/20} \cdot Q_l \tag{2.3}$$

$$Q_u = \frac{1}{\frac{1}{Q_l} - \frac{1}{Q_{ext}}} \tag{2.4}$$

The simulated value of Q_u was calculated to be 623 at 60.25 GHz.

2.3.4 3-Pole Cavity BPFs

A vertically stacked LTCC 3-pole cavity BPF [17] is developed for 3D integrated, 59–64 GHz, industrial, scientific, and medical (ISM) band transceiver front-end modules. The center frequencies of 60.25 GHz and 62.75 GHz in the band are selected for the Rx channel and the Tx channel, respectively.

First, the cavity BPF for the Rx channel selection is designed with a center frequency of 60.25GHz, a <3-dB insertion loss, a 0.1-dB ripple, and a 4.15% (≈2.5 GHz) fractional bandwidth based on a Chebyshev lowpass prototype. The filter schematic is implemented with ten substrate layers of LTCC tape. Its 3D overview, side view, and top view of the feeding structure and the inter-resonator coupling structure are illustrated in Figure 2.12. The top five substrate layers (substrates 1–5 in Figure 2.12b) are occupied by the Rx filters, and the remaining layers are reserved for the antenna and the RF active devices, that could be integrated into front-end modules. The microstrip lines on metal 1 and metal 6 are utilized as the feeding structures to excite the 1st and 3rd cavities, respectively. Three identical cavity resonators (1st cavity, 2nd cavity, and 3rd cavity in Figure 2.12b) designed in section 1.2.3 are vertically stacked and coupled through slots to achieve the desired frequency response with a high level of compactness. This filter is also an effective solution for connecting the active devices on the top of the LTCC board and the antenna integrated on the back side.

Two external slots (Figure 2.12b) on metal layers 2 and 5 are dedicated to magnetically coupling the energy from the input/output (I/O) microstrip lines into the 1st and 3rd cavity resonators, respectively. To maximize magnetic coupling by maximizing the current, the microstrip feed lines are terminated with a $\lambda_g/4$ open stub beyond the center of each external slot. The fringing field generated by an open-end discontinuity can be modeled by an equivalent length of transmission line, which is determined to be about $\lambda_g/20$. Therefore, the optimum length of the stub is approximately $\lambda_g/5$ (*MS* in Figure 2.12c) [18]. The position and size of the external slots

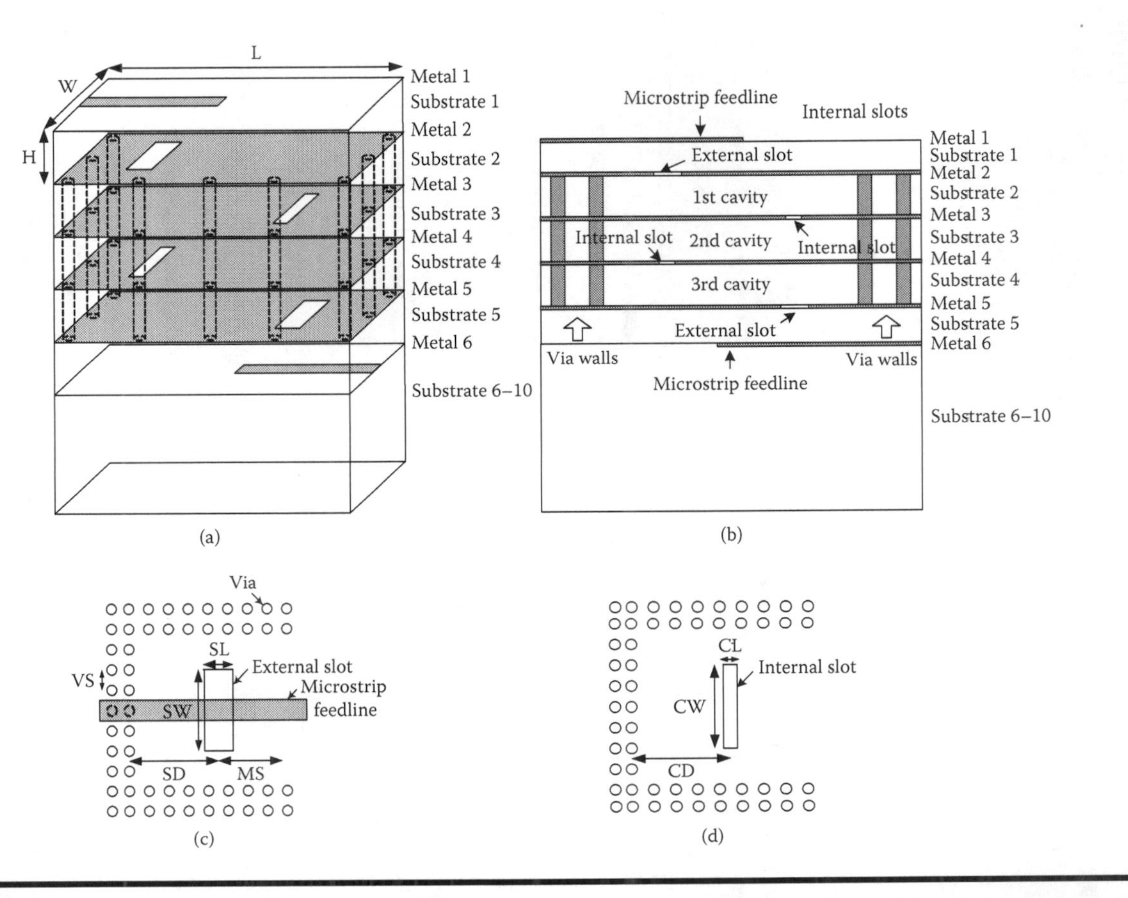

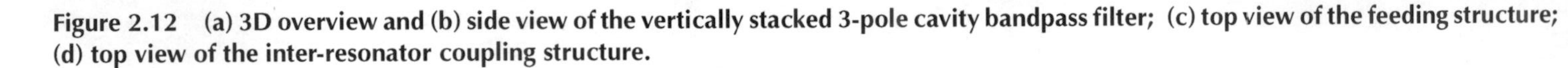

Figure 2.12 (a) 3D overview and (b) side view of the vertically stacked 3-pole cavity bandpass filter; (c) top view of the feeding structure; (d) top view of the inter-resonator coupling structure.

are the main design parameters to provide the necessary Q_{ext}. The external quality factor (Q_{ext}), which controls the insertion loss and ripple over the pass band, can be defined from the specifications as follows [19]:

$$Q_{ext} = \frac{g_i g_{i+1} f_{res}}{BW} \tag{2.5}$$

where g_i are the element values of the low-pass prototype, BW is the bandwidth of the filter and f_{res} is the resonant frequency.

The calculated Q_{ext} is 24.86. The external slot is initially positioned at $L/4$ from the edge of the cavity, and the width (SW in Figure 2.12c) of the slot is fixed to $\lambda_g/4$. Then, the length (SL in Figure 2.12c) of the slot is tuned until the simulated Q_{ext} converges to the prototype requirement. Figure 2.13 shows the relationship between the length variation of the external slots and the Q_{ext} extracted from the simulation using [19]

$$Q_{ext} = \frac{f_{res}}{\Delta f_{\pm 90}} \tag{2.6}$$

where $\Delta f_{\pm 90^\circ}$ is the frequency difference between $\pm 90^\circ$ phase response of S11.

The latter internal slots on metal layers 3 and 4 (Figure 2.12b) are employed to couple energy from the 1st and 3rd cavity resonators into the 2nd resonator, and their design procedure is similar to that of the external slots. The internal slots are located one-quarter cavity length from the sides. The desired inter-resonator coupling coefficients ($k_{12} = k_{23} = 0.0381$) are

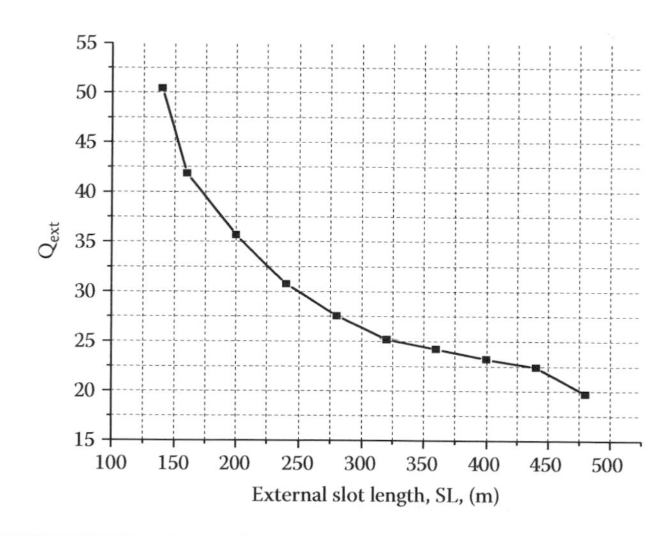

Figure 2.13 External quality factor (Q_{ext}) evaluated as a function of external slot length (SL).

obtained by [19]

$$k_{jj+1} = \frac{BW}{f_{res}} \sqrt{\frac{1}{g_j g_{j+1}}} \tag{2.7}$$

This desired prototype, k_{jj+1}, can be physically realized by varying the slot length (*CL* in Figure 2.12d) with a fixed slot width ($CW \approx \lambda_g/4$ in Figure 2.12d). Full-wave simulations are employed to find the two characteristic frequencies (f_{p1}, f_{p2}) that are the resonant frequencies in the transmission response of the coupled structure [19], and its plot versus frequency is shown in Figure 2.14a. These characteristic frequencies are associated with the inter-resonator coupling between the cavity resonators

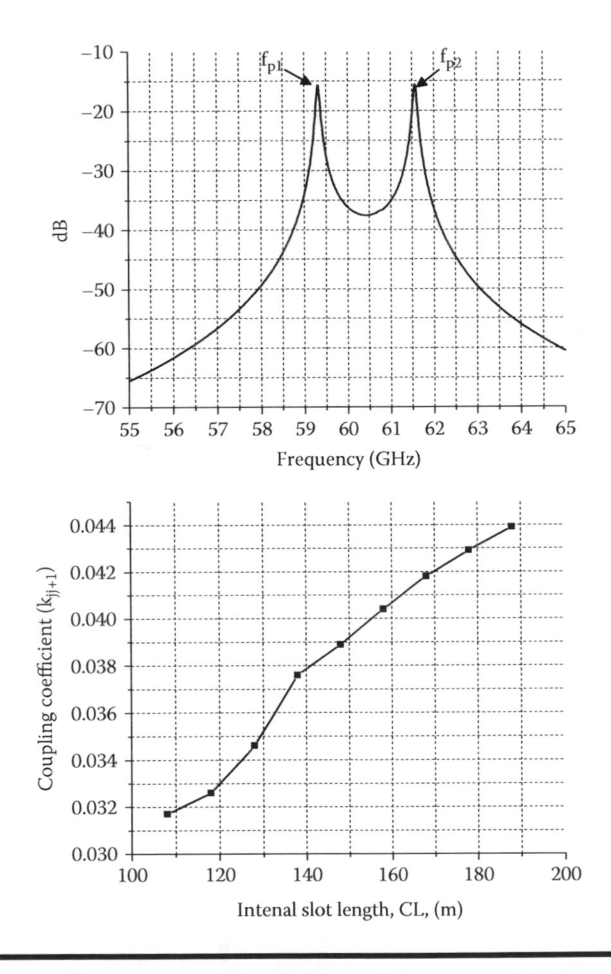

Figure 2.14 (a) Two characteristic frequencies (f_{p1}, f_{p2}) of the coupled cavities to calculate the internal coupling coefficients (k_{jj+1}). (b) Inter-resonator coupling coefficient (k_{jj+1}) as a function of internal slot length (CL).

as follows [19]:

$$k_{jj+1} = \frac{f_{p2}^2 - f_{p1}^2}{f_{p2}^2 + f_{p1}^2} \tag{2.8}$$

Figure 2.14(b) shows the internal coupling as a function of the variation of the internal slot length (*CL* in Figure 2.12(d)). By adjusting the slot length, we can determine the optimal size of an internal slot for a given prototype value. Using the initial dimensions of the external (*SW, SL*) and internal slot (*CW, CL*) size as the design variables, we optimized the design variables to realize the desired frequency response. Considering the minimum and maximum of the fabrication tolerances, the design can be finely tuned. Then, the final variable values that match the desired frequency response can be determined.

The filters, including CPW pads and a vertical transition, were fabricated in LTCC and measured on an HP8510C Vector Network Analyzer using SOLT calibration. WinCal software gives us the ability to de-embed capacitance effects of CPW open pads and inductive effects of short pads from the measured S-parameters so that the loading shift effect is negligible. Figure 2.15 shows the photograph of the fabricated filter with CPW pads and a transition. The cavity size is determined to be 1.95 × 1.284 × 0.1 mm^3 ($L \times W \times H$ in Figure 2.12a.

From the measurement, it is observed that the center frequency of the filter shifts from 60.2 to 57.5 GHz. This might be attributable to the dielectric constant variation at these high frequencies and the fabrication accuracy of vias positioning caused by XY shrinkage. The HFSS simulation is performed again in terms of two aspects. First, the dielectric constant of 5.4 was extracted using cavity resonator characterization techniques [20] at 35 GHz. The dielectric constant is expected to increase to 5.5 across 55–65 GHz [16]. In addition, the tolerance of XY shrinkage is expected to be ±15%. XY shrinkage can significantly affect the via positioning, which is the major factor in determining the resonant frequency of a cavity filter. From our investigation, the averaged relative permittivity was evaluated to be 5.5 across 55–65 GHz [16], and the cavity size was modified to

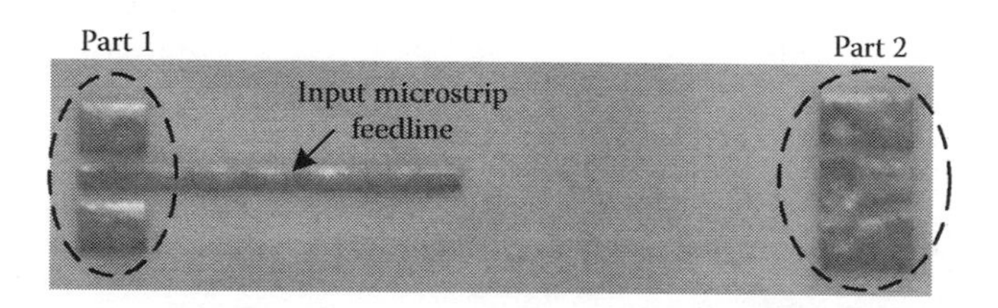

Figure 2.15 The photograph of the cavity bandpass filter fabricated on LTCC.

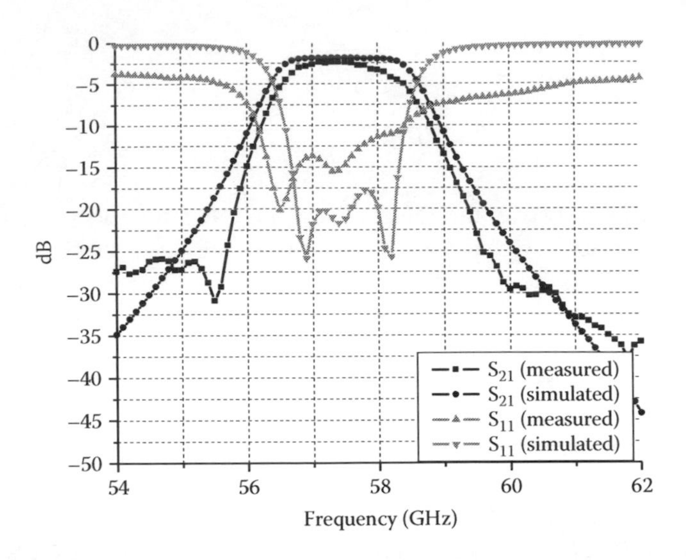

Figure 2.16 Comparison between measured and simulated S-parameters (S11 and S21) of the Rx 3-pole cavity band filter with $\epsilon_r = 5.5$ and modified cavity size (2.048 × 1.348 × 0.1 mm^3).

2.048 × 1.348 × 0.1 mm^3, with 5% of XY shrinkage effect. Then, the exact coincidence between the measured center frequency (57.5 GHz) and the simulated (57.5 GHz) is observed in Figure 2.16. The filter exhibits an insertion loss <2.37 dB and a 3-dB bandwidth about 3.5% (≈ 2 GHz).

Then, the same techniques were applied to the design of the cavity BPF for the Tx channel (61.5–64 GHz). The Chebyshev prototype filter was designed with a center frequency of 62.75 GHz, <3 dB insertion loss, 0.1 dB band ripple, and 3.98% 3-dB bandwidth. To meet the specified center frequency requirement, the cavity width (W) was decreased. Then the cavity size was determined to be 1.95 × 1.206 × 0.1 mm^3 ($L \times W \times H$ in Figure 2.12a). The external and internal coupling slot sizes are used as the main design parameters to obtain the desired external quality factors and coupling coefficients, respectively. The measured results of the Tx filter exhibit an insertion loss of 2.39 dB, with a 3-dB bandwidth of 3.33% (≈2 GHz) at the center frequency of 59.9 GHz. The center frequency is downshifted approximately 2.72 GHz, similarly to the Rx filter. A new theoretical simulation was performed with $\epsilon_r = 5.5$ and the 5% increase in the volume of the cavity (2.048 × 1.266 × 0.1 mm^3). The measured and simulated results are presented in Figure 2.17. The simulation showed a minimum insertion loss of 1.97 dB, with a slightly increased 3-dB bandwidth of 4% (≈2.4 GHz). The center frequency of the simulated filter was 59.9 GHz. The center frequency shift is consistent with all devices using this LTCC process because of the fabrication tolerances mentioned.

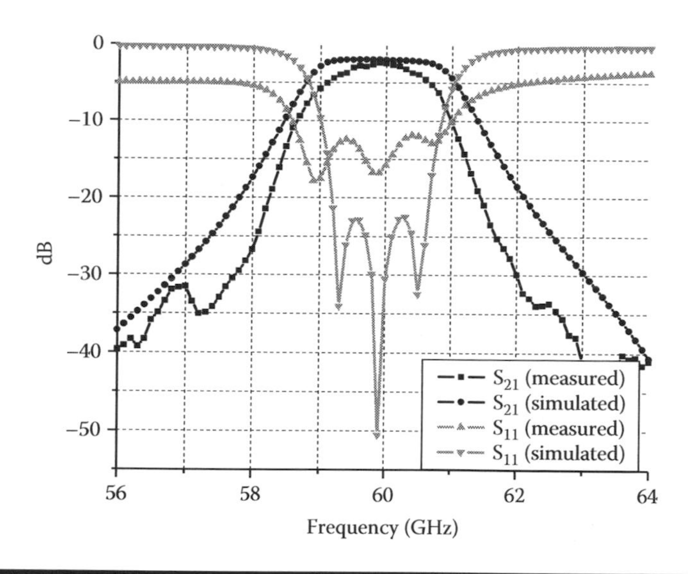

Figure 2.17 Comparison between measured and simulated S-parameters (S11 and S21) of Tx 3-pole cavity band filter (simulation with ϵ_r = 5.5 and modified cavity size [2.048 × 1.266 × 0.1 mm^3] vs. measurement).

2.4 Integrated (Filters and Antenna)

The optimal integration of antennas and filters into a 3D, 59–64 GHz transceiver front-end module is significantly desirable because it not only reduces cost, size, and system complexity but also achieves a high level of band selectivity and spurious suppression. Although cost, electrical performance, integration density, and packaging capability are often at odds in RF front-end designs, the performance of the module can be significantly improved with the 3D integration of filters and antennas using the flexibility of multilayer architecture in LTCC. In this section, the fully integrated Rx and Tx filters and the dual-polarized antenna that covers Rx (1st) and Tx (2nd) channels are proposed to employ the presented designs of the filters [17]. The filters' matching (>10 dB) toward the antenna and the isolation (>45 dB) between Rx and Tx paths comprise the excellent features of this compact 3D design. The stringent demand of high isolation between two channels induces the advanced design of a duplexer and an antenna as a fully integrated function for the V-band front-end module.

The 3D overview and the cross-section view of the topology chosen for the integration are shown in Figure 2.18a and b, respectively. A cross-shaped patch antenna [14], designed to cover two bands between 59 and 64 GHz (1st channel: 59–61.5 GHz; 2nd channel: 61.75–64 GHz), is located at the most bottom metal layer (M11 in Figure 2.18b). The cross-shaped geometry was utilized to decrease the cross-polarization, which contributes

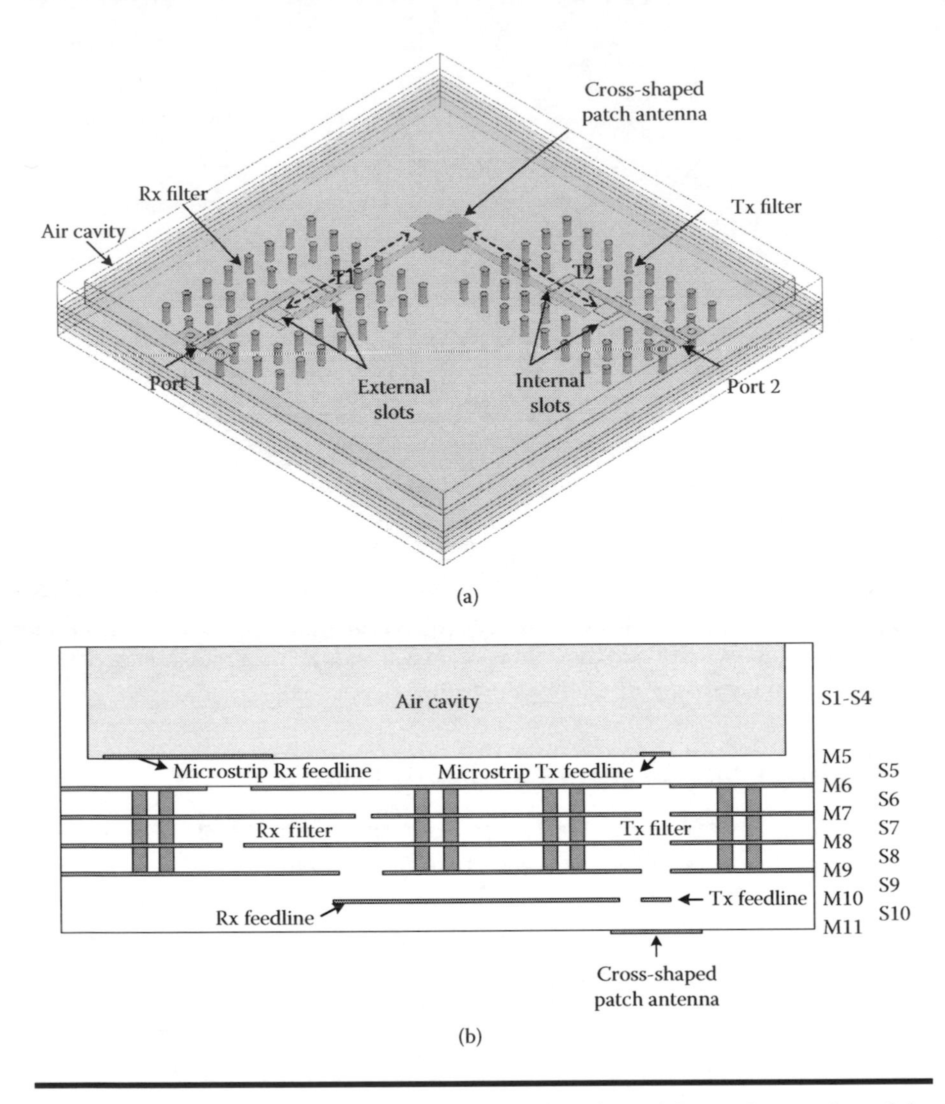

Figure 2.18 (a) 3D overview and (b) cross-section view of the 3D integration of the filters and antennas using LTCC multilayer technologies.

to unwanted side lobes in the radiation pattern [14]. The cross-channel isolation can be improved by receiving and transmitting signals in two orthogonal polarizations. The feed lines and the patch are implemented into different vertical metal layers (M10 and M11, respectively), and then the end-gap capacitive coupling is realized by overlapping the end of the embedded microstrip feed lines and the patch. The overlap distance for the Rx and Tx feed lines is approximately 0.029 and 0.03 mm, respectively. The common ground plane for the feed lines and the patch is placed one layer above the feed lines as shown in Figure 2.18b. The two antenna feed lines (Rx feedline and Tx feed line in Figure 2.18b) are commonly utilized as the filters' feed lines, which excite the Rx and Tx filters accordingly through external slots placed at M9 in Figure 2.18b. The lengths of the Rx and Tx feed lines (T1 and T2 in Figure 2.18a) connecting the cross-shaped antenna to the Rx and Tx filters, respectively, are initially set up to be one guided wavelength at the corresponding center frequency of each channel and are optimized using the HFSS simulator (T1: 2.745 mm, T2: 2.650 mm). The 3D Rx and Tx filters (see Figure 2.12) designed in section 1.2.4 are directly integrated with the antenna. The integrated filters and antenna function occupy six substrate layers (S5–S10:600 μm). The remaining four substrate layers (S1–S4 in Figure 2.18(b)) are dedicated to the air cavities reserved for burying RF active devices (RF receiver and transmitter MMICs [monolithic millimeter-wave integrated circuit]) that are located beneath the antenna on purpose so as not to interfere with the antenna performance and be highly integrated with the microstrip (Rx/Tx) feed lines, leading to significant volume reduction, as shown in Figure 2.18.

The cavities are fabricated by removing the inner portion of the LTCC material outlined by the successively punched vias. The deformation factor of a cavity, defined as the physical depth difference between the designed one and the fabricated one, is stable in the LTCC process when the depth of the cavity is less than two-thirds of the height of the board. Since we have chosen the air cavity depth of 400 μm, which is suitable for Rx/Tx MMIC chipsets, to enable the full integration of MMICs and passive front-end components, we can limit the fabrication tolerance's effect of an air cavity to the other integrated circuitries. Figure 2.19 shows a photograph

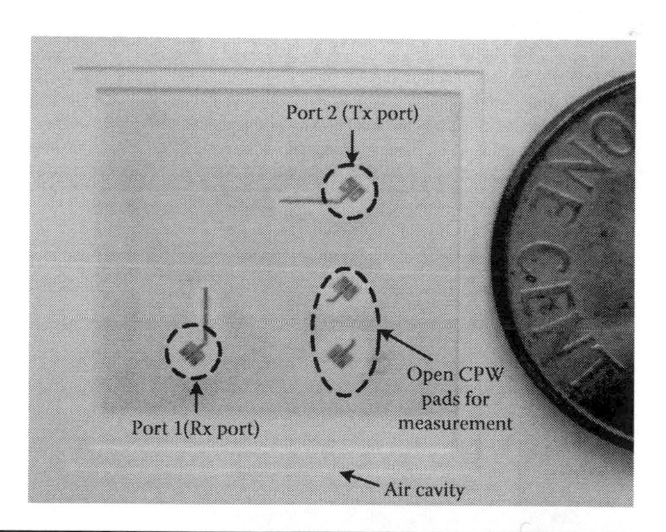

Figure 2.19 Photograph of the top view of the integrated function of Rx/Tx cavity filters and a cross-shaped patch antenna with the air cavity at the top.

of the integrated device equipped with one air cavity at the top layers. The device occupies an area of 7.94 × 7.82 × 1 mm^3, including the CPW measurement pads.

Figure 2.20 shows the simulated and measured return losses (S11/S22) of the integrated structure. In the simulation, the higher dielectric constant ($\epsilon_r = 5.5$) and the 5% increase in the volume of the cavity were applied. It is observed from the 1st channel that the 10-dB return loss bandwidth is

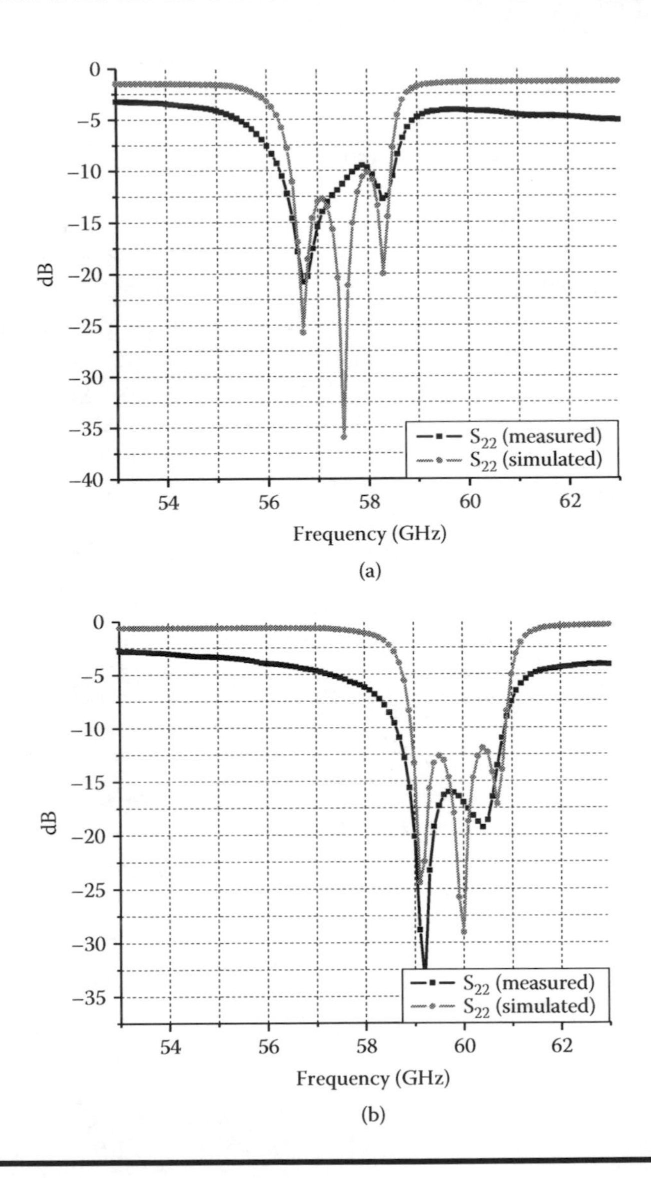

Figure 2.20 The comparison between measured and simulated return losses of the 1st channel (a) and of the 2nd channel (b).

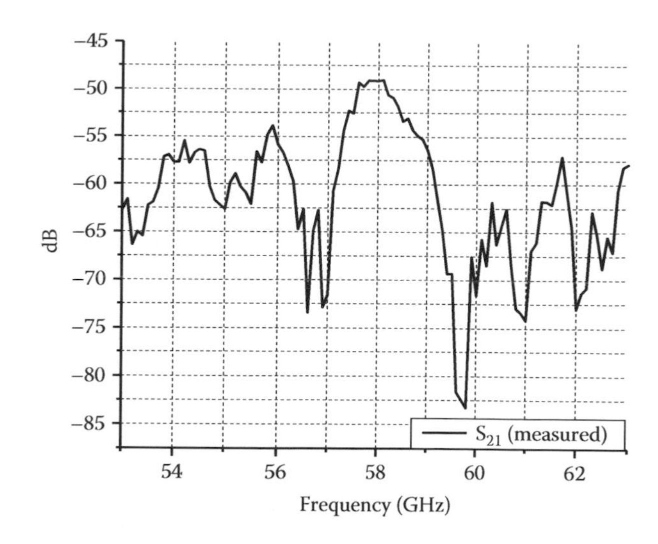

Figure 2.21 The measured channel-to-channel isolation (S21) of the integrated structure.

approximately 2.4 GHz (≈4.18%) at the center frequency of 57.45 GHz, which is slightly wider than the simulation of 2.1 GHz (≈3.65%) at 57.5 GHz, as shown in Figure 2.20a. The slightly increased bandwidth may be attributed to parasitic radiation from the feed lines or the measurement pads. In Figure 2.20b, the return loss measurement from the 2nd channel also exhibits a wider bandwidth of 2.3 GHz (≈3.84%) at the center frequency of 59.85 GHz compared to the simulated value of 2.1 GHz (≈3.51%) at that of 59.9 GHz. The measured channel-to-channel isolation is illustrated in Figure 2.21. The measured isolation is better than 49.1 dB across the 1st band (56.2–58.6 GHz) and better than 51.9 dB across the 2nd band (58.4–60.7 GHz).

References

[1] C. H. Doan, S. Emami, D. A. Sobel, A. M. Niknejad, and R. W. Brodersen, "Design considerations for 60GHz CMOS radios," *IEEE Communications Magazine,* vol. 42, no. 12, pp. 132–140, Dec. 2004.

[2] H. H. Meinel, "Commercial applications of millimeterwaves history, present status, and future trends," *IEEE Transactions on Microwave Theory and Techniques,* vol. 43, no. 7, pp. 1639–1653, July 1995.

[3] M. M. Tentzeris, J. Laskar, J. Papapolymerou, S. Pinel, V. Palazzari, R. Li, G. DeJean, N. Papageorgiou, D. Thompson, R. Bairavasubramanian, S. Sarkar, and J.-H. Lee, "3-D-integrated RF and millimeter-wave functions and modules using liquid crystal polymer (LCP) system-on-package

technology," *IEEE Transactions on Advanced Packaging*, vol. 27, no. 2, pp. 332–340, May 2004.

[4] K. Lim, A. Obatoyinbo, A. Sutuno, S. Chakraborty, C. Lee, E. Gebara, A. Raghavan, and J. Laskar, "A highly integrated transceiver module for 5.8 GHz OFDM communication system using multi-layer packaging technology," in IEEE MTT-S Int. Microwave Symp., Dig., vol. 1, 2001, pp. 65–68.

[5] W. Diels, K. Vaesen, K. Wambacq, P. Donnay, S. De Raedt, W. Engels, and M. Bolsens, "A single-package integration of RF blocks for a 5 GHz WLAN application," *IEEE Trans. Components Packaging Technol. Adv. Packaging* pt. B, vol. 24, pp. 384–391, Aug. 2001.

[6] M. J. Vaughan, K. Y. Hur, and R. C. Compton, "Improvement of microstrip patch antenna radiation patterns," *IEEE Trans. Antennas Propagation.*, vol. 42, no. 6, pp. 882–885, June 1994.

[7] R. Gonzalo, P. de. Maagt, and M. Sorolla, "Enhanced patch-antenna performance by suppressing surface waves using photonic-bandgap substrates," *IEEE Trans. Microwave Theory Tech.,* vol. 47, no. 11, pp. 2131–2138, Nov. 1999.

[8] R. Coccioli, F.-R. Yang, K.-P. Ma, and T. Itoh, "Aperture-coupled patch antenna on UC-PBG substrate," *IEEE Trans. Microwave Theory Tech.,* vol. 47, no. 11, pp. 2123–2130, Nov. 1999.

[9] R. L. Li, G. DeJean, J. Papapolymerou, J. Laskar, and M. M. Tentzeris, "Radiation-pattern improvement of patch antennas on a large-size substrate using a compact soft surface structure and its realization on LTCC multilayer technology," *IEEE Trans. Antennas Propagation,* vol. 53, no. 1, pp. 200–208, Jan. 2005.

[10] A. M. Tentzeris, R. L. Li, K. Lim, M. Maeng, E. Tsai, G. DeJean, and J. Laskar, "Design of compact stacked-patch antennas on LTCC technology for wireless communication applications," *Proceedings of IEEE Antenna and Propagation Society International Symposium,* vol. 2, pp. 500–503, June 2002.

[11] V. Kondratyev, M. Lahti, and T. Jaakola, "On the design of LTCC filter for millimeter-waves," in IEEE MTT-S Int. Microwave Symp. Dig., pp. 1771–1773, June 2003.

[12] J.-H. Lee, G. DeJean, S. Sarkar, S. Pinel, K. Lim, J. Papapolymerou, J. Laskar, M. M. Tentzeris, "Highly integrated millimeter-wave passive components using 3-D LTCC system-on-package technology," *IEEE Trans. Microwave Theory Tech.,* vol. 53, no. 6, pp. 2220–2229, June 2005.

[13] D. M. Pozar and D. H. Schauber, *Microstrip Antennas,* Piscataway, NJ, IEEE Press, 1995.

[14] R. E. Collin, *Foundations for Microwave Engineering,* New York, McGraw-Hill, 1992.

[15] M. J. Hill, R. W. Ziolkowski, and J. Papapolymerou, "Simulated and measured results from a duroid-based planar MBG cavity resonator filter," *IEEE Microwave and Wireless Components Letters,* vol. 10, no. 12, pp. 528–530, Dec. 2000.

[16] J.-H. Lee, S. Pinel, J. Papapolymerou, J. Laskar, and M. M. Tentzeris, "Low loss LTCC cavity filters using system-on-package technology at 60 GHz,"

IEEE Transactions on Microwave Theory and Techniques, vol. 53, no. 12, pp. 231–244, Dec. 2005.

[17] J.-H. Lee, K. Nobutaka, G. DeJean, S. Pinel, J. Laskar, and M. M. Tentzeris, “A v-band front-end with 3-D integrated with cavity filters/duplexers and antennas in LTCC technologies,” *IEEE Transactions on Microwave Theory and Techniques*, vol. 54, no. 7, pp. 2925–2936, July 2005.

[18] L. Harle and L. P. B. Katehi, “A vertically integrated micromachined filter,” *IEEE Trans. on Microwave Theory and Techniques,* vol. 50, no. 9, pp. 2063–2068, Sept. 2002.

[19] J.-S. Hong and M. J. Lancaster, *Microstrip Filters for RF/Microwave Applications,* New York, John Wiley and Sons, Inc., 2001.

[20] D. C. Thomson, O. Tantot, H. Jallageas, G. E. Ponchak, M. M. Tentzeris, and J. Papapolymerou, “Characterization of liquid crystal polymer (LCP) material and transmission lines on LCP substrates from 30 to 110 GHz,” *IEEE Transactions on Microwave Theory and Techniques*, vol.52, no.4, pp. 1343–1352, April 2004.

Chapter 3

Antennas and Channel Modeling in Millimeter-Wave Wireless PAN, LAN, and MAN

Shao-Qiu Xiao, Jijun Yao, Bing-Zhong Wang, Jianpeng Wang, and Huilai Liu

Contents

3.1 Introduction

In the future mobile communication networks, there are four types of access networks, namely, wireless wide area cellular (WWAC), wireless personal area network (WPAN), wireless local area network (WLAN), and wireless metropolitan area network (WMAN). WWAC is developed for communicating in a large or global area. WPAN, WLAN, and WMAN are proposed to realize communications in a smaller area [1,2]. Some special requirements, such as high data rate, large communication capacity, limited communication area, and high angular resolution communication, are required in emerging WPAN, WLAN, and WMAN. Millimeter waves are capable of providing these desirable performances due to their unique characteristics; i.e., wide-spectrum bandwidth, short wavelength, and large air propagation loss. Therefore, millimeter waves are important candidates for future WPAN, WLAN, and WMAN [1–10].

As one of the most critical components in wireless communication systems, antennas can be regarded as a means for radiating or receiving radio waves. In other words, it is the transitional device between free space and a guiding device, as shown in Figure 3.1 [3]. In the transmitter, the antenna, called the transmitting antenna, radiates the electromagnetic waves, and in the receiver, the antenna, called the receiving antenna, receives the electromagnetic waves. Antennas with excellent design can improve the communication performance. Due to the short wavelengths of the millimeter waves, it is easy to obtain an electrically larger antenna aperture. It means that high gain and angular resolution are easily obtained. However, it is also difficult to develop a millimeter-wave antenna because of the rigorous manufacturing requirement and large metal and dielectric loss.

Various forms of the millimeter-wave antennas have been developed to meet particular requirements [3–16]. From the configuration standpoint,

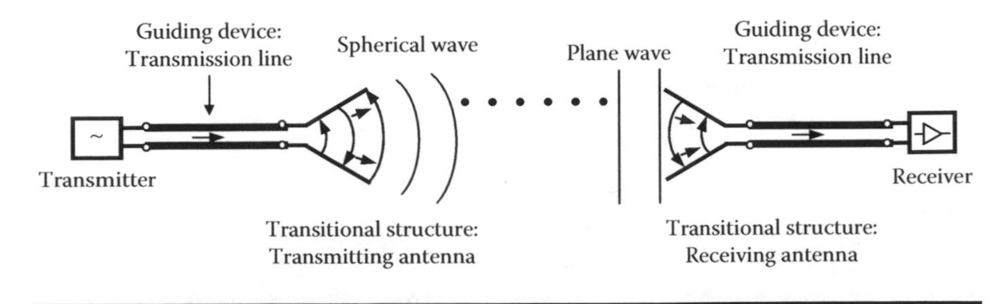

Figure 3.1 The transmitter and receiver antennas in a wireless system.

most antenna forms in millimeter waves are similar to those in microwaves. However, some forms are more suitable and popular in millimeter-wave applications, such as horn antenna and reflector antenna.

Another key issue related to radio wave is the wireless channel. In order to estimate the communication performances accurately for millimeter-wave wireless systems, it is necessary to estimate signal propagation characteristics through a specific environment. The concept of channel modeling is adopted to describe all possible propagation effects, including absorption, attenuation, reflection, motion, and fading. Various channel modeling has been established for various operational environments and application systems. They range from simple geometric models to more complex statistical and combined geometric/statistical and very complex ray tracing models. In the designs of WPAN, WLAN, and WMAN systems, the channel modeling must be considered accurately in the link budget [9].

This chapter focuses mainly on antennas, which have been applied or are potential candidates in millimeter-wave WPAN, WLAN, WMAN, and the channel modeling for millimeter-wave propagation.

3.2 Basic Principles for Radiation

The radiations of the antenna are caused by electric and magnetic current sources. These sources are either real electromagnetic current sources, $\boldsymbol{J}_r$ and $\boldsymbol{M}_r$, on the antenna volume, or equivalent electromagnetic currents $\boldsymbol{J}_e$ and $\boldsymbol{M}_e$, on the surface of the antenna's closed envelope, as shown in Figure 3.2. The latter can be considered as secondary wave source; i.e., Huygens source. Based on field equivalence principles, once the electric or magnetic fields on the surface of the antenna's closed envelope are known, the radiation fields can be determined by replacing the antenna with the equivalent surface electric and magnetic currents. The relationships between equivalent electromagnetic currents and the electromagnetic fields

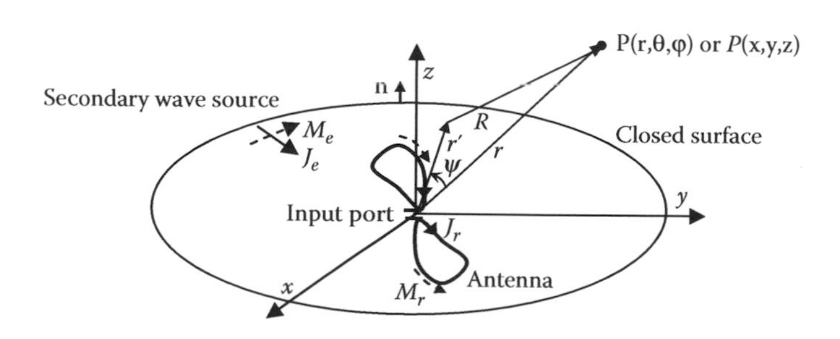

Figure 3.2 Antenna radiation and its equivalent problem.

on the envelope are

$$\boldsymbol{J}_e = \hat{\boldsymbol{n}} \times \boldsymbol{H} \tag{3.1}$$

$$\boldsymbol{M}_e = -\hat{\boldsymbol{n}} \times \boldsymbol{E} \tag{3.2}$$

where $\hat{\boldsymbol{n}}$ is the outward normal vector on the surface of the envelope. Therefore, in order to analyze the antenna characteristics, the crucial work is to determine the electromagnetic current on the antenna volume or the electromagnetic field over the envelope. The equivalent sources, $\boldsymbol{J}_e$ and $\boldsymbol{M}_e$, behave just like the real sources, $\boldsymbol{J}_r$ and $\boldsymbol{M}_r$.

The solutions for the radiation problem are based on Maxwell's equations. The auxiliary functions, such as vector potential $\boldsymbol{A}$ for electric current source $\boldsymbol{J}$ and vector potential $\boldsymbol{F}$ for magnetic current source $\boldsymbol{M}$, are introduced usually to simplify the solution of Maxwell's equations. If $\boldsymbol{J}$ and $\boldsymbol{M}$ are specified, the auxiliary function $\boldsymbol{A}$ and $\boldsymbol{F}$ can be determined as

$$\boldsymbol{A} = \frac{\mu}{4\pi}\int\int\int_v \frac{\boldsymbol{J}e^{-jkR}}{R}dv' \tag{3.3}$$

and

$$\boldsymbol{F} = \frac{\varepsilon}{4\pi}\int\int\int_v \frac{\boldsymbol{M}e^{-jkR}}{R}dv' \tag{3.4}$$

where $k = \omega\sqrt{\mu\varepsilon}$ (ω is angle frequency) and R is the distance from any point in the source to the observation point, respectively. The electronic field and the magnetic field generated by $\boldsymbol{J}$ are

$$\boldsymbol{E}_A = -j\omega\boldsymbol{A} - j\frac{1}{\omega\mu\varepsilon}\nabla(\nabla \bullet \boldsymbol{A}) \tag{3.5}$$

and

$$\boldsymbol{H}_A = \frac{1}{\mu}\nabla \times \boldsymbol{A} \tag{3.6}$$

respectively. In the far-field region, they can be re-expressed approximately as

$$\begin{aligned} E_{Ar} &= 0 \\ E_{A\theta} &= -j\omega A_\theta \\ E_{A\varphi} &= -j\omega A_\varphi \end{aligned} \tag{3.7}$$

and

$$H_{Ar} = 0$$

$$H_{A\theta} = j\omega A_{\varphi}/\eta_0 \tag{3.8}$$

$$H_{A\varphi} = -j\omega A_{\theta}/\eta_0$$

where η_0 is the intrinsic impedance of free space. Equations (3.7) and (3.8) indicate that the ***E***- and ***H***- fields radiated by ***J*** have only θ-direction and φ-direction components in polar coordinates. Similarly, the electric field and the magnetic field generated by ***M*** are

$$\boldsymbol{E}_F = -\frac{1}{\varepsilon}\nabla \times \boldsymbol{F} \tag{3.9}$$

and

$$\boldsymbol{H}_F = -j\omega \boldsymbol{F} - j\frac{1}{\omega\mu\varepsilon}\nabla(\nabla \bullet \boldsymbol{F}) \tag{3.10}$$

respectively. In the far-field region, they can be re-expressed approximately as

$$H_{Fr} = 0$$

$$H_{F\theta} = -j\omega F_{\theta} \tag{3.11}$$

$$H_{F\varphi} = -j\omega F_{\varphi}$$

and

$$E_{Fr} = 0$$

$$E_{F\theta} = -j\omega\eta_0 F_{\varphi} \tag{3.12}$$

$$E_{F\varphi} = -j\omega\eta_0 F_{\theta}$$

The ***E***- and ***H***- fields radiated by ***M*** have also only θ-direction and φ-direction components. Then the total ***E***- and ***H***- fields radiated by ***J*** and ***M*** at the observation position P are

$$\boldsymbol{E} = \boldsymbol{E}_A + \boldsymbol{E}_F \tag{3.13}$$

and

$$\boldsymbol{H} = \boldsymbol{H}_A + \boldsymbol{H}_F \tag{3.14}$$

The far-zone $\boldsymbol{E}$- and $\boldsymbol{H}$-field components radiated by antenna are orthogonal to each other and form TEM (to $\hat{r}$) mode fields. The ratio between the amplitude of the $\boldsymbol{E}$- and $\boldsymbol{H}$-fields is equal to the intrinsic impedance of free space, that is,

$$\left|\frac{\boldsymbol{E}}{\boldsymbol{H}}\right| = \eta_0 \tag{3.15}$$

Based on the introduced principle, the first step of analyzing a radiation problem is to specify the source and define the auxiliary functions. The fields radiated by sources could be calculated in the next step.

Some important parameters are defined by IEEE standard definitions of term for antenna to describe the radiation performance of an antenna, such as directivity, gain, efficiency, half-power beamwidth, bandwidth, polarization, effective aperture, and so forth. However, not all of them need to be specified for the antenna since some of them are mutually dependent. The antenna radiation patterns in the far-field region ($R_a \geq 2L_m{}^2/\lambda_0$, where R_a is the distance between the antenna and the observer's position, L_m is the largest dimension of the antenna, and λ_0 is the wavelength in free space) are determined by Equations (3.13) and (3.14). Once the radiation patterns and the input characteristics (such as input impedance) of the antenna are known, all of the other parameters can be obtained.

A typical wireless link in wireless communication system is shown in Figure 3.3. The distance between transmitting and receiving antennas is r_{TR}. Power P_t generated by the transmitter is perfectly fed into the transmitting antenna whose gain and effective area are G_t and A_{et}, respectively. The receiving antenna has a gain and effective area, G_r and A_{er}, respectively. If the intervening medium is linear, passive, and isotropic, the power density at the receiving antenna is

$$S_r = \frac{P_t}{4\pi r_{TR}{}^2} G_t \tag{3.16}$$

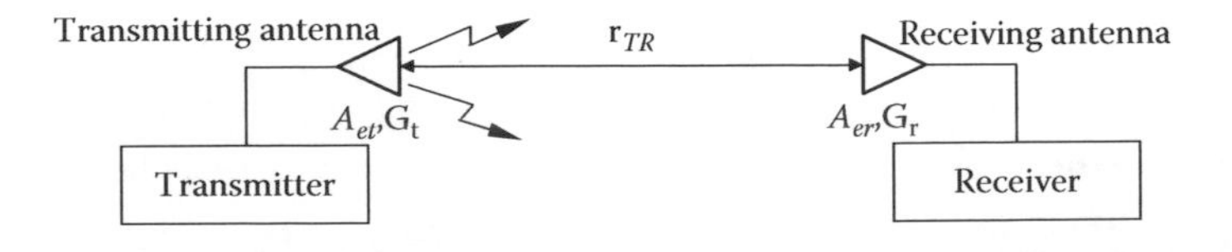

Figure 3.3 The wireless link between the transmitter and the receiver antennas.

Assuming the receiving antenna is lossless, and matched perfectly (i.e., impedance-matched to the load and polarization-matched to the incoming wave), the received power of the receiving antenna is

$$P_r = S_r A_{er} \tag{3.17}$$

The relationship between the effective area A_e and the directivity D of an antenna is

$$D = 4\pi \frac{A_e}{\lambda^2} \tag{3.18}$$

where λ is the wave length. When the antenna is matched perfectly, the gain equals to the directivity based on equation $D = \eta G$ (η is the efficiency factor). Therefore, combining Equations (3.16), (3.17), and (3.18), the relationship between transmitting and receiving power can be expressed as

$$\frac{P_r}{P_t} = \frac{A_{er} A_{et}}{{r_{TR}}^2 \lambda^2} \tag{3.19}$$

or

$$\frac{P_r}{P_t} = \left(\frac{\lambda}{4\pi r_{TR}}\right)^2 G_r G_t \tag{3.20}$$

which are called Friis transmission formula. The term $(\lambda/4\pi r_{TR})^2$ in Equation (3.20) is regarded as the free-space loss factor. The Friis transmission formula is significant to guide the design of the wireless link in a millimeter-wave system. For example, if the receiving power sensitivity P_{rmin} of the receiver and the gains of transmitting and receiving antennas are known, we can determine the maximal communication distance under the limited transmitting power. Also, we can determine the minimal transmitting power under the limited communication distance. However, generally speaking, the propagation environment of radio waves is very complex compared with free space, so the free-space loss factor in Equation (3.20) should be replaced by practical propagation loss in actual wireless communication system designs. A special research field, known as propagation channel model, has been developed for this important issue. Friis transmission formula of radio wave is presented to reveal simply the influence of antenna performance on wireless communications.

3.3 Types of Antennas in Millimeter-Wave WPAN, WLAN, and WMAN

The choice of antennas for millimeter-wave WPAN, WLAN, and WMAN depends on the applications and the propagation environment, but clearly both a broad-beam antenna and a steerable-beam antenna are required. When a long transmission distance or suppression of multipath interface is needed, the directional antenna is desirable. The omnidirectional antenna is required in case of omnidirectional communication. Usually, aperture antenna, lens antenna, leaky-wave and surface wave antenna, and reflector antenna are easily used as directional and beam-steerable antenna in millimeter-wave WPAN, WLAN, and WMAN. The microstrip antenna and slot antenna are suitable for broad-beam antenna. However, a category of antenna may be either the directional antenna or the omnidirectional antenna based on designing. The array configured by microstrip and slot antenna elements is also a beam-steerable antenna. Recently, the technology of planar integrated antenna has been developed for millimeter-wave applications due to the trend of the integration in radio frequency front-end circuits and systems. In this section, these main antennas form in millimeter-wave WPAN, WLAN, and WMAN and their recent developments are presented.

3.3.1 Aperture Antennas

Aperture antenna is an important form for millimeter-wave WPAN, WLAN, and WMAN because it can provide good gain performance. The most common aperture antenna is the horn. The horn may be square, rectangular, circular, elliptical, biconical, and so on. Several simple horn configurations are shown in Figure 3.4. The high gain performance of the horn operating in millimeter waves comes from the electrically large aperture. The horns with high angular resolution (such as the horns in Figure 3.4a–d) are suitable for applications of high-data-rate point-to-point communications. Combining simple mechanical steerable technology, the millimeter-wave wireless systems can support nomadic or slow mobile users. A scanning antenna with high gain gives improved link margin and multipath rejection.

The field equivalence principle is an effective technique for aperture antenna analysis. To develop an exact equivalence for horn antenna, it is necessary that the tangential electric and magnetic field components over a closed surface are known. The selected closed surface is usually an infinite plane that coincides with the horn aperture and passes through the horn mouth. However, the fields outside the aperture are not known if the horn aperture is not mounted on an infinite ground plane. An effective approximate method is to assume that the fields outside the horn aperture

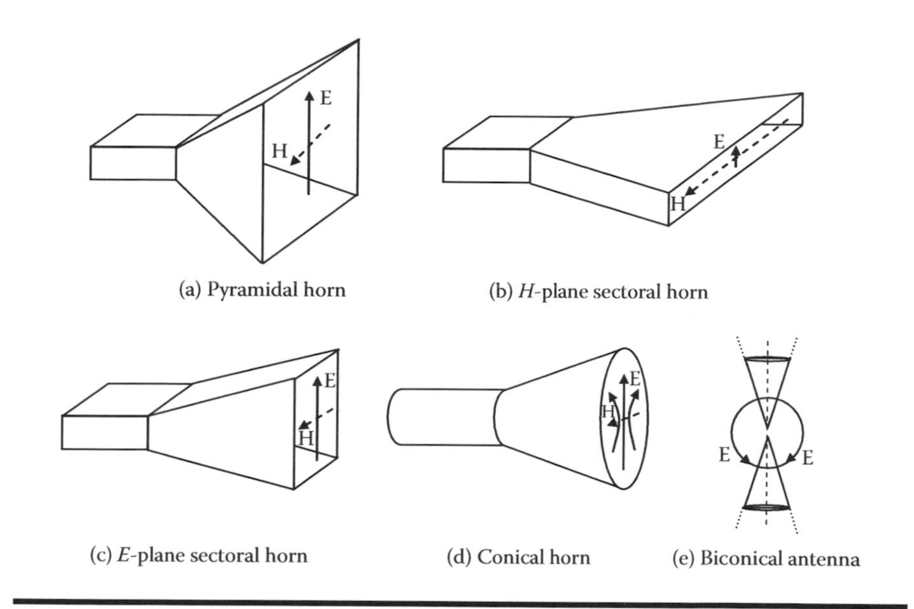

Figure 3.4 The typical configurations of the horn antenna.

are zero. The validity of this approximate method has been validated by lots of experiments. Here E-plane sectoral horn in Figure 3.4c is illustrated as an example. The feeding waveguide operates with the fundamental mode, i.e., TE_{10} mode, and the horn length is large compared with the aperture dimensions. In the coordinate system of Figure 3.5, the electromagnetic fields on the horn aperture are described approximately as follows

$$E_z = E_x = H_y = 0 \tag{3.21}$$

$$E_y(x, y) \approx E_0 \cos\left(\frac{\pi}{a}x\right)e^{-jk\delta} \tag{3.22}$$

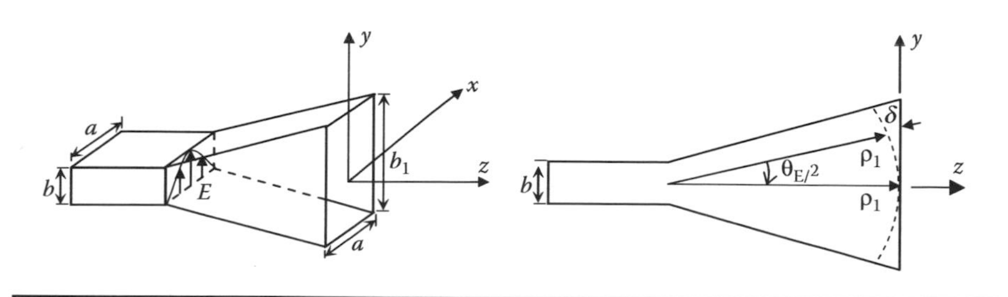

Figure 3.5 *E*-plane horn and coordinate system.

$$H_z(x, y) \approx jE_0\left(\frac{\pi}{ka\eta_0}\right)\sin\left(\frac{\pi}{a}x\right)e^{-jk\delta} \tag{3.23}$$

$$H_x(x, y) \approx -\frac{E_0}{\eta_0}\cos\left(\frac{\pi}{a}x\right)e^{-jk\delta} \tag{3.24}$$

$$\delta = y^2/(2\rho_1) \tag{3.25}$$

where E_0 is a constant, and $k\delta$ is the relative phase delay of the fields on the horn aperture to the field at the aperture center point. Based on Equations (3.1) and (3.2), the approximate equivalent electric and magnetic current sources over the infinite plane that coincides with the horn aperture and passes through the horn mouth can be given by

$$\begin{array}{ll} \left.\begin{array}{l} J_{ey} = -\frac{E_0}{\eta}\cos\left(\frac{\pi}{a}x\right)e^{-jk\delta} \\ M_{ex} = E_0\cos\left(\frac{\pi}{a}x\right)e^{-jk\delta} \end{array}\right\} & \begin{array}{l} -a/2 \le x \le a/2 \\ -b_1/2 \le y \le b_1/2 \end{array} \\ J_e = M_e = 0 \quad \textit{elsewhere} & \end{array} \tag{3.26}$$

Based on Equations (3.3) and (3.4) and the coordinate system in Figure 3.5, $\boldsymbol{A}$ and $\boldsymbol{F}$ generated by the sources on closed surface can be rewritten with

$$\mathbf{A} = \frac{\mu}{4\pi}\int\int_s \frac{\boldsymbol{J}e^{-jkR}}{R}ds' \approx \frac{\mu e^{-jkr}}{4\pi r}\boldsymbol{N} \tag{3.27}$$

and

$$\mathbf{F} = \frac{\varepsilon}{4\pi}\int\int_s \frac{\boldsymbol{M}e^{-jkR}}{R}ds' \approx \frac{\varepsilon e^{-jkr}}{4\pi r}\boldsymbol{L} \tag{3.28}$$

where

$$\boldsymbol{N} = \int\int_s \boldsymbol{J}e^{jkr'\cos\Psi}ds' \tag{3.29}$$

$$\boldsymbol{L} = \int\int_s \boldsymbol{M}e^{jkr'\cos\Psi}ds' \tag{3.30}$$

The far region fields can be expressed approximately by

$$E_r = 0 \tag{3.31}$$

$$E_\theta = -\frac{jke^{-jkr}}{4\pi r}(L_\phi + \eta N_\theta) \tag{3.32}$$

$$E_\phi = \frac{jke^{-jkr}}{4\pi r}(L_\theta - \eta N_\phi) \tag{3.33}$$

$$H_r = 0 \tag{3.34}$$

$$H_\theta = \frac{jke^{-jkr}}{4\pi r}\left(N_\phi - \frac{L_\theta}{\eta}\right) \tag{3.35}$$

$$H_\phi = -\frac{jke^{-jkr}}{4\pi r}\left(N_\theta + \frac{L_\phi}{\eta}\right) \tag{3.36}$$

Using the rectangular-to-spherical component transformation, L_θ, L_φ, N_θ, N_φ may be calculated by

$$N_\theta = \int\int_s [J_x \cos\theta \cos\varphi + J_y \cos\theta \sin\varphi - J_z \sin\theta] e^{jkr' \cos\Psi} ds' \tag{3.37}$$

$$N_\phi = \int\int_s [-J_x \sin\phi + J_y \cos\varphi] e^{jkr' \cos\Psi} ds' \tag{3.38}$$

$$L_\theta = \int\int_s [M_x \cos\theta \cos\varphi + M_y \cos\theta \sin\varphi - M_z \sin\theta] e^{jkr' \cos\Psi} ds' \tag{3.39}$$

$$L_\phi = \int\int_s [-M_x \sin\phi + M_y \cos\varphi] e^{jkr' \cos\Psi} ds' \tag{3.40}$$

The integrations in Equations (3.37) to (3.40) are usually performed numerically by computer. The radiation patterns of the E-plane sectoral horn antennas with $\rho_1 = 8\lambda$ are shown in Figure 3.6. Figure 3.6a–e demonstrates the influence of the flare angle on E-plane patterns. But the patterns in H-plane vary slightly when the flare angles change and Figure 3.6f shows the typical patterns in H-plane. The E-plane sectoral horn antenna is useful in obtaining fan-shaped beams in E-plane. The horn antenna in Figure 3.4a is very useful in obtaining specified beamwidths independently in two principal planes. The H-plane sectoral horn in Figure 3.4b can obtain fan-shaped beams in a plane containing the flare, while the pattern is very broad in another plane. The conical horn antenna in Figure 3.4d can handle any polarization of the dominant TE_{11} mode. The biconical horn antenna in Figure 3.4e gives an omnidirectional radiation pattern in the plane normal to the axis, and is useful in millimeter-wave omnidirectional communications.

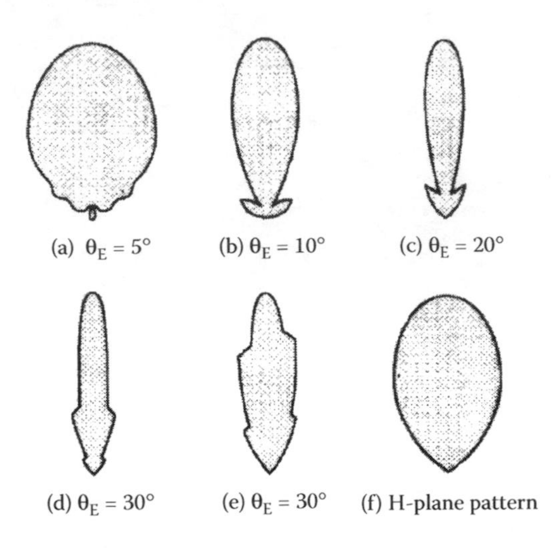

Figure 3.6 The radiation patterns of *E*-plane sectoral horn antenna with $\rho_1 = 8\lambda_0$.

3.3.2 Lens Antennas

The lens antenna is frequently used in millimeter-wave communication systems. It is a secondary antenna and needs a primary antenna to excite it. The lens antenna can be designed to produce highly shaped beams that significantly improve the system performance. It enables transformation between spherical and plane waves, and scans its pattern by moveable feeding antenna. The lens antenna is built from inhomogeneous or homogeneous materials. Lens profiles are either conical or shaped. Shaped lenses are usually designed with synthesis methods based on geometrical optics (GO) principles, and the resulting profiles are axisymmetric, quasi-axisymmetric, or arbitrary. The design of lens antenna is also relative to the given primary source. Note that multiple-beam antennas with good off-axis performance can be obtained using bifocal or zoned lenses. Conical shapes (cylinders, spheres, ellipses, hemispheres, or extended hemispheres) have been extensively studied in millimeter waves.

Figure 3.7 shows some typical configuration of millimeter-wave lens antennas. As an example, a typical axisymmetric dielectric lens antenna shown in Figure 3.7a is analyzed simply. It is assumed that λ_d is the wavelength in dielectric lens, L_0 is the distance between the focus point O and the top point Q of the lens surface, and L is the distance between the focus point O and any point on the lens surface. When the following relationship

$$\frac{L}{\lambda_0} = \frac{L_0}{\lambda_0} + \frac{L\cos\theta_L - L_0}{\lambda_d} \tag{3.41}$$

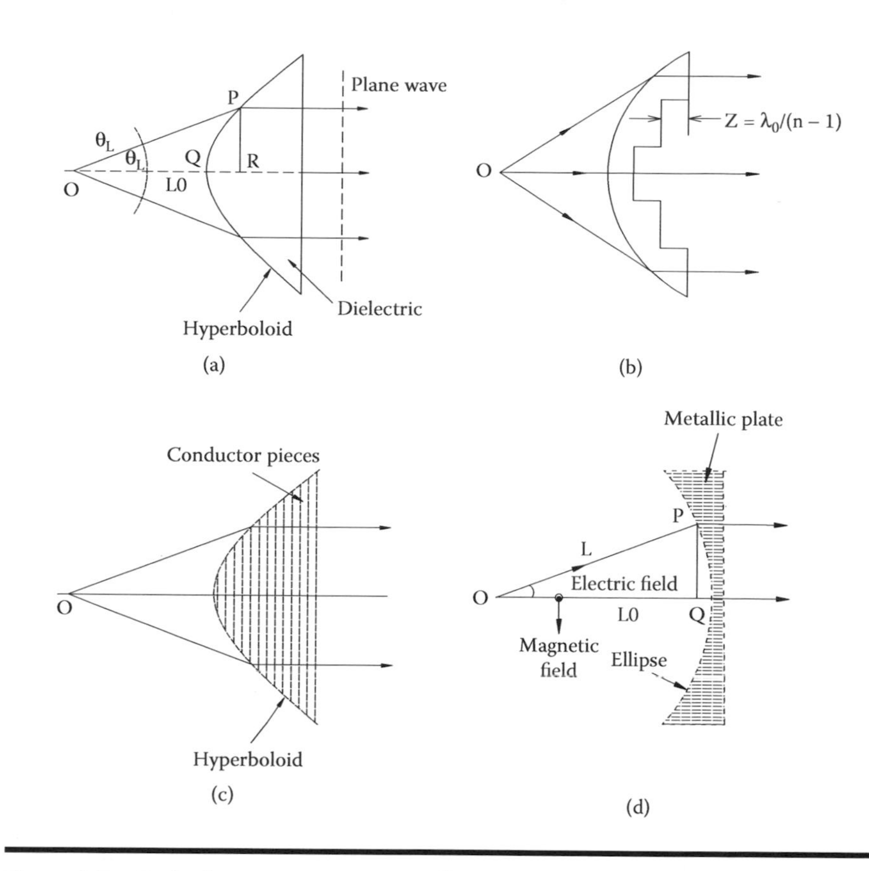

Figure 3.7 Typical antenna structures of lens antenna. (a) Traditional dielectric lens; (b) the zoned dielectric lens; (c) lens formed by artificial dielectric; (d) lens formed by *E*-plane metallic plate.

is satisfied, the spherical wave from focus O can be transferred into plane wave. θ_L is the angle between line OP and line OQ. Based on the relationship between the refractive index n and the dielectric constant ε_r of the dielectric lens

$$\varepsilon_r = n^2 = (\lambda_0/\lambda)^2 \tag{3.42}$$

Equation (3.41) can be rewritten as

$$L = L_0 + n(L \cos\theta_L - L_0) \tag{3.43}$$

Solving L from Equation (3.43), it can be expressed by a hyperboloid function

$$L = \frac{(n-1)\,L_0}{n\,\cos\theta_L - 1} \tag{3.44}$$

For an isotropic point source, the flare angle θ_L to the edge of the lens should be less than 20° to obtain a nearly uniform illumination over the lens aperture. However, in practice, the feed source is not isotropic and the illumination becomes even more nonuniform, unless the pattern of the feed antenna has less intensity in the axial direction than in the direction off the axis. Fortunately, a tapered illumination over the lens aperture contributes to suppress the minor lobes. Thus, a large flare angle of the lens is desirable sometimes. But a large angle results in large body and weight. The scheme in Figure 3.7(b) has been developed to reduce the volume and weight of the antenna by removing sections or zones. The thickness Z of the removed sections or zones should satisfy the following equation:

$$\frac{Z}{\lambda_d} - \frac{Z}{\lambda_0} = N \ (N = 1, 2, \ldots) \tag{3.45}$$

Usually, N=1 is selected. Equation (3.45) can be rewritten as

$$Z = \lambda_0 / (n - 1) \tag{3.46}$$

Obviously, the performance of the zoned lens antenna is sensitive to frequency.

Essentially, the lens antenna is structured by the mediums with a different phase velocity from free space. Artificial dielectric lens techniques are often used for various purposes. For example, the lens formed by burying some metal pieces in the medium, shown in Figure 3.7, is used to reduce the weight. Here, metal pieces play a key role and the light medium is applied. A lens formed by the E-plane metal plate is shown in Figure 3.7d. It is also artificial dielectric lens. Instead of a dielectric, the parallel plate waveguide provides different phase velocities. When the electromagnetic fields pass through the parallel plate waveguide in TE_{10} mode, the phase velocity is faster than that in free space. Therefore, the lens antennas can be categorized into two main types based on the propagation characteristics of the electromagnetic waves in the lens medium. One is a delay lens antenna in which the electrical path length is increased by the lens medium, such as those in Figures 3.7a–c. The other one can make the electrical path length decreased, such as the antenna in Figure 3.7d.

There are some typical applications of lens antenna in millimeter-wave communications. Figure 3.8a shows a photograph of a Plexiglas ($\varepsilon_r = 2.53$ and $\tan\delta_l = 0.012$) lens antenna for indoor millimeter-wave WLAN [17,18]. The antenna may contribute to significantly enhanced system performances. The antenna is fed by an aperture of a circular metallic waveguide, which is embedded in the lens body. TE_{11} mode of the circular waveguide is used to excite the lens body. The lens surface is conveniently shaped to transform the feed aperture radiation pattern into the desired output beam.

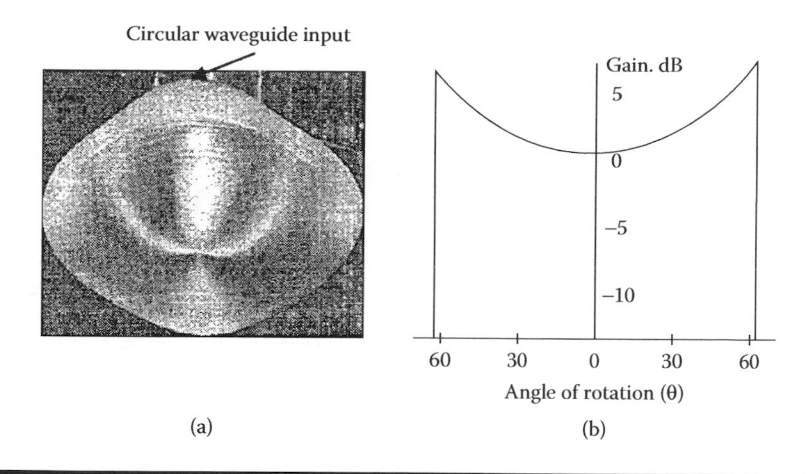

Figure 3.8 (a) Photograph of a prototype Plexiglas dielectric lens antenna. (b) Pattern.

The antenna structure is very flexible, allowing the design for different target patterns; from secant squared to hemispherical type, with linear or circular polarization. The return loss of the antenna in the frequency range of 61 to 62 GHz is better than -20 dB. The radiation pattern at frequency 62 GHz is shown in Figure 3.8b, which indicates that the radiated energy is restricted within the angle range of $\theta = 0°$ to $\pm 63°$.

3.3.3 Leaky-Wave Antennas

The leaky-wave antenna belongs to the traveling-wave antenna based on its operation principle. Various leaky antenna configurations have been studied and two typical leaky-wave antennas are shown in Figure 3.9. The electromagnetic field's amplitude is attenuated exponentially along the leaky-wave structure due to radiation. The propagation constant and attenuation constant along the extension direction of the configuration (x-axis in the coordinate system of Figure 3.9) are β_x and α_x, respectively, and the electrical field amplitude in the antenna can be expressed as

$$E(x) = E_0 e^{-(\alpha_x - j\beta_x)x} \tag{3.47}$$

The beam direction in xoz-plane is given as

$$\theta_0 = \sin^{-1}(\beta_x / k_0) \tag{3.48}$$

where k_0 $(2\pi/\lambda_0)$ is the wave number in free space. The beam direction θ_0 is changeable with the variation of the frequencies; i.e., the leaky-wave

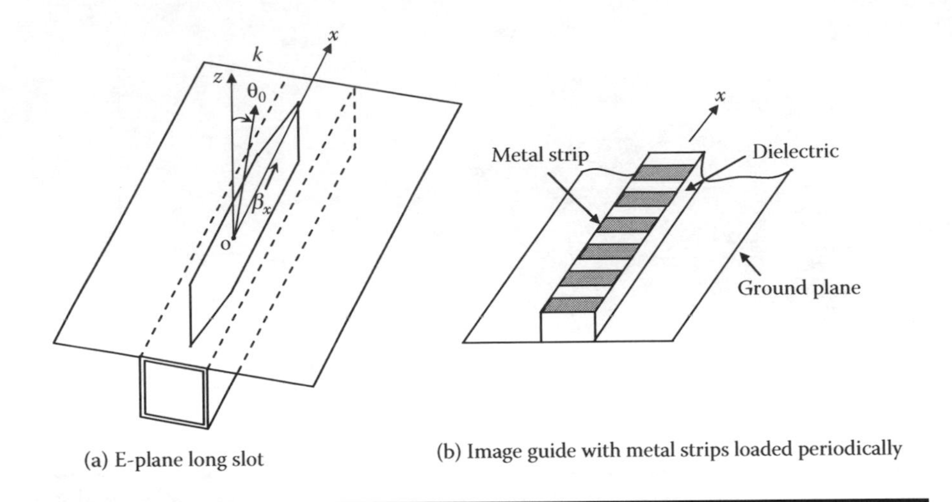

Figure 3.9 Some simple leaky-wave antenna structures.

antenna is the one with frequency scanning characteristics. This is one of the main merits of the leaky-wave antenna. Additionally, the leaky-wave antenna can provide a high gain due to its large aperture.

Equation (3.48) indicates that $\beta_x < k_0$ is necessary for radiation, which means that the antennas can support the fast wave propagation alone extension direction of configuration. Because of metal waveguide structure, transverse resonance occurs in the antenna structures of Figure 3.9a when the frequencies are higher than the waveguide cutoff frequency. Assuming $k_t(>0)$ is the wave number corresponding to transverse resonance, according to the equation

$$k_0{}^2 = k_t^2 + \beta_x^2 \tag{3.49}$$

$\beta_x < k_0$ can be satisfied.

The periodic structure antenna in Figure 3.9b is a slow wave structure; however, it can also radiate electromagnetic energy. Based on Floquet's theorem, the transmitted electromagnetic fields can be expressed as the sum of an infinite number of space harmonics in a periodic structure with period P

$$\mathrm{E}(z) = e^{-\alpha_x x} \sum_{m=-\infty}^{\infty} \mathrm{E}_{Pm} e^{-j\beta_{xm} x} \tag{3.50}$$

with

$$\beta_{xm} = \beta_{x0} + \frac{2\pi m}{P} \tag{3.51}$$

where E_{Pm} and β_{xm} are the amplitude and phase constant of the m-th space harmonic, respectively. β_{x0} is the phase constant in structure free from periodic perturbation. When $|\beta_{xm}| < k_0$, the m-th space harmonic becomes a fast wave and radiates a leaky wave in the direction

$$\theta_m = \arcsin\left(\frac{\beta_{x0}}{k_0} + \frac{m\lambda_0}{P}\right) \tag{3.52}$$

Generally, m is a negative number to satisfy $|\beta_{xm}| < k_0$, and many space harmonics can radiate energy simultaneously. In order to avoid grating lobes, it is desirable that only one fast wave mode is launched. For this purpose, besides $|\beta_{xm}| < k_0$, the period P should be limited by the following equations:

$$\frac{-2m-1}{2\left(\beta_{x0}/k_0 - 1\right)} < \frac{P}{\lambda_0} < \frac{-2m+1}{2\left(\beta_{x0}/k_0 + 1\right)} \tag{3.53}$$

$m = -1$ order space harmonic is usually used due to its larger amplitude.

From Equation (3.52), we can find that the periodic structure leaky antenna can scan its main beam if the period P can be changed in real time. A pattern reconfigurable coplanar waveguide (CPW) antenna in millimeter wave has been developed [19]. The antenna configuration is shown in Figure 3.10. One hundred and twenty five pairs of slits are etched on the ground symmetrically from the origin, O. In each perturbation slit, one micro-electro-mechanical system (MEMS) switch or PIN diode switch is installed and closed to the CPW slots. A period of $P = p \times \Delta P$ (p is a positive integer) can be obtained by making every p switch open and the others closed. So it is easy to control P by adjusting the states of the switches installed in the slits. The minimum step of the reconfigurable period is ΔP. The operation frequency of the antenna is 35.0 GHz. The $m = -1$ order space harmonic is chosen to form a single main beam. The antenna radiates its main beam in a backward direction. When Port 1 is the feed port, the single main beam can be scanned from $\theta_m = -90°$ to $-0°$ by adjusting switch states. But the antenna cannot scan a beam with a small angle step by fixed ΔP. Reconfigurable states with the mixed-period structures are presented to reduce the scanning angle step. The mixed-period structure is constructed by the cascade of compound cells. The compound cells are formed by a serial connection between M cells with a period of P_1 and N cells with a period of P_2. The equivalent period of the mixed-period structure can be calculated by

$$P_{eff} = \frac{MP_1 + NP_2}{M + N} \tag{3.54}$$

The design results indicate that the antenna can scan the beam from $\theta_m = -90°$ to $-0°$ with ten states (four uniform states and six mixed-period

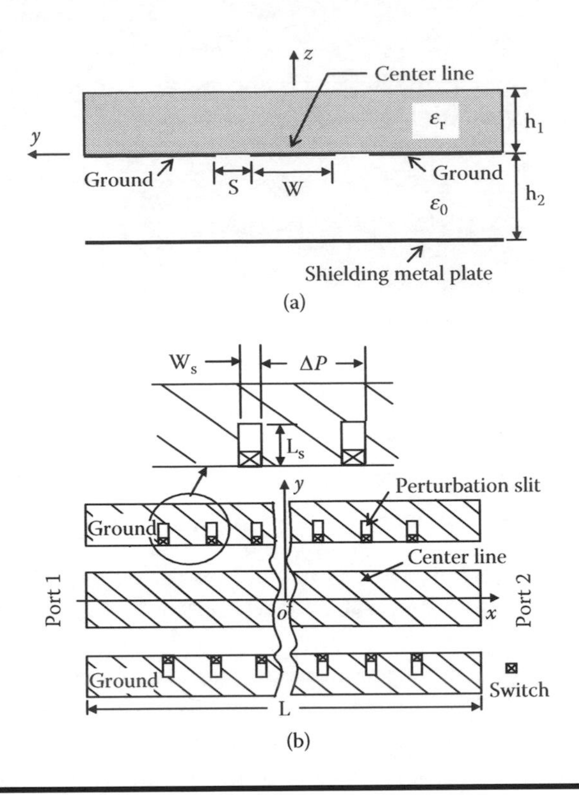

Figure 3.10 Configuration of the CPW pattern reconfigurable antenna. (a) Cross-section view. (b) CPW plane view.

states). Due to symmetry, the single main beam can be scanned from $\theta_m = 0°$ to $90°$ by using Port 2 as the feed port.

Another kind of leaky-wave antenna is proposed to radiate its $m = 0$ order space harmonic; i.e., the fundamental mode [20]. The antenna radiates its main beam in a forward direction. As an example, one CPW-based leaky-wave antenna is shown in Figure 3.11. The period of the structure

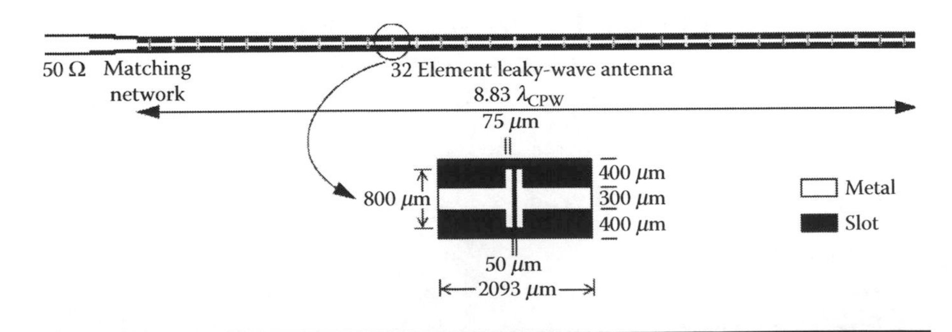

Figure 3.11 CPW-based leaky-wave antenna structures for forward radiation.

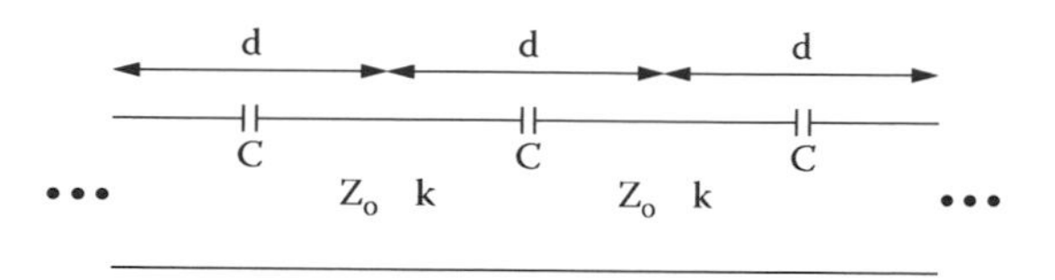

Figure 3.12 The simple transmission line mode for CPW-based leaky-wave antenna.

is denoted by d. A simple transmission line model in Figure 3.12 is used for simplifying analysis. The loaded transmission line system exhibits the following dispersion relation:

$$\cos(\beta_{x0}d) = \cos(kd) + X\sin(kd)/2Z_0 \tag{3.55}$$

where kd and Z_0 are phase delay and characteristic impedance of interconnecting transmission line segments, respectively. $\beta_{x0}d$ is the phase delay per periodic unit cell and $X = 1/(\omega C)$ is the reactance due to the series loading capacitor. This dispersion relation can be examined graphically by means of the Brillouin diagram, shown in Figure 3.13a.

The dotted line in Figure 3.13a marks the transition between slow- and fast-wave propagation related to free space in the first Brillouin zone. The $m = 0$ order space harmonic becomes fast wave ($\beta_{x0} < k_0$) and the antenna becomes leaky. The antenna operates at the frequency of 30.0 GHz. The coordinate system and the principle planes of pattern are shown in Figure 3.13b. The measured patterns in H-plane and E-plane are shown in Figure 3.13c and d, respectively. The measured results indicate that the radiated beam points at an angle of $\theta_m = 55°$ while exhibiting a VSWR <2 bandwidth of 26%. In addition, the antenna demonstrates a gain of 9.2 dB while suffering only minimal conductor feed-line losses of 0.4 dB. The transverse size of the structure is electrically small, which allows the compact integration of several antennas side-by-side in order to form an array.

3.3.4 Surface Wave Antennas

When electromagnetic wave propagates in a dielectric layer or on the surface of a metal structure with a smaller phase velocity than that in free space, the transmitted wave is called a surface wave. From Equation (3.48), a surface wave cannot radiate energy. However, if the surface waveguide structure is truncated to a finite length, electromagnetic waves are radiated into space from the end of the waveguide. Thus, a surface wave antenna is formed. Typical surface wave structures are dielectric slab, dielectric rod, corrugated metal or dielectric structures, and long Yagi.

The surface wave antenna is often used in millimeter waves owing to its small physical size. An example of a dielectric rod antenna at $f = 62$ GHz

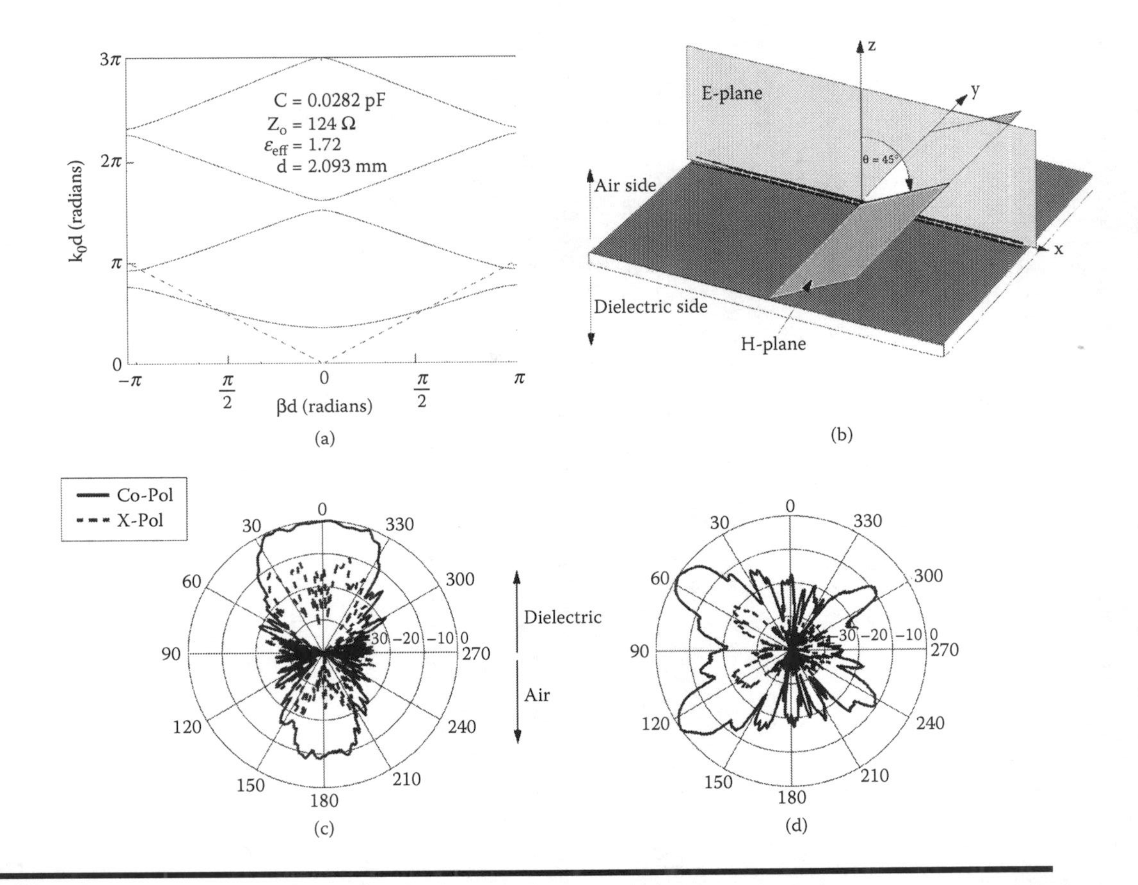

Figure 3.13 The dispersion relation of the CPW-based leaky-wave antenna and the antenna radiation patterns.

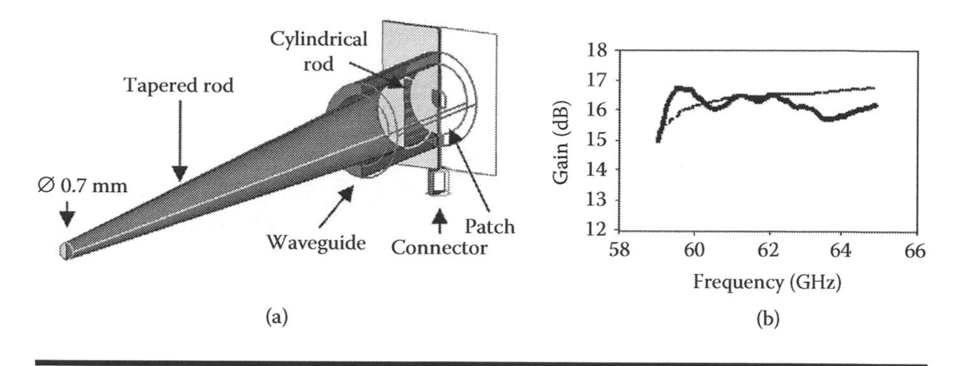

Figure 3.14 The patch-fed rod antenna and its gain performance.

is shown in Figure 3.14a [21]. The rod antenna consists of a cylindrical part and a tapered part. The rod is fed by a patch, which is sequentially energized by a microstrip line connected to a coaxial connector. The antenna configuration can be easily built and integrated with other millimeter-wave functional modules or planar circuits. The antenna is simulated by CST Microwave Studio. While the antenna is radiating, energy from the patch is transferred to the tapered rod through a small cylindrical rod and a circular waveguide. The height of the cylindrical rod is set to 3 mm while that of the circular waveguide is 7 mm. The upper part of the rod is tapered linearly to a terminal with 0.7-mm diameter and 30-mm height in order to have high antenna gain. The tapered rod can be treated as an impedance transducer and it reduces the reflection caused by an abrupt discontinuity. The rod antenna with circular cross-section has a symmetrical shape and therefore it generates the same energy in both right-hand circular polarization (RHCP) and left-hand circular polarization (LHCP). The rod antenna is a directional antenna and it radiates along the central axis of the rod. The direction of the beam can be adjusted by changing the direction of the central axis of the rod antenna. The matching is approximately −15 dB between 61 and 63 GHz. The measured gain is shown in Figure 3.14b. The half power beamwidth is 20°. This antenna is suitable for millimeter-wave high-data-rate wireless communications.

3.3.5 Reflector Antennas

Reflector antennas, including plate reflector, corner reflector, and curved reflector (shown in Figure 3.15), are applied often in high frequency. The millimeter-wave reflector antenna has three main merits: (1) it has a wide bandwidth because of independent structure from frequency, (2) its simple feed system results in less feed loss, and (3) it provides a high gain due to a large aperture. The polarization of the source and the feed position

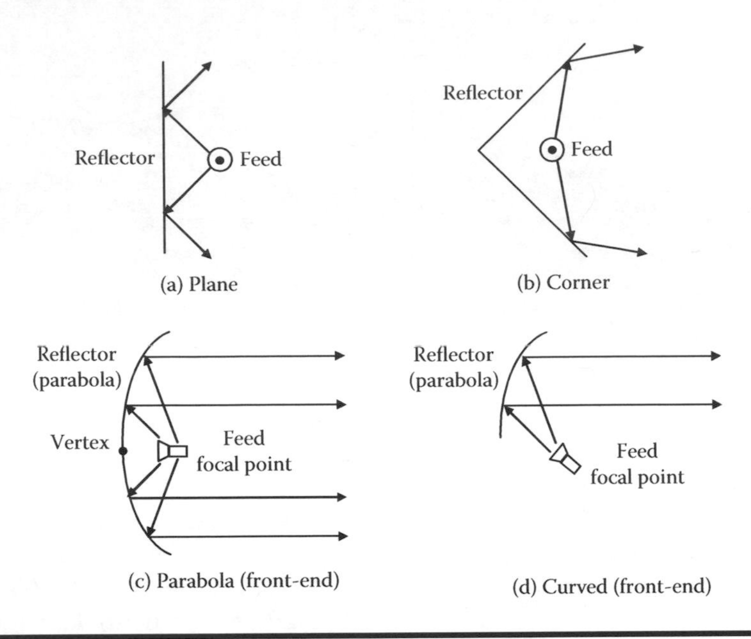

Figure 3.15 The basic forms of the reflector antenna.

relative to the reflecting surface can be used to control the radiating properties (pattern, impedance, directivity). Corner reflectors can provide better directional radiation performance than plate ones. The best overall radiation characteristic is obtained by using a reflector with a parabolic configuration. Paraboloidal reflectors can be fed by a horn. The performances of high-gain and low-noise paraboloidal reflectors are suitable for communications with a longer distance. However, the paraboloid reflector antenna in Figure 3.15c has some disadvantages; namely, (1) the wave reflected from the parabola back to the feed antenna produces interaction and mismatching, and (2) the feed antenna acts as an obstruction, which results in increasing side lobe and decreasing gain. An improved structure, shown in Figure 3.15d, has been developed. Only a portion of the paraboloid is used to avoid these drawbacks. But the modified schemes worsen the radiation performance, including the polarization performance and the radiation beam. Just like lens antenna, the radiation beam of the parabolic reflector antenna can be controlled by shifting the feed horn from the focal point in the focal plane.

In order to increase efficiency and decrease side lobe, dual-reflector antennas are utilized usually because it is easy to control the amplitude distribution over the aperture. Two main forms of dual-reflector antennas are the Cassegrain antenna and the Gregorian antenna. Their structure includes a main reflector (parabola) and a subreflector (hyperbolic reflector). The difference from the omni-reflector antenna in Figure 3.15c is that the feed horn is moved to a location between the vertex and the focus of the

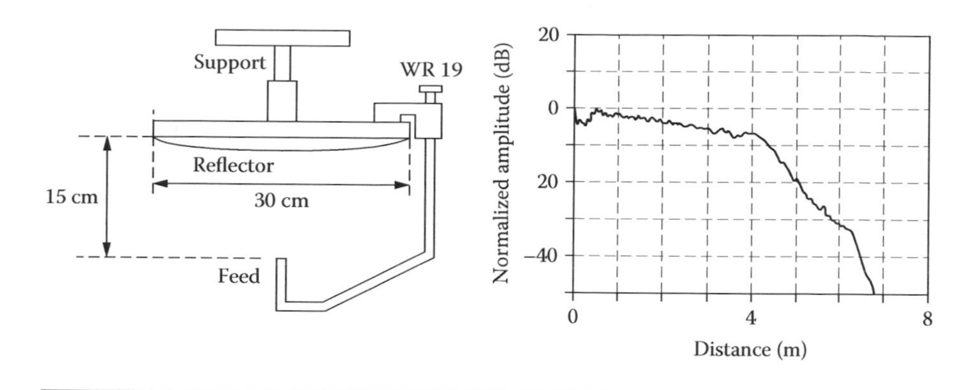

Figure 3.16 (a) The geometry of the shaped reflector antenna, and (b) its near-field distribution.

parabola reflector, and a subreflector is located near the focus. However, the setting of the subreflector has to be arranged precisely in this manner: one focus of the hyperbolic reflector matches the phase center of the feed horn and the other matches the focus of the parabola reflector.

The main techniques in analyzing the reflector antenna are the aperture distribution method and the current distribution method. Geometrical theory of diffraction (GTD) can be used to analyze the diffraction of the reflector edge.

Besides being used as a directional device, the reflector antenna is used as an omnidirectional antenna, which is frequently applied as a radiator for indoor wireless LAN access points. A shaped reflector antenna in Figure 3.16a is presented for a 60-GHz indoor wireless LAN with the converge area of 8×8 m^2 [22]. The feed is a corrugated conical horn that is designed to provide a circular symmetrical radiation pattern $2(n+1)cos\, n\psi$ (ψ denotes angle from feed axis). $n = 6.6$ is determined for reflector edge illumination -10 dB compared with that of center. The feed is connected via a transition to standard WR-19 waveguide. The transition transforms the TE_{10} mode in WR-19 waveguide into the TE_{11} mode in circular waveguide. The circular polarization radiation field is achieved by using a polarizer in the feed entrance. Figure 3.16b shows the normalized near-field distribution below the antenna as three meters, which indicates that the amplitude across the footprint area remains fairly constant within 6 dB of uniformity.

3.3.6 Planar Integrated Antennas

With the development of integrated circuits, the microstrip and printed circuit antennas have been widely used for the microwave and millimeter-wave band due to their low profile, low cost, light weight, conformity to

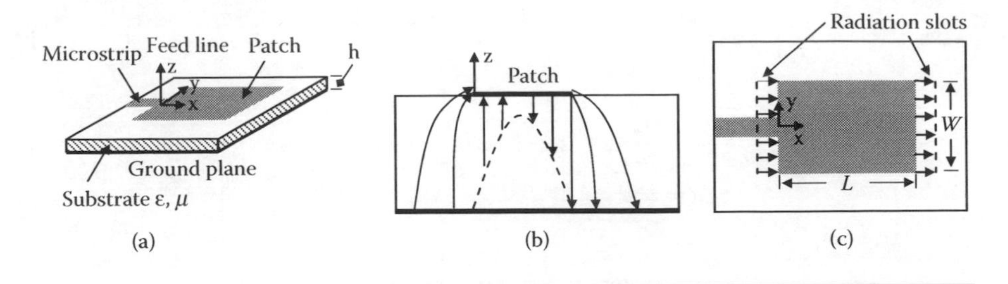

Figure 3.17 Rectangle patch antenna and its field distribution on the section and top views. (a) Basic configuration; (b) the field distribution on section view; and (c) the field distribution on top view.

surface, mass production, and direct integrability with microwave circuitry [23–25]. The rectangular microstrip antenna is shown in Figure 3.17a. It consists of a metallic patch on a grounded substrate. The field distribution of the fundamental mode from the section view and top view are shown in Figure 3.17b and c, respectively. The metallic patch can take various configurations, as shown in Figure 3.18. Essentially, the microstrip antenna radiates energy by slots formed by patch edges and ground plane.

The microstrip antenna has a main disadvantage in narrow band. In order to enhance its bandwidth, several approaches have been used, such as using impedance matching network, thick substrate with low dielectrics constant, parasitic patches stacked on the top of the main patch or close to the main patch in the same plane, and chip-resistor and integrated reactive loading. Some of these technologies are indicated in Figure 3.19. Another disadvantage of the microstrip antenna is its poor power capacity, which leads the microstrip antenna to be applied in the case of lower power radiation. Compared with the microstrip antenna in microwave, the miniaturizing technologies are not very significant for millimeter-wave applications due to the smaller wavelength.

The simplest method to analyze the microstrip antenna is the empirical models. There are two kinds of empirical models: transmission line model (TLM) and cavity model. In TLM, each of the radiating edges of the antenna is simulated by a radiating slot with a complex admittance. However, the TLM can handle only rectangular or square patches. The cavity model, in principle, can handle arbitrary-shaped patches. The mathematics involved

Figure 3.18 Some typical shapes of patch elements.

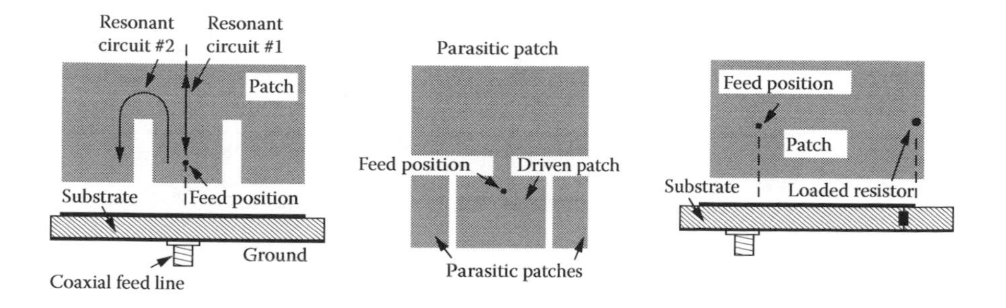

Figure 3.19 Some simple technologies for enhancing microstrip antenna bandwidth.

is tractable only when rectangle, square, circular, or triangular patch shapes are considered. The model basically treats the patch antenna as a filled dielectric cavity that is bounded by electric conductors (above and below it) and by magnetic walls (to simulate an open circuit) along the perimeter of the patch. The electromagnetic fields in the cavity are the sum of various resonance modes. A given mode in the rectangle microstrip antenna can be denoted by *mnp* mode, where m, n and p correspond to x-, y-, and z-axis, respectively. k_x, k_y, and k_z is the wavenumber along the x, y, and z directions, respectively. The wave number of *mnp* mode can be determined by

$$k^2 = \omega^2 \mu\varepsilon = k_x^2 + k_y^2 + k_z^2 \tag{3.56}$$

with

$$\begin{aligned} k_x &= m\pi/L \quad m = 0, 1, 2\ldots \\ k_y &= n\pi/W \quad n = 0, 1, 2\ldots \\ k_z &= p\pi/h \quad p = 0, 1, 2\ldots \end{aligned} \tag{3.57}$$

where m, n, and p are the number of half-cycle field variations along the x, y, and z directions, respectively, and cannot equal zero simultaneously. The resonant frequencies are calculated by

$$\begin{aligned} f_{mnp} &= \frac{1}{2\pi\sqrt{\varepsilon\mu}}\sqrt{k_x^2 + k_y^2 + k_z^2} \\ &= \frac{1}{2\pi\sqrt{\varepsilon\mu}}\sqrt{\left(\frac{m\pi}{L}\right)^2 + \left(\frac{n\pi}{W}\right)^2 + \left(\frac{p\pi}{h}\right)^2} \end{aligned} \tag{3.58}$$

In general, h is much smaller than L and W. Under the condition $L > W$, the mode with the lowest resonant frequency, also called dominant mode,

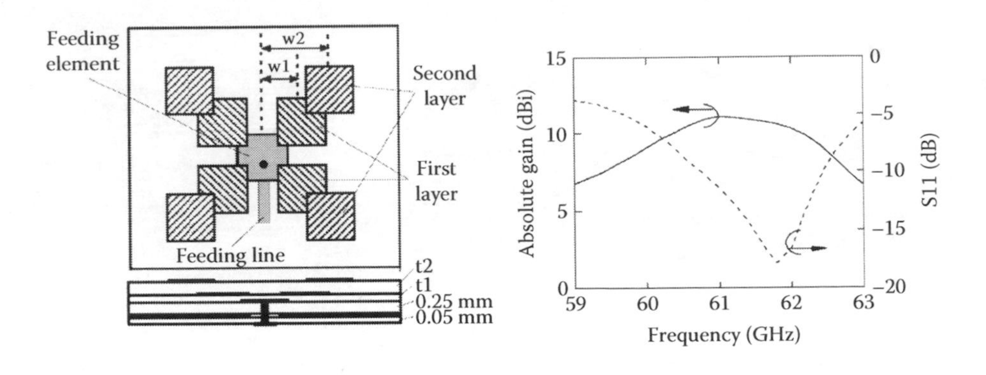

Figure 3.20 Microstrip antenna integration design example and its performance.

is the TM^z_{100} mode. Its resonance frequency is

$$f_{100} = \frac{1}{2L\sqrt{\mu\varepsilon}} \tag{3.59}$$

The full-wave methods, including finite-difference time-domain (FDTD) method, method of moment (MOM), and finite element method (FEM), can provide more comprehensive and accurate predictions for microstrip antenna performance. Currently, commercial software based on full-wave methods can be obtained. For example, XFDTD and QFDTD based on FDTD method, IE3D based on MOM, and HFSS based on FEM have been developed. The semiempirical model is a hybrid of empirical and full-wave analysis. The involved analytical and computational complexity is more than that of the empirical models and less than that of the full-wave analysis.

Figure 3.20 shows a recent design of the millimeter-wave microstrip antenna [26,27]. It is a compact and highly efficient multilayered parasitic microstrip array antenna. The antenna array is constructed on a multilayer Teflon substrate for millimeter-wave system-on-package modules. The substrate has a relative dielectric constant $\varepsilon_r = 2.2$ with $tan\delta = 0.0007$ at 10 GHz. The antenna array employs three layers with a 2×2 parasitic array on each layer and is well suited to achieve high gain and a wide bandwidth. The via-fed method is used. The size of the feeding element is $0.3\lambda_0 \times 0.3\lambda_0$ and the feeding point is 0.01 λ_0 away from the center of the patch. The size of the parasitic elements mounted on the first parasitic layer is 0.32 $\lambda_0 \times 0.32\ \lambda_0$ and the patch size of the second parasitic layer is 0.28 $\lambda_0 \times 0.28\ \lambda_0$. The four parasitic elements mounted on each layer are arranged so that they are equidistant from the center of the feeding element and the distance of the parasitic elements on the first parasitic layer and that on the second parasitic layer are w_1 and w_2, respectively. The substrate thicknesses of the first and second parasitic layers are t_1

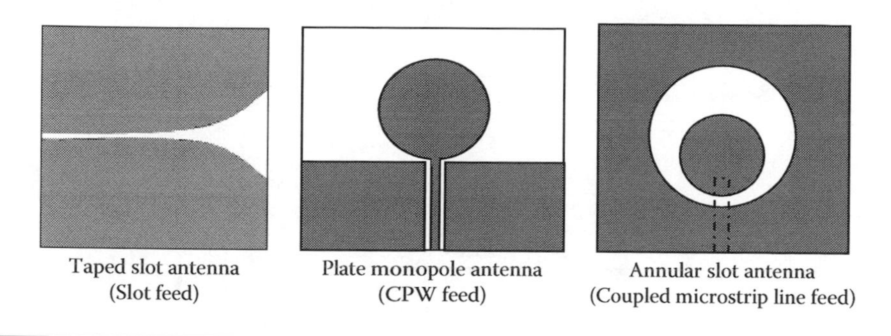

Figure 3.21 Some typical structures of planar slot antennas.

and t_2, respectively. The influence of the configuration parameters on the antenna performance has been studied. The developed antenna with optimized configuration achieves a radiation efficiency greater than 91 percent and an associated antenna gain of 11.1 dB at 60 GHz. The antenna size is only 10 mm × 10 mm. A similar design based on low temperature cofired ceramic (LTCC) technology has also been realized.

Another planar antenna suitable for integration is slot antenna, which radiates energy by the etched slots on the ground or tapered slots. The basic configuration of the tapered slot antenna is shown in Figure 3.21. Besides the similar merits of microstrip antenna, the slot antenna can obtain a wide bandwidth, moderately high gain (7–10 dB), and symmetrical *E*- and *H*-plane patterns. However, the analysis method of the tapered slot antenna is not yet built and the designs are basically based on empirical techniques. The tapered slot antenna also belongs to endfire traveling-wave antennas. In some cases, the slot antenna is formed by drilling some slots with various configurations on the wall of a waveguide. In order to radiate the electromagnetic wave with good efficiency, the slots need to cut the current on the wall and have a length of one-half the wavelength.

A design of the tapered slot antenna in millimeter waves is shown in Figure 3.22a–d [28]. The multilayer LTCC structure is used to integrate this antenna to the RF front-end module. This substrate consists of nine layers that are each 100 μm thickness. The dielectric constant of the material is 5.4 and its loss tangent is 0.0015. The antenna consists of three parts: a radiator, a feeding structure, and an air cavity. The radiator is a linear tapered slot, which is located in the second layer. The feeding structure is on the first layer to allow for easy measurement and characterization. The radiator is in the second layer to obtain tight coupling with the feeding structure. The feeding structure is a transition from a microstrip line to a slot line. The air cavity back is located from the fourth layer to the ninth layer to lower the effective dielectric constant. The size of the air cavity is 5.0 mm × 2.5 mm. The depth of air cavity has been fixed on 600 μm to ensure a sufficient

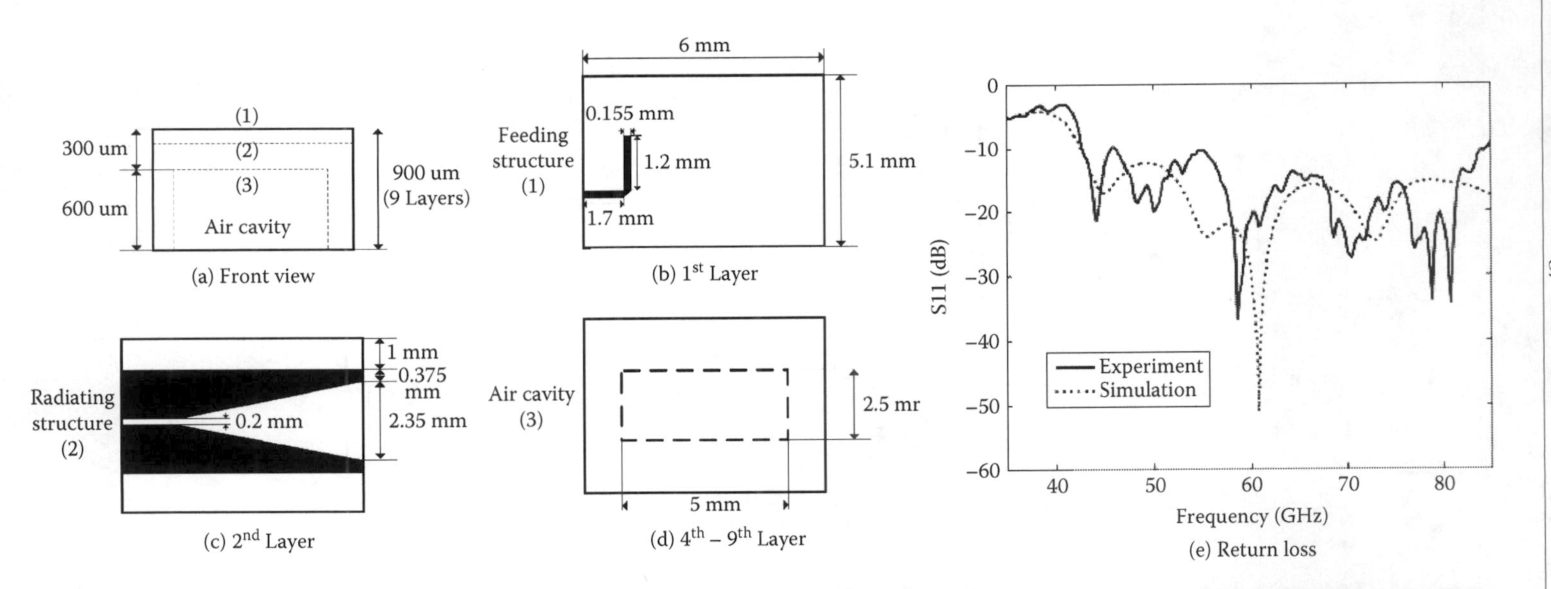

Figure 3.22 The design of a millimeter-wave tapered slot antenna on LTCC substrate.

mechanical strength and stability. The antenna is designed by HFSS. The return loss and the gain of the antenna is shown in Figure 3.22e. It can be seen that the bandwidth with a return loss of less than −10 dB is about 32 GHz (43–75 GHz).

Phased arrays are an important technology in millimeter-wave communications. It is regretted that phased arrays have not been included in this book because the planned chapter was withdrawn by the authors.

3.4 Channel Modeling for Millimeter-Wave Propagation

3.4.1 *Basic Propagation Characteristics*

It is well known that the power density radiated by antennas is attenuated along the direction of electromagnetic wave propagation, according to the rule demonstrated in Equation (3.20). The loss between two isotropic antennas is expressed in absolute numbers by the following Equation [29]:

$$L_{free} = (4\pi r_{TR}/\lambda)^2 \tag{3.60}$$

After converting to units of frequency and putting them in decibel form, the equation becomes

$$L_{free} = 92.4 + 20\log_{10} f + 20\log_{10} r_{TR} \tag{3.61}$$

where f is the frequency in gigahertz and r_{TR} is the line-of-sight (LOS) range between antennas in kilometers. Equation (3.61) indicates that the higher the frequency, the larger the free-space loss. This suggests that, even in a short distance, the free-space loss can be quite high, and only a short-distance communication link is supported for the millimeter-wave spectrum.

In microwave systems, transmission loss is accounted for principally by the free-space loss. However, in millimeter-wave band, besides the propagation attenuation, the absorptive attenuation of electromagnetic wave occurs due to the presence of atmospheric molecules, especially oxygen and water vapor. The absorption lines of the electromagnetic wave can be found in Chapter 1. The absorption contribution by oxygen results from its magnetic interaction with incident wave and that by water vapor results from the electric polarity of the water molecule. In each case, there are resonant frequency regions where the absorption is abnormally large. For oxygen, the resonance occurs at 60 GHz and 120 GHz. In the case of water vapor, the resonance occurs at 22 GHz and 183 GHz. At the same time, it can be found that there are four windows with fewer propagation losses;

i.e., 35 GHz, 94 GHz, 140 GHz, and 220 GHz. At 60 GHz oxygen absorption peak, the working range for a typical fixed service communication link is very short, in the order of 2 km. It implies the frequency reuse possibilities; however, antenna directivity and other factors must be considered in determining actual frequency reuse. The absorption attenuation due to oxygen and water vapor molecules can be calculated accurately by the model obtained from spectroscopic experiments. An effective model was established by ITU-R in 1996. The model is effective at frequencies lower than 350 GHz.

Weather can influence the propagation characteristic of millimeter waves. When millimeter wave propagates through rainfall, the attenuation and depolarization occur because of scattering or absorption due to raindrops. The specific attenuation due to rain was studied by Ihara in CRL (Japan) in 1984. Figure 3.23 shows the attenuation per kilometer as a function of rain rate. The studied results indicate that the specific attenuation increases rapidly as the frequency becomes higher and tends to become saturated beyond 100 GHz for a given rainfall and the attenuation increases as the rainfall intensity becomes stronger. Likewise, millimeter waves can be attenuated when they propagate through snowfall. However, it is difficult to study the attenuation by snowfall because the property of snowfall depends on water vapor content. Usually, the attenuation due to dry snow is negligible. Fog and clouds can be regarded as water droplets with small diameters compared with raindrops. The attenuation caused by fog and

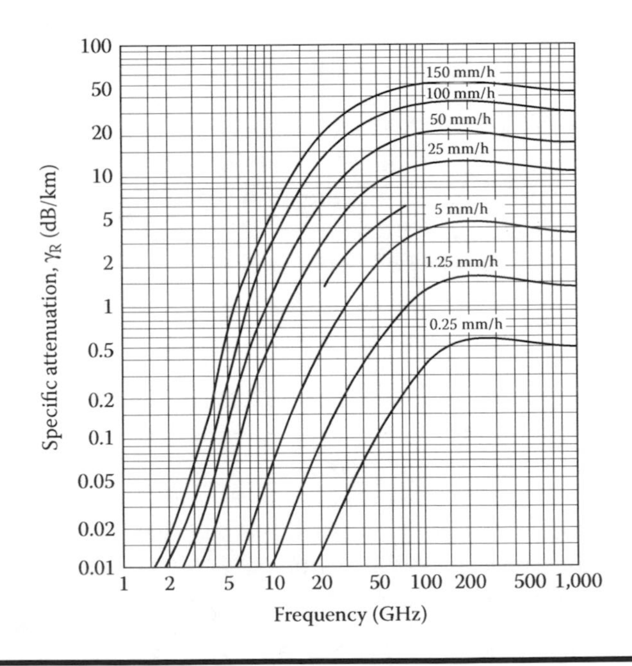

Figure 3.23 Specific attenuation due to rain.

clouds was studied by ITU-R in 1995. Among all of these weather cases, rainfall has the most serious influence on millimeter-wave propagation.

Another familiar case that results in significant losses for millimeter-wave propagation is woods. For the case where the foliage depth is less than 400 m, the loss is given by

$$L_{fol} = 0.2 f^{0.3} R_{fol}^{0.6} \text{ dB} \tag{3.62}$$

where f is the frequency in megahertz and R_{fol} is the depth of foliage transversed in meters. This expression is applicable for frequencies in the range 0.2–95 GHz. Based on Equation (3.62), when millimeter waves with 40 GHz frequency penetrate a large tree or two in tandem (approximately 10 m), the foliage loss is about 19 dB.

3.4.2 Channel Modeling

3.4.2.1 The Traditional Channel Modeling

In order to estimate signal parameters accurately in millimeter-wave wireless communication systems, it is also necessary to estimate signal propagation characteristics in different terrain environments, such as free space, urban, suburb, country, and indoor. To some extent, the communication quality is influenced mainly by applied terrain environment. Propagation analysis provides a good initial estimate of the signal characteristics. Channel modeling is defined as the communication path between transmitting and receiving antennas. In the designs of the modern wireless communication systems, the channel modeling must be considered accurately as a part of the whole system.

The channel must consider all possible propagation effects, including absorption, attenuation, reflection, motion, diffracting, and fading [30,31]. In different operation environments, the propagation modeling is dissimilar. Under outdoor propagation conditions, the simplest channel modeling is the flat Earth model. In this model, only two propagation paths are considered, just as shown in Figure 3.24. The received signal is a sum of the signals from two paths. When the reflector characteristic of the ground plane is given, the propagation modeling can be extracted.

However, usually numerous paths can be created by the random distribution of large numbers of reflecting, diffracting, refracting, and scattering objects, such as in urban areas. Then, the multipath signal amplitudes, phases, and time delays become random variables and one must revert to a statistical model for estimating signal and channel behavior. Fading is used to describe the fluctuation in a received signal as a result of multipath components. The fading can be classified as fast fading, slow fading, flat fading, and frequency selective fading.

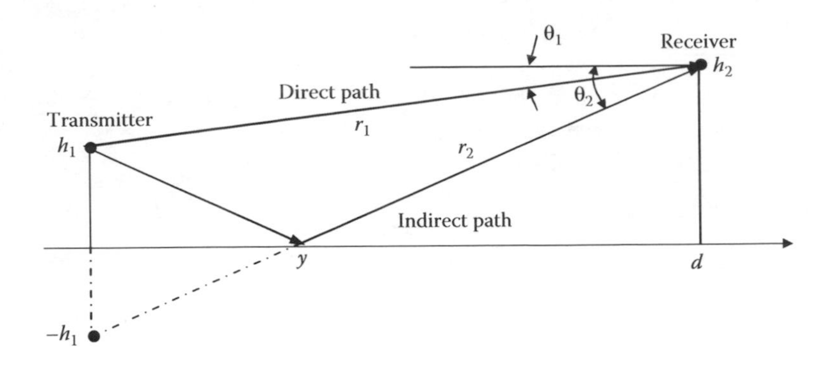

Figure 3.24 The flat Earth model.

Many channel modelings have been established for various applications. Application systems include fixed and mobile millimeter-wave communications. They range from simple geometric models to more complex statistical and combined geometric/statistical and very complex ray tracing models. The indoor millimeter-wave channel modeling has been widely studied recently because indoor millimeter-wave communications play a dominant role in the present market. In indoor wireless systems, the prediction of microwave and millimeter-wave propagation characteristics is usually applied to a specific location. The FDTD method and MOM are used to perform this prediction in some cases. However, many channel modelings are found by experimental method today.

In order to suppress the multipath propagation in high-speed wireless transmission, circular polarization and directional antennas may be adopted. The experiments at 60 GHz by Manabe in 1995 and 1996 have indicated that the circular polarization and directional antenna can improve the propagation performance significantly.

3.4.2.2 The TR Channel for UWB Communications

Recently, an important method, called time-reversal (TR) technique, has been developed to transmit ultra-wideband (UWB) wireless signals and overcome the problems arising from multipath propagation [32–34]. Essentially, it is a kind of adaptive antenna technique. Time reversal has been studied for a long time as a method to focus an ultrasonic wave in both time and space. Based on the same principle, it can be used to focus an electromagnetic wave on both time and space. Basically, the field radiated by an electromagnetic source is initially recorded at a transducer array. Then it is time reversed and retransmitted through the medium by the same array acting as a time-reversal mirror (TRM). The resulting wave travels back and refocuses on the initial source position (spatial focusing). Even when the medium is highly dispersive due to reverberation or scattering, the

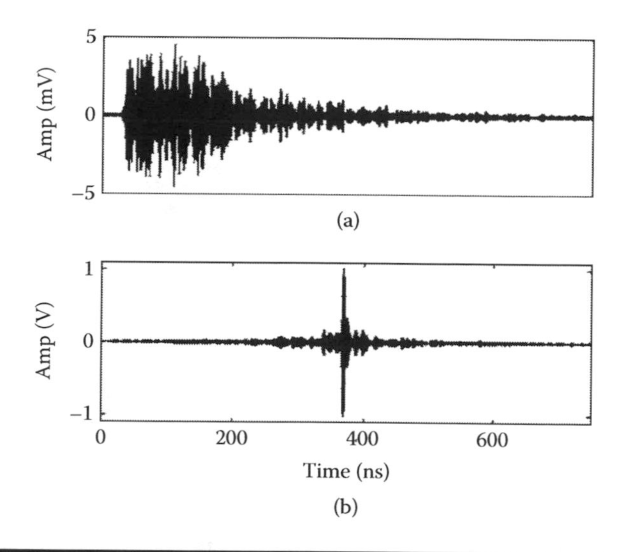

Figure 3.25 (a) A typical pulse response of the multipath propagation. (b) The time compression effect by the time-reversal process.

time-reversed signals are compressed into a pulse as short as the transmitted one (time compression). Spatial focusing and time compression have been studied both theoretically and experimentally. Figures 3.25 and 3.26 illustrate the typical effect of signal transmission based on the time-reversal technique.

When the time-reversal technique is applied in wireless communications, the main advantages are (1) mitigating intersymbol interference (ISI)

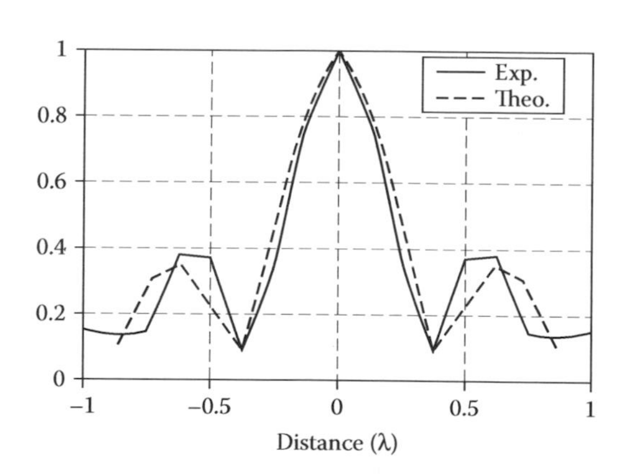

Figure 3.26 The spatial focusing effect based on the time-reversal technique. The amplitude is normalized to energy at the focus point.

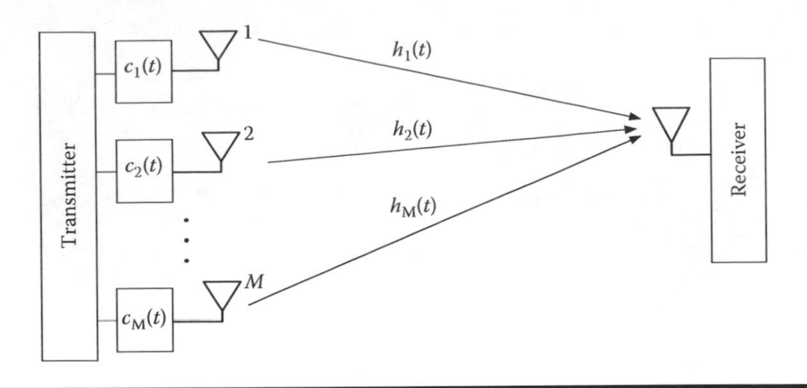

Figure 3.27 MISO wireless communication system based on the time-reversal technique.

without using an equalizer and (2) focusing the signal on the point of interest thereby reducing the cochannel interference. The simple communication process of the time-reversal UWB multiple-input, single-output (MISO) system is shown in Figure 3.27. The antenna array can be viewed as a TRM that records a waveform $f(t)$, time reverses it, and then retransmits $f(-t)$. There are three steps in TR communications. First, a signal $x(t)$ is transmitted on the receiver side to sound the channel with its CIR (from R_x to T_x), $h_m(t)$, between the receiving element and the transmitting element. Next, the transmitter records the received signal, $y_m(t) = x(t) * h_m(t)$. Finally, the transmitter transmits data and uses the TR signal as the preceding filters; i.e., $C_m(t) = y_m{}^*(-t)$. The channel reciprocity implies that the CIRs from T_x to R_x are identical to the CIRs from R_x to T_x and both are represented by $h_m(t)$. Assuming the channel reciprocity (that has been experimentally confirmed), the received signal can be expressed as

$$y(t) = x^*(-t) * \left(\sum_{m=1}^{M} h_m^*(-t) * h_m(t) \right) + n(t) \tag{3.63}$$

If subsequently one seeks to focus a signal $S(t)$ such that its spectral support $W_s(t)$ is contained in that of $x(t)$, and the power spectral density of $x(t)$ is flat over $W_s(t)$, then it is possible to reuse the implicit environmental information gained from $x(t)$ by transmitting $y(t) = S(t) * x^*(-t) * h_m^*(-t)$. The process above can be also applied to a multireceiver case. Due to wide bandwidth and focus ability in time and space, the communication capacity offered by time-reversal UWB systems may exceed other existent communication systems. Unlike using circular polarization and directional antenna to suppress the multipath propagation, the time-reversal technique utilizes the multipath signal to obtain a diversity gain.

3.5 Conclusion

This chapter introduces the antenna technologies that have been applied or are potential candidates in millimeter-wave WPAN, WLAN, and WMAN. The basic radiation principle is displayed simply for understanding the discussed antennas, and Friis transmission formula of radio wave is presented to reveal the influence of antenna performance on wireless communication systems. The operation characteristics of the aperture antenna, the lens antenna, the leaky-wave antenna, the surface wave antenna, the reflector antenna, and the planar integrated antenna and array have been demonstrated for showing their unique value for millimeter-wave wireless communication systems. As an important part of modern wireless communication systems, the propagation channel and channel modeling are introduced roughly. The concept of time-reversal UWB communication shows a recent development in wireless communications.

References

[1] For 802 activities, see www.ieee802.org
[2] Standard 802.15.3-2003 can be found at http://standards.ieee.org/getieee802/
[3] C. A. Balanis. *Antenna Theory Analysis and Design*, John Wiley, New York, USA, 1997.
[4] J. D. Kraus and R. J. Marhefka. *Antennas: For All Applications* (3rd Edition), McGraw-Hill Companies, New York, USA, 2002.
[5] Hans Schantz. The Art and Science of Ultrawideband Antennas, Artech House, Norwood, MA, USA, 2005.
[6] K.-L. Wong. *Compact and Broadband Microstrip Antennas*, John Wiley, New York, USA, 2002.
[7] S. J. Johnston. *Millimeter Wave Radar*, Artech House, Norwood, MA, USA, 1980.
[8] T. Teshirogi and T. Yoneyama, Eds. *Modern Millimeter-Wave Technologies*, Ohmsha Ltd., Tokyo, Japan, 2001.
[9] D. M. Dobkin. *RF Engineering for Wireless Networks Hardware, Antenna and Propagation*, Elsevier, Cambridge, MA, USA.
[10] J.-F. Kiang. Novel Technologies for Microwave and Millimeter-Wave Applications, Kluwer Academic Pub., Norwell, MA, USA, 2004.
[11] K. F. Lee and W. Chen. *Advances in Microstrip and Printed Antennas*, John Wiley, New York, USA, 1997.
[12] P. Bhartia, K. V. S. Rao, and R. S. Tomar. *Millimeter-Wave Microstrip and Printed Circuit Antennas*, Artech House, Norwood, MA, USA, 1991.
[13] K. L. Wong. *Planar Antenna for Wireless Communications*, John Wiley, New York, USA, 2003.
[14] J. C. Sletten. *Reflector and Lens Antennas*, Artech House, Norwood, MA, USA, 1988.

[15] R. E. Collin. *Field Theory of Guided Wave* (2nd edition), Wiley-IEEE Press, New York, USA, 1990.
[16] G. Passiopoulos, I. D. Robertson, and E. Grindrod. Integrated endfire sectored antennas for microwave and millimeter-wave LANs, 1997 IEEE 10th International Conference on Antenna and Propagation, pp. 1394–1398.
[17] T. S. Bird, J. S. Kot, N. Nikolic, G. L. Hames, and S. J. Barker. Millimeter-wave antenna and propagation studies for indoor wireless LANs, 1994 IEEE International Symposium on Antennas and Propagation, pp. 336–339.
[18] A. Kumar. Antennas for wireless indoor millimeter-wave application, 2003 IEEE Canadian Conference on Electrical and Computer Engineering (CCECE), Vol. 3, pp. 1877–1880.
[19] S. Q. Xiao, Z. Shao, M. Fujise, and B.-Z. Wang. Pattern reconfigurable leaky-wave antenna design by FDTD method, *IEEE Trans. on Antennas and Propagation*, Vol. 53, pp. 1845–1848, May 2005.
[20] A. Grbic and G. V. Eleftheriades. Leaky CPW-based slot antenna arrays for millimeter-wave applications, *IEEE Trans. on Antennas and Propagation*, Vol. 50, No. 11, pp. 1494–1504, 2002.
[21] K. C. Huang and Z. Wang. V-band patch-fed rod antennas for high data-rate wireless communications, *IEEE Trans. on Antennas and Propagation*, Vol. 50, No. 11, pp. 297–300, 2002.
[22] P. F. M. Smulders and M. H. A. J. Herben. A shaped reflector antenna for 60GHz radio access points, *IEEE Trans. on Vehicular Technology*, Vol. 54, No. 1, pp. 584–591, 2005.
[23] J. J. Wang, Y. P. Zhang, Kai Meng Chua, and Albert Chee Wai Lu. Circuit model of microstrip patch antenna on ceramic land grid array package for antenna-chip codesign of highly integrated RF transceivers, *IEEE Trans. on Antennas and Propagation*, Vol. 53, No. 12, pp. 3877–3883, 2005.
[24] S. Q. Xiao, B.-Z. Wang, W. Shao, and Y. Zhang. Bandwidth-enhancing ultra low profile compact patch antenna, *IEEE Trans. on Antennas and Propagation*, Vol. 53, pp. 3443–3447, Nov. 2005.
[25] S. Q. Xiao, Z. Shao, B.-Z. Wang, M. T. Zhou, and M. Fujise. Design of low profile microstrip antenna with enhanced bandwidth and reduced size, *IEEE Trans. on Antennas and Propagation*, vol. 54, pp. 1594–1599, May 2006.
[26] T. Seki, N. Honma, K. Nishikawa, and K. Tsunekawa. A 60-GHz multi-layer parasitic microstrip array antenna on LTCC substrate for system-on-package, *IEEE Microwave and Wireless Components Letters*, Vol. 15, No. 5, pp. 339–341, 2005.
[27] T. Seki, N. Honma, K. Nishikawa, and K. Tsunekawa. Millimeter-wave high-efficiency multilayer parasitic microstrip antenna array on Teflon substrate, *IEEE Trans. on Microwave Theory and Techniques*, Vol. 53, No. 6, pp. 2101–2106, 2005.
[28] I. K. Kim, N. K. S. Pinel, J. Papapolymerou, J. Laskar, J.-G. Yook, and M. M. Tentzeris. Linear tapered cavity-backed slot antenna for millimeter-wave LTCC modules, *IEEE Antenna and Wireless Propagation Letters*, Vol. 5, pp. 175–178, 2006.

[29] M. Marcus and B. Pattan. Millimeter wave propagation: Spectrum management implications, *IEEE Microwave Magazine*, Vol. 6, No. 2, pp. 54–62, June 2005.

[30] M. Gaddar, L. Talbi, and T. A. Denidni. Millimeter wave propagation modeling for indoor high-speed PC systems, 2003 IEEE 12th International Conference on Antenna and Propagation, Vol. 2, pp. 767–770.

[31] N. Moraitis and P. Constantinou. Millimeter wave propagation measurements and characterization in an indoor environment for wireless 4G systems, 2005 IEEE 16th International Symposium on Personal, Indoor and Mobile Radio Communications, pp. 594–598.

[32] G. Lerosey, J. de Rosny, A. Tourin, A. Derode, and M. Fink. Time reversal of wideband microwaves, *Applied Physics Letters*, Vol. 88, 1–3, 154101, 2006.

[33] H. T. Nguyen, J. B. Andersen, and G. F. Pedersen. The potential use of time reversal techniques in multiple element antenna systems, *IEEE Communications Letters*, Vol. 9, No. 1, pp. 40–42, Jan. 2005.

[34] R. C. Qiu, C. Zhou, N. Guo, and J. Q. Zhang. Time reversal with MISO for ultrawideband communications: Experimental results, *IEEE Antenna and Wireless Propagation Letters*, Vol. 5, pp. 269–273, 2006.

Chapter 4

MAC Protocols for Millimeter-Wave Wireless LAN and PAN

T. Chen, H. Woesner, and I. Chlamtac

Contents

4.1 Introduction

Intensive efforts in research, development, regulation, and standardization have paved the way for a massive usage of millimeter-wave bands in short-range communications. In 2001, the U.S. Federal Communications Commission (FCC) set aside a continuous block of 7 GHz spectra between 57 and 64 GHz for license-free wireless communications. Japan has released the 59–66 GHz band for unlicensed usage. In Europe, there is a consideration by the European Conference of Postal and Telecommunications Administrations (CEPT) to allocate 54–66 GHz for the same purpose. Research activities in signal processing, system packaging, and front-end design make low-cost SiGe-based millimeter-wave front ends feasible. The standard activities in

IEEE 802.15.3c working group and the recent announcement of 60-GHz single-chip radio front ends show the age of millimeter-wave wireless local area networks (WLANs) and wireless personal area networks (WPANs) is coming.

The Media Access Control (MAC) protocol is an essential part of every WLAN/WPAN. The function of MAC is to coordinate the channel access among competing nodes of a network in an orderly and efficient manner. It determines the behavior and performance of a network. On one hand, the MAC protocol of a millimeter-wave system matches the application scenario and satisfies general design rules such as efficiency, fairness, flexibility, and scalability. On the other hand, it should take into account the unique characteristics exhibited in millimeter-wave bands.

Millimeter-wave bands have their own characteristics different from lower frequency bands. While signals at lower frequency bands can easily traverse through building materials, millimeter-wave signals cannot penetrate solid materials very well. Consequently, millimeter waves permit a dense packing of communication frequencies, thus providing very efficient spectrum utilization, and increasing the security of transmissions. Moreover, like light waves, millimeter-wave signals result in low diffraction, but are subject to more shadowing and reflection. For non-line-of-sight (NLOS) propagation, the greatest contribution at the receiver is the reflected power. Shorter wavelengths cause the reflecting material to appear relatively rougher, which results in greater diffusion of the signal and less direct reflection. Since diffusion provides less power at the receiver than directly reflected power, millimeter-wave systems usually rely on line-of-sight communication condition. Directional antennas are normally required in these systems to achieve reliable communication.

Apart from this, millimeter-wave systems share common features with other wireless systems. These include: (1) high error rate and bursty errors; (2) location-dependent and time-varying wireless link capacity; (3) half-duplex communication; (4) user mobility; and (5) power constraints of mobile users. All of the above characteristics challenge the development of effective and efficient MAC protocols.

4.2 Classification of Wireless MAC

Wireless MAC protocols can be broadly classified into two categories: distributed and centralized protocols. In distributed protocols, competing nodes in the network contend for the medium access without any centralized coordination. On the other hand, in centralized protocols, there is a special node responsible for channel allocation. That node can be a dedicated node, like base station (BS), or a normal node elected from a group of nodes, like the piconet coordinator (PNC) in Bluetooth.

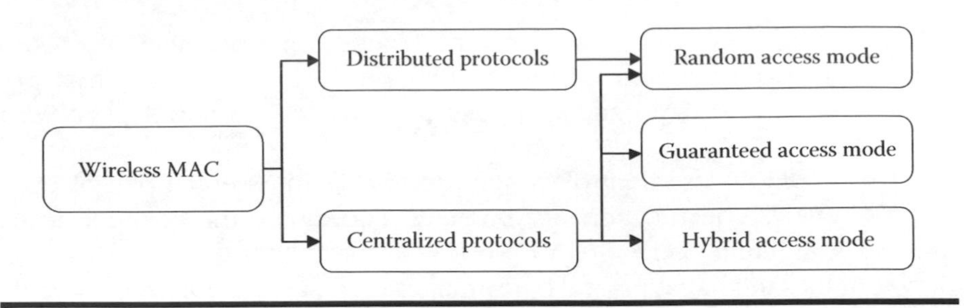

Figure 4.1 The classification of wireless MAC protocols.

Based on the method of operation, the MAC protocols are further divided into three access modes: random access, guaranteed access, and hybrid access (see Figure 4.1). In the random access mode, all nodes contend for the medium access. When several nodes simultaneously access the channel, a collision occurs and leads to the failure of the transmission. A contention resolution algorithm (CRA) is adopted in the random access mode to resolve collisions. Three families of CRAs are widely used in wireless MAC protocols: the binary exponential backoff (BEB), the p-persistence backoff, and splitting algorithms like n-ary tree [1].

In the guaranteed access mode, competing nodes access the medium in an orderly manner, usually in a round robin fashion. There are two approaches to implement the guarantee access mode. One uses master–slave configuration, in which the master polls slaves for channel access. The other exchanges a token in a distributed way. Only the node holding the token is allowed to transmit. The token is passed to the next node after finishing the transmission. Because of its implementation complexity (measures have to be taken to re-create *exactly one* token after it is lost) it is rarely used in wireless systems.

The third access mode combines the benefits of the above two access modes. Most hybrid access protocols are based on request–grant mechanisms; i.e., each node sends a request to the BS indicating the bandwidth it requires on a contention basis. The BS then allocates upstream time slots for the actual data transmission and sends a grant to the requesting node indicating the allocation.

4.3 State of the Art in Millimeter-Wave Based MultiAccess Systems

Among current wireless multiaccess systems on the market few are operated at millimeter-wave bands. There are several reasons for this situation. First, the front-end and baseband signal processing technologies capable of operating at millimeter-wave bands are not mature enough for mass deployment. The III-V radio frequency (RF) components, which have typically been used for millimeter-wave front-end in the past, are too expensive for portable devices. Second, the demand for bandwidth is tightly related to applications. Truly bandwidth-intensive applications such as HDTV have emerged only in recent years. As long as legacy wireless systems are capable of providing adequate capacity for conventional applications, there is no driving force to move to millimeter-wave bands. In the following, we briefly introduce the state of the art of millimeter-wave based multiaccess systems.

LMDS (Local Multipoint Distribution System) is a fixed broadband wireless access system operating at the 28-GHz band in the United States and the 40-GHz band in Europe. Its achievable data rates depend on distance and modulation format; typical figures are 40 megabits-per-second (Mb/s) for the downlink and 10 Mb/s for the uplink for links of a few kilometers. There are two specifications for LMDS: the Digital Video Broadcasting (DVB) specification from the European broadcasting union, and the Digital Audio Video Council (DAVIC) specification from DAVIC, which is an international body formed by major network operators, service providers, and industry vendors. The DVB specification only focuses on broadcast services. The DAVIC defines a MAC protocol for uplink channel access, which is similar to 802.14 [1] and DOCSIS MAC. The later two protocols are deployed in hybrid fiber coaxial systems. We find that LMDS and 802.16 have a strong relationship with the DOCSIS standard. The DAVIC MAC uses the request and grant procedure for bandwidth allocation. The contention slots are used for registration, bandwidth request, and short message exchange. Ali et al. [2] provided a comparison of DAVIC, 802.14, and DOCSIS MAC.

In 2002, IEEE published 802.16.1 to provide high-speed wireless Internet access between buildings with exterior antennas. It operates at 10–60 GHz bands. The MAC of 802.16.1 is based on a point-to-multipoint (PMP) topology. The uplink and downlink of the channel are structured into frames and the channel access is Time Division Multiple Access (TDMA) based. The BS governs the bandwidth allocation of both up/downlink channels. The fine granularity QoS is guaranteed through a connection-oriented MAC protocol. Due to the LOS propagation required in millimeter-wave bands, the 802.16.1 system had a coverage issue in urban areas. The IEEE then worked out a new PHY interface operating at 2–11 GHz. This new amendment is referred to as 802.16a.

The current research on millimeter-wave bands is heavily focused on the 60-GHz band, with a goal to support multiple gigabits-per-second (Gb/s) data transmission. The IEEE 802.15.3c working group is working on a 60-GHz PHY alternative for 802.15.3 WPAN. It is expected that the standardizing process will be completed in 2008. The objective of the 802.15.3c is to support bandwidth-intensive applications, and consumer device interconnection in WPAN. Due to the characteristics of the 60 GHz band, there is a need to support directional antenna or beamforming in its MAC protocol.

Several research projects have targeted 60-GHz multiaccess systems. The Europe IST broadway project proposed the use of the 60-GHz band as an ad hoc extension for the HiperLAN/2 system [3]. It is a 5-GHz and 60-GHz hybrid system, where the 60-GHz subsystem provides a data rate of at least 100 Mb/s for peer-to-peer connections. The MAC protocol is based on HiperLAN/2 but with modification to accommodate the 60-GHz ad hoc extension. Currently, the Wireless Gigabit With Advanced Multimedia support (WIGWAM) project has been set up by 27 partners in Europe and coordinated by TU Dresden. It is aimed at designing a 1 Gb/s system concept for home/office, public-access, and high-velocity scenarios. A distributed MAC is used in home/office scenarios, and a centralized MAC is used in public access scenarios. The frequency bands to be used include 5 GHz, 17 GHz, 24 GHz, 38 GHz, and 60 GHz.

As we can see, currently there is no widely used millimeter-wave WLAN/WPAN system. LMDS and 802.16.1 are established products, but not multiaccess systems, as there is one channel exclusively assigned to each node. All other systems are prototypes that arose from research projects. In the following section we will look into the reasons for this situation.

4.4 Challenges of Millimeter-Wave Based MAC Design

According to the nature of multiple access systems, and the characteristics of millimeter-wave bands, the challenges of millimeter-wave based MAC exist in four areas: physical constraints, resource allocation, QoS provisioning, and handover issues.

4.4.1 Physical Constraints

The Friis equation [4] indicates that the path loss of a radio signal is proportional to the square of its carrier frequency. For instance, with equal antenna gain, the 60 GHz band has an additional 21 dB of path loss compared to the 5 GHz band. As the path loss combines with other channel impairments such as delay spread, there is a need to employ directional antennas in millimeter-wave systems in order to achieve reliable communications. For instance, a highly directive horn antenna as used in [5] compensates for this loss entirely, but for the price of an opening angle of 7°, thus basically prohibiting mobility.

To enable user mobility, the beamforming process through the use of MIMO antennas seems to be a required process in millimeter-wave multiaccess systems. The beamforming is achieved by adaptively adjusting the weight of each antenna branch in an antenna array. A typical beamforming process works as follows: the initiator sends a steering request to the responder, including a training sequence. The responder estimates the channel state and sends back to the initiator the channel state information or computed steering matrix. Thereafter, the initiator uses the feedback to steer the beam of the MIMO antennas accordingly. The use of directional antenna brings several challenges to the MAC protocol.

4.4.1.1 Carrier Sensing

Most MAC protocols assume the use of the omnidirectional antenna at the PHY layer for broadcasting control messages. If a directional antenna is used either at the transmitter or at the receiver, a new mechanism at the MAC layer is required to detect neighbor nodes.

As the data rate of a system increases, the overhead of the MAC protocol becomes the bottleneck to the system throughput. As reported by Xiao [6], the upper-bound throughput of 802.11a is only 75.24 Mb/s when the data rate of the system becomes infinite. The MAC overhead in 802.11a includes a contention period, guard intervals between transmission operation, an acknowledgment (ACK) process, and a frame header for each packet. The most important parameter for carrier-sensing based protocols is the Rx/Tx turnaround time, which is physically determined by the time to power up and down the antenna. This means that the number of changes between listening to the channel and transmitting for a given number of bits should be minimum.

The block transmission and block acknowledgment are common approaches to reducing the MAC overhead, in which a transmitter is allowed to continuously transmit a number of packets with a single ACK message. To further reduce the overhead, the 802.11n introduces a concept of traffic aggregation; i.e., a number of packets from the upper layer can be aggregated into a single frame at the MAC layer. The guard interval and duplicated frame headers are therefore reduced. It is certain that large packets can improve the system throughput. However, there is a trade-off between the packet size and channel condition. The loss of large packets degrades performance. The challenge on this problem is to determine the optimum frame size according to the channel coherence time and collision rate, or to adapt the maximum frame size to the channel states.

4.4.1.2 Hidden Terminals

As shown in Figure 4.2A, the use of directional antennas creates a new hidden terminal problem. In contrast to omnidirectional transmission where hidden terminals are somewhat of an exception, with directed antennas the ongoing transmissions in the vicinity of a node are almost never detectable,

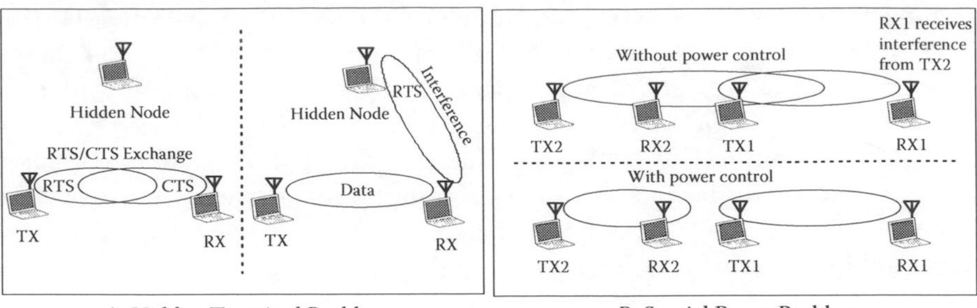

Figure 4.2 MAC issues in using directional antennas.

as the narrow beam of the antenna is purposely suppressing signals from the side. This means that if there is a collision of two transmitters inside the receiver's reception beam/angle, it will be of the hidden terminal type. The exchange of RTS/CTS (request-to-send/clear-to-send) messages prior to transmission should help here.

4.4.1.3 Spatial Reuse

Directional antennas increase spatial reuse by means of narrowing the beamwidth. However, as shown in Figure 4.2B, with the same transmission power, the directional antennas may have a longer transmission range, which leads to a specific kind of exposed terminal problem (in that the receivers and transmitters are exchanged compared to the usual sketch of the problem). Here, a potential transmitter node TX1 would hear an RTS from TX2, but not the CTS from RX2, as this is being directed toward TX2. According to the normal procedure, a node hearing an RTS but no CTS may conclude that its own transmission does not interfere with another ongoing transmission. Yet, if immediate acknowledgments are sent out, these may collide with ongoing data transmissions from the respective exposed terminal. Power control at the MAC layer can alleviate this problem.

4.4.2 Resource Allocation

Main resources to be allocated in a wireless network are bandwidth, power, and space, which are allocated through bandwidth allocation, rate control, and power control algorithms. Bandwidth allocation algorithms share bandwidth among multiple users. Rate control adapts bitrate to channel condition and the queue state of the nodes. Power control aims to prolong battery life, control interference level, and increase spatial reuse. Moreover, handover allows nodes to change the cell and use the shorter transmission links, which leads to a reduced area that is covered by the transmission and, thus, to an increased overall capacity.

4.4.2.1 Bandwidth Allocation

The challenge of a millimeter-wave system is to allocate the bandwidth according to the channel condition. The channel-state–dependent (CSD) scheduling exploits the time varying channel to achieve better system performance. It is highly suitable for a millimeter-wave system since a channel in a millimeter-wave band may experience more fluctuation than a low-frequency band due to fading and shadowing effects. Different CSD scheduling approaches are reported in [7,8]. The essential idea of CSD scheduling is to schedule the node with good channel quality and defer the one with bad channel quality, therefore the overall system performance improves. The challenges in CSD scheduling include efficient detection of the channel condition, and the fairness of the scheduling algorithms.

4.4.2.2 Power Control

Power consumption is a critical issue for millimeter-wave systems due to their high carrier frequency. A typical approach for energy saving is to shift the node into the sleep mode when it has no data to send [9,10]. The sleeping node wakes up periodically to receive the data from the access point (AP) or other nodes. Other approaches use a power control scheme during transmission [11,12]. In the case of using MIMO, which is highly likely in millimeter-wave systems, a transceiver can further save the power by switching other antenna chains off and leaving only one working antenna chain to monitor the network. The power saving mode needs the support of the MAC protocol.

The dilemma of power control is the following: When a transmitter observes there is high interference in the channel, it recognizes more power is needed to make a successful transmission. Therefore it would be better to back off, buffer the traffic, and wait for the interference to lessen. However, the new traffic may arrive during the back-off to fill up the buffer and delay raises, which pushes the transmitter to become power-aggressive in order to reduce its backlog. This problem has been studied by Bambos under the framework of dynamic programming [13].

Power control is also helpful in alleviating the hidden terminal problem in ad hoc networks [14,15]. As seen from Figure 4.2B, the transmission power can be effectively controlled so as to control the transmission range of a node. With channel knowledge from neighbor nodes, a hidden node can adjust its transmitting power in a way that the ongoing transmission will not be disrupted. A joint optimization process to maximize the battery life, make the network stable, and improve the system performance and capacity is a highly challenging task.

4.4.2.3 Rate Control

The rate control is discussed on two levels. One is to control the data rate by means of different modulation and coding schemes (MCS). The motivation behind this is that current wireless systems usually support a set of MCS. The adaptive data rate control refers to the dynamic change of the transmission data rate according to the channel condition. The second refers to the adaptive coding rate control at the channel coding level, which adapts the data transmission rate with more or less redundancy to compensate channel fading and interference.

An adaptive data rate control scheme needs the estimation of the channel conditions. It can be done at the transmitter or the receiver [16,17]; however, the best place to do it is at the receiver. In a receiver initial scheme, the receiver feeds back the channel state or the rate of the choice to the transmitter. The transmitter chooses the appropriate data rate according to the feedback. For instance, Holland [17] proposes a Receiver-Based AutoRate (RBAR) protocol based on RTS/CTS-based protocols, in which the rate selection is performance on a packet-by-packet basis during the RTS/CTS exchange, just prior to the packet transmission. The use of a data rate control MAC scheme is reasonable in high-data-rate millimeter-wave systems. It provides the opportunity for the systems to adapt to the complex channel conditions. The challenge in a rate control scheme lies in the mechanism to estimate the channel quality timely and accurately.

The power control scheme cannot solve the performance problem when the channel condition degrades to a certain level. The coding rate control is deployed in this case to further guarantee the performance. In a typical adaptive code rate scheme, the channel quality estimation is made at the receiver and sent back to the transmitter through the feedback message. To reduce the transmission overhead, incremental redundancy is usually applied, in which a special channel coding scheme is adopted such that the high-rate code is the subset of the lower-rate code. If the current code cannot provide sufficient protection for decoding at the receiver, only the redundant bits, which are different bits between two channel codes, are transmitted. An example of incremental redundancy code is rate-compatible punctured convolutional (RCPC) code [18]. In RCPC codes, a high-rate code is generated by puncturing the lowest-rate block of code bits. Since the encoder needs to generate only the code with the lowest coding rate, one encoder/decoder pair is enough to encode and decode the codes with all coding rates. The incremental redundancy code scheme belongs to the type II hybrid automatic repeat request (ARQ) schemes.

It is noted that the adaptive data rate and coding rate scheme both use the channel estimation to determine the appropriate rate. It is interesting to combine two rate control schemes for better system performance.

4.4.3 QoS Provisioning

In wireless networks, fairness becomes a complex problem. A wireless link may turn to the error state when a flow is transmitted. To maximize the overall system throughput, it is better to defer the transmission until the link changes to the good state. However, there is a trade-off between channel utilization and fairness. To ensure fairness, the flow or node should be compensated for the loss when the link recovers. However, the definition and objectives of fairness become ambiguous in wireless networks. The appropriate interpretation of fairness for wireless networks depends on the service model, traffic type, and channel characteristics. For ad hoc networks, the challenges lie in the following factors: (1) the location-dependent contention of wireless channel; (2) the trade-off between the channel utilization and fairness; and (3) the inaccurate state and decentralized control. The challenges of fairness in millimeter-wave–based systems include the trade-off between channel utilization and fairness and the guarantee of short-term and long-term fairness through scheduling algorithms.

Moreover, admission control is an essential component in MAC for strict QoS guarantee. It is difficult to implement it in a distributed MAC since there is no global information available. The research efforts based on measurement have been devoted to these issues [19,20]. However, proposed distributed admission control schemes only address coarse QoS. The remaining open issues of distributed admission control include [21]:

1. It should support heterogenous traffic.
2. It should jointly consider the resources at the network, link, and physical layers.
3. It depends on channel access methods.
4. It requires an efficient information exchange mechanism among nodes.

4.4.4 Handover

The use of cell-based millimeter-wave systems makes the handover problem prominent. As mentioned, the cell of a millimeter-wave–based WLAN is largely confined to a room. The small cell size drastically increases the frequency of handover. As the data rate of the system increases, the overhead of handover becomes a critical performance issue. The fast handover at the MAC layer becomes a challenge in such a system.

Handover is broadly divided into two types: soft and hard handover. In soft handover, the call is uninterrupted when the node is moving from one cell to another. It is supported in the Universal Mobile Telecommunications System (UMTS). The hard handover tears down the old connection before setting up a new connection. IEEE 802.11F is an example of hard handover. It is obvious that soft handover provides better performance; however, the internal procedure of the soft handover is more complex since the connection state of the handover node needs to be exchanged between the BSs involving the handover. The handover procedure consists of measurements, decision, and execution of the handover. In cellular networks, the measurements are typically done via signal strength detection. The handover decision is made if a node detects that the signal strength of the neighbor station outperforms the current one at a predefined level. In 802.11, the handover is done by mobiles. After losing the connection, a mobile scans channels and tries to detect a new AP by using active probing or passive listening to beacon signals emitted periodically from the APs. Research found the discovery (probe) phase is the dominant contributor in handover latency [22]. The total handover latency in 802.11b was shown to be 75 and 350 ms, which is far too long to the delay requirement of voice or video traffic (max 50 ms). It is clear that to enable fast handover, a cross-layer approach with handover indication from the physical layer is necessary. The fast handover solution is also related to the MAC implementation; i.e., the connection oriented or connectionless.

4.5 Candidate Millimeter-Wave MAC Protocols

Based on the challenges of MAC protocol design for millimeter-wave systems, we identify several common features of millimeter-wave based MAC protocols: be efficient and scalable to bear an ultra-high data rate up to multiple Gb/s; be flexible to support QoS for a wide range of applications; be capable of supporting directional antenna and advance antenna technologies like beamforming.

After investigating current and under-development WLAN/WPAN systems, we chose the MAC protocols of IEEE 802.11n, HiperLAN/2, and IEEE 802.15.3 as the candidates for millimeter-wave based WLANs. Although the IEEE 802.16 is a standard for wireless metropolitan area networks (WMANs), its centralized architecture, operation mechanism, and QoS support make it a competitive reference model for millimeter-wave based WLAN/WPAN systems. We also propose a novel system for 60 GHz indoor WLAN, named wireless gigabit Ethernet extension (WiGEE), and design a MAC protocol that is suitable for small cell communications with a large number of cells. In the rest of this section, we introduce the candidate protocols on their channel access, bandwidth allocation, QoS, and error control schemes. At the end, we provide a qualitative comparison.

4.5.1 802.11n

IEEE 802.11n is the high-throughput version of IEEE 802.11 standards, which aims at providing a minimum data rate of 100 Mb/s, and a peak data rate as high as 600 Mb/s. The first draft was approved by IEEE in March 2006. It is expected that the final standard will be published in October 2007. This new amendment boosts the system capacity primarily in three aspects: the use of MIMO to increase the peak data rate; the introduction of double-width 40 MHz channels for extra throughput; the enhancement of MAC to improve the MAC efficiency. The enhancements of 802.11n MAC can be summarized as follows: traffic aggregation, block acknowledgment, link adaptation, and protection from legacy 802.11 services. The first two enhancements aim at reducing the MAC overhead, which is necessary when the data rate of the system is very high. The link adaptation functions provide 802.11n with capabilities like beamforming.

4.5.1.1 Channel Access

As in legacy 802.11, two operation modes are defined in the 802.11n draft: the infrastructure mode where an AP coordinates the communication between stations; and the ad hoc mode where stations directly communicate to each other without centralized coordination. The ad hoc mode provides the flexibility to form a network, however, the infrastructure mode used is more popular since in most application scenarios stations need to access infrastructure networks.

The legacy 802.11 introduced two modes of channel access: distributed coordination function (DCF) for random access and point coordination function (PCF) based on DCF for coordinated access. The period operating in random access mode is called contention period (CP), and the period in coordinated access mode is called contention-free period (CFP). The CP and CFP are announced in a special control subframe called a beacon. As shown in Figure 4.3, the beacon, CP, and CFP form a structure called a superframe.

The DCF mode is the basic access mode that all other access modes rely on. It uses Carrier Sensing Multiple Access/Collision Avoidance (CSMA/CA) mechanism for channel access. Truncated BEB is employed to resolve collisions.

In addition to DCF, 802.11n uses hybrid coordination function (HCF) to support QoS in channel access. The HCF was first defined in the 802.11e

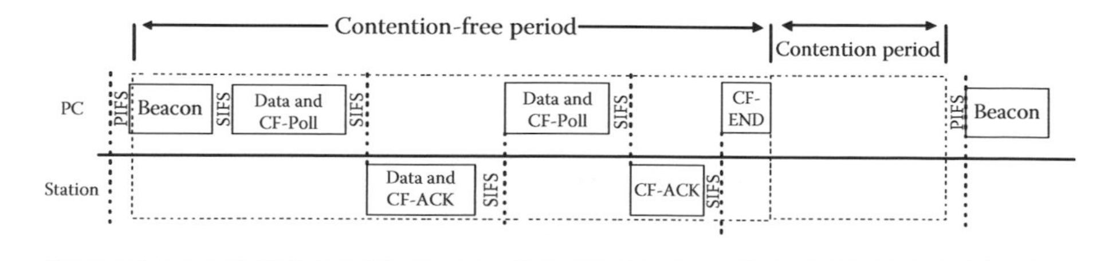

Figure 4.3 The contention-free and contention period in 802.11.

standard, which is the QoS amendment of legacy 802.11 standards. 802.11e extends the DCF to Enhanced Distributed Channel Access (EDCA), and the PCF to HCF Controlled Channel Access (HCCA) for better channel access control. Both modes need the participation of the AP to control access parameters on the fly. Therefore, they can operate only in the infrastructure mode.

The EDCA enhances the DCF in two aspects. An important concept, the transmission opportunity (TXOP), is introduced to limit the channel access time of each station. A TXOP is a time interval in which the station is permitted to transmit its MAC service data units (MSDU). Once seizing the channel, a station is permitted to transmit many frames as long as the TXOP is not exceeded. Both EDCA and HCCA use TXOP to control channel access time of each station. For EDCA, the duration of TXOP is announced in the beacon. For HCCA, it is specified by the AP and sent in polling messages. The TXOP brings two benefits: the channel access time can be well controlled; and the channel access overhead can be reduced. To seize the channel in legacy 802.11, each node has to wait for a time interval called DCF interframe space (DIFS) after sensing the channel to be idle. In EDCA, a new set of interframe spaces, named arbitrary interframe space (AIFS), is introduced to provide channel access priorities for differential services. As shown in Figure 4.4, a shorter AIFS gives a node higher priority to access the channel.

In the HCCA access mode, there is a hybrid coordinator (HC) residing in the AP, fully controlling the HCCA operation. The HC uses polling-based scheduling. It polls a station using a contention-free poll (CF-Poll) frame, or the combination of data and CF-Poll frame if it has data to send to that station. The granted time for channel access is given by a form of TXOP specified in the CF-Poll frame. The polled station transmits its data frames and ACK during the granted TXOP. If the HC receives no response from a polled station after a predefined interval, it polls the next station

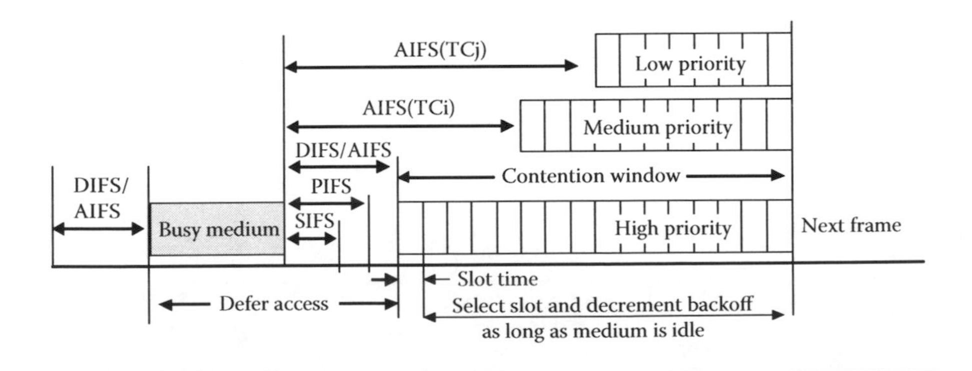

Figure 4.4 Differential services supported by AIFS in 802.11e.

or ends the CFP by broadcasting a CF-end frame. Note that although the EDCA is only permitted in CP, the HCCA can be used in both CP and CFP. In CFP, the HC uses the shortest AIFS to seize the channel earlier than normal stations.

4.5.1.2 Frame Structure

802.11n specifies new frames and fields for high throughput operation, which includes frame structures for traffic aggregation, and control frames for block acknowledgment, MIMO manipulation, and link adaptation. We introduce traffic aggregation because it can be generalized and used in millimeter-wave based MAC protocols. Used properly, it can significantly reduce MAC overhead.

Traffic aggregation is the opposite process of fragmentation. The purpose of traffic aggregation in 802.11n is to reduce the overhead of frame header and guard intervals. Two kinds of traffic aggregation are defined in the 802.11n draft: aggregated MAC service data unit (A-MSDU), which assembles multiple MSDUs into one MAC frame, and aggregated MAC protocol data unit (A-MPDU), which combines multiple MPDUs into one Physical Layer Convergence Protocol (PLCP) frame.

An A-MSDU allows multiple MSDUs sent to the same receiver to be aggregated into a single MPDU. The efficiency of the MAC layer is improved particularly when there are lots of small MSDUs such as TCP acknowledgments. The structure of a data MPDU containing an A-MSDU, which is a sequence of several MSDU subframes, is shown in Figure 4.5. Each subframe contains a subframe header followed by a MSDU and 0–3 bytes of padding. The subframe header consists of three fields: the destination address (DA), the source address (SA), and the length field, which contains the length of the MSDU in bytes. Note that only those MSDUs transmitted to the same receiver can be aggregated. An A-MSDU is treated like any other MSDU. The maximum length of an A-MSDU is limited to 8000 bytes.

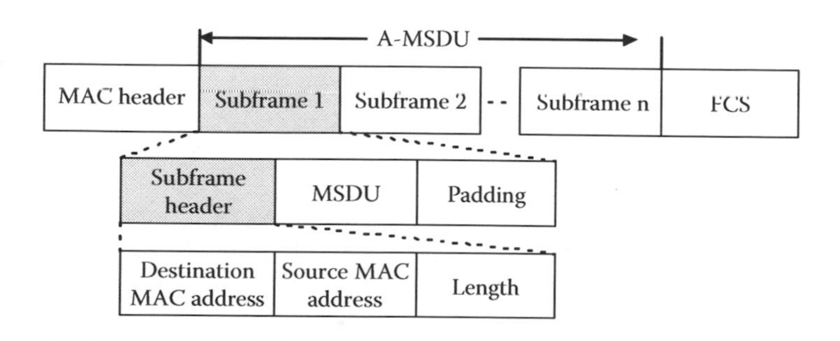

Figure 4.5 A-MSDU in 802.11n.

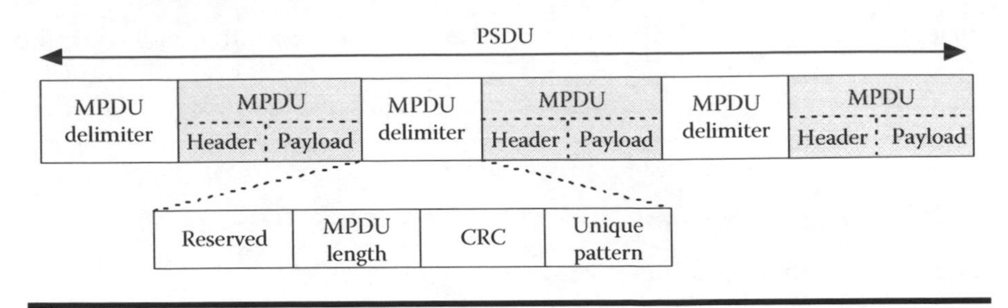

Figure 4.6 A-MPDU in 802.11n.

The problem of A-MSDU is that a large frame size may degrade the system performance when the channel is in the bad state. The A-MPDU is proposed to alleviate this problem. The structure of an A-MPDU is shown in Figure 4.6. An A-MPDU consists of a number of MPDU delimiters each followed by an MPDU. The purpose of the MPDU delimiter is to robustly delimit the MPDUs within the aggregate even when one or more MPDU delimiters are collapsed. An MPDU delimiter contains an 8-bit pattern field that provides a unique pattern to delimit the start of each MPDU. MPDUs are put into the aggregate without modification. If one frame out of an aggregate is collapsed, it is possible to successfully decode other frames. The maximum length of an A-MPDU is 65,535 bytes.

4.5.1.3 Bandwidth Allocation and QoS

As mentioned, the 802.11 MAC is essentially a contention-based protocol. However, 802.11e and 802.11n provide a centralized bandwidth allocation scheme in their controlled channel access mode to provide fine granularity QoS. In 802.11n, the HCCA mode is dedicated for centralized bandwidth allocation. A signaling mechanism is provided to let stations report their bandwidth requests to HC. Two kinds of requests are defined: the queue state indicator, which reflects the queue state of a station, or the traffic specification, which defines the bandwidth and delay requirements of a traffic stream in a station. Requests are sent to HC in two ways: (1) transmit during CFP period when using HCCA; (2) contend during the CP period when using EDCA. Once the HC admits the request, it issues a CF-Poll frame to the requesting station with channel access time and duration defined in a TXOP field.

In addition to support guaranteed QoS in the HCCA mode, 802.11n provides differential services in the EDCA mode. Four types of access categories (ACs) are introduced and mapped to voice, video, best effort, and background traffic, respectively. The concept of ACs was introduced in 802.11e. As shown in Figure 4.3, each AC has its own backoff entities and

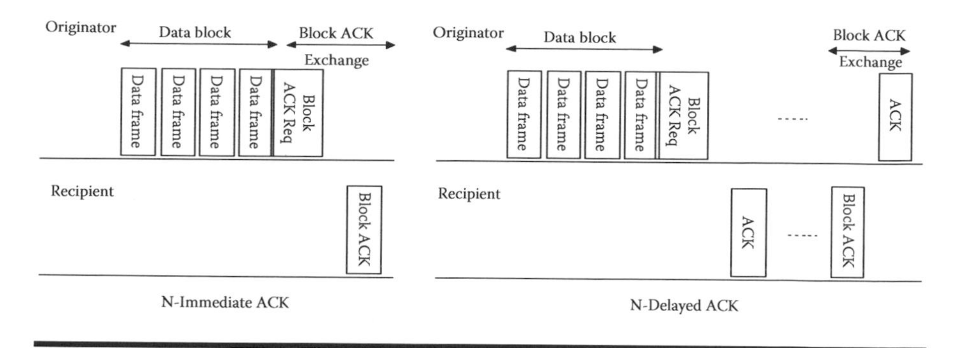

Figure 4.7 Block acknowledgment in 802.11n.

uses AIFS for channel access. The AIFS provides channel access priorities for each AC.

4.5.1.4 Error Control Mechanism

The legacy 802.11 MAC requires an explicit ACK for each unicast data frame. More flexible acknowledgment policies are provided in 802.11n to support differential services and improve throughput. 802.11n allows for two acknowledgment options: normal ACK, and no ACK. The new no ACK option allows a recipient to take no action upon receipt of the frame. In addition, a block acknowledgment (BA) is introduced to further reduce transmission overhead. The BA process is shown in Figure 4.7. Two kinds of BAs are used: N-immediate BA, which acknowledges the block of data right after the transmission, and N-delayed BA, which delays the BA until the next TXOP.

4.5.2 HiperLAN/2

HiperLAN/2 is a WLAN standard developed by the European Telecommunications Standards Institute (ETSI). Although it is technically superior to 802.11 standards in several aspects, such as throughput and QoS support, it failed in competition with 802.11 in the marketplace. However, the design principles of HiperLAN/2 provide us with valuable insights for the MAC protocol design of millimeter-wave systems.

HiperLAN/2 uses a TDMA-based connection-oriented MAC protocol. Similar to 802.11, it provides two operation modes: cellular access network configuration for business scenarios, and ad hoc LAN configuration for residential scenarios. The former uses AP to provide access over a certain area. An AP interconnects all mobile terminals (MT) associated with it, and all communication goes through the AP. APs are interconnected by external networks such as Ethernet to extend coverage. Ad hoc LAN configuration is operated in an ad hoc manner where MTs directly exchange data with each other. However, unlike the ad hoc mode in 802.11, a node is dynamically elected among MTs to provide control functions within its subnet. This special MT is called the central controller (CC). Accordingly, HiperLAN/2 uses two access modes: centralized or direct mode. In the centralized mode all traffic goes through the AP. In the direct mode the traffic between MTs is exchanged directly without passing through the CC, while channel access is still controlled by the CC.

HiperLAN/2 defines three layers in its protocol stack, from upper to lower listed as follows: the convergence layer to provide an interface between upper layer and data link layer; the data link control layer (DLC); and the physical layer. The DLC layer is the target of our interest. It consists of three main entities: the radio link control (RLC), MAC, and error control (EC). The functions of DLC are divided into two categories: user data plane functions that are responsible for delivering data between the upper layer and the physical layer, and providing the error control mechanism; and control plane functions provide the radio resource control (RRC), association control function (ACF), and DLC connection control (DCC). The RRC is responsible for detecting and efficiently utilizing available radio resources. It manages handover, frequency selection, power saving, and power control functions. The ACF manages the association process for association, disassociation, authentication, key management, and encryption. The DCC takes charge of connection setup, maintenance, and release. As we can see, the HiperLAN/2 uses a cross-layer approach between the PHY and DLC layers for efficiently utilizing radio resources.

4.5.2.1 Frame Structure

Instead of using the CSMA/CA model, HiperLAN/2 MAC uses dynamic TDD and dynamic TDMA for channel access. The channel is structured into frames. A unique feature of HiperLAN/2 is that the duration of each frame is fixed to 2 ms. Since the length of a frame header is fixed, as the data rate increases, a frame can accommodate more data payload and provide more throughput. Xiao [6] showed that the MAC protocol of HiperLAN/2 outperforms that of 802.11 in terms of throughput.

HiperLAN/2 introduces the concepts of logical channel and transport channel in the DLC. The concept of channels here is used to classify control and user data messages that originate from the DLC or higher layer. Control and data messages are first mapped to proper logical channels based on their contents, and then passed to the corresponding transport channels, where MAC frames are constructed and appropriate channel access modes are assigned. The logical channels are identified by message types. The transport channels are identified by message formats and channel access modes. The standard defines ten logical channels and six transport channels.

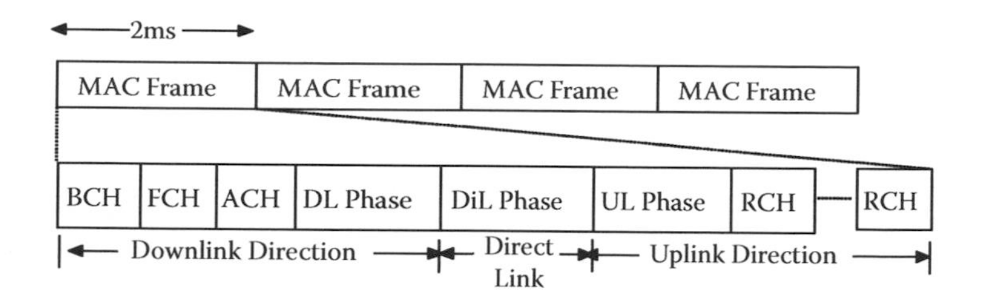

Figure 4.8 HiperLAN/2 frame structure.

As shown in Figure 4.8, each frame starts with a broadcast channel (BCH) duration, followed by frame control channel (FCH) duration, access feedback channel (ACH) duration, data phase, and at least one random access channel (RCH) duration. Let us define the transmission direction from MT to AP/CC as the uplink and the opposite direction as the downlink. The BCH transport channel is used in the downlink direction to carry the information about the entire cell, such as network identity (ID), AP ID, and transmission power for AP. The FCH is used in the downlink direction to carry the information describing the structure of the frame. The information specifies the transmission time slots, and the PHY mode used in each time slot. The RCH is used by MTs to send control information to AP/CC on a contention basis. A data phase is divided into several subphases. If transmissions occur between AP/CC and MTs, a data phase includes the downlink (DL) phase and/or uplink (UL) phase. If transmissions occur between MTs in the direct mode, a data phase consists of several direct link (DiL) phases.

4.5.2.2 Channel Access, Bandwidth Allocation, and QoS

HiperLAN/2 uses a request and grant channel access scheme, which is controlled by the AP/CC. An MT with pending data to send requests the bandwidth from the AP/CC using the bandwidth allocation mechanism defined in the standard. The bandwidth allocation is performed per connection basis. Each connection is identified by the DLC user connection ID (DUC ID). In the centralized mode, the DUC ID is composed of the DLC connection ID (DLCC ID) and the MAC ID, which is assigned by the RLC of the AP during the association process. In the direct mode, the DUC ID is composed of the source MAC ID, the destination MAC ID, and the DLCC ID. The DUC ID acts as a tag for classification and enables per-flow QoS.

The resource request (RR) message and resource grant (RG) message are used by MTs for bandwidth requests and by AP/CC for bandwidth allocations on a connection basis. The RR contains the queue status of the requesting MT and is initiated by an MT in two ways: a response to a polling issued by AP/CC, or a standalone request sent in a special transport channel. The RG is carried in the FCH transport channel, which specifies the location of the transmission/reception in the frame.

HiperLAN/2 is capable of providing a strict QoS guarantee due to its TDD/TDMA-based channel access mode and centralized-bandwidth allocation scheme.

4.5.2.3 Error Control Mechanism

The EC in the DLC layer manages error detection and retransmission. It ensures in-order delivery of packets. The selective repeat ARQ (SR-ARQ) mechanism is employed in the standard. Three modes are specified for different levels of transmission reliability: the acknowledge mode, in which the EC retransmits acknowledgments from the receiver; the repetition mode, in which the EC simply repeats the transmission without acknowledgment; and the unacknowledge mode, for unreliable, low-latency services without the need of retransmission. Unicast data are sent in either acknowledge or unacknowledge mode; broadcast data are transmitted in either repetition or unacknowledge mode; and multicast data are sent in unacknowledge mode.

4.5.3 802.15.3

The IEEE 802.15 working group focuses on developing standards for WPAN. A set of 802.15 standards has been developed so far. IEEE 802.15.1 adopts Bluetooth MAC and physical layer specifications for data and audio communications among portable devices. It operates at a 2.4 GHz frequency band and provide a data rate up to 1 Mb/s. The IEEE 802.15.3 draft aims at providing high data rates in WPAN. The initial version provides data rates from 11 to 55 Mb/s by means of a single carrier scheme. The IEEE 802.15.3c working group is working on a new PHY interface based on the 60-GHz unlicensed band. The initial objective of 802.15.3c is to provide a data rate up to 2 Gb/s. The applications of WPANs include audio, digital video, HDTV, media-rich interactive games, and data applications running on portable devices, laptops, home theater, and other consumer electronics. Although air interfaces are different, the MAC protocols of 802.15.3 standards are similar.

4.5.3.1 Architecture

The basic topology unit in 802.15.3 is piconet, a concept derived from Bluetooth. As shown in Figure 4.9, a piconet is formed in an ad hoc manner

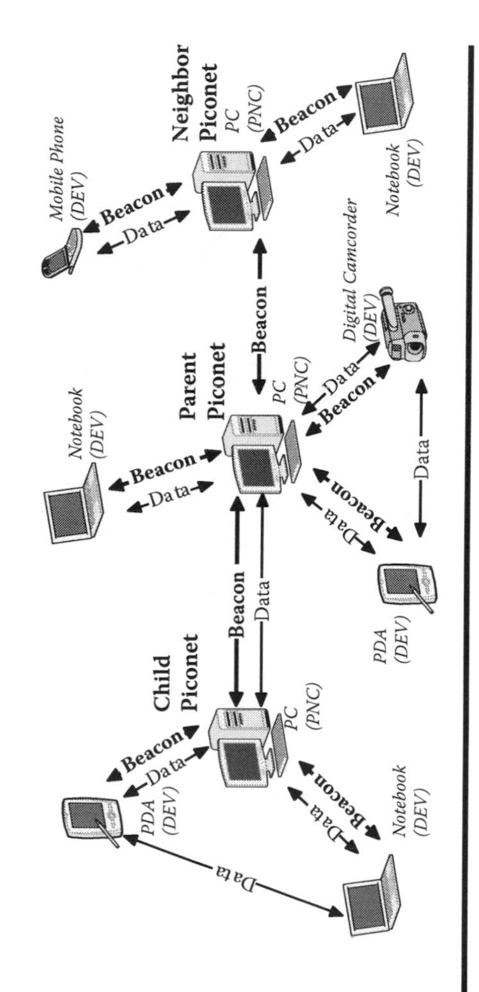

Figure 4.9 PICONET Concept in 802.15.3.

and controlled by a PNC, which is a node dynamically elected among member devices (DEVs). As its name suggested, the piconet is confined to a small area, typically covering a range from 10 to 70 m. This is the range a millimeter-wave system usually covers in an indoor environment. A DEV is allowed to join and leave the piconet with a short association time. Once the piconet is formed, the PNC issues control messages in beacons periodically, through which the PNC maintains network synchronization, controls channel access, manages QoS provisioning, and performs admission control. To join a piconet, a DEV needs to associate with the PNC and follow the channel access information provided in beacons. A PNC can handover its PNC role to other DEVs before it leaves the piconet or changes its role.

To extend the coverage, increase the number of DEVs to be supported, and distribute control functions among DEVs, the concept of the child piconet has been introduced. The child piconet has the same topology as the parent piconet. The PNC of the child piconet fully controls its own piconet. However, the PNC of a child piconet is also the member of the parent piconet, and the parent PNC allocates time slots for the channel access inside the child piconet. DEVs in a child piconet are only allowed to transmit within the time slots assigned by the parent PNC.

In order to share frequency resources among different piconets, neighbor piconets are used to provide coexistence. Similar to a child piconet, a neighbor piconet associates with a parent piconet, and the channel access time of a neighbor piconet is allocated by the parent PNC. But unlike a child piconet, the PNC of a neighbor piconet is not a real member of the parent PNC. It is only permitted to exchange certain commands with the parent PNC, such as an association/disassociation request, a channel time request, and authentication. The topology of the piconet, child piconet, and neighbor piconet is shown in Figure 4.9.

4.5.3.2 Frame Structure

The 802.15.3 MAC is based on a time-slotted superframe structure that consists of three phases: beacon, contention access period (CAP), and channel time allocation period (CTAP). The structure of a superframe is shown in Figure 4.10. At the beginning of each superframe, the PNC broadcasts a

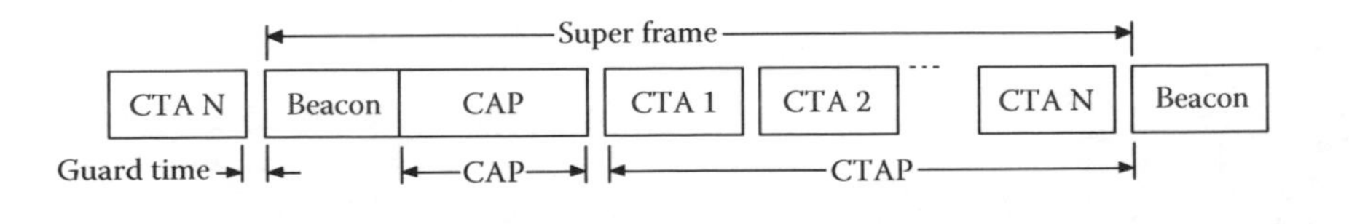

Figure 4.10 Superframe structure in 802.15.3.

beacon, specifying time synchronization, control, and resource allocation information of the piconet. It follows a CAP that is used to exchange a small amount of data and control messages such as bandwidth requests on a contention basis. The CTAP is a contention-free period, which consists of two types of time slots: the management time slot (MTS), reserved for command exchange between PNC and DEVs, and the guaranteed time slot (GTS), used for data transmission among DEVs. The length of CAP and CTAP is specified in the beacon.

4.5.3.3 Channel Access

The channel access at the CAP is on a contention basis. The collision is solved by a CSMA/CA mechanism similar to the one used in 802.11 MAC. Like 802.11, interframe spaces (IFS) are used to guarantee the protocol operation. The actual values of IFSs are PHY layer dependent. The traffic allowed to be transmitted in a CAP period includes asynchronous data traffic and control commands. Isochronous traffic is allowed only in the CTAP.

The CTAP is based on TDMA, in which the PNC specifies the starting time and the duration of the time slot for each traffic stream. The PNC divides the CTAP into a number of channel time allocations (CTAs). A CTA is specified by a CTA information element (IE) in the beacon. The CTA information includes the source and destination address of the DEV, the stream index that identifies the connection, and the starting time and duration of reserved time slots. There are two types of time slots: GTS and MTS. GTS is used for asynchronous and isochronous traffic. Two kinds of GTSs are defined: dynamic GTS and pseudostatic GTS. The PNC can dynamically change the location of the dynamic GTS within the superframe on a superframe-by-superframe basis. Pseudostatic GTSs, which are only allocated to isochronous traffic, have relative fixed locations within the CTAP. However, it can be changed by PNC on a long-term basis in order to optimize the channel utilization. MTSs are used for exchanging commands between PNC and DEVs. There are two ways to use MTSs: the direct uplink MTS for a dedicated DEV or the open MTS for multiple DEVs on a contention basis. For the latter, a slotted ALOHA protocol is used for the contention resolution.

4.5.3.4 Bandwidth Allocation

Each DEV uses an explicit bandwidth allocation mechanism to request CTA in CTAP. A source DEV requiring time slots first sends a channel time request (CTR) command to the PNC, indicating the recurring duration and the number of required time slots. The parameters of CTR for asynchronous and isochronous traffic are different. The CTR for the former only contains the total amount of required time. For the latter, the CTR includes the number of time slots or GTSs needed per superframe, and the minimum and desired duration of each time slot. Each GTS is actually a time slot reserved for a specified DEV. It is the DEV's duty to determine how to use the GTS; i.e., determine which command, stream, and asynchronous traffic will be transmitted.

4.5.3.5 QoS

The CAP does not provide any service assurance. It only offers a best-effort service. The QoS is guaranteed in CTAP, in which the channel access is on a reservation basis. After receiving a CTR command, the PNC performs admission control, and allocates GTSs to the requesting DEVs when resources are available. For flexibility, the admission control and scheduling algorithms are open in the standard. The standard specifies the stream index to identify the connection and traffic stream. A stream index is assigned by the PNC during the connection establishment procedure. There are three reserved stream indices: 0X00 for all asynchronous traffic, 0XFD for MCTA traffic, and 0XFE for unassigned streams. A stream index, except the reserved stream indices, is uniquely assigned for each isochronous stream in a piconet. Asynchronous traffic is assigned to one stream index. Therefore there is no service differentiation among asynchronous connections in a DEV.

4.5.3.6 Error Control Mechanism

Three ARQ policies are specified in 802.15.3: No-ACK, Immediate ACK, and Delayed ACK. If the ACK policy of a frame is set to No-ACK, upon the reception no ACK is sent by the intended recipient. The broadcast and multicast frames must use the No-ACK policy upon transmission. If the Immediate ACK policy is used, the intended recipient is required to send an Immediate ACK right after the reception. The Delayed ACK allows for a single ACK acknowledging a block of frames. It is used only for isochronous streams. The parameters for Delayed ACK are negotiated between the source and destination DEV.

The retransmission in CAP and CTAP are different. The former uses a backoff mechanism; i.e., the source DEV starts a backoff process to retransmit a frame when the expected ACK is not received in a given time interval. The maximum waiting time is limited by a predefined parameter. In CTAP, the source DEV awaits the ACK for a period defined by retransmission interframe space (RIFS) before starting the retransmission.

4.5.4 IEEE 802.16

The IEEE develops IEEE 802.16 standards for WMAN. The objective of WMANs is to provide high-speed wireless Internet access over a large geographic area, similar to wired access technologies such as Digital Subscriber

Line (DSL), Ethernet, and fiber optic. The first standard of 802.16 was published in 2002, which is referred to as IEEE 802.16.1. It uses 10–60 GHz frequency band to provide fixed wireless access in a point-to-multipoint (PMP) topology. Because the LOS propagation condition limits the coverage of 802.16.1, especially in urban areas, the IEEE developed a new air interface at 2–11 GHz band, known as 802.16a. The 802.16a also adds optional capabilities at the MAC layer to support mesh networks so as to increase its coverage. Both 802.16.1 and 802.16a only support fixed wireless access. The mobility support is enabled in standard version 802.16 2005, which was published in December 2005. The 802.16 standards are different in air interfaces but similar in MAC protocols; i.e., they are all connection-oriented MAC protocols with the capability to support continuous or bursty traffic.

4.5.4.1 Frame Structure

802.16 uses a dynamic TDMA scheme for channel access, where the channels are structured into frames. The communication path from BS to SS is defined as downlink and the opposite direction as uplink. Both uplink and downlink can operate in time division duplex (TDD) or frequency division duplex (FDD) mode, as shown in Figures 4.11 and 4.12, respectively. In TDD, a frame is divided into an uplink and downlink subframe, where the former follows the latter; in FDD, the uplink and downlink subframes are transmitted simultaneously over different frequency bands. Full-duplex SS and half-duplex SS are both supported in standards. Each subframe consists of a number of time slots. BS and SSs must synchronize and transmit data in predefined time slots.

It is the BS's duty to assign time slots in both uplink and downlink subframes. The channel assignment information is issued at the head of each downlink subframe. The downlink map message (DL-MAP) is used to specify the downlink channel usage in the current downlink subframe. The uplink map message (UL-MAP) is used to allocate the uplink channel to SSs. The standards support adaptive data burst profiling, in which transmission parameters such as modulation and coding settings can be modified on a

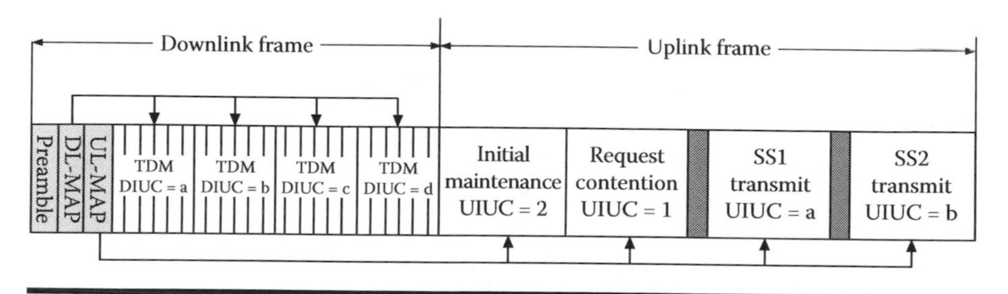

Figure 4.11 TDD access mode in 802.16.

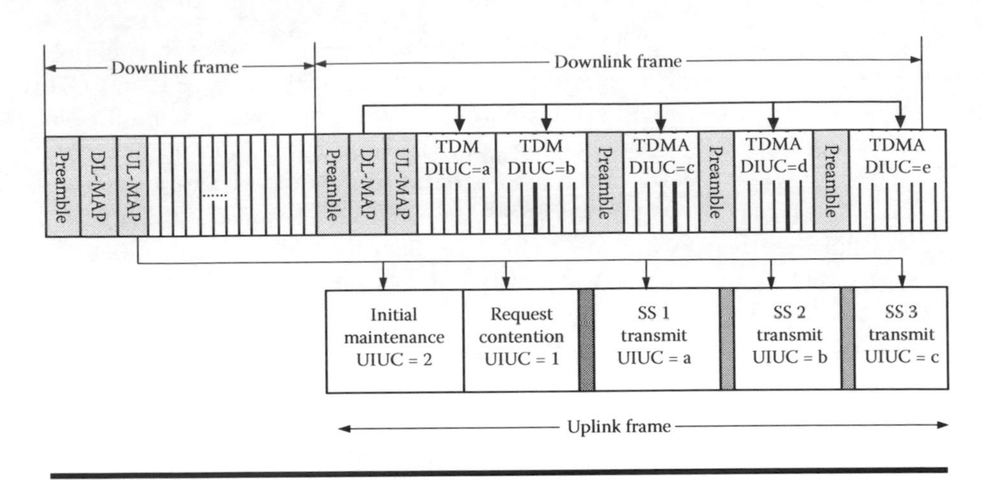

Figure 4.12 FDD access mode in 802.16.

frame-by-frame basis in both uplink and downlink transmission. The data burst profiles are identified by a code named interval usage code (IUC). The IUC for the downlink is called DIUC, and for the uplink is called UIUC. The DL-MAP and UL-MAP uses the DIUC and UIUC to specify the data burst profiles used for each time slot, respectively.

4.5.4.2 Channel Access

The downlink channel allocation is simple in 802.16. The downlink subframes are different for FDD and TDD. The subframe in FDD is divided into TDM and TDMA portions as shown in Figure 4.12, where the TDMA portions follow the TDM portions. The TDMA portions are separated by preambles. This design is used for half-duplex SS operating at the FDD mode. The half-duplex SS can transmit its data earlier in the subframe and synchronize back to the downlink using the preamble for receiving data. This allows a SS to decode a specific portion of the downlink without the need to decode the entire downlink subframe. The TDD downlink subframe only contains TDM portions.

The uplink subframe contains time slots for SS association, bandwidth request, and data transmission. As shown in Figure 4.12, it includes three periods: initial maintenance, bandwidth (BW) request contention, and data grants. Different periods are identified by their UIUC. The BS announces these periods in a UL-MAP, and can specify such periods in any order and length. In the initial maintenance period, SSs send initial maintenance–related messages; e.g., ranging requests for BS to determine the network delay. A new station may join the network in this period. In a BW request contention period, SSs request bandwidth based upon multicast and broadcast polls issued by the BS. The data grant period is dedicated for

SSs to transmit their data. The initial maintenance period and BW request contention period are accessed on a contention basis. The truncated BEB algorithm is used to solve collisions.

4.5.4.3 Bandwidth Allocation

Uplink bandwidth is always requested on a connection basis. The connection ID (CID) assigned by the BS is used to identify each connection. The BW request can be sent in an explicit packet or piggybacked on another packet. The requested bandwidth is expressed in two modes: either incremental, meaning how much additional bandwidth is required, or aggregate, meaning how much total bandwidth is needed. Both modes are allowed in explicit BW requests, while only incremental mode is also allowed in piggyback BW requests. A BW request can be initiated directly by a connection of a SS or in response to a polling message given by the BS. A BW request is transmitted during the uplink subframe in one of three ways: contending in a BW request period; transmitted in a predefined time slot indicated by UL-MAP; or piggybacked on another packet.

BS issues unicast as well as multicast and broadcast polls. The polling process does not use an explicit poll message from thc BS to a SS. Rather, the BS allocates bandwidth in UP-MAP for potential requests from SSs. The BS polls stations in multicast or broadcast method in case the BS finds there is not enough bandwidth to support unicast polling.

The BS grants uplink bandwidth to SSs based on one of two modes: grant per subscriber station (GPSS) or grant per connection (GPC). In GPSS mode, the BS allocates bandwidth for individual SSs. It is the SS's duty to allocate bandwidth among its connections. In GPC mode, the BS allocates bandwidth for individual connections. The BS allocates bandwidth based on the following facts: the amount of bandwidth requested by the connections; the QoS parameters of delay and bandwidth needed by current applications in the SS; and the available network resources.

4.5.4.4 QoS

The principal mechanism to enable QoS in 802.16 is to associate each connection with a service flow. A service flow is a unidirectional flow of packets provided with a particular QoS. A service flow is characterized by a set of QoS parameters such as latency, delay jitter, and throughput. Each network application is associated with a service flow by assigning a unique service flow ID (SFID). All packets must be tagged with SFID and CID in order for the network to provide appropriate QoS. Four types of service flows are defined in the standard: unsolicited grant service (UGS), real-time polling service (rtPS), non-real-time polling service (nrtPS), and best effort service (BE). The UGS supports real-time service flows that generate fixed-size periodic data packets. A connection with UGS flow is prohibited from using any contention-based BW request. The rtPS supports real-time service flows that generate periodic data with various sizes. The BS needs to poll an SS periodically in order to get unicast BW requests from the SS. An SS is prohibited from using contention-based BW request in order to avoid unpredicted delay. The nrtPS supports non-real-time service flows that have data packets of various sizes. The BS needs to poll an SS on a regular basis and allows a SS to send its unicast BW request. An SS is allowed to use contention-based BW requests for nrtPS services. Finally, the BE is used to support service flows that do not need QoS. An SS sends its BW requests in the request contention periods.

4.5.4.5 Error Control Mechanism

The ARQ mechanism is a part of the MAC, which is an option for implementation. When implemented, ARQ is enabled on a connection basis. A connection cannot have a mixture of ARQ and non-ARQ traffic. The standards use SR-ARQ as the ARQ mechanism. The transmission of data is based on blocks. An MSDU is logically partitioned into several blocks. Each block is identified by its block sequence number (BSN). Sets of blocks selected for transmission or retransmission are encapsulated into a PDU. To retransmit blocks in a PDU, there are two options: with or without rearrangement of blocks. The former retransmits the original PDU; the latter rearranges the blocks into different PDUs. The ARQ feedback information can be sent in a standalone message or piggybacked on an existing connection.

Two types of acknowledgments are defined: the accumulative ACK and selective ACK. The accumulative ACK acknowledges all blocks up to the BSN specified in the ACK. The selective ACK comes with a bitmap. The blocks with their BSN in the bitmap are acknowledged. A hybrid ACK that combines accumulative ACK and selective ACK is allowed in an ARQ feedback message.

4.5.5 MPCP MAC Protocol for 60 GHz System

For millimeter-wave systems deployed in indoor environment for wireless Internet access or business usage, the number of APs will be huge due to the characteristics of millimeter-wave frequency band. To support hundreds of cells with a peak data rate up to Gb/s in each cell, it is necessary to design a new system architecture.

As shown in Figure 4.13, a novel 60 GHz indoor WLAN system, named WiGEE, has been proposed in [23]. It combines an optical fiber infrastructure and 60 GHz radio to provide Gb/s communications in indoor environment. The optical fiber infrastructure is an extension to Ethernet passive

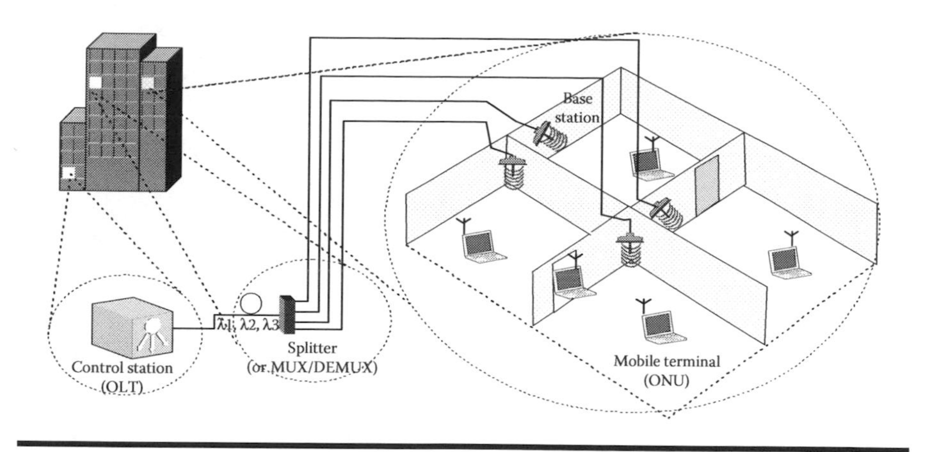

Figure 4.13 WiGEE System.

optical network (EPON), an optical access network that provides an economical and reliable gigabit-speed alternative for wired-access networks [24]. In EPON, an optical line termination (OLT) at the central office connects a number (typically 16 or 32) of optical network units (ONU) in clients' premises through a passive optical splitter. It uses a PMP topology in which ONUs can directly communicate only with the OLT. Standardized by IEEE 802.3ah [25], a centralized MAC protocol, named Multi-Point Control Protocol (MPCP), is used for channel access in EPON. The current EPON standard only supports one wavelength. In order to support a large number of cells the optical fiber infrastructure used in WiGEE extends EPON to support multiple wavelengths.

The WiGEE can be regarded as an extension to an EPON system in a sense that a piece of fiber is replaced by 60 GHz radio to extend the connections to MT. Similar to EPON, the central station (CS) broadcasts downlink signals to MTs through the relay of APs. In the uplink, MTs access the channel on a TDMA basis, which is controlled by the CS. The system takes advantage of the multiple GHz-wide spectrum at 60 GHz band, and adequate capacity of optical fiber infrastructure to provide reliable Gb/s transmission on cells basis. Several characteristics of the system make it different from other WLAN/WPAN systems: (1) The number of APs is huge in order to cover a large area in the buildings. (2) It uses FDD for uplink and downlink operation in order to fully explore the infrastructure capacity. (3) There are two collision domains in the system, one in optical fiber and the other in wireless medium. The MAC protocol jointly solves collisions in two domains. (4) The optical fiber in the infrastructure produces significant propagation delay. A sliding window-based error control scheme is necessary in WiGEE.

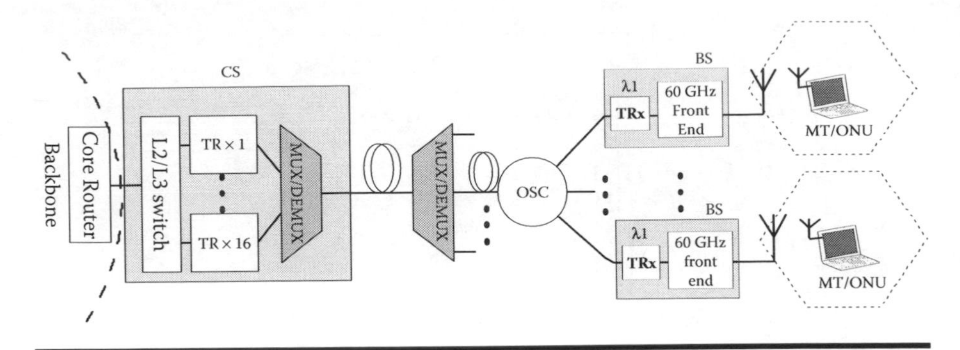

Figure 4.14 Architecture of WiGEE system, Configuration I.

4.5.5.1 WiGEE Architecture

As shown in Figure 4.14, the system consists of five components: CS, AP, MT, multiplexor/demultiplexor (MUX/DEMUX), and optical splitter and combiner (OSC). The CS, which connects to a large number of APs in a tree topology through optical components such as OSC and MUX/DEMUX, is responsible for controlling the whole system. APs extend signals from the optical to the wireless domain, and provide MTs wireless access in their coverage.

Three configurations of building blocks are shown in Figure 4.14, Figure 4.15 and Figure 4.16, respectively. They differ in the layout of the radio cells and trade off system cost against dynamic reconfigurability.

The first two configurations shown in Figure 4.14 and 4.15 use MUX/DEMUX and OSC in the optical infrastructure. In Figure 4.14 the wavelengths from CS are first demultiplexed and then each wavelength is fed into a single OSC, [26]. The APs connecting to the same splitter share the

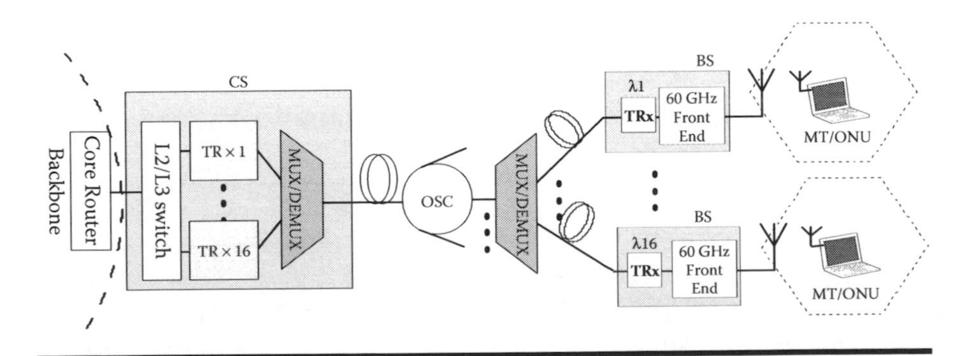

Figure 4.15 Architecture of WiGEE system, Configuration II.

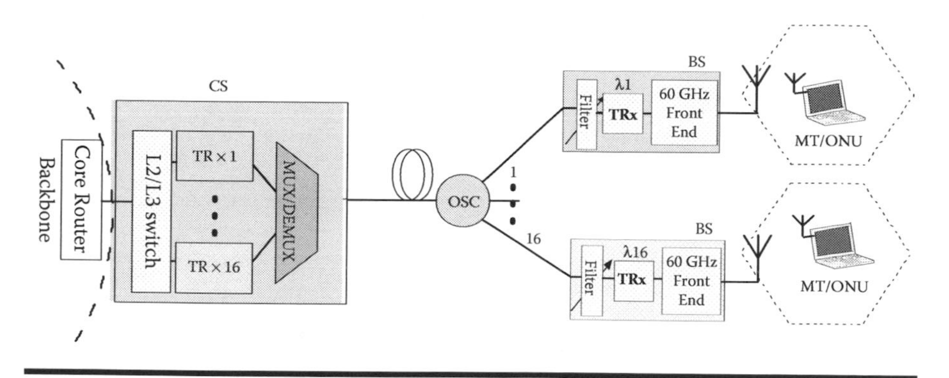

Figure 4.16 Architecture of WiGEE system, Configuration III.

same wavelength and hence the same uplink and downlink bandwidth. Assuming the APs served by the same wavelength are adjacent in their physical locations, e.g. the same floor of the building, a large virtual cell is formed and the MAC protocol is able to schedule the uplink channel on the virtual cell basis. The virtual cell is a desirable feature in network planning.

In Figure 4.15, the wavelengths are first split and then demultiplexed to APs. APs connecting to the same MUX/DEMUX use different wavelengths. It is suitable to provide adequate bandwidth in dense communication areas where MTs aggregate together, e.g., offices or conference rooms.

Aforementioned architectures assign wavelengths in a fixed method. As a result, the residual bandwidth of a wavelength can not be shared by other traffic-intensive cells if they do not share that wavelength. As shown in Figure 4.16, an alternative architecture is proposed to improve the bandwidth utilization, in which, only OSCs are used between CS and APs. Multiple wavelengths are fed into each OSC simultaneously and hence received by each AP. Each AP is equipped with a tunable filter to pick up its working wavelength, and a tunable laser to send back data to the CS. The tunable components can be dynamically configured by the CS according to the wavelength utilization of the system. The third configuration provides superior flexibility on bandwidth utilization and QoS provisioning. Virtual cells can be formed on demand based on the traffic load and service requirement. However, the downside of this configuration is its complexity and cost.

4.5.5.2 Medium Access Control

A centralized MAC protocol, named Multiple Wavelength MPCP (MW-MPCP), is developed in WiGEE. It extends the IEEE 802.3ah MAC protocol to support multiple wavelengths. For each wavelength, there is an MPCP entity

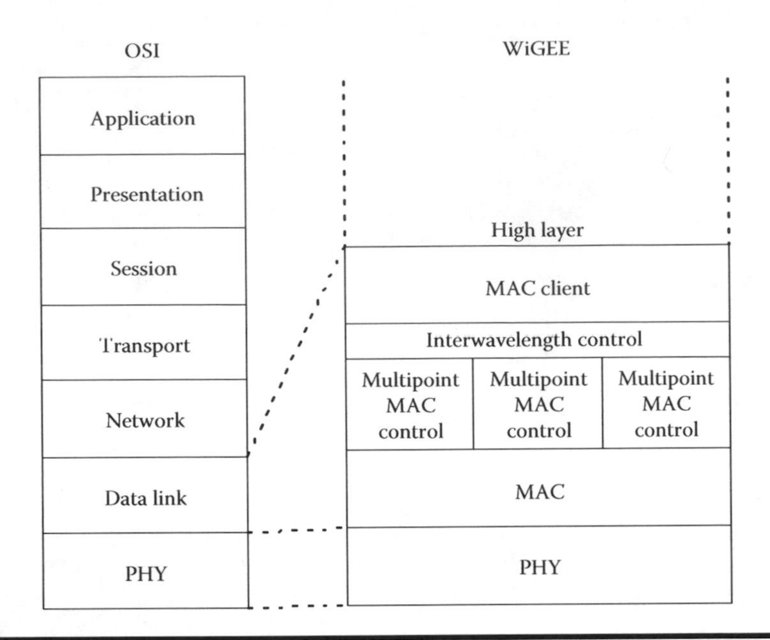

Figure 4.17 The relationship of WiGEE protocol stack and OSI network reference model.

residing in the MAC layer of the CS, functioning as the MPCP entity in the OLT of EPON. However, an additional component behind all MPCP entities of the CS is introduced to perform inter-wavelength scheduling and handover control, named the inter-wavelength control (IWC) unit. The relationship of the MW-MPCP and the OSI protocol stack is shown in Figure 4.17.

The MW-MPCP has three main processes: the association process, which deals with registration, deregistration, authentication, and inter-wavelength handover; the report process, which provides the bandwidth request mechanism between the CS and MTs; and the grant process, which provides the bandwidth assignment mechanism between the CS and MTs.

- Association process. The CS broadcasts DISCOVERY control messages periodically to MTs. A DISCOVERY message defines the time period in which the MTs can send registering, deregistering, and inter-wavelength handover requests on a contention basis.
- Report process. The report message is sent by an MT to the CS for uplink bandwidth request. In a report message, the queue sizes of different priority queues are included. It is up to the CS to determine the actual amount of grant according to available resources.
- Gate process. In this process, the uplink bandwidth is assigned to MTs through gate messages generated by the CS according to a

bandwidth allocation algorithm. The gate message specifies an uplink transmission opportunity of an MT at a given time period, during which only the MT is permitted to transmit data.

4.5.5.3 Channel Access, Bandwidth Allocation, and QoS

WiGEE uses a request-and-grant mechanism at the MAC layer to allocate bandwidth among competing MTs. The bandwidth request of an MT, which is usually the queue states of the MT, is sent through a report message. WiGEE only provides contention-free bandwidth requests; i.e., the report message is sent only during the grant period of that MT.

The dynamic bandwidth allocation (DBA) algorithm at the CS makes the bandwidth allocation decision based on scheduling policies and queue states of MTs obtaining report messages. The implementation of the DBA relies on network scenarios. WiGEE leaves the DBA algorithm open for different network scenarios. An overview of the DBA algorithms used in EPON networks is provided in [27]. They provide references to design DBA algorithms in WiGEE accordingly.

WiGEE provides a fine-granularity QoS guarantee through the DBA mechanism. In report messages, MTs can indicate their uplink bandwidth needs per 802.1Q priority queues. The states of up to eight priority queues are permitted in a report message. The grant for queues of an MT is issued by the CS as a whole. The MT is responsible for allocating the grant among different priority queues.

4.5.5.4 Error Control Mechanism

The considerable propagation delay introduced by the fiber infrastructure demands a sliding window-based error control scheme in WiGEE, in which SR-ARQ is used. It is worth noting that the uplink and downlink employ different sliding window size. In the downlink, because each MT is permitted to transmit only at a granted time period, the ACK time, which is the duration of a frame waiting for acknowledgment, is tightly bound to the DBA algorithm. Consequently, the ARQ window size for the downlink is a function of the maximum cycle time of the uplink.

4.5.6 Comparison of Candidate MAC Protocols

The design of an MAC protocol depends on the network scenario, service requirements, and the number of nodes to be supported in a network. We identify three common features of millimeter-wave based MAC protocols: to support a very high data rate; to support a wide range of services; and to provide functionalities dedicated to millimeter-wave bands (e.g., directional antenna or beamforming support). Based on these features, we compare the MAC candidates on the aspects of channel access, data rate, service support, millimeter-wave oriented functions, and other important aspects related to high-speed WLAN/WPAN. Table 4.1 provides a detailed comparison of these protocols.

As seen from Table 4.1, the centralized channel access mode is used by the majority of the candidates. It outperforms the distributed access mode of fine-granularity QoS support and channel access overhead reduction. Considering emerging bandwidth intensive applications, a centralized access scheme is a better choice for high-speed WLAN/WPAN. Among the centralized-access based candidates, three of them use hybrid access mode, in which the bandwidth request or other control messages can be sent in a contention or contention-free period. It increases the robustness of the wireless systems. However, the distributed access mode is more robust than centralized access mode because it has no single point of failure problem.

Among the candidate systems, 802.15.3c and WiGEE operate at data rates of multiple gigabits per second. 802.11n is going to support a data rate up to 600 Mb/s. The remaining two support a maximum data rate less than 100 Mb/s. However, the maximum data rate a system can support is limited by its physical layer and is not equal to the maximum rate the MAC layer can support. For MAC protocols, its efficiency determines the maximum throughput a system can afford. All candidates provide a centralized access scheme to reduce channel access overhead. All of them provide block transmission and acknowledgment to reduce transmission overhead. In HiperLAN/2, 802.16 2005, and WiGEE, it is implemented by SR-ARQ. In 802.11n and 802.15.3, it is realized by block acknowledgment mechanisms. Moreover, 802.11n provides traffic aggregation to further reduce the MAC layer overhead. In this sense, all candidates are suitable for high-data-rate wireless systems.

All candidates support QoS in different degrees. Centralized protocols apparently outperform distributed protocols on QoS provisioning. The 802.11n supports differential services in the EDCA mode, and fine granularity services in the HCCA mode. Other candidates support fine granularity QoS by means of a centralized control scheme, in which scheduling algorithms are all open for flexibility.

One unique feature of millimeter-wave based WLAN/WPAN is the support of directional antennas or beamforming in their MAC protocols. 802.11n has options to support MIMO and beamforming process in its MAC protocol. HiperLAN/2 is capable of using sector antennas in its AP to increase spatial reuse. The main problem to use directional antennas is the implementation of broadcasting function since most MAC protocols rely on broadcasting to enable their control and management functions.

Table 4.1 Comparison of Millimeter-Wave MAC Candidates

	802.11n	*HiperLAN/2*	*802.15.3c*	*802.16 2005*	*WiGEE*
Channel access	Distributed	Centralized/hybrid	Centralized/hybrid	Centralized/hybrid	Centralized/guaranteed
Max. data rate (Mb/s)	600	54	2000	70	1000
Duplex	TDD	TDD	TDD	TDD/FDD	FDD
Frame structure	Superframe	Superframe	Superframe	Downlink & uplink frame	None
QoS	Access priority	Scheduling	Scheduling	Scheduling	Scheduling
Connection oriented	No	Yes	Yes	Yes	No
Error control scheme	BA	SR-ARQ	BA	SR-ARQ	SR-ARQ
CRA	BEB	BEB	BEB	BEB	S-ALOHA
Duplex	TDD	TDD	TDD	TDD/FDD	FDD
Power saving	Yes	Yes	Yes	Yes	No
Mobility	Hard handover	Handover	Partial	Hard handover	Soft handover
Multiple MCS	Yes	Yes	Yes	Yes	No

In HiperLAN/2 the broadcasting problem is solved by sending a copy of control messages to each sector antenna. In 802.11n, the beamforming process in only performed in the data transmission phase. In other phases, the antenna is working in the omnidirectional mode. Therefore, the broadcasting based control functions are not affected. For centralized schemes like WiGEE and 802.16, the sector antennas used in BS can be a simple solution. However, considering the flexibility, beamforming is a better choice. For distributed systems such as 802.15.3, beamforming may be the only choice. MAC candidates need to be modified accordingly in order to accommodate the beamforming process.

Except for WiGEE, all other candidates' systems support multiple MCS. It is easy for them to develop an adaptive data rate scheme with a minor modification of MAC protocols. Again, except for WiGEE, all other candidates support power-saving operations. It is an important feature for a millimeter-wave system since it has more power consumption compared to lower-frequency based systems. Note that WiGEE is only in its infancy phase; its advanced features are still under investigation.

In addition, all candidates support a certain level of mobility. 802.11n supports mobile initialized hard handover. HiperLAN/2 supports handover in business scenarios. The handover is performed by the RRC unit in the RLC block. The 802.16 2005 supports hard handover. In WiGEE, because the CS has the knowledge of all BSs, it has the capability to support soft handover. Even for 802.15.3c, the handover of PNC in a piconet guarantees a certain level of mobility.

4.6 Directions of Millimeter-Wave WLAN and WPAN

As high-bitrate applications are emerging and the bandwidth demand of mobile users increases, it can be expected that millimeter-wave systems will dominate the area of WLAN/WPAN due to the huge bandwidth offered by millimeter-wave bands. There are, as discussed in the previous sections, some common features of the millimeter-wave WLAN and WPAN.

First, the millimeter-wave systems will support a data rate from several hundred Mb/s to multiple Gb/s. In such a data rate, the MAC protocols of these systems must be efficient to provide a reasonable performance. Techniques such as *block transmission* and *block acknowledgment* will be a standard part of those MAC protocols. Moreover, high-data-rate transmission requires a higher SNR at the receiver. In the millimeter-wave band, the channel condition is more susceptible to the environment. It is therefore hard to guarantee the desired SNR all the time. As a result, an *adaptive data rate* scheme supported by the MAC layer will be more attractive.

Second, the millimeter-wave systems aim to support a variety of applications with different QoS requirements. Bandwidth-intensive applications such as HDTV are more sensitive to the bandwidth change. The fine-granularity QoS on a flow or connection basis becomes a better choice. Cross-layer approaches, such as channel state-dependent scheduling, power and rate control, and advanced signaling process, will be promoted with the goal to improve the QoS and the system performance.

Third, the use of directional antennas will be necessary in millimeter-wave systems. A more advanced option will be the use of beamforming techniques via MIMO antennas. Advanced antenna techniques provide millimeter-wave systems extra gain to combat against multipath fading, increase the data rate, and extend the communication range. There is a requirement in the MAC layer to fully exploit those techniques. This is an especially interesting topic for distributed systems. It will therefore hold tables with antenna parameters for each destination node and thus perform routing for each packet, leading to the integration of layers 1 and 3 into the MAC.

And finally, the cognitive radio (CR) technologies [28] are imaginable in millimeter-wave systems. The basic idea of CR is to efficiently use the frequency resources via implementing intelligence at transceivers. In a typical CR scenario, the transmitter is able to sense and choose frequency bands for operation. The CR involves the cooperation of PHY, MAC, and upper layers. For coexistence, connectivity, and better service quality, there is a need to use CR radio techniques in millimeter-wave systems. For instance, if a system can operate at UWB and 60-GHz band, it can use 60 GHz for high-speed transmission and UWB to keep the connection alive when a blockage at 60 GHz is detected. It should be noted that so far the research on CR is in the infancy phase. How to bring CR into millimeter-wave systems is still a challenge.

4.7 Conclusion

The characteristics of millimeter-wave bands bring both challenges and opportunities to the MAC protocol design. In this chapter, we identify the characteristics of millimeter-wave bands that affect the MAC design, provide an overview of current millimeter-wave based systems, investigate the MAC issues in millimeter-wave based WLAN/WPAN, and introduce those MAC protocols of current high-data-rate WLAN, WPAN, and WMAN, which have the potential to be operated in millimeter-wave based multi-access systems. All candidate protocols are capable of working at a high data rate with a decent support on QoS. However, millimeter-wave systems are more susceptible to environmental changes than lower-frequency–based systems. Enhancements are needed to cope with the characteristics of millimeter-wave bands. To release the full power of a millimeter-wave system, cross-layer approaches considering power and rate control, beamforming, and channel state-dependent scheduling between multiple layers of the network protocol stack are expected.

References

[1] Y. Lin, "On IEEE 802.14 medium access control protocol," *IEEE Communications Surveys*, vol. 1, no. 1, pp. 2–10, 1998.

[2] M. Ali, R. Grover, G. Stamatelos, and D. Falconer, "Performance evaluation of candidate MAC protocols for LMCS/LMDS networks," *IEEE Journal on Selected Areas in Communications*, vol. 18, no. 7, pp. 1261–1270, 2000.

[3] M. de Courville, S. Zeisberg, M. Muck, and J. Schoenthier, "BROADWAY-the way to broadband access at 60GHz," *International Conference on Telecommunication*.

[4] T. Rappaport, *Wireless Communications: Principles and Practice*. Prentice Hall, Upper Saddle River, NJ, 1996.

[5] G. Grosskopf, A. Norrdine, D. Rohde, and M. Schlosser, "Transmission experiments with gigabit-Ethernet signals in the 60 GHz frequency band," 14th International POF Conference, 2005.

[6] Y. Xiao, "IEEE 802.11 N: Enhancements for higher throughput in wireless LANs," *IEEE Wireless Communications* [see also *IEEE Personal Communications*], vol. 12, no. 6, pp. 82–91, 2005.

[7] P. Bhagwat, P. Bhattacharya, A. Krishna, and S. Tripathi, "Enhancing throughput over wireless LANs using channel state dependent packet scheduling," *INFOCOM 1996*, vol. 3, pp. 1133–1140, 1996.

[8] X. Liu, E. Chong, and N. Shroff, "A framework for opportunistic scheduling in wireless networks," *Computer Networks*, vol. 41, no. 4, pp. 451–474, 2003.

[9] H. Woesner, J. Ebert, M. Schlager, and A. Wolisz, "Power-saving mechanisms in emerging standards for wireless LANs: The MAC level perspective," *IEEE Personal Communications,* [see also *IEEE Wireless Communications*], vol. 5, no. 3, pp. 40–48, 1998.

[10] E. Jung and N. Vaidya, "An energy efficient MAC protocol for wireless LANs," *INFOCOM 2002*, vol. 3, 2002.

[11] J. Wieselthier, G. Nguyen, and A. Ephremides, "On the construction of energy-efficient broadcast and multicast trees in wireless networks," *INFOCOM 2000*, vol. 2, 2000.

[12] E. Jung and N. Vaidya, "A power control MAC protocol for ad hoc networks," *Wireless Networks*, vol. 11, no. 1, pp. 55–66, 2005.

[13] N. Bambos and S. Kandukuri, "Power controlled multiple access (PCMA) in wireless communication networks," *INFOCOM 2000*, vol. 2, pp. 386–395, 2000.

[14] J. Monks, V. Bharghavan, and W. Hwu, "A power controlled multiple access protocol for wireless packet networks," *INFOCOM 2001*, vol. 1, pp. 219–228, 2001.

[15] T. ElBatt and A. Ephremides, "Joint scheduling and power control for wireless ad hoc networks," *IEEE Transactions on Wireless Communications*, vol. 3, no. 1, pp. 74–85, 2004.

[16] A. Kamerman and L. Monteban, "WaveLAN-II: A high-performance wireless LAN for the unlicensed band," *Bell Labs Technical Journal*, Summer, pp. 118–133, 1997.

[17] G. Holland, N. Vaidya, and P. Bahl, "A rate-adaptive MAC protocol for multi-Hop wireless networks," *Proceedings of the 7th Annual International Conference on Mobile Computing and Networking*, pp. 236–251, 2001.

[18] J. Hagenauer, "Rate-compatible punctured convolutional codes (RCPC codes) and their applications," *IEEE Transactions on Communications*, vol. 36, no. 4, pp. 389–400, 1988.

[19] Y. Xiao and H. Li, "Local data control and admission control for QoS support in wireless ad hoc networks," *IEEE Transactions on Vehicular Technology*, vol. 53, no. 5, pp. 1558–1572, 2004.

[20] F. Cuomo, C. Martello, A. Baiocchi, and F. Capriotti, "Radio resource sharing for ad hoc networking with UWB," *IEEE Journal on Selected Areas in Communications*, vol. 20, no. 9, pp. 1722–1732, 2002.

[21] X. Shen, W. Zhuang, H. Jiang, and J. Cai, "Medium access control in ultra-wideband wireless networks," *IEEE Transactions on Vehicular Technology*, vol. 54, no. 5, pp. 1663–1677, 2005.

[22] A. Mishra, M. Shin, and W. Arbaugh, "An empirical analysis of the IEEE 802.11 MAC layer handoff process," *ACM SIGCOMM Computer Commun. Review*, vol. 33, pp. 93–102, 2003.

[23] T. Chen, H. Woesner, Y. Ye, and I. Chlamtac, "Wireless gigabit ethernet extension," in *BROADNETS 2005*, 2005.

[24] G. Kramer and G. Pesavento, "Ethernet passive optical network (EPON): Building a next-generation optical access network," *IEEE Communications Magazine*, no. 2, pp. 66–73, 2002.

[25] IEEE Standard for information technology—Telecommunications and information exchange between systems—local and metropolitan area networks—specific requirements part 3: Carrier sense multiple access with collision detection (CSMA/CD) access method and physical layer specifications amendment: Media access control parameters, physical layers, and management parameters for subscriber access networks, IEEE 802.3ah, 2004.

[26] D. J. Shin, D. K. Jung, H. S. Shin, J. W. Kwon, S. Hwang, Y. Oh, and C. Shim, "Hybrid WDM/TDM-PON with Wavelength-Selection-Free transmitters," *Journal of Lightwave Technologies*, vol. 23, pp. 187–195, 2005.

[27] M. McGarry, M. Maier, and M. Reisslein, "Ethernet PONs: A survey of dynamic bandwidth allocation (DBA) algorithms," *IEEE Communications Magazine*, vol. 42, no. 8, pp. S8–15, 2004.

[28] I. Akyildiz, W. Lee, M. Vuran, and S. Mohanty, "NeXt generation/dynamic spectrum access/cognitive radio wireless networks: A survey," *Computer Networks*, vol. 50, no. 13, pp. 2127–2159, 2006.

Chapter 5

Millimeter Waves for Wireless Networks

James P. K. Gilb and Sheung L. Li

Contents

5.1 Introduction

Although wireless local area networks (WLANs) and wireless personal area networks (WPANs) have advanced in recent years, many applications still are not served by current standards and technology. In particular, there is not currently a wireless solution for applications that require greater than 1 Gb/s at ranges of 10 m or more.

Some of the applications that need these high data rates include:

- A wireless equivalent of Gigabit Ethernet (GigE)
- Eliminating the cable connecting video sources and projectors
- Delivery of high-definition uncompressed baseband audio and video (A/V) without wires
- Practical high-bandwidth wireless links to either high-capacity or low-power devices such as hard drives or digital video cameras

These applications can be served with the high transmit power, wideband frequency allocations approved and available worldwide for unlicensed or license-exempt systems around 60 GHz.

The band around 60 GHz has been limited in use based on the difficulties associated with at that radio propagation frequency and the cost of producing commercial products. This band had been opened up by various regulatory authorities because it is seen as having the potential to be well suited for high-speed communications in dense environments. Many radio designs that deliver the link budgets necessary for high data rate coverage do so by increasing the modulation complexity. However, this also increases the ratio of the range at which interference with other radio frequency systems is possible to the range at which connections are possible. Because of the peak in the attenuation characteristics due to oxygen absorption around 60 GHz, approximately 1.5 dB per 100 m, a high degree of spectrum reuse is possible. The additional attenuation combined with lower order modulations made possible by the large allocated frequency band makes the ratio of undesirable interference to desirable coverage area much lower than it is at the lower unlicensed frequencies. Furthermore, with recent developments in wireless technology, the economies of supporting applications in the 60-GHz band have changed dramatically.

Low-cost, highly integrated radios will need to be developed to take advantage of the 60-GHz frequency allocations and serve a commercial market. Previously, millimeter-wave (mm-wave) designs required exotic and expensive semiconductor processes, such as gallium arsenide (GaAs) and indium phosphide (InP). However, recent advances in bot complementary metal oxide semiconductor (CMOS) and silicon germanium (SiGe) technologies have the potential to deliver lower cost solutions that allow the integration of a significant amount of digital circuits with the millimeter-wave components. In addition, the ability to combine the digital and RF portions of a millimeter-wave radio in CMOS may yield even lower cost integrated solutions. Along with the advances in semiconductor technology, new developments in packaging and antenna design all point to the future development of low-cost, highly integrated millimeter-wave solutions.

With the market ready for solutions and new lower cost technology becoming available, the final piece to the millimeter-wave PAN puzzle is standards. Three organizations are in the process of developing the first millimeter-wave standards and specifications: the IEEE 802.15.3c working group, the WirelessHD special interest group, and the Wireless Gigabit With Advanced Multimedia Support (WIGWAM) project.

5.2 Applications, Description, and Requirements

Although the current WLAN and WPAN solutions (e.g., 802.11 and 802.15.1/Bluetooth) satisfy a large number of applications, many more applications cannot be served with the current standards and specifications. In particular, the standards for gigabit-level applications are still in development. The applications cover a wide range of areas, including data, high-definition (HD) media, and transportation. A high-level summary of the characteristics of these Gb/s applications is given in Table 5.1.

A common requirement for all of these applications is that the solution cost must be similar to the wired solution that it is replacing. The consumer may be willing to pay a little extra for the convenience of a wireless solution, but this premium will likely be small.

5.2.1 Gb/s Wireless Networking

In the wired world, 100 Mb/s networking is old hat and many new computers are being equipped by default with 1 Gb/s LAN (GigE) connections.

Table 5.1 Summary of the Requirements for Various Gb/s Applications

Application	*Throughput*	*Latency*	*Typical Motion*
Uncompressed HD A/V	3.0 Gb/s	Low	Stationary
Wireless projector	2 Gb/s	Very low	Stationary
Uncompressed audio	40 Mb/s	Low	Mobile
Lossy compressed A/V	40 Mb/s	Moderate	Mobile
Gigabit networks	1 Gb/s	High	Mobile
Data transfer	2.4 Gb/s	Low	Mobile

In the office environment, the expectation is that GigE connectivity will become the preferred technology in the office and in the server room. It is a natural evolution for this wired connectivity to be replicated in the wireless domain.

A wireless Gb/s networking solution would allow businesses to set up new computers or move them around the office without having to pull GigE cable to the new locations. However, the real advantage for the office environment is that it would allow laptops to connect at the same speed wirelessly as they do when they are connected to the wired network.

While GigE connections are not yet common in the home, laptops that are purchased for businesses are also used away from the office, for example in the home, hotel, or hotspot. Users have become used to having Internet connectivity wherever they are, whether it is the office, kitchen table, bedroom, or local coffee shop. Thus, the support of limited mobility for this application is needed as well.

For WPANs, the high-rate connectivity need is different. In this case, the goal is to connect devices that have large amounts of storage locally. Typical applications would include synchronizing data between a portable device and fixed storage, quickly sending data to a printer from a mobile device, and connecting peripherals to a PC. Although the specific applications for WPAN are somewhat different from WLAN, the requirements to support these applications are the same.

The requirements of the Gb/s WPAN/WLAN applications are

- Asynchronous data transfer
- A wide variety of data rates are useful
- Quality of service (QoS) support is not required
- Ad hoc connectivity, especially for WPAN applications
- Limited mobility

5.2.2 Video Source to Projector Connection

A common desire of business people is to have wireless connectivity between laptops and a projector. There are two main reasons why this is very attractive to businesses. The first reason is that it allows the projector to be mounted on the ceiling without having to run a long cable to the conference table. The second reason is that it allows multiple people to share the projector in a meeting without having to pass around the video cable connection. The throughput required for this application depends on the color depth, resolution, and refresh rate. For example, for a 1280 × 1024 screen with 24-bit color (16 million colors) and 60-Hz refresh rate, the raw data rate would be around 1.8 Gb/s. Larger displays or higher refresh rates could easily double or triple the required throughput.

In a home environment, the equivalent application is the connection of video sources such as DVD players, set-top boxes (STBs), and digital video recorders to cinema-quality video projectors. The reasons why it is attractive to use a wireless connection in the home are very similar to the reasons for the business case. The optimal position for a projector in a room is elevated and out of the way, primarily on the ceiling. Potential video sources are not normally ceiling mounted as they require user access (DVD trays) or are otherwise linked to connections that are not in the ceiling; e.g., cable or satellite-based STBs. The data rates for these home applications are comparable to those of business applications, starting at approximately 500 Mb/s for progressive standard resolution video (480p or 576p) and increasing to 3 Gb/s to support HD video.

These applications are very sensitive to latency issues because the user interacts with, and pays close attention to, displayed data in the business case and both video menus and audio in the home application. When the user moves the mouse on the computer, the expectation is that the mouse on the screen will move at essentially the same time in the correct direction. Any buffering of the data will result in a bad experience for the user. While it is possible to buffer a presentation where the user only shows the slides, very often a presentation will involve the user modifying the presentation, taking down notes, or producing a joint work product interactively with the audience. Thus, in this application, minimizing the latency is critical.

The requirements of this application include:

- Isochronous data delivery
- Very high data rates; 1.6 Gb/s, with 3–4 Gb/s in the near future
- The data rate required is fixed by the application and doesn't change during operation
- QoS is very important
- Low latency for data delivery

There are now some projectors available that implement a solution based on IEEE 802.11a and/or 802.11g. However, due to the limited throughput of these wireless technologies (<30 Mb/s for UDP and <20 Mb/s for TCP), the entire screen is not sent to the projector to be displayed. Rather a special application captures the drawing commands that are being dispatched to the operating system and sends these to the projector as well. The projector then renders these drawing commands to display a copy of the screen. This greatly reduces the bandwidth required for many office applications. However, this approach will not work with rapidly changing content, such as video data, because the bandwidth required for these drawing commands would exceed the throughput supported by the link. In addition, this solution is operating system specific (indeed, it can also be version specific as well) and so it is not applicable for CE devices that do not use vector-based display architectures such as Microsoft Windows DirectDraw.

5.2.3 Uncompressed Baseband High-Definition Video

HD digital video has begun to penetrate the home market and soon it will be commonplace in all homes. Cabled connections for HD video are effective and easily carry very high data rates. However, the user does not want to hook up cables and would prefer a wireless solution if it was easy to use and setup. A common example of the need for a wireless video solution is a flat-panel display (FPD) hanging on a wall. In this case, the consumer wants to have the minimum number of wires attached to the display. In addition, most walls have power outlets, but not video delivery outlets. A wireless solution that connects the content source to the display is highly desired by consumers. An example of this application is illustrated in Figure 5.1.

The typical consumer will have more than one source of HD content, but will want it shown on any display that is available. Currently, wired solutions either require switching the cables (as most displays only have one or possibly two digital input ports) or using an expensive central switch

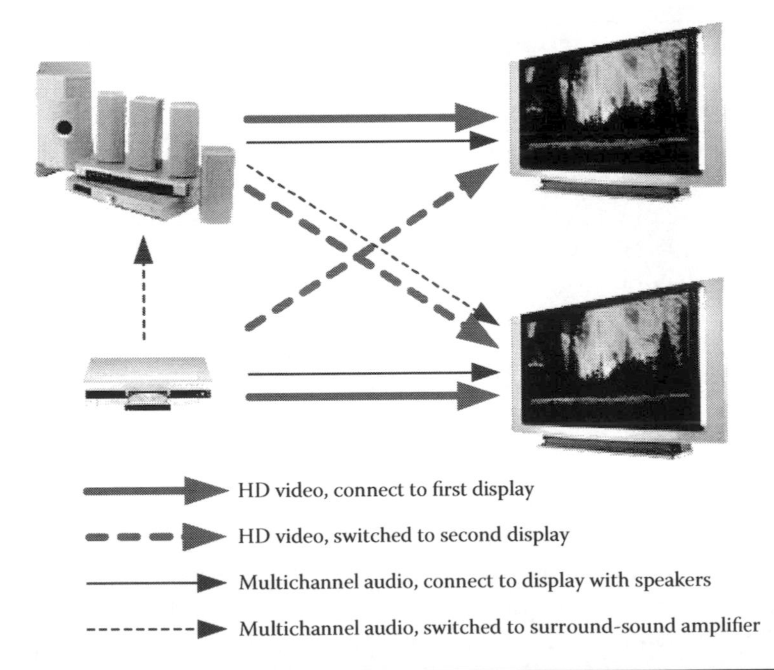

Figure 5.1 Uncompressed video distribution application with two sources and two displays.

box to connect the input to the output. With wireless technology, it is easy to have any source connect directly to any sink. For example, a user might have an STB that brings in HD content from a cable, satellite, or Internet connection, a source of prerecorded HD content, such as a Blu-Ray disc (BD) player or HD DVD player, and a gaming console. Any one of these three needs to be able to connect quickly and easily to the display. In addition, the display could combine video streams from the STB and the BD player to do picture-in-picture (PiP) display.

While it is possible to do video distribution in this manner with compressed video, there are challenges and costs with this approach. One issue is that many digital displays do not have the decompression hardware necessary to display the compressed data and so this would have to be added to the wireless solution, increasing the cost. Compression and decompression of the video stream requires extra time, which will introduce latency in the video display. This latency is already an issue in wired solutions where the audio and video take separate paths from the content source and are not played back at the correct time interval; i.e., the lip-sync problem.

Uncompressed video places little burden on the display device because the video is already formatted correctly for the screen. Additional features, such as PiP and on-screen display (OSD), are more readily added to the uncompressed video data. With compressed data, this information needs to be sent as a separate stream of data and needs to be composited by the display. The low latency due to handling uncompressed video data also allows the source to better synchronize the audio and video presentation. Otherwise, the source needs to determine the end-to-end latency of all of the delivery mechanisms so that it can delay the audio data to arrive at the same time as the video data.

The major HD video modes and the approximate data rates are summarized in Table 5.2. Note that the values in the table are only approximate; the actual values are available in CEA-861-D [1]. In addition, the rates shown in the table are for the video only and do not include the requirements for audio or control messaging.

Table 5.2 Approximate Throughput Required for Various Uncompressed HD Video Content

	Required Throughput		
Format	*8-bit RGB Color*	*10-bit RGB Color*	*12-bit RGB Color*
480p	0.5 Gb/s	0.625 Gb/s	0.75 Gb/s
720p	1.4 Gb/s	1.75 Gb/s	2.1 Gb/s
1080i	1.5 Gb/s	1.9 Gb/s	2.25 Gb/s
1080p	3.0 Gb/s	3.75 Gb/s	4.5 Gb/s

Table 5.3 Approximate Throughput Required for Various Audio and Video Content, Including Compression

Data Type	*Required Throughput*
Uncompressed 192 kHz 7.1 surround-sound audio	40 Mb/s
MPEG2 compressed broadcast 1080i A/V	20–24 Mb/s
VC-1 compressed 1080p prerecorded A/V	36 Mb/s
Uncompressed 5.1 surround-sound audio	20 Mb/s
Compressed 5.1 surround-sound audio	1.5 Mb/s

The approximate throughput required for compressed A/V, compressed audio, and uncompressed audio is summarized for selected formats in Table 5.3.

In summary, the requirements for carrying uncompressed A/V data are:

- Isochronous data delivery
- Very high data rates, from 0.5 to 4.5 Gb/s
- Excellent support for QoS
- Low latency
- Support for multiple simultaneous connections

5.2.4 Uncompressed High-Quality Audio

Another application for very high rate WPANs is the distribution of high-quality audio. As indicated in Table 5.3, the throughput required for high-end audio is in the tens of Mb/s. While the raw data rate of 802.11a and 802.11b (54 Mb/s) would appear to support uncompressed audio applications, the actual throughput is much lower (less than 30 Mb/s) and even lower still when QoS considerations are taken into account. In addition, if the uncompressed audio is not the only stream being transmitted, then additional throughput is required to support both applications at the same time.

One example of wireless delivery of uncompressed high-quality audio is illustrated in Figure 5.2. In this application, the FPD is receiving HD broadcast TV with its antenna and it is forwarding the audio portion to the audio receiver so that it can be distributed to the 5.1, 7.1, or 13.1 Dolby TrueHD or DTS-HD surround-sound system. In this case, the latency of the audio decoding is potentially an issue, especially with low-cost audio systems. Being able to stream high-quality, uncompressed audio is a real advantage.

The high-quality audio application has the following requirements:

- Very low latency
- Synchronization of the playback of multiple destinations

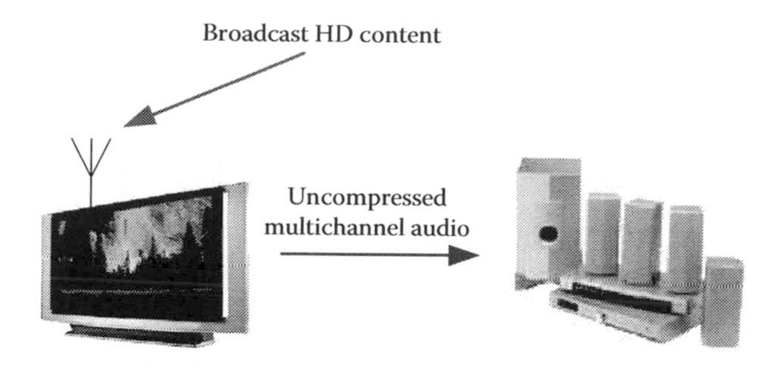

Figure 5.2 Example of streaming uncompressed audio from a display with broadcast receiver to a surround-sound system.

- Moderate to high data rates
- Isochronous data delivery

5.2.5 High-Bandwidth Data Transfer (USB/1394/SATA)

In contrast with the GigE networking applications, high-bandwidth data transfer applications involve the transparent bridging of various short-range wired connection technologies with wireless technology. These wired connection technologies are primarily used for single point-to-point data transfer and data storage applications such as copying stored videos from a digital video camera, digital still camera, or streaming media from an external hard drive.

At the lower end, the actual data throughput demands are similar to those of data networking technologies (400 Mb/s link rate for IEEE 1394a and 480 Mb/s link rate for USB 2.0). At the higher end, the throughput requirements push to and beyond the requirements of data networking (800 Mb/s and 1.6 Gb/s link rate for IEEE 1394b and 1.2–2.4 Gb/s for SATA).

One function that all of these data transfer applications share is the ability to readily bridge (from both a protocol and an electrical perspective) between the legacy wired technology and the new wireless solution. Devices that implement these types of connections have limited capability, due to cost constraints, and so any potential wireless replacement cannot require the implementation of a full networking stack. Instead, the original protocol needs to be supported by the wireless solution, creating a link that is indistinguishable from a wired link. Replicating a wired protocol over wireless requires additional overhead as well as low latency, pointing to the need for very high data rates like those promised by mm-wave solutions.

High-bandwidth data transfer requires the following:

- Asynchronous data transfer
- Convergence with existing wired protocols
- Maximum possible net data throughput
- Low packet error rate

5.2.6 Low-Power Mobile Applications

It may seem counterintuitive to use a high-frequency, high-throughput wireless technology in a lower power device. However, this application relies on an often neglected factor, the fact that it is not just the instantaneous power consumption, but rather the energy efficiency that usually determines the suitability of a design for mobile applications.

To a casual observer, a typical 60-GHz wireless design, optimized for sustained gigabit-level transfers over higher distances may appear to consume more power than other wireless technologies. This conclusion is misleading, because it does not take into account the energy that any design would need to physically process gigabit or higher levels of data.

Having a faster communications technology available, however, does not in itself increase the amount of data a mobile or fixed application needs to send. Furthermore, a higher speed solution may reduce the total power consumption required to transfer a given amount of data due to its ability to operate at a much lower duty cycle than slower radio technologies. In this case, the figure of merit is the energy required per bit transmitted that determines power efficiency. The net result is that the faster the underlying radio technology, the lower the actual transmit/receive duty cycle and the lower the amount of energy necessary to transfer each bit of information. In addition to reduced power for the radio, the host processor and storage are used for a shorter period of time, reducing their impact on the overall power drain. That's just what every mobile application needs.

Another advantage of high throughput is that the time required to complete a transaction is decreased. A user wants to touch a button and have his or her personal media player instantly sync with the digital video recorder. Any data transfer must take some fixed amount of time, but the less time the user has to wait, the better the experience. In addition, because the application involves a mobile device, the user may want to move out of range with the device soon after initiating the data transfer process. High-throughput connections enable the user to quickly accomplish the data transfer.

Low-power mobile applications have the following requirements:

- Both asynchronous data transfer and isochronous streaming
- Minimal energy per bit transferred
- Low peak energy consumption

5.2.7 Intelligent Transportation Systems

Intelligent transportation systems encompass a variety of safety, informational, and entertainment communications applications either between multiple moving vehicles or between an in-vehicle system and infrastructure such as roadside equipment (RSE). These applications are relevant because of the popularity in some regulatory domains of allocating portions of the 60-GHz band for intelligent transportation systems (ITS). These co-allocations range from a designated primary use of some fraction of the 60-GHz band by licensed ITS to the designation that outdoor ITS is available for use in a license-exempt manner.

In practice, these ITS applications cover a wide range of applications, including:

- Accident or collision avoidance
- Infrastructure-based communication of local points of interest
- The ready download of media entertainment from roadside units
- Dual-purpose designs such as multimedia-enabled refueling pumps

Under one of the proposed methods for ITS, sign- or streetlight-based RSE would take advantage of directional, 60-GHz signals to continuously scan and broadcast to each lane on a highway. The RSE's transmitter functions would include broadcasting information on upcoming traffic and potential safety conditions (such as an impending rapid slowdown) as well as commercial functions such as local advertisements. The RSE's receiver functions include collecting information on potential accidents and vehicle identification for toll collection, road usage monitoring, and other applications. In the toll application, the use of high-frequency directional signals enables tiering of traffic depending on highway lane and direction of travel.

Another application is the rapid download of multimedia content from RSE or refueling pumps to the vehicle for later viewing and/or listening. In this case, it is very important that the data download happen as quickly as possible, as this may be an impulse purchase. For example, when the user has finished filling up his or her vehicle, the refueling pump may offer a movie, video program, or audio program for download. The transfer of data, perhaps 1–9 GB of data, needs to take place in only a few seconds. This requires low latency and very high data rates.

Transportation applications share these characteristics:

- Pseudo-isochronous communication
- Directionality for lane isolation and position sensing
- Extremely high frequency reusability
- High availability and reliability

5.3 Regulatory Environment

The 60-GHz band has not been heavily utilized because of the attenuation of RF signals due to oxygen absorption in this frequency band. This has allowed regulators around the world to create huge frequency allocations, as much as 7 GHz, for unlicensed use. To put it in perspective, 7 GHz includes all AM radio stations, FM radio stations, all broadcast television allocations, all cellular telephony, both 802.11 frequency bands, and still has room to spare. A summary of the 60-GHz regulations is listed in Table 5.4. A graphical representation of the allocated frequencies is illustrated in Figure 5.3.

The relevant documents from the geographic regions are summarized in Table 5.5. The list of documents is not exhaustive and there are generally other documents that govern the practical certification of radio devices and the limits of unlicensed operation in various regulatory domains.

In addition to the requirements in Table 5.4, each of the geographic regions has some additional requirements. In Japan, for example, the maximum bandwidth for a device is 2.5 GHz and the frequency accuracy is required to be less than 500 ppm. The South Korean regulations are in draft form and are expected to be finalized in early 2007.

Given the potential for high-radio-frequency power output, it is useful to consider the practical impact of power density limits from an exposure perspective. A watt of power at a higher frequency has lower epidemiological impact than a watt of power at a lower frequency, because of the higher frequency's greater path loss and material reflectivity characteristics. In fact, the actual (and to one accustomed to lower frequencies, seemingly high) power limits are typically already set by the field strength limits for human exposure to electromagnetic fields [7] in any particular regulatory domain. A maximal 40-dBm transmitter only achieves a 10 W per square meter field strength at a range of 30 cm from the body core in the boresight direction, well outside the normal limits of any 60-GHz usage models.

5.4 Why Millimeter Waves for Gb/s WPAN?

The new applications for WPAN require very high data rates with isochronous data delivery. Because of the high expectations of the user, a very high level of QoS will be required to satisfy their desires. The millimeter-wave frequencies, particularly the 60-GHz unlicensed spectrum, provide a unique opportunity to wirelessly enable these applications. In particular, the 60-GHz unlicensed band provides the following advantages:

- *Huge frequency allocation.* Typically 7 GHz in most regions, with 5 GHz of overlapping allocations
- *Much higher power available.* Unlike the strict transmit power restrictions on ultrawide band unlicensed operation, the millimeter-wave

Table 5.4 Summary of 60-GHz Regulations in Selected Geographic Regions

Parameter	*Japan*	*North America*	*South Korea (draft)*	*EU (draft)*	*Australia*
Frequency	59–66 GHz	57–64 GHz	57–64 GHz	59–66 GHz	59.4–62.9 GHz
Conducted TX power	+10 dBm	N/A	+10 dBm	+27 dBm	+10 dBm
EIRP	+57 dBm	+40 dBm ave. +43 dBm peak	+57 dBm	+40 dBm ave. +43 dBm peak	+51.8 dBm
TX spurious	100 μW	N/A	N/A	−10 dBm[a]	N/A
RX spurious	100 μW	500 μW/m at 3 m	N/A	−10 dBm[a]	N/A

[a] Measured per ERC74-01.

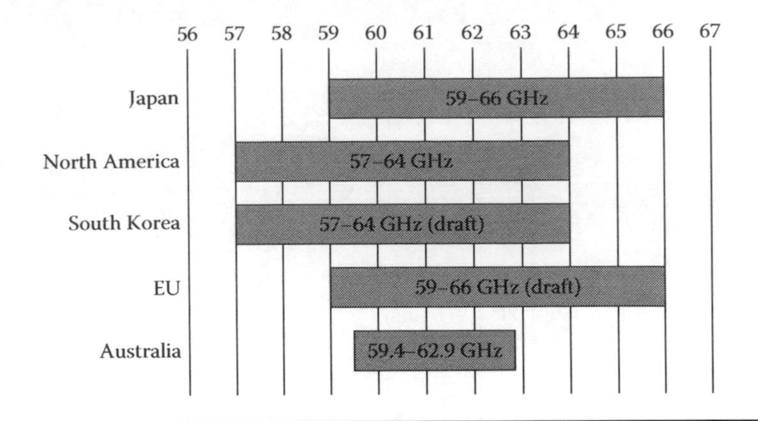

Figure 5.3 Graphical representation of unlicensed 60-GHz frequency allocations in selected geographical regions.

band allows an EIRP (effective isotropic radiated power) that is significantly greater, 40–50 dB greater

- *Clean spectrum, no incumbents.* There are not any widely deployed 60 GHz radiators in the home or office, so there is less chance for interference
- *High frequency allows small, high-gain antennas.* A 25-dB gain antenna has an effective aperture approximately one square inch. High-gain antennas allow high EIRP with low-power RF amplifiers
- *High gain allows overlapping networks that don't interfere.* Because the antennas are highly directional at these frequencies, spatial reuse is enabled for colocated systems

There are some challenges to creating millimeter-wave gigabit-per-second WPANs. Some of the biggest challenges that remain unsolved at this time are

Table 5.5 Summary of Regulatory Documents for Various Geographic Regions

Region	*Document*
Australia	Radiocommunications class license 200 [2]
Canada	RSS-210, Issue 6, September 2005 [3]
Japan	Regulations for Enforcement of the Radio Law 6-4-2 Specified Low Power Radio Station (11) 59–66 GHz band [4]
USA	CFR Title 47 Part 15.255 [5]
European Union	ETSI DTR/ERM-RM-049 [6]

- Low-cost, high-yield implementations
- Multi-gigabit-per-second wireless modems
- Seamless connectivity with highly directional antennas

Some of the challenges, in particular cost, are being addressed with advances in process technology and design methodologies. Others, like how to take advantage of highly directional antennas without requiring an antenna engineer to place the devices, are still topics of ongoing research.

5.5 Survey of Technology

Key to the creation of millimeter-wave WPANs is the development of cost-effective technologies that support very high throughput as well as millimeter-wave RF circuits. The technologies required include RF semiconductor processes, high-frequency packaging, and advanced antenna technology. All of these technologies need to support not only low-cost, but also high-volume production to support the consumer electronics market.

The acceptable cost of a wireless solution is determined by the value it adds to the solution, the price of the cable it replaces, and the price of the device for which it is intended. For example, a $2000 FPD can support a much more expensive wireless solution than a $50 DVD player. Likewise, if the wireless solution replaces a $75 cable and two $5 connectors, one at each end point, then a solution price of $85 could be justified. In some cases, the mobility and convenience of a wireless solution will provide additional value that can justify an increased price. The price of a cordless telephone is significantly greater than the price of a corded telephone, but the typical consumer values the mobility and convenience that it provides. Because of this, the acceptable cost of a millimeter-wave WPAN solution will depend strongly on the application.

5.5.1 Old Technology: GaAs, InP

Until recently, the design of millimeter-wave circuits required the use of exotic technologies. In particular, GaAs heterojunction bipolar transistors (HBTs) and InP-based semiconductors were required to design 60-GHz circuits. While these exotic processes are still unmatched in terms of the RF performance that is possible, the cost of the resulting circuits make the designs practical only for the least cost-sensitive applications.

There are a variety of reasons why compound semiconductor processes such as GaAs and InP are not cost effective for high-volume applications. One reason is that these processes are not in large-scale production, as is the case with SiGe, RF CMOS, and CMOS, which leverage the enormous volumes of standard digital CMOS processes. Because of this, very few of the fabrication lines for compound semiconductors are using 150-mm wafers. Most use 100-mm or smaller wafers, as opposed to the 200-mm or 300-mm wafer diameters that are common for modern CMOS-based processes. Larger wafers give higher volumes of parts at significantly lower costs. The higher volumes of silicon technology also allow a significantly greater monetary investment in process development to support cost reductions and yield enhancements.

Another factor that raises the cost of systems based on compound semiconductors is that it is not possible to do even moderate scale integration of the radio functions on a single die. Using more than one die increases the system cost as well as creates a design challenge in bringing the millimeter-wave signals onto and off the dice. With CMOS-based processes, a significant portion of the radio can be integrated into a single die and only one millimeter-wave connection is required, the attachment to the antenna.

An example of a compound semiconductor 60-GHz radio is one based on 0.15-μm GaAs pHEMT (pseudomorphic high electron mobility transistor) process and designed at the Department of Microelectronics at Chalmers University of Technology [8]. The key building blocks include a low-noise amplifier (LNA), 5-GHz intermediate frequency (IF) amplifier, power amplifier (PA), mixer, 7-GHz voltage-controlled oscillator (VCO), and a frequency multiplier. Another example used a square-law detector to simplify the receiver architecture [9]. While these radio designs accomplish the task of generating and receiving 60-GHz signals, they are not able to hit the cost targets or volumes required for WPAN applications.

5.5.2 SiGe Is Ready Now

With the recent advances in silicon technology, there are now multiple companies that offer foundry service with a form of SiGe that can be used to design millimeter-wave circuits. Some of these foundries are listed in Table 5.6 along with key figures of merit for the process. The table does not list all SiGe foundries, rather it is a sampling of publicly available information from selected foundries.

Using the advanced SiGe processes that are now available, a variety of building blocks as well as some elementary transceivers have been designed, fabricated and tested. A fully integrated phase-locked loop (PLL), including VCO, that can tune from 54.5 to 57.8 GHz was demonstrated in SiGe:C BiCMOS [10,11]. This PLL was then used to fabricate a transmit and receive chipset for 250 Mb/s operation [12]. The receive chip was comprised of an LNA, down mixer, and PLL, while the transmit chip implemented the up mixer, buffer amplifier, and PLL. Both chips used a 5-GHz IF and a single-ended Vivaldi antenna integrated in the board.

IBM research has used the IBM 8HP process to design a variety of circuits at 60 GHz. The team initially designed a variety of individual circuits (LNA, PA, VCO, and direct conversion mixer) in its 8HP process to operate

Table 5.6 Selected SiGe Foundries

Company	*Process*	f_t	f_{max}	BV_{vceo}
IHP	SG25H1[a]	190 GHz	220 GHz	1.9 V
IHP	SG25H2	170 GHz	170 GHz	1.9 V
Jazz	SBC18H2	200 GHz	200 GHz	
Jazz	SBC18HX	150 GHz	170 GHz	2.2 V
IBM	8HP	200 GHz	250 GHz	1.2 V[b]

[a]The IHP process is SiGe:C.
[b]At 1.7 V, f_t is 200 GHz and f_{max} is 180 GHz.

at 60 GHz [13]. The team then used those building blocks to successfully design and fabricate transmit and receive chains that integrated these blocks and which were combined with an antenna that directly connected to die using flip-chip type technology [14,15]. The receive chip integrated an LNA, 60-GHz down mixer, variable gain IF amplifier, I/Q down mixer, baseband amplifiers, and a 3X multiplier for the local oscillator (LO). The TX chip integrated the I/Q up mixers, variable gain IF amp, 60-GHz up mixer, driver amp, and PA.

Other 60-GHz receiver designs in SiGe have also been presented [16] with an LNA, mixer, VCO, and wideband IF amplifier that used integrated transformers to provide the match between the LNA and the mixer. A transmitter was developed in 180-nm SiGe BiCMOS that included a VCO, subharmonic mixer, PA, and a tapered-slot antenna [17].

SiGe processes offer significant cost advantages over compound semiconductor processes and also allow integration of large amounts of digital logic. However, they are more expensive than standard digital CMOS processes for the same technology node. In general, the price of SiGe is comparative to the cost for RF CMOS processes, which add special features to standard digital CMOS processes to improve the analog and RF performance.

SiGe has been traditionally thought of as more power efficient than CMOS, particularly due to its higher breakdown voltage. However, the high-speed SiGe processes used for millimeter-wave design have relatively low breakdown voltages, as shown in Table 5.6, similar to those of CMOS processes with equivalent frequency performance.

5.5.3 Is CMOS the Future?

To further reduce the cost of millimeter-wave solutions, some research groups have been investigating using standard digital CMOS for 60-GHz circuit design. Initial work in CMOS design was the development of an LNA [18,19], and mixer [20], in a 130-nm standard digital CMOS process.

An example of a complete CMOS receiver design was also presented by Razavi [21] that used 130-nm CMOS technology and implemented the LNA, quadrature mixers, baseband amplifiers, and a single-ended-to-differential converter for the LO. Even higher frequencies have been designed in CMOS, for example, a 70-GHz distributed amplifier [22] has been demonstrated using 90-nm CMOS. In addition, a 114-GHz VCO was realized in 130-nm CMOS technology, but with very low output power (−22.5 dBm).

Currently, all of the work on 60-GHz CMOS is being done as university research projects. At this point, the researchers are attempting to determine the limits of this approach to circuit design. If they are successful in designing not only millimeter-wave functional blocks, but also entire transceivers, the next step will be to see if it can be done reliably in high volumes to support consumer electronics applications. If standard digital CMOS can be used for mm-wave WPAN radios, it will likely be the cheapest solution.

5.5.4 Packaging Is Key to Implementation

Even if it is possible to lower the cost of the RFICs used in millimeter-wave radios, packaging 60-GHz circuits remains a significant technical and cost challenge. The key to millimeter-wave packaging is twofold. First, the package needs to ensure good grounding by creating short paths to ground. Second, the design must reduce the number of high-frequency connections that need to go off-chip. Ideally, the only high-frequency connection should be the connection to the antenna.

Short ground connections are achieved by using flip-chip technology instead of wire bonding. This also allows short RF connections that are easier to match over a wider range of frequencies. Additionally, because the antennas are so small, it is best to make them part of the package, or even to integrate them as part of the die [23,24].

5.6 Standardization of Millimeter Waves

There are currently three main efforts in the standardization of millimeter-wave technology, IEEE 802.15.3c, WirelessHD, and the WIGWAM project.

5.6.1 IEEE 802.15.3c

In the 2003 July Plenary meeting of IEEE 802, the 802.15 working group approved a motion to form an interest group to study the possibility of developing a standard for millimeter-wave WPANs. The millimeter-wave frequencies were seen as appropriate for the relatively short range that is traditional for WPANs. The 60-GHz band, in particular, was identified as appropriate due to the oxygen absorption band. More information can be found at http://www.ieee802.org/15/pub/TG3c.html

The 802.15 millimeter-wave interest group met four times and in March of 2004 was approved to become study group 3c under 802.15 (sg3c). The purpose of a study group is to write a project authorization request (PAR) and describe how the new standard would satisfy the 802 committee's five criteria (5C) (broad market potential, compatibility, distinct identity, technical feasibility, and economic feasibility). The study group worked for another year on the PAR and 5C and was successful in getting approval in March of 2005 to operate as a task group, 802.15.3c, to develop an amendment to IEEE Std 802.15.3-2003 for an alternate PHY. As an amendment, the 802.15.3c task group will produce a PHY that works with the Media Access Control (MAC) from IEEE Std 802.15.3-2003, much as 802.11a, 802.11b, and 802.11g produced PHYs that operated with the 802.11 MAC. The 802.15.3c task group has identified three key attributes that are expected from the proposals:

1. Unlicensed operation in the 60-GHz band
2. Very high data rates, greater than 1Gb/s
3. Support for multimedia

From their first meeting as a task group in May of 2005, 802.15.3c has been working on various documents that the group wants prior to reviewing proposals and selecting a solution. The documents include:

- *System requirements document*, which describes the desired system performance for the amendment; e.g., the range, data rate, power consumption, form factor, complexity, and so forth
- *Selection criteria document*, which lists specific characteristics that should be provided by a proposer; e.g., sensitivity, range, performance in a multipath environment, throughput, power consumption, and so forth
- *Channel model*, which defines the radio channel based on measured data so that it can be used for simulating the performance of a proposal in a multipath environment
- *Usage model document*, which provides a variety of use cases and applications that the standard is expected to address
- *Down selection procedure*, which determines the process by which the 802.15.3c task group will review and select a proposal to begin drafting the standard

The progress of 802.15.3c had been held up by the development of the channel model. Unlike other channel modeling efforts for wireless standards, the 802.15.3c task group has specifically requested a channel model that enables the simulation of systems with highly directional antennas. To do this, new measured data was required that include angle-of-arrival (AoA) information. A more complex mathematical model was also selected that could use the AoA information so that proposers would be free to propose different antenna gains, but simulate against the same channel model. Despite these challenges, a variety of companies and organizations have submitted many measurement results and these have been combined to create a set of channel models suitable for simulation work.

Because of the delay in developing the channel model, the down selection process has been delayed from its initial date of November 2005 to July 2007.

The 802.15.3c task group is very active and its status can change at least as often as it meets, which is every other month. The 802.15 working group Web site (http://grouper.ieee.org/groups) is the best place to find the latest information on 802.15.3c and to download documents related to 802.15.

5.6.2 WirelessHD Special Interest Group

WirelessHD is a special interest group of consumer electronics companies focused on creating a common network interface specification that supports multi-gigabit-per-second connections. This specification is intended to enable the low-cost delivery of high-definition baseband A/V streaming, high-speed content transmission, and control signals for consumer electronic devices. The A/V streaming will support both existing EIA/CEA and ITU standards as well as upcoming high-definition formats. The group intends to present the format for adoption as soon as the specifications are completed in Summer 2007. More information can be found on the WirelessHD Web site, (http://www.WirelessHD.org).

The WirelessHD special interest group seeks to leverage its use of the 60-GHz band with a variety of smart antenna technologies that invert the traditional mix of high-power amplifier output/low antenna gain used in lower frequency consumer electronics applications. Instead, WirelessHD solutions target low power amplifier output combined with high antenna gain to get similar link margins. By leveraging these high-efficiency antenna technologies, WirelessHD plans to support designs suitable for both fixed and mobile products as well as provide a means of scaling to even more demanding media requirements in the future.

WirelessHD also reflects its consumer electronics roots with an integration of control technologies that make the setup and operation of wireless products easier for typical consumers. In addition, the control technologies enable convergence with existing wired mechanisms for providing content protection, copy management, and personal privacy.

The WirelessHD specification uses several new technologies that enable multi-gigabit-per-second data rates required to support high-definition, uncompressed video streaming. These new technologies will enable low-cost solutions, better image quality, and higher performance wireless A/V systems. The key characteristics of the WirelessHD technology are

- High interoperability supported by major CE device manufacturers
- Uncompressed HD video, audio, and data transmission, scaleable to future high-definition A/V formats
- High-speed wireless, multigigabit technology in the unlicensed 60-GHz band
- Smart antenna technology to overcome line-of-sight constraints of 60 GHz
- Secure communications
- Device control for simple operation of consumer electronics products
- Error protection, framing, and timing control techniques for a quality consumer experience

5.6.3 WIGWAM

WIGWAM (Wireless Gigabit With Advanced Multimedia Support) is an initiative funded by the German Ministry of Education and Research (BMBF) and consisting of European partners from industry, universities, and independent research institutes to create a wireless communications system with a maximum data rate of 1 Gb/s. To achieve this, WIGWAM has targeted the 5 GHz band for the initial implementation but is also considering extensions in the 17-, 24-, and 60-GHz bands. The home page for the WIGWAM project is http://www.wigwam-project.com

The WIGWAM project seeks to develop an entire wireless system and so it covers all aspects of a communications system, from use cases to protocols to hardware, both baseband and RF. The project began in October 2003 and is scheduled to create deliverables by March 2007.

The key application that is targeted by WIGWAM is the delivery of multimedia content in a variety of locations, including [25]:

- Home
- Large offices
- Public access (i.e., hotspots)
- High-velocity mobile applications, including trains and automobiles

The 60-GHz portion of the WIGWAM project is located in a subproject, "Design and Implementation of an Ultra High-Speed WLAN: Concept, SoC-Implementation and Demonstrator (WIGWAM-IHP)," that is being led by IHP (Innovations for High Performance Microelectronics), a nonprofit organization supported by the German state of Brandenburg and the German federal government. The subproject is investigating

- 60-GHz radio front-end architecture and implementations
- High-speed baseband processor for 1 Gb/s data rate
- Ultra-high-throughput protocol processor
- Integration of radio components on a single chip
- Development of a demonstration platform at 60 GHz

The 60-GHz design work has focused on using SiGe:C to develop the necessary components for the RF front end. SiGe:C uses a carbon-doped SiGe base layer to improve the performance of the transistor [10]. The project has successfully used SiGe:C to develop the following parts at 60 GHz:

- Frequency divider
- LNA
- Mixer
- VCO
- RF switch

The researchers involved in the 60-GHz portion of the WIGWAM project have been in communication with IEEE 802.15.3c and have presented ideas based on their work for consideration by the task group [27]. Members of WIGWAM have also declared their intent to submit a proposal for consideration in the down selection process of 802.15.3c.

The WIGWAM project continues to develop new technology in all of the disciplines required for developing a high-speed wireless link. Current information and publications on the WIGWAM project can be found at the home page for the WIGWAM project, (http://www.wigwam-project.com).

References

[1] CEA-861-D, "A DTV Profile for Uncompressed High Speed Digital Interfaces," July 2006.
[2] Australia Radiocommunications Class License 2000.
[3] Canada RSS-210, Issue 6, September 2005.
[4] Japan Regulations for Enforcement of the Radio Law 6-4-2 Specified Low Power Radio Station (11) 59–66 GHz Band.
[5] FCC 47 CFR Part 15.255, 57–64 GHz, September 19, 2005.
[6] Electromagnetic compatibility and Radio spectrum Matters (ERM); System reference document; Technical characteristics of multiple gigabit wireless systems in the 60 GHz range, ETSI DTR/ERM-RM-049, July 2006.

[7] Generic standard to demonstrate the compliance of electronic and electrical apparatus with the basic restrictions related to human exposure to electromagnetic fields (10 MHz to 300 GHz), General public, CENELEC EN 50 392.

[8] J. Noreus, M. Flament, H. Zirath, and A. Alping, "System Considerations for Hardware Parameters in a 60 GHz WLAN," in *Proc. of the GHz 2000 Symposium on Gigahertz Electronics*, pp. 267–271, Gothenburg, Sweden, March 2000.

[9] Y. Shoji, K. Hamaguchi, and H. Ogawa, "Millimeter-Wave Remote Self-Heterodyne System for Extremely Stable and Low-Cost Broad-Band Signal Transmission," *IEEE Trans. on Microwave Theory and Techniques*, vol. 50, pp. 1458–1467, June 2002.

[10] W. Winkler, B. Heinemann, and D. Knoll, "Application of SiGe:C BiCMOS to Wireless and Radar," 12th GAAS Symposium, 2004.

[11] W. Winkler, J. Borngräber, B. Heinemann, and F. Herzel, "A Fully Integrated BiCMOS PLL for 60 GHz Wireless Applications," in *IEEE International Solid-State Circuits Conference Digest*, pp. 406–407, 2005.

[12] Y. Sun, S. Glisic, F. Herzel, K. Schmalz, E. Grass, W. Winkler, and R. Kraemer, "An Integrated 60 GHz Transceiver Front-End for OFDM in SiGe: BiCMOS," Wireless World Research Forum (WWRF) 16 Apr. 2006, http://www.wireless-world-research.org/

[13] B. Gaucher, "Silicon Millimeter Wave Integrated Circuits for Wireless Applications," IEEE 802.15 document number 15-04-0665-01-003c, November, 2004.

[14] B. Floyd, S. Reynold, U. Pfeiffer, T. Beukema, J. Grzyb, and C. Haymes, "A Silicon 60 GHz Receiver and Transmitter Chipset for Broadband Communications," in *IEEE International Solid-State Circuits Conference Digest*, pp. 184–185, 2006.

[15] B. Gaucher, "Completely Integrated 60 GHz ISM Band Front End Chip Set and Test Results," IEEE 802.15 document number 15-06-0003-00-003c, January, 2006.

[16] M. Q. Gordong, T. Yao, and S. P. Voinigescu, "65-GHz Receiver in SiGe BiCMOS Using Monolithic Inductors and Transformers," *Si Monolithic Integrated Circuits in RF Systems*, Technical Digest, pp. 265–268, Jan. 2006.

[17] C.-H. Wang, et al., "A 60 GHz Transmitter with Integrated Antenna in 0.18 μm SiGe BiCMOS Technology," in *IEEE International Solid-State Circuits Conference Digest*, pp. 186–187, 2006.

[18] C. Doan, S. Emami, A. Niknejad, and R. Bordersen, "Design of CMOS for 60 GHz Applications," in *IEEE International Solid-State Circuits Conference Digest*, pp. 1–3, 2004.

[19] C. Doan, S. Emami, A. Niknejad, and R. Brodersen, "Millimeter-wave CMOS design," *IEEE J. Solid-State Circuits*, vol. 40, pp. 144–155, Jan. 2005.

[20] S. Emami, C. Doan, A. Niknejad and R. Brodersen, "A 60-GHz Down-Converting CMOS Single-Gate Mixer," *IEEE RFIC Symp. Dig.*, pp. 163–166, June 2005.

[21] B. Razavi, "A 60 GHz Direct-Conversion CMOS Receiver," in *IEEE International Solid-State Circuits Conference Digest*, pp. 400–401, 2005.

[22] M.-D. Tsai, H. Wang, J.-F. Kuan, and C.-S. Chang, "A 70 GHz Cascaded Multi-Stage Distributed Amplifier in 90 nm CMOS Technology," in *IEEE International Solid-State Circuits Conference Digest*, pp. 402–403, 2005.

[23] A. Babakhani, X. Guan, A. Komijani, A. Natarajan, and A. Hajimiri, "A 77-GHz 4-Element Phased Array Receiver with On-Chip Dipole Antennas in Silicon," in *IEEE International Solid-State Circuits Conference Digest*, pp. 180–181, 2006.

[24] A. Natarajan, A. Komijani, X. Guan, A. Babakhani, Y. Wang, and A. Hajimiri, "A 77 GHz Phase-Array Transmitter with Local LO-Path Phase-Shifting in Silicon," in *IEEE International Solid-State Circuits Conference Digest*, pp. 182–183, 2006.

[25] G. Fettweis, T. Hentschel, and E. Zimmermann, "WIGWAM A Wireless Gigabit System with Advanced Multimedia Support," http://www.wigwam.de

[26] W. Winkler, J. Borngräber, B. Heinemann, and F. Herzel, "60 GHz Transceiver Circuits in SiGe:C BiCMOS Technology," *European Solid State Circuits Conference*, pp. 83–86, Sept. 2004.

[27] E. Grass, M. Piz, F. Herzel, and R. Kraemer, "Draft PHY Proposal for 60 GHz WPAN," IEEE 802.15 document number 15-05-0634-01-003c, November 2005.

Chapter 6

The WiMedia Standard for Wireless Personal Area Networks

Joerg Habetha and Javier del Pardo Pavon

Contents

6.1 Introduction

The WiMedia Alliance is a nonprofit association of more than 200 industry corporations and research institutions for the specification, promotion, and multivendor interoperability of ultrawideband (UWB) communication solutions. The six largest consumer electronics companies of the world and nine of the ten largest semiconductor companies are members of WiMedia. The first UWB standard of WiMedia enables data rates of up to 480 Mb/s and can therefore be used for the wireless transmission of USB2.0 frames between a personal computer and its peripheral devices. Another strength of UWB is its low power consumption, which makes it suitable for mobile devices, such as cameras, MP3 players, and so forth.

The WiMedia Alliance specifies a "common radio platform" that includes the physical (PHY) layer and the Media Access Control (MAC) sublayer of the protocol stack. The WiMedia UWB common radio platform is designed to operate with application stacks developed by the 1394 Trade Association, the Certified Wireless USB Promoter Group, and the Bluetooth-SIG.

The present chapter will provide an overview of the WiMedia specifications for PHY and MAC.

The physical layer of WiMedia is based on a proposal that was first submitted to the IEEE 802.15.3a task group. IEEE 802.15 is the working group for wireless personal area networks (WPANs). In December 2002 a call for proposals was issued by IEEE 802.15.3a. Of the 24 proposals received in March 2003, 23 were UWB based. Between May and July 2003 many companies merged their proposals into a multiband orthogonal frequency division multiplexing (MB-OFDM) proposal. The main opponent to this proposal was a proposal based on direct-sequence UWB (DS-UWB). In the following two years neither of the two remaining proposals could reach the required majority of 75 percent of the votes. This situation led to the creation of the Multi-Band OFDM Alliance (MBOA), which became WiMedia in March 2005.

Shortly after the foundation of MBOA, the alliance started to develop not only a standard for the PHY layer but also a new standard for the MAC layer. The reason MBOA began to develop a new MAC layer was twofold: On the one hand, the centralized MAC architecture of IEEE 802.15.3 with a piconet coordinator as its central element was not optimally suited for peer-to-peer communication in an ad hoc network. On the other hand, piconets could not operate simultaneously, due to the fact that some channels in the MB-OFDM physical layer were not fully orthogonal.

MBOA issued a call for proposals for the MAC layer in early 2004. Several companies submitted proposals in February 2004 and the final proposal selection took place in March. The authors of this chapter are the main authors of the winning proposal.

The MBOA standards have been adopted as international standards ECMA-368 [1] and ECMA-369 [2] in the meantime. The companies supporting a DS-UWB solution have also pursued product development of the DS-UWB solution. Because of the deadlock situation, the IEEE 802.15.3a working group was dissolved in January 2006. Several associations, among them the Bluetooth Special Interest Group, have adopted the WiMedia standard, and the first products according to this standard are available on the market.

6.2 MultiBand OFDM vs. Conventional Pulse-Based Ultra-Wide-band PHY Layer/System

Conventional UWB communication systems [3–4] are based on the transmission of very short (in picoseconds) pulses that spread the signal over a single (ultra) wide frequency band (measured in gigahertz). Modulation schemes such as on–off keying (OOK), binary phase shift keying (BPSK), or pulse position modulation (PPM) can be used with this single-band UWB system. The bandwidth is determined by the pulse width, whereas the data rate is determined by the pulse repetition rate. In such a system, it is easy to resolve the multipath components at the receiver due to the short duration of the pulses, and hence a Rake combiner [5] could be used. Such a system, however, has some limitations in addressing the requirements of a high-speed wireless network.

First, for low-data-rate applications, where the pulse repetition interval will usually be larger than the delay spread, it is feasible to avoid intersymbol interference (ISI). On the other hand, high-data-rate systems are susceptible to multipath interference because the pulse repetition interval is shorter than the delay spread.

Second, a single-band UWB system also has limited flexibility in terms of the number of simultaneously operating networks. Since all the WPAN networks operate in the same single-frequency band, it is not possible to use frequency division multiple access (FDMA) or frequency-hopping spread spectrum (FHSS) multiple access to accommodate them. Other multiple access schemes, such as direct-sequence spread spectrum (DSSS) multiple access, could be used to improve the capacity, but they require additional computational resources at the receiver.

Finally, it is not easy to avoid narrowband interferences; a pulse UWB system would have to use a notch filter to suppress the interference. This approach is not very attractive because the location of the interferer may not be fixed.

To overcome some of these limitations, an approach based on multiple frequency bands was developed [6, 7]. This system has been specified and finalized by the WiMedia Alliance. In this scheme, the UWB frequency band is divided into subbands of 528 MHz wide (which is considered UWB by FCC requirements). The modulated pulses are transmitted in these bands in a time-interleaved fashion. In this manner, multiple networks can operate simultaneously by using different frequency sequences. Further, narrowband interferences can be avoided by simply not transmitting in the band of interference.

The computational complexity of a pulsed UWB receiver is typically quite high, as it requires a Rake combiner in each of its paths in order to fully capture the signal energy [5]. This would be even more complex for multiband systems. On the other hand, we could consider an OFDM-based system, which has very good signal energy capture properties. A system combining the features of multiband and OFDM has evolved as an attractive option for high-data-rate WPAN applications. A proposal based on a frequency-hopping OFDM system has gained wide acceptance [7].

6.3 Overview of the WiMedia Physical Layer

The WiMedia physical layer (PHY) provides data rates from 53.3 Mb/s to 480 Mb/s. It uses a 128-point IFFT (inverse fast Fourier transform) and a symbol rate of 3.2 MHz. It hops among three bands and the hopping is done per symbol. The subcarrier spacing is 4.125 MHz and the signal occupies a bandwidth of 528 MHz in each hop. This system provides a link margin of 6.0 dB at 10 m separation using a 106.6 Mb/s mode.

Figure 6.1 shows a simplified block diagram of a WiMedia transmitter. The input data is randomized by a scrambler that is reset on each frame header and payload based on the scrambler initialization seed identifier carried in the PHY header. A constraint length 7, rate - 1/3 convolutional code, with a generator polynomial [133_8 145_8 175_8], is used for forward error

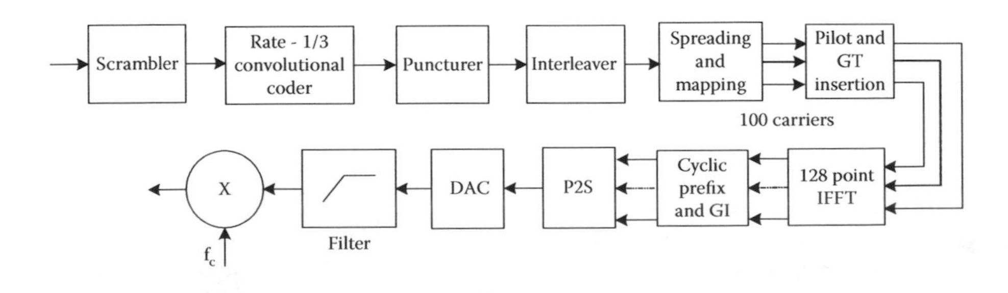

Figure 6.1 Simplified block diagram of a WiMedia transmitter.

Table 6.1 Rate-Dependent Parameters

Data Rate (Mb/s)	*Coding Rate*	*Conjugate Symmetry Input to IFFT*	*Time Spread Factor*
53.3	1/3	Yes	2
80	1/2	Yes	2
106.7	1/3	No	2
160	1/2	No	2
200	5/8	No	2
320	1/2	No	1
400	5/8	No	1
480	3/4	No	1

correction. A three-stage bit interleaver is used to protect the data against burst errors. The coded bits are mapped using quadrature phase shift keying (QPSK) constellation onto 100 data carriers. A combination of frequency spreading, time spreading, and puncturing is used to achieve different data rates. Table 6.1 shows some of the rate-dependent parameters.

Of the remaining 28 subcarriers, 12 are specified as pilot subcarriers, 10 are specified as guard tones, and 6 are specified as null carriers. The null subcarriers (except for one subcarrier at DC) are specified at the bandedges in order to relax filter requirements. A 32-sample zero prefix is added to the 128-point IFFT output, to mitigate ISI. This enables the system to tolerate delay spreads of up to 60.6 ns. A five-sample zero-postfix guard interval (GI) is added to the 160-point vector to enable band–carrier frequency switching. Figure 6.2 shows the format of an OFDM symbol.

An OFDM baseband symbol is represented by the following equation:

$$r_k(t) = \begin{cases} 0 & t \in [0,\ T_{ZP}] \\ \sum\limits_{n=-N_{ST}/2}^{N_{ST}/2} C_n \exp(j2\pi n\Delta_f)(t - T_{ZP}) & t \in [T_{ZP},\ T_{FFT} + T_{ZP}] \\ 0 & t \in [T_{FFT} + T_{ZP},\ T_{SYM}] \end{cases} \quad (6.1)$$

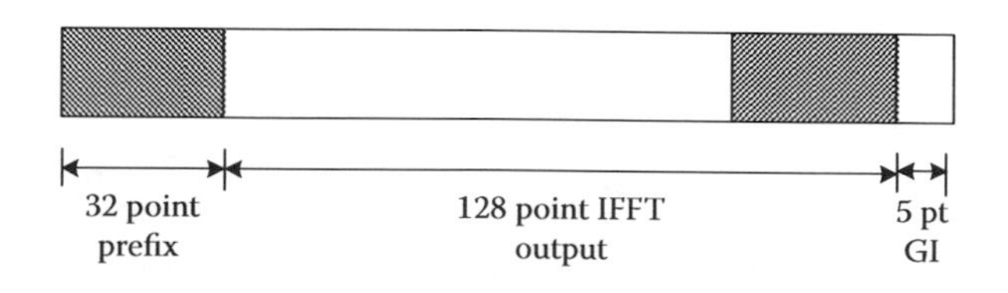

Figure 6.2 Format of an OFDM symbol.

Table 6.2 Timing-Related Parameters

Parameter	Description	Value
N_{ST}	Number of total subcarriers	122
N_{SD}	Number of data subcarriers	100
N_{SDP}	Number of pilot subcarriers	12
N_{SG}	Number of guard subcarriers	10
Δ_F	Subcarrier frequency spacing (= 528 MHz/128)	4.125 MHz
T_{FFT}	IFFT/FFT period (= $1/\Delta_F$)	242.42 ns
T_{ZP}	Zero-prefix duration (= 32/528 MHz)	60.61 ns
T_{GI}	Guard interval duration (= 5/528 MHz)	9.47 ns
T_{SYM}	Symbol period (= $T_{FFT} + T_{ZP} + T_{GI}$)	312.5 ns

The coefficients, C_n, are derived from the data, pilots, or the training sequences. Table 6.2 defines the parameters and the specific values to be used for the WiMedia PHY.

The complex baseband OFDM symbols, $r_k(t)$, are modulated onto the carrier frequency, f_k, as shown in Equation (6.2), to form the RF transmitted signal.

$$r_{RF}(t) = \mathrm{Re}\left\{\sum_{k=0}^{N-1} r_k(t - kT_{SYM}) \exp(j2\pi f_k t)\right\} \tag{6.2}$$

The carrier center frequency, f_k, of the signal is modified on each OFDM symbol and is determined by the hopping pattern for the devices that use the same band plan. Each band plan will have a unique hopping pattern (referred to as time-frequency code, or TFC) and can hop up to a maximum of three bands (each band is 528 MHz wide). The following subsection provides a detailed description of the WiMedia channelization scheme.

At the transmit side, the MAC controls the operation of the PHY through the transmit vector (TXVECTOR) parameter, which specifies the packet length, the data rate, the scrambler initialization seed identifier to be used for the payload, and the transmit power. At the receiver, the MAC communicates with the PHY through the receive vector (RXVECTOR) parameter, which includes the packet length, data rate, and received signal strength indicator (RSSI) values.

6.3.1 PLCP Frame Format

The Physical Layer Convergence Protocol (PLCP) frame consists of a a PLCP preamble, a PLCP header, and the payload portions. Figure 6.3 shows the WiMedia frame format.

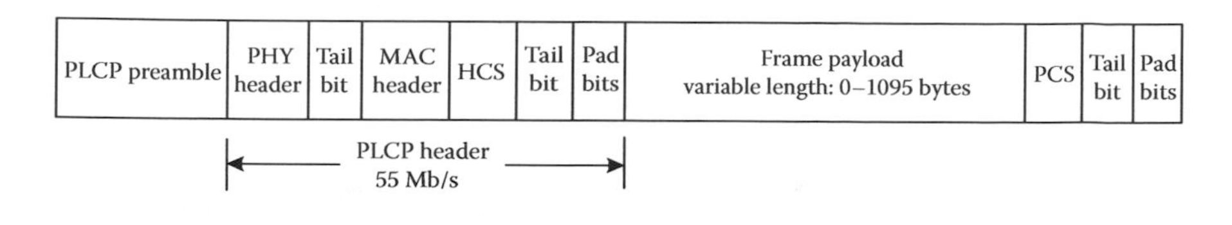

Figure 6.3 Format of a WIMEDIA frame.

The WiMedia PHY specification defines two PLCP preamble formats: standard mode and streaming mode. The standard-mode PLCP preamble is used for all the regular packet transmissions and for the first packet of a burst transmission. The duration of the preamble is 9.375 μs and it consists of 24 time-domain (TD) preamble symbols and 6 frequency-domain (FD) preamble symbols. The TD preamble consists of 18 symbols for packet synchronization and 6 symbols for frame synchronization.

The streaming-mode PLCP preamble is used for all the packets, except for the first packet, of a burst transmission. A burst transmission is a sequence of data frames, without intermediate acknowledgments, separated by a minimum interframe space (MIFS) (equal to 1.875 μs). The streaming-mode preamble consists of 12 TD preamble symbols and 6 FD preamble symbols, for a total duration of 5.625 μs. The TD preamble consists of six packet synchronization symbols and six frame synchronization symbols.

6.3.1.1 Generation of TD and FD Preamble Symbols

Each TFC has a unique TD preamble pattern associated with it. This pattern is read from a table preappended with 32 zero samples and appended with 5 GI samples to form one symbol. The symbol thus formed is repeated 18 times and 6 times for standard mode and streaming mode, respectively, to form the packet synchronization part of the TD preamble. A combination of the symbol and a 180-degree-rotated version of this symbol are repeated three times to form the frame synchronization part of the TD preamble for both the modes.

In order to generate the FD preamble, a predefined pattern is read from a table and is processed by an IFFT block. The output of the IFFT block is then preappended with 32 zero samples and appended with 5 GI samples. The symbol thus formed is transmitted six times. All the TFCs use the same FD preamble pattern. Figure 6.4 shows the format of the standard-mode and streaming-mode PLCP preambles.

The robustness of the preamble is crucial for the overall performance of the system as well as for improving the efficiency of the MAC. The WiMedia preamble has been designed to work at signal levels well below the sensitivity levels of the lowest data rate modes. In a single BP environment

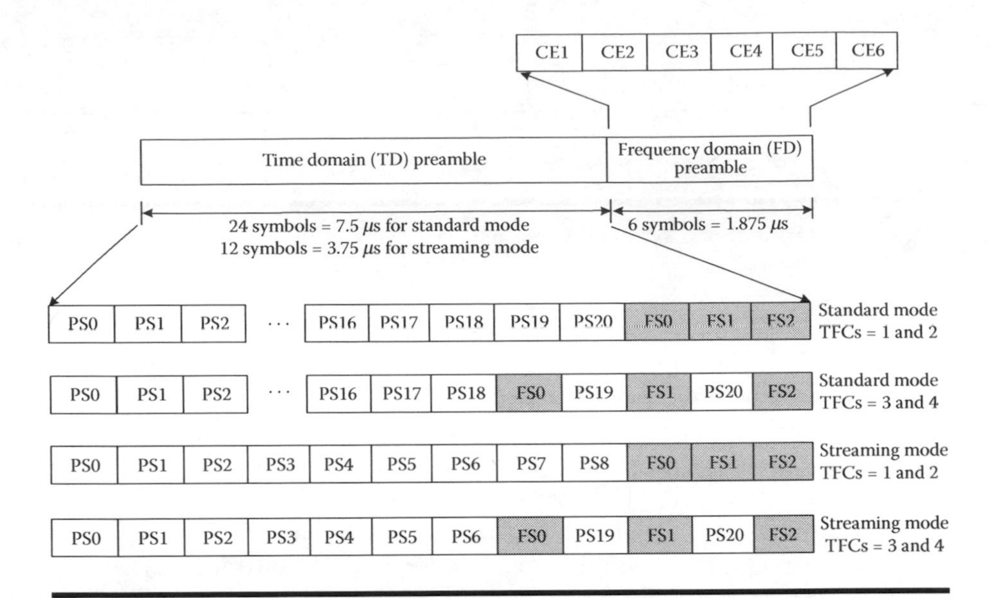

Figure 6.4 Format of the PLCP preamble for the standard and streaming modes. (The shaded symbols represent symbols with inverted polarity; PS, packet synchronization; FS, frame synchronization; CE, channel estimation symbols.)

and in typical channel conditions the probability of missed detection is less than 2.0×10^{-5} and the probability of false detection is less than 6.2×10^{-4} for the mandatory data rate modes.

The PLCP header follows the PLCP preamble. It includes the PHY header, the MAC header, a header check sequence (HCS), tail bits, and pad bits. The HCS is generated based on the PHY and MAC header fields [7] using a CCITT CRC-16 code. The PHY header includes the rate, the packet length, and the scrambler initialization seed identifier information. The PLCP header is transmitted using a 53.3-Mb/s data rate mode.

The payload portion of the PLCP frame consists of the MAC payload, the frame check sequence (FCS), the tail bits, and the pad bits. The size of the MAC payload can range from 0 to 4095 bytes. The FCS is 4 bytes in duration and is calculated on the MAC payload. The tail bits are used to bring the coder to zero state, and the pad bits are specified in order to make the total number of bits an exact multiple of the number of coded bits per symbol (NCBPS).

6.3.2 Channelization

The WiMedia PHY provides channelization through a combination of the following methods:

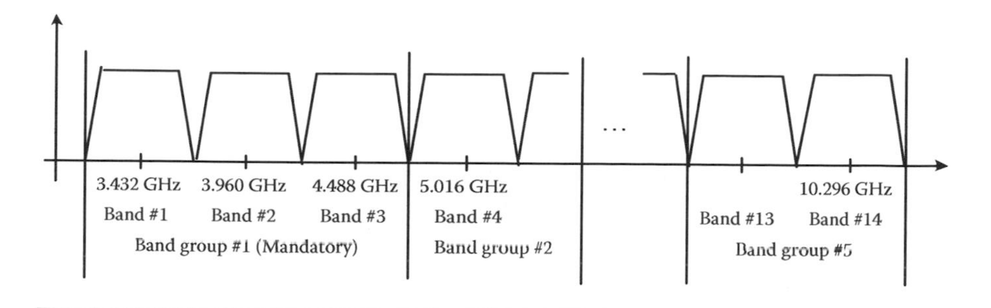

Figure 6.5 Classification of bands into band groups.

- *Band groups*: The UWB (3.1–10.6 GHz) is divided into 14 bands, each about 528 MHz wide. Further, these bands are grouped into 4 band groups of 3 bands each and 1 band group of 2 bands, as shown in Figure 6.5. Band group #1, consisting of bands 1, 2, and 3, is mandatory.
- *Time-frequency codes* (TFCs): Within each band group, further isolation among devices can be provided by assigning a unique hopping pattern or TFC to the devices that share the same band plan. Devices in other band plans may use a different TFC. There are two types of TFCs: (1) time-frequency interleaving (TFI), where the symbols are interleaved over the three subbands, and (2) fixed-frequency interleaving (FFI), where the symbols are transmitted on a single subband. Within the first four band groups, four TFI codes and three FFI codes are defined, and hence up to seven channels are supported in each band group. In the fifth band group there are two FFI codes defined. For example, devices with TFC #1 transmit in band #1 in the first symbol period, in band #2 in the second symbol period, and so on. Similarly, devices using TFC #2 (shown in Figure 6.6) transmit in band #1 in the first symbol period, in band #3 in the second symbol period, and so on. The pattern repeats every six symbols.

6.4 WiMedia PHY Performance

The performance of a system depends significantly on the propagation channel. Therefore, in order to design a reasonable/best system, a good understanding of the channel is vital. An effort to better understand the UWB channel through field measurements has been taken up by different groups [8]. From these studies, it was observed that the geometry of the location and the building architecture had a significant impact on the received signal. Another phenomenon that was observed in these studies was that the multipath components arrived in clusters.

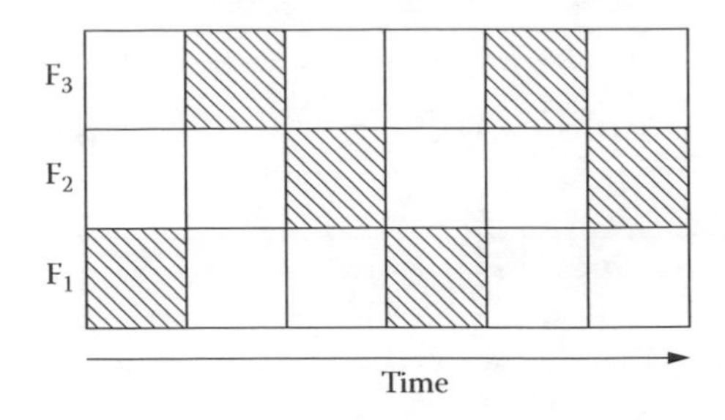

Figure 6.6 Time-frequency pattern for TFC #2. The devices use the same TFC for beacon transmission as well as payload transmission. Because the devices do not have prior knowledge of the hopping pattern, they have to scan all the channels (band groups and TFCs) at start-up and whenever they join a beacon period to determine the TFC.

For the multipath channel, the subcommittee has recommended the use of the Saleh–Valenzuela (S–V) [9] model with a couple of modifications. A lognormal distribution is used instead of a Rayleigh distribution to generate the gain magnitudes. In addition, independent fading is assumed for each cluster as well as each ray within the cluster. A more detailed description of the channel model is available [8,9].

The following key parameters define the channel model: the cluster arrival rate (Λ), the ray arrival rate (λ), the cluster decay factor (Γ), the ray decay factor (γ), standard deviation of the cluster lognormal fading term (σ_1), the standard deviation of the ray lognormal fading term (σ_2), and the standard deviation of the lognormal shadowing term for the total multipath realization (σ_x). These parameters are derived based on channel characteristics, such as mean excess delay, RMS delay spread, and the number of significant (within 10 dB of the peak) multipath components. Figure 6.7 shows example multipath realizations using the model specified with line of sight (LOS), or channel model 1 (CM1), and no line of sight (NLOS), or channel model 2 (CM2).

OFDM systems have very good energy collection properties. Further, the OFDM systems minimize ISI effects by using a cyclic prefix or zero prefix. A simple one-tap equalizer can be used to compensate for the channel effects.

The performance of the PHY in the presence of white noise can be evaluated by analyzing the transfer function of the convolutional code [10,11]. For QPSK modulation, an upper bound on the first-event error probability, P_e, and on the bit-error probability, P_b, is given by the following equations:

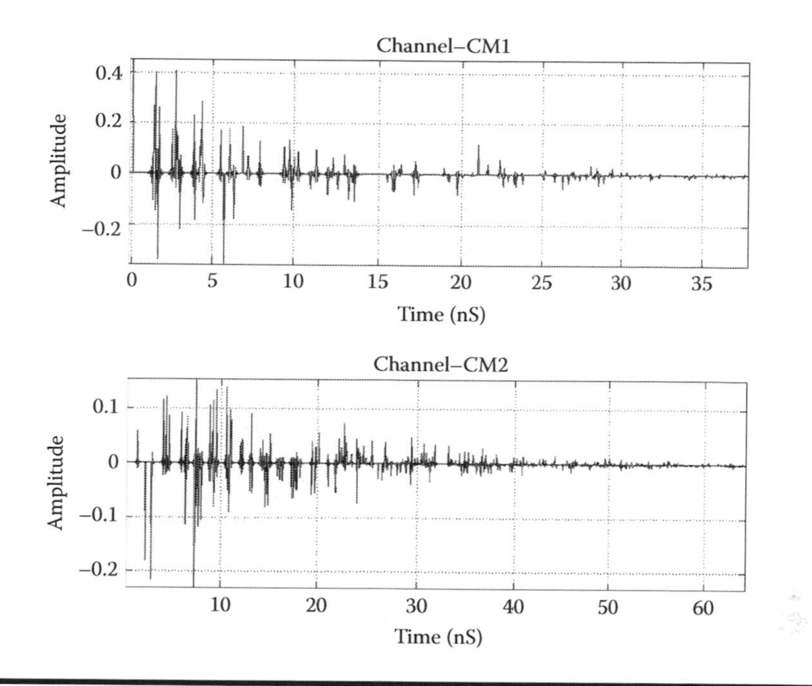

Figure 6.7 One realization each of channel models CM1 (LOS) and CM2 (NLOS).

$$P_e \leq \sum_{j=d_{free}}^{\infty} a_j Q\left(\sqrt{2jRE_b/N_o}\right) \tag{6.3}$$

$$P_e \leq \sum_{j=d_{free}}^{\infty} c_j Q\left(\sqrt{2jRE_b/N_o}\right) \tag{6.4}$$

where E_b/N_o is the energy per bit to noise density, R is the code rate, and $Q(x)$ is the complementary distribution function

$$Q(x) = \int_{x}^{\infty} \frac{1}{\sqrt{2\pi}} \exp\left(-\frac{1}{2}t^2\right) dt \tag{6.5}$$

In these expressions, d_{free} is the free distance of the code, a_j is the number of incorrect paths of distance j, and c_j is the number of errors in all the incorrect paths of distance j. The first few terms in the weight spectrum of the rate-1/3 convolutional code and the different puncturing patterns defined in the WiMedia specification are given by Ojard [11]. Figure 6.8

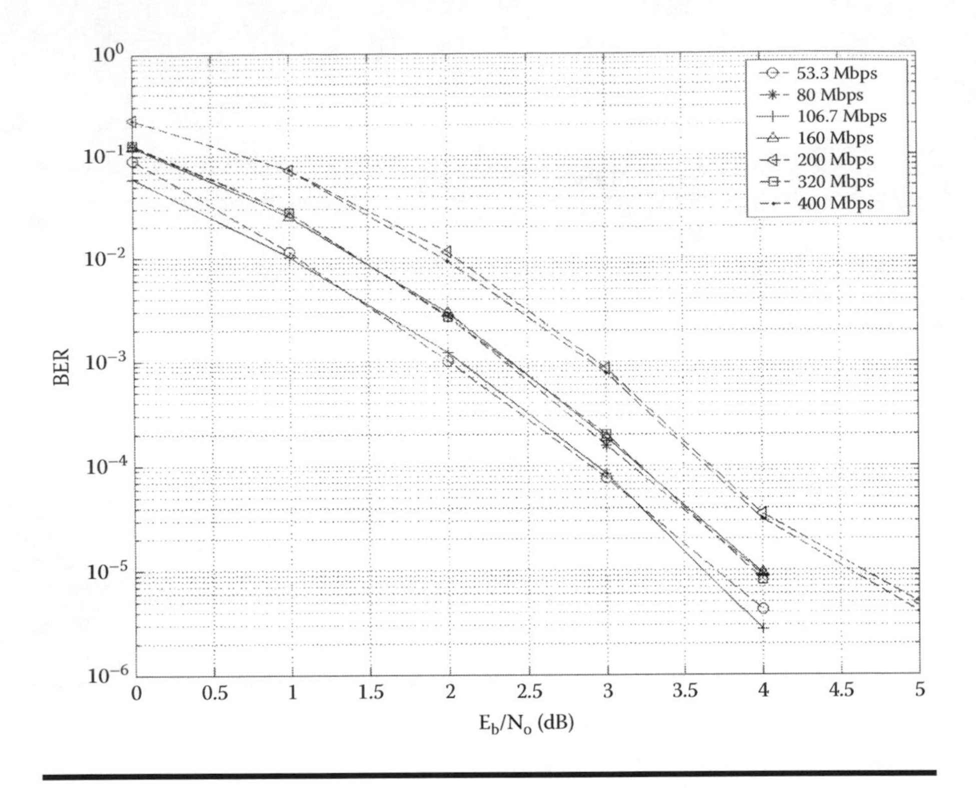

Figure 6.8 BER (bit error rate) vs. E_b/N_o for the different coding rates of WiMedia PHY.

shows the performance of some of the data rate modes of the WiMedia system in an AWGN (additive white Gaussian noise) channel.

6.5 Media Access Control Standard

This chapter provides an overview of WiMedia's MAC standard, which has been published as the ECMA-368 standard. The standard is mainly described from a functional perspective in order to give insights into the innovative protocols, which have been conceived by the authors of the specification.

6.5.1 *Overview*

The WiMedia MAC protocol is decentralized. The main reason for this design choice is that most of the application scenarios of the standard imply peer-to-peer communication of devices. In other words, no base

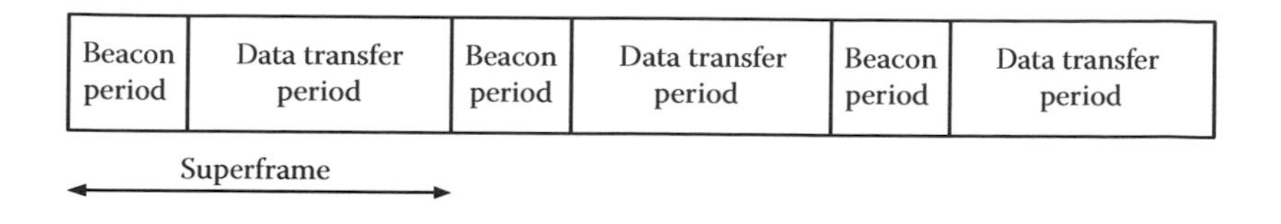

Figure 6.9 Superframe structure.

station or access point is needed in order to establish the communication. Two communicating devices have equal rights and responsibilities. This avoids complicated selection mechanisms of master devices and coexistence management of the clusters of multiple masters.

Time is structured in periodic "superframes." As shown in Figure 6.9, a superframe is divided into a beacon period (BP) and a data transfer period (DTP).

In contrast to wireless local area networks (WLANs), according to the standard IEEE 802.11, where only the access point transmits beacon frames, every WiMedia device transmits its own beacon frame in the beacon period. The rules for beacon transmission and protocols to avoid and resolve beacon collisions are described in Section 6.5.2.

During the data transfer period the actual (application) data transmission takes place. Two different media access mechanisms are combined in the WiMedia MAC protocol: distributed reservation and prioritized contention. Despite the fact that the MAC protocol is decentralized, reservation of transmit capacity is possible. This is achieved by the Distributed Reservation Protocol (DRP), which is presented in Section 6.5.3.1. All capacity that has not been reserved by any device can be accessed in contention with other devices, ruled by the Prioritized Contention Access (PCA) protocol. Section 6.5.3.2 contains a brief overview of PCA.

Fragmentation and reassembly of data frames as well as frame aggregation and corresponding acknowledgment schemes are described in Section 5.3.3. An overview of the power management is given in Section 6.5.4 and Section 6.5.5 contains a brief description of the security mechanisms.

6.5.2 Beacon Period

The beacon period (BP) solves various issues that arise in a distributed communications environment, such as

- Mutual device discovery
- Frame synchronization
- Distributed reservation of transmit capacity
- Power management

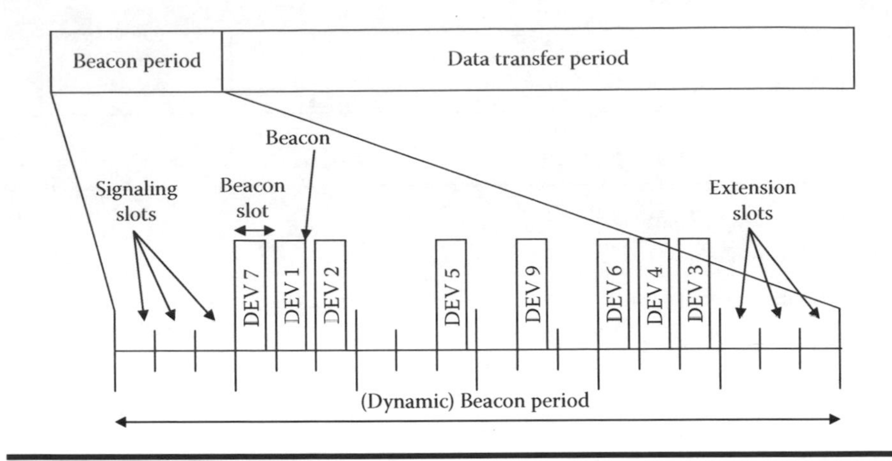

Figure 6.10 Beacon period structure.

The BP is slotted into beacon slots, as shown in Figure 6.10. Each beacon slot has a constant duration, during which a device can transmit its beacon with variable length at 55 Mbps. The maximum length of a beacon is obviously limited by the size of a beacon slot (minus a so-called short interframe space [SIFS] and minus a guard time).

The BP contains a variable number of beacon slots, depending on the number of devices in the system, but cannot exceed a certain maximum length. The beginning of the beacon period is marked by the first beacon that is transmitted in the beacon period. Some (normally empty) signaling slots are located before this first transmitted beacon. The end of the beacon period can be deduced from a field in every beacon that is specifying the length of the beacon period. Among other information, a beacon contains an identifier of the device that transmits the beacon, the number of the beacon slot in which the beacon is transmitted, as well as several information elements (IEs). Different IEs are defined for various purposes. Three of these IEs will play a role in the protocols described in the following sections.

6.5.2.1 Beacon Collision Resolution Protocol

The Beacon Collision Resolution Protocol (BCRP) defines the basic mechanisms for how beacon collisions are detected and resolved.

When powered up, a device first scans the medium for beacons that are transmitted in existing, beacon periods. If no other beacon is received (i.e., the device is the first in the network), it starts to transmit its own beacon and thereby defines a new beacon period. If, on the other hand, an existing beacon period is detected, the device chooses a free beacon slot within this period and starts to transmit its beacon in that slot. This

Octets: 1	1	1	K	2	...	2
Element ID	Length (= 1 + K + 2 × N)	BP length	Beacon slot info bitmap	DevAddr 1	...	DevAddr N

Figure 6.11 Fields of the beacon period occupancy information element (BPOIE) [1].

is where beacon collisions may occur. Two devices may have started up in the same superframe and randomly chosen the same beacon slot for transmission. Another reason for a collision may be a so-called hidden-station scenario, in which a receiving device is detecting a beacon collision but the two transmitting devices are not in range of each other and consider the respective beacon slot as free.

In order to detect such beacon collisions, every beacon includes something called the beacon period occupancy information element (BPOIE), which reflects the occupancy of beacon slots in the previous superframe. The fields of the BPOIE are shown in Figure 6.11.

The beacon slot info bitmap (from Figure 6.11) indicates with 2 bits for every beacon slot whether the respective slot is occupied and whether the beacon is movable. The movability of a beacon will be explained in Section 6.5.3. If the slot is occupied, the device address (DevAddr in Figure 6.11) of the occupying device is also included in the BPOIE (in ascending beacon slot order). A device determines whether a beacon slot is occupied and by whom by listening to the beacon transmissions of all devices in reception range. Based on this information, the device composes the BPOIE and includes it in its own beacon.

The BPOIE plays a key role in detecting beacon collisions. A device uses the BPOIEs that are included in the beacons of the neighboring devices to determine whether its own beacon was correctly received by these neighbors. If its own beacon slot is signaled as occupied and its own address is listed in the BPOIE, the beacon has been received correctly. However, if there is one device that is signaling the respective beacon slot as occupied but associates a different device address with it, a beacon collision has occurred. In a similar way, the reception of a BPOIE in a certain number of consecutive superframes, in which the broadcast address is associated with the respective slot, also signals a beacon collision.

In addition to this mechanism, a device aperiodically skips the transmission of its own beacon and listens to other transmissions in the respective beacon slot. If a beacon from another device is received in that slot or if the beacon slot is reported as occupied in the following BPOIE of any neighbor, another device must be active in the same slot and a beacon collision is detected.

If a device has detected a beacon collision by any of the possibilities described above, it has to switch its beacon transmission to a beacon slot

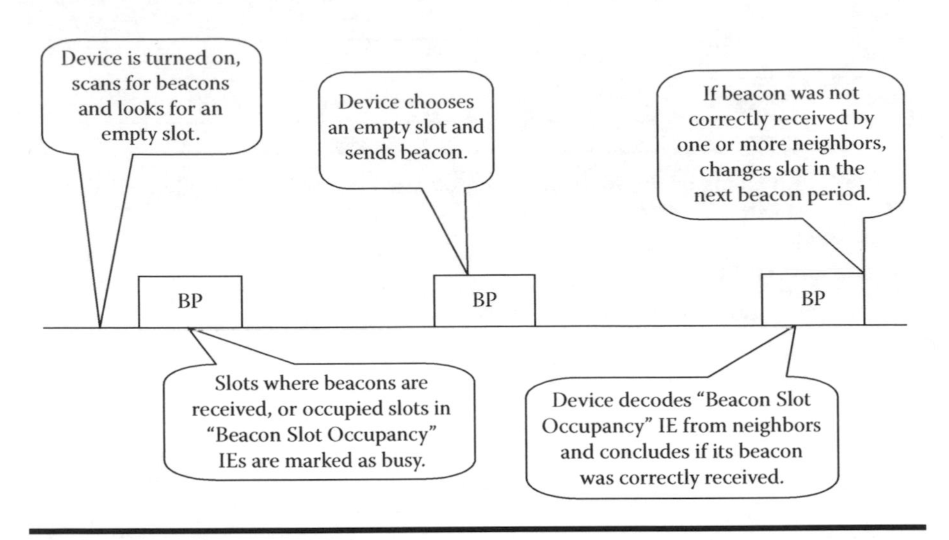

Figure 6.12 Beacon collision resolution protocol.

that has been unoccupied for at least a certain number of superframes. The procedure is illustrated in Figure 6.12.

6.5.2.2 Beacon Period Extension and Contraction

For the purpose of beacon period extension, a certain number of slots at the beginning of the beacon period are reserved and called signaling slots. These beacon slots can be used by a device that would have chosen a beacon slot that exceeded the beacon period length of a neighboring device (because it did not find any other free slot). The beacon period length of the neighbor can be depicted from a field in the neighbor's beacon. The device has to transmit its beacon in the signaling slot until the respective neighbor has extended its beacon period length and all other neighbors also signal a sufficiently large beacon period. In this way the transmission of a beacon in a signaling slot has resulted in an extension of the beacon period. Similarly, a certain number of beacon slots at the end of the beacon period are reserved for devices having to switch their beacon slot as well as for devices joining the network. These slots are referred to as extension slots. A device that has transmitted its beacon in a signaling slot until the beacon period length has been extended will also select an extension slot for the permanent beacon allocation.

The fact that joining devices can only select a slot among the extension slots would result in a frequent extension and thereby ever-growing beacon period. Furthermore, devices that were switched off or moving outside of the network would leave "holes" in the beacon period, which would result in a very inefficient use of the beacon period after some time. In order to

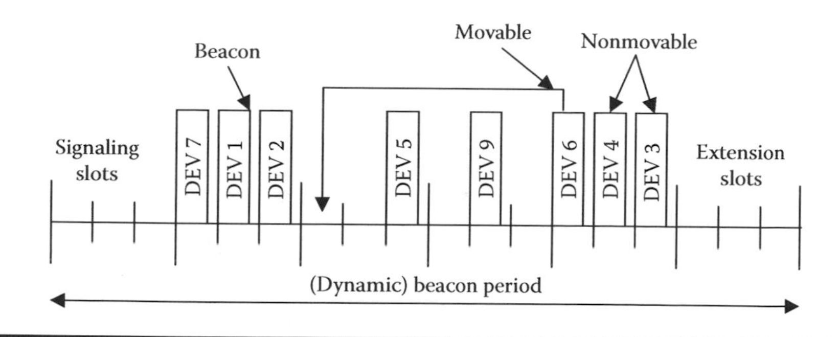

Figure 6.13 Process of beacon contraction and movability concept.

avoid such a growth and inefficient use of the beacon period, a permanent and automatic beacon contraction is foreseen. If a hole appears in the beacon period (i.e., a beacon slot has become available), the last movable beacon within the beacon period is moved to the available slot. The concept of movability has been foreseen in order to allow certain devices, such as devices in hibernation mode, not to move their beacon into a free slot. The process is illustrated in Figure 6.13.

6.5.2.3 Merging of Multiple Beacon Periods

Due to user mobility or changing propagation conditions, it may happen that groups of devices come into range of others that have formed separate networks beforehand. In such a case either merging or coexistence of such networks could be attempted. Which strategy should be preferred depends, e.g., on the speed of the users. In the case of networks that are crossing each other for a short period of time, a coexistence strategy may be the best solution, whereas in a situation of longer colocation, a merger would probably be the better solution. In the WiMedia standard, the issue has been addressed in the following way. Two scenarios are distinguished: one in which the beacon periods of two networks are overlapping and one in which the beacon period of one network falls fully into the data transfer period of the other network.

In the first scenario, a merger is always carried out immediately. The beacon period with the earlier beacon period start time (BPST) is maintained and the beacons of the other beacon period are relocated into the maintained period. The length of the maintained beacon period has to be increased to at least the sum of the occupied beacon slots of both periods. For this purpose, a device that has to relocate its beacon because it has detected a beacon period with an earlier BPST adjusts its beacon slot number to the sum of its current slot number plus the number of the highest occupied beacon slot in the other period plus one minus the number of signaling slots in its old period. Alternatively, a device may also join the maintained beacon period as if it was joining the network for the first time, according to the corresponding rules.

In the second scenario, of nonoverlapping beacon periods, a merger of the networks is started after a certain waiting time. The waiting time is mainly used to determine which beacon period is maintained and which beacon period is given up. This is achieved by assuming a waiting time of mBPMergeWaitTime if the BPST of the alien beacon period falls within the first half of its own superframe and a waiting time of 1.5 × mBPMergeWaitTime if the alien BPST falls within the second half of the superframe. This basically means that one network assumes a shorter waiting time than the other, and it will be the one that gives up its beacon period and relocates its beacons into the beacon period of the other network.

During the waiting time, the two networks coexist by protecting each other's beacon period. For this purpose a special type of medium reservation has been defined for the data transfer period, which is called alien beacon period reservation. Every device that has detected an alien beacon period includes an alien beacon period reservation in its beacon, which covers at least the entire duration of the alien beacon period and prevents devices of the network from transmitting and interfering with the other beacons.

If a device has started the waiting time for relocating its beacon, it includes a so-called beacon period switch information element (BP switch IE) into its beacon to announce the impending relocation of its beacon. The BP switch IE is received by neighboring devices in the alien network, which stop their own relocation process if their remaining waiting time is longer than the waiting time (BP move countdown) that is announced in the respective BP switch IE. In this way, the devices in the network with the shorter waiting time will be the only ones that relocate their beacons.

The BP switch IE also contains information on the planned offset of the BPST, which is the time difference between the starting times of the two beacon periods, as well as information on the planned beacon slot offset, which is set to the number of the highest occupied beacon slot in the other period plus one minus the number of signaling slots in the old period.

In a typical case of networks coming into range of each other, not all devices in the two networks receive each other. Usually the devices located closest to the other network will become aware of the other network first, as shown in Figure 6.14.

For proper network operation, all devices in the same network have to share the same beacon period (even though beacon slots may be reused after two radio hops). This means that if some devices start to relocate their beacons into a new beacon period, all other devices in the network have to do the same. Devices that are not receiving the alien beacon period will nevertheless become aware of it and of the planned beacon relocation by the BP switch IE of their neighbors. They will consequently also start

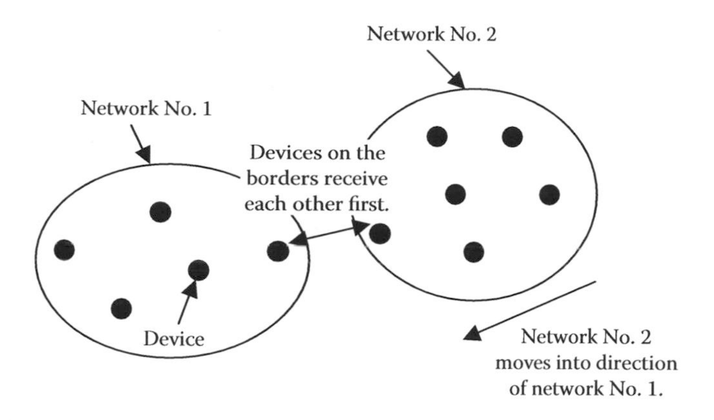

Figure 6.14 Example scenario for the beginning of a beacon period collision.

to relocate their beacon and basically copy the information from the BP switch IE of their neighbors, except for the beacon slot number, which will be chosen according to its own beacon slot position.

6.5.3 Data Transfer Period

As mentioned before, the data transfer period is divided into media access slots (MASs), which can be accessed by two different media access protocols.

6.5.3.1 Distributed Reservation Protocol (DRP)

One of the two access mechanisms is the DRP. This protocol allows a device to exclusively reserve a certain time in the future (in the form of a certain number of MASs) for communication with one or several other devices. Because time is divided into superframe periods and many applications require a bandwidth guarantee over the duration of many superframes, a DRP reservation has to be issued only once and applies to the current as well as all following superframes until its termination. The key to the reservation process is its decentralized character: devices involved in the planned frame exchange determine themselves at which position in the superframe a reservation is possible and which is the most suitable one. For this purpose any reservation has to be announced to all neighbors of the involved devices. It is quite straightforward that reservations are announced in the beacon frames of the devices. The neighbors consequently mark the respective MASs internally as occupied and can deduce free spots for their own reservations. In order to make this decentralized process work, it is absolutely necessary that a successfully established reservation is respected by all neighbors, which means that neighbors do not announce reservations for the respective periods and also do not access the medium at the start of the reserved period of time.

Five different types of reservations are distinguished:

- *Hard*, to reserve the medium exclusively
- *Soft*, to have first access to the medium
- *Alien Beacon Period*, to reserve the medium for a detected alien BP
- *Private*, to reserve the medium exclusively for higher layer protocols
- *PCA*, to reserve time for access according to the PCA method (see Section 6.5.3.2)

Explicit DRP Negotiation A reservation has to be negotiated between the devices that want to exchange information. The device that is planning to initiate the frame exchange during the reservation is also the device that initiates the negotiation process. It is called the reservation owner, whereas the other involved devices are called the reservation targets. Two alternatives for negotiating a reservation are foreseen: an explicit and an implicit negotiation.

In the explicit negotiation process, command frames are exchanged between reservation owner and target during the data period of a superframe, as illustrated in Figure 6.15.

The reservation owner initiates the process by sending a DRP reservation request frame. A reservation target evaluates whether the medium is free on its side during the proposed MASs; e.g., by checking locally stored information on already established reservations of its neighbors. The reservation target answers with a DRP reservation response frame,

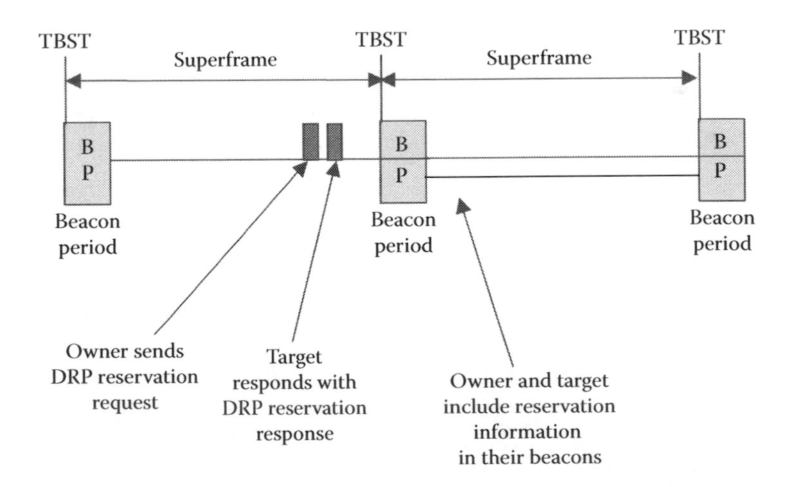

Figure 6.15 Overview of DRP operation.

in which a reservation status bit is set to 1, if the reservation proposal is accepted, and to 0 otherwise. A reason code field in the DRP reservation response is used to give the reservation owner more information on the reasons for the response and can be set to values for "conflict," "denied," or "pending."

In case of a conflict with existing reservations, the reservation target includes a DRP availability IE in the DRP reservation response frame, which gives an overview of the slot occupancy by the reservation target and its neighbors. The reservation owner can use this information to make a new, modified proposal for the reservation that is better suited for the reservation target. In the case of multiple reservation targets (i.e., a multicast reservation), the targets will always include a DRP availability IE in their response, in order to allow the reservation owner to derive the best, potentially modified proposal that would suit a maximum number of reservation targets.

Once the negotiation is completed successfully, reservation owner and targets include a DRP IE in their beacon. As explained before, this is necessary in order to inform all neighbors of owner and targets about the established reservation. Owner and targets have to include the DRP IE in the beacon of every superframe until the respective reservation is terminated.

Implicit DRP Negotiation The implicit negotiation is an alternative to the explicit negotiation and does not require the exchange of command frames in order to establish a reservation. The reservation owner will just include a DRP IE in its beacon, in which the device addresses of the potential reservation targets are included. The reservation targets will notice the attempt of the owner to establish a reservation, because every device that is supporting DRP has to scan all received beacons for its device address present in a DRP IE. A reservation target will respond by including a DRP IE in its own beacon. Every DRP IE includes a reservation status bit as well as a reason code. To signal the acceptance of the DRP proposal, the reservation target sets the reason code in its respective DRP IE to "accepted." When accepting a unicast reservation the reservation status bit is set to 1, whereas in the case of a multicast reservation the reservation status bit is set to the same value as in the DRP IE of the reservation owner.

If the reservation is not granted, the reservation target sets the reservation status bit to 0. In this case the reason code may be set to "conflict," "denied," or "pending." If the reason code is "conflict" or "pending," the reservation target includes a "DRP availability IE in its beacon (apart from the DRP IE) in order to signal alternative MASs to the reservation owner. This is especially important in case of a multicast reservation, because it may be otherwise difficult to find a set of MASs that fits all members of the multicast group.

Upon receipt of a DRP IE with reason code "Accepted," the reservation owner will also set the reservation status bit of its own DRP IE to 1 and the reason code to "Accepted." The reason code and reservation status fields are present in every DRP IE and, in the case of an explicit DRP negotiation, where the fields are set in a corresponding manner.

For an established reservation, the "reservation code" may also be set to "Modified." This is used in the case that the reservation owner wants to decrease or increase the number of MASs in the reservation. A removal of MASs does not require a new negotiation, whereas an inclusion of additional MASs requires a negotiation with an additional DRP IE. After successful negotiation, existing and new DRP IEs are combined.

Independent of whether a reservation has been established with the explicit or implicit negotiation process, a reservation owner can terminate any reservation by just removing the respective DRP IE from its beacon. A reservation target can initiate a termination by setting the reservation status bit to 0 and the reason code to a corresponding value. Upon receipt, the reservation owner will terminate the reservation by removing the DRP IE, which will be followed by a DRP IE removal of the reservation targets.

Resolution of Reservation Conflicts A reservation conflict occurs if two groups of reservation owners and targets have reserved the same or an overlapping set of MASs. This may happen because of a parallel reservation negotiation, because of mobility, or because of changing propagation conditions.

The resolution of such a conflict depends on the type of the reservation. The type of the reservation is announced within the DRP IE of the involved devices.

A reservation for an alien beacon period always has highest priority and the conflicting reservation has to be changed. This has been defined because otherwise the entire communication in the alien network could be heavily disturbed by the conflicting reservation. For all other reservation types the resolution is guided by the values of the reservation status bit, a so-called conflict tie-breaker bit, and the beacon slot number. The conflict tie-breaker bit is set randomly.

Different combinations of reservation status and conflict tie-breaker bit in the conflicting reservations may occur.

In the case of two different values of the reservation status bit, the reservation with the status bit set to 1 will prevail and the other reservation has to be changed. The rationale is that the reservation with the status set to 1 has already completed the negotiation process, whereas the other reservation has not.

If the reservation status bits of the conflicting reservations have the same value, the combination of tie-breaker bit and beacon slot number is used to resolve the conflict. If the tie-breaker bits have the same value, the device with the lower beacon slot number can maintain its reservation, whereas in the case of different tie-breaker values, the device with the higher beacon slot number prevails. The other device has to stop the negotiation process (in case the status bit has not yet been set to 1) or remove the conflicting MASs from the reservation (in case of an already established reservation with status bit set to 1). Before restarting the reservation negotiation with different MASs or before reserving additional MASs, the device has to carry out a backoff process, which results in a random waiting time of a multiple of superframes. This is useful in order to avoid repeating reservation conflicts.

Releasing Unused Reservations The difference between hard and soft reservations is that no other devices can access the medium during a hard reservation, whereas MASs of a soft reservation can be accessed in PCA mode (cf. Section 6.5.3.2). In a soft reservation, the reservation owner is just given the highest priority arbitrary interframe space (AIFS) and the right not to perform a backoff. If the reservation owner does not use its right for first medium access, other devices will use the MASs during the reservation.

This is different in the case of a hard reservation, where the other devices are not allowed to access the medium and the MASs would be consequently wasted if the owner did not use them. This is why a reservation owner may explicitly grant access to unused parts of a reservation by broadcasting an unused DRP announcement (UDA) control frame. The UDA frame consists of a list of devices that should respond to the UDA frame with an unused DRP response (UDR) in the order in which the devices are listed in the UDA frame. The transmission of UDR frames by the neighbors of the reservation owner is required in order to also inform the devices that are two radio hops away from the reservation owner about the freed MASs. The MASs are considered accessible in PCA mode after the projected end of the last UDR frame.

6.5.3.2 Prioritized Contention Access

The PCA protocol originates from IEEE 802.11e, where it is called Enhanced Distributed Channel Access (EDCA) [12]. The underlying access scheme is Carrier Sense Multiple Access with Collision Avoidance (CSMA/CA) with a prioritized backoff procedure.

Two key characteristics of CSMA/CA are the "listen before talk" concept as well as a backoff procedure after a frame collision. "Listen before talk" implies that every device has to sense the medium for ongoing transmissions before accessing it. Only if the medium is idle can an access attempt be started. The duration of sensing the medium is fixed and given by the standard as AIFS. Four AIFSs are distinguished, depending on the priority of the frames that are to be transmitted:

- Background (AC_BK)
- Best effort (AC_BE)
- Video (AC_VI)
- Voice (AC_VO)

With increasing priority from background to voice, the AIFSs are of shorter duration. This has the effect that a device with data frames of higher priority to transmit will have a higher chance of accessing the medium earlier than a device with lower priority frames.

The reason the order of transmission is not deterministic (even between higher and lower priority frames) is that every device has to sense the medium for an additional random backoff time. This random time is a multiple of a fixed slot time, where the multiple is a random number drawn from a uniformly distributed interval of (0, CW).

Without the additional random time, all devices with pending data frames of the same priority would access the medium at the same time and the respective frames would collide on the medium. The random sensing time avoids this situation as far as possible, even though it may happen that two or more devices have randomly chosen the same number of backoff slots. The frames of these devices will therefore collide on the medium. The resolution of such packet collisions is achieved by means of the backoff process.

The backoff process foresees that devices involved in a collision will draw a new random number from an interval (0, CW), after having doubled the value CW (after each collision) up to a maximum value CW_{max}. This is why the process is called exponential backoff. CW_{min} is the initial value of CW. The duration of CW_{min} depends on the priority of the pending frames. This is a second way of prioritizing different types of traffic (besides the different AIFS values).

Whenever a device senses the channel as idle, it decreases an internal slot counter by one. If the slot counter reaches zero, the device transmits one or several pending data frames, as illustrated in Figure 6.16, in a so-called transmission opportunity (TXOP). A TXOP has a maximum duration (to ensure fairness between all devices), which is dependent on the priority of the data frames. This is a third and final mechanism to prioritize different traffic classes.

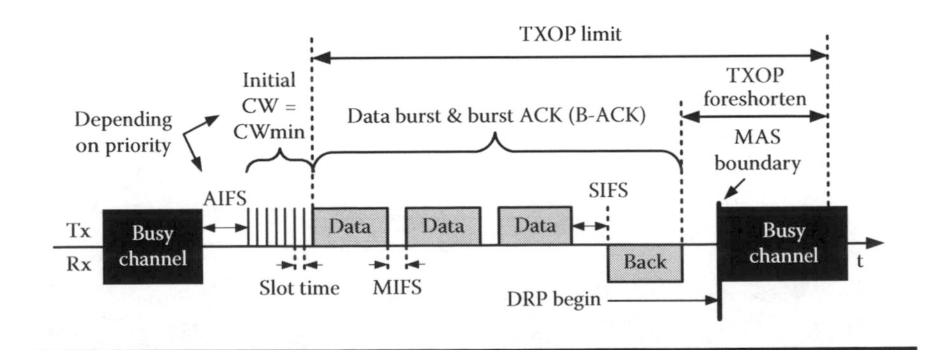

Figure 6.16 PCA mechanism.

One or several data frames are acknowledged by the receiver with an ACK frame (see Section 6.5.3.3). The sender will recognize a packet collision by a missing acknowledgment.

If the station senses the channel as busy, it freezes its slot counter. After the channel is sensed as idle for an AIFS period again, the backoff entity starts to count down the remaining slots.

There are a few differences between the point coordination function (PCF) of the WiMedia standard and the EDCA of the IEEE 802.11 standard. An important difference is that the data phase is slotted into MASs and that only those MASs can be accessed by the PCFs that have not been blocked by DRP reservations of type hard, private, or alien. As described in Section 6.5.3, a soft DRP reservation may be accessed by devices in the PCF mode with the special rule that the DRP owner can access the medium after the shortest waiting time and without additional random backoff time.

Because of the DRP and the beacon period, a TXOP may have to be shortened to a value shorter than the TXOPLimit (as illustrated in Figure 6.16), in order not to interfere with the beacon period or a DRP reservation.

The carrier sensing for determining whether the medium is idle or busy is more challenging with the UWB physical layer of WiMedia than with the IEEE 802.11 PHY because of the low spectral density, rsp. transmit power of UWB frames. Because it is difficult to distinguish ongoing transmissions from background noise, the carrier sensing of WiMedia mainly relies on detecting the preamble of a transmitted frame (which contains known signals). This mechanism is sometimes also referred to as Preamble Sense Multiple Access (PSMA).

With PSMA, the concept of network allocation vectors (NAVs) is of special importance. Upon reception of any frame, a device locally sets an NAV based on the so-called duration field in the header of the frame. The duration field is set by the sender of the frame to a value that covers the entire frame exchange sequence (e.g., until the expected end of the acknowledgment by the receiver). If an NAV is (internally) set, a device considers the medium as busy. In other words, it is not absolutely necessary that devices can decode the entire frames of neighboring devices, but just the duration fields in the frame headers.

6.5.3.3 Fragmentation, Reassembly, Aggregation, and Acknowledgment Policies

Data frames on the MAC level, called MAC service data units (MSDUs), can be split by the sending device into several fragments, called MAC protocol data units (MPDUs). Fragmentation may become necessary if the remaining time within a reservation would not be sufficient to transmit the entire frame.

Each MPDU can be identified by the receiver by its MSDU number and a fragment number that are included in the header of the frame. This allows the receiver to reassemble all fragments once all fragments of an MSDU have been received. It is the responsibility of the receiver to arrange the fragments back in the right order of increasing fragment numbers. The order may have been changed during the transmission of the fragments if one or several fragments had to be retransmitted by the sending device.

The inverse operation to the splitting up of data frames, the aggregation of frames, is also possible. The sending device may aggregate multiple MSDUs that are intended for the same receiver(s) as payload of a single frame. Frame aggregation is useful to increase the data throughput of the system (see Section 6.5).

Three different types of acknowledgment (ACK) policies are possible, which are set in the header of the respective frame:

- No ACK
- Immediate ACK
- Block ACK

Frames transmitted under the no-ACK policy are not acknowledged by the receiver. This policy may be applied to streaming applications, where retransmissions would violate delay boundaries.

The immediate-ACK (Imm-ACK) policy is the classical acknowledgment policy, where every MPDU is acknowledged by the receiver. However, this acknowledgment scheme is quite inefficient, because it results in transceiver turnaround and media access times between consecutive frames (in addition to the acknowledgment itself).

This is why a block-ACK (B-ACK) policy has been defined. With this acknowledgment scheme it is possible to transmit a block of several MPDUs that do not have to be acknowledged by the receiver individually but with a single B-ACK for the entire block. This is obviously much more efficient than the Immediate ACK scheme (see Figure 6.18a).

The maximum number and size of MPDUs in a block is negotiated between sender and receiver (depending on the buffering capabilities of the sender and receiver(s)). The negotiation is part of the data transfer itself; i.e., the sender sets the ACK policy field in the header of the first transmitted frame to B-ACK request, which is answered by the receiver within a so-called SIFS time, with a B-ACK frame including as payload the buffer size available for the next B-ACK sequence. If the payload is empty, the receiver signals that it rejects the B-ACK request. In all succeeding frame exchange sequences, the ACK policy field of the data frames is set to B-ACK, except for the last frame of the block, in which the ACK policy field is set to B-ACK request (to negotiate the next sequence).

The acknowledgment mechanism is of the type selective repeat automatic repeat request (SR-ARQ). This means that the sender does not have to retransmit an entire block if one or several frames within the block have not been correctly received, but just those incorrect frames. For this purpose the B-ACK contains a bitmap to signal correct and incorrect frames. An acknowledgment window (specified by a sequence control field and the frame bitmap field) specifies (as usual in ARQ protocols) which frames are currently being acknowledged and buffered for eventual retransmission.

6.5.4 Power Management

Devices can be in two different types of modes: active mode or hibernation mode. Within the active mode two substates are distinguished: awake and sleep.

The hibernation mode was conceived to allow devices to save (battery) power. In the hibernation mode devices can fall asleep for more than one superframe in a row without waking up for the intermediate beacon phases (in contrast to the sleep state in the active mode). For going into hibernation, a device includes a hibernation IE (Figure 6.17) in its beacon.

The hibernation IE contains a hibernation countdown and a hibernation duration field, as shown in Figure 6.17. The hibernation countdown value signals how many superframes the device will enter into hibernation. The hibernation duration signals how many superframes the device will leave in hibernation mode.

Neighboring devices that receive the hibernation IE will store this information and will not attempt any data transmissions directed to the hibernating device during the hibernation phase (except for multicast traffic with additional receivers). All DRP connections of the hibernating device are considered terminated.

Even though the hibernating device stops beaconing during the hibernation phase, the neighboring devices will still mark the respective beacon slot as busy and include the device address of the hibernating device in the BPOIE of its own beacon. This is to prevent other devices from occupying the beacon position of the hibernating device.

Devices in active mode may also go into a sleep state, but only within the data transfer period of a superframe. However, any device has to be awake prior to DRP reservations in which it is involved as a sender or

Octets: 1	1	1	1
Element ID	Length (=2)	Hibernation countdown	Hibernation duration

Figure 6.17 Structure of hibernation information element [1].

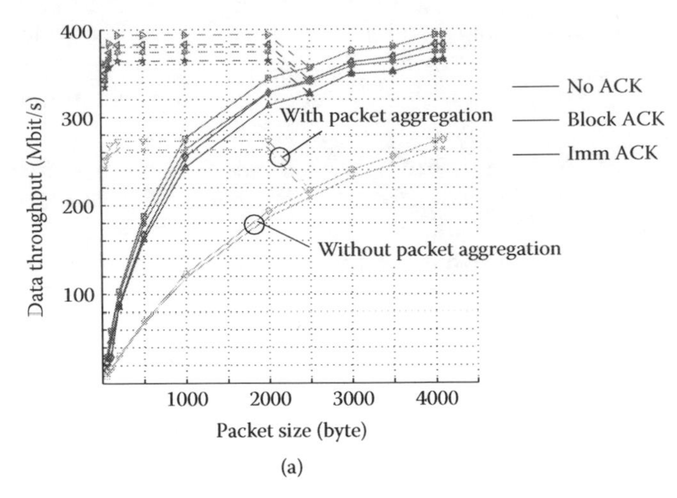

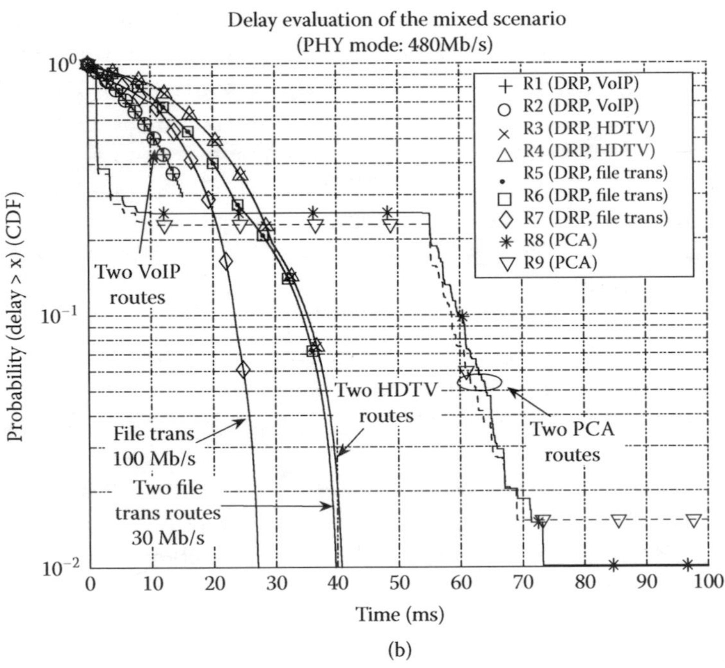

Figure 6.18 (a) Maximum data throughput between two devices [14]. (b) Delay evaluation for seven DRP routes and two PCA routes on a 480-Mb/s channel.

receiver of frames. The same holds true for communication in PCA mode. For the latter purpose, a device with pending PCA traffic can include a traffic indication map IE (TIM IE) in its beacon, with the device addresses of the receivers. This signals to the respective receivers to stay awake during this superframe.

6.5.5 *Security*

The WiMedia standard foresees three different modes of operation, which will allow communication without any security protection, with and without protection, or with security protection only. The security mechanisms that are provided in the two latter modes are mainly based on the advanced encryption standard (AES), specified in Federal Information Processing Standards Publication 197 [13].

The different security features are

- Authentication
- Privacy/encryption
- Integrity
- Replay attack prevention

The authentication establishment of a secure relationship between devices is based on a preshared key. How the devices obtain the unique preshared key is not specified within the standard. Two devices use the preshared key during a four-way handshake to mutually authenticate each other. During the handshake the devices establish a pair-wise temporal key (PTK), which is used, e.g., for the encryption of all the following frames that are exchanged between the two devices. For multicast or broadcast communication between multiple devices a group temporal key (GTK) is exchanged between the devices over unicast links that are encrypted with the different PTKs. Once the GTK has been exchanged, all multicast packets of the respective group are encrypted with the GTK of this group. The encryption mechanism is the AES-128 counter mode.

The integrity of frames is ensured by appending a message integrity code (MIC) to each frame, which enables a receiver to detect whether devices other than the sender have modified a message. The MIC is a cryptographic checksum, which makes use of the PTK or GTK and can therefore be calculated only by the sender and the intended receiver(s). The MIC is calculated according to the AES-128 cipher block chaining message authentication code (CBC-MAC).

Replay attacks are prevented by secure frame counters and replay counters that are also making use of the PTK or GTK.

For more details regarding the security mechanisms, the reader is referred to the WiMedia specification 24.

6.6 WiMedia MAC Performance

We are reporting simulation results from a former analysis [14] in order to determine typical values for data throughput and packet delay.

The maximum data throughput is analyzed with a simple scenario of one transmitting and one receiving station. DRP and PCA have been compared with each other for three different acknowledgment policies in separate simulation runs. A beacon period of 8 MASs and a data period of 248 MASs have been assumed. The TXOP length of PCA has been set to 1024 μs. All results with ACK policy set to burst ACK have been derived with a burst buffer size of 16 frames. The aggregation function has been evaluated with a timeout value of 100 μs. For transmissions in stream mode, the MIFS and stream mode preamble have been taken into account.

Figure 6.18a shows that the throughput of DRP always outperforms that of PCA with the same ACK policy by several Mb/s. This is because in PCA mode the sender has to wait for an AIFS duration and perform a backoff before transmitting, which decreases the throughput. As expected, in both DRP and PCA modes the no-ACK policy achieves the highest throughput, followed by burst ACK, which reaches a slightly lower throughput than no ACK due to the overhead of transmitting acknowledgment frames. As illustrated here, frame aggregation is very important for an efficient packet-oriented MAC. The throughput that is achieved with frame aggregation is comparable to the throughput of long packets.

In order to assess typical packet delays we have used a different scenario Figure 6.18(b) [14]. One pair of stations communicates via Voice-over-IP (VoIP) (150 kb/s each direction, 120-B packets, R1–2). A wireless streaming server provides HDTV to two different clients (24 Mb/s each, 1500-B packets, R3–4). Two stations handle file transfers at 30 Mb/s (R5–6) and one station handles a file transfer of 100 Mb/s (R7), each using 1500-B packets. The above-mentioned streams all use DRP. Two additional PCA connections complete the scenario (best effort, 1500-B, R8–9). DRP enables constant support for QoS to the high-priority transmissions, whereas PCA access fits the needs of low-priority background services. As depicted in Figure 6.17b 11 DRP results in well-bounded delays. However, for traffic carried by PCA, large delays can occur for some packets, even though other packets experience a lower delay than DRP packets. The reason for the shape of the PCA delay distribution is the slot reservation scheme. Sometimes packets immediately access the channel via PCA when an MAS is not reserved. Thus the delay is small. However, reserved MASs can lead to large delays for PCAs of more than 60 ms.

References

[1] *Standard ECMA-368— High Rate Ultra Wideband PHY and MAC Standard*, ECMA International, December 2005.

[2] *Standard ECMA-369— MAC-PHY Interface for ECMA 368*, ECMA International, December 2005.

[3] M.Z. Win and R.A. Scholtz, "Impulse radio: How it works," *IEEE Communications Letters*, Vol. 2, No. 2, Feb. 1998.

[4] M.Z. Win and R.A. Scholtz, "Ultra-wide bandwidth time-hopping spread-spectrum impulse radio for wireless multiple-access communications," *IEEE Trans. On Comm.*, Vol. 44, Apr. 2000.

[5] M.Z. Win and R.A. Scholtz, "On the energy capture of ultra wideband signals in dense multipath environments," *IEEE Comm. Letters*, Sept. 1998.

[6] A. Batra et al., "TI physical layer proposal: Time-frequency interleaved OFDM," March 2003, *IEEE 802.15.3/141/r0.*

[7] WiMedia, "Multi-band OFDM physical layer proposal for IEEE task group 3a," Sept. 2004, http://www.multibandofdm.org/papers/MultiBand_OFDM_Physical_Layer_Proposal_for_IEEE_802.15.3a_Sept_04.pdf

[8] IEEE 802.15 SG 3a channel modeling sub-committee, "Channel modeling sub-committee report final," Feb. 2003, IEEE P802.15-02/490r1-SG3a.

[9] A.A. Saleh and R. Valenzuela, "A statistical model for indoor multipath propagation," *IEEE Journal on Selected Areas in Comm.*, Vol. 5, Feb. 1987.

[10] D. Haccoun and G. Begin, "High-rate punctured convolutional codes for Viterbi and sequential decoding," *IEEE Trans. On Comm.*, Vol. 37, Nov. 198.

[11] E. Ojard, "Proposed improved convolutional code for the MB-OFDM PHY," Submission to WiMedia system definition sub-group, Jan. 2004.

[12] IEEE Standard for Information technology—Telecommunication and information exchange between systems—Local and metropolitan area networks—Specific requirements, Part 11: Wireless Medium Access Control (MAC) and Physical Layer (PHY) specifications: Amendment 7: Medium Access Control (MAC) Quality of Service (QoS) Enhancements, IEEE Draft Amendment P802.11e/D11.0, Oct. 2004.

[13] Federal Information Processing Standards Publication 197, Specification for the Advanced Encryption Standard (AES), November 26, 2001.

[14] Y. Zang, G. Hiertz, J. Habetha, and M. Zeybek, Multiband OFDM Alliance—The next generation of Wireless Personal Area Networks. In Proc. IEEE Sarnoff Symposium on Advances in Wired and Wireless Communications, Stanford, USA, April 19, 2005.

Chapter 7

Millimeter-Wave–Based IEEE 802.16 Wireless MAN

Jun Zheng, Yan Zhang, and Emma Regentova

Contents

7.1 Introduction

Broadband wireless access (BWA) systems, as reported by the International Telecommunication Union (ITU) [1], are one of the most promising solutions for broadband access. BWA systems provide an alternative to current wired broadband access options such as cable modem using coaxial systems, leased lines based on fiber-optic links, and Digital Subscribe Line (DSL) access networks [2]. It has several advantages, such as rapid deployment, high scalability, low maintenance and upgrade costs, and granular investment to match the market growth [3]. Installation of such systems can be beneficial in either highly crowded geographical areas such as cities or rural areas without wired infrastructure. To provide an industry-wide standard for BWA systems, the IEEE 802 committee set up the IEEE 802.16 working group to develop broadband wireless standards. The IEEE 802.16 standard defines the air interface and Media Access Control (MAC) specifications for BWA systems supporting metropolitan area network (MAN) architecture [4,5]. An alternative standard in Europe is the high-performance radio metropolitan area network (HiperMAN) standard created by the European Telecommunications Standards Institute (ETSI).

The 802.16 activities were initiated at an August 1998 meeting called by the National Wireless Electronic Systems Testbed (N-WEST) of the U.S. National Institute of Standards and Technology (NIST) [6]. The IEEE 802.16 working group was then formed to develop BWA standards [7]. IEEE 802.16 is an evolutionary standard. Much of the evolution has occurred in the physical layer (PHY) while the MAC specification of the standard can support all PHY options. The first version of the IEEE 802.16 standard, IEEE 802.16.1, was approved in 2001. It addresses the frequencies in millimeter-wave bands (10–66 GHz) with up to 134 Mbps data rate [4]. Due to the short wavelength in millimeter-wave bands, line of sight (LOS) required for transmission and multipath interference is negligible. A single-carrier modulation air interface, called ***WirelessMAN-SC***, is designed with a high degree of flexibility to allow service providers to optimize the system deployments. The system operating in this environment uses a point-to-multipoint (PMP) architecture to support applications from small office/home offices

(SOHOs) through medium to large offices. The next temporary version, IEEE 802.16.2, attempted to minimize the interference between coexisting wireless MAN systems.

In April 2003, the IEEE 802.16 working group released IEEE 802.16a to support the BWA systems in microwave bands (2–11 GHz) with a data rate up to 75 Mbps and maximum range 50 km [8]. Due to the longer wavelength in microwave bands, LOS is no longer necessary and multipath may be significant, which allows inexpensive and flexible consumer deployment and operation. To support near-LOS and non-LOS (NLOS) scenarios, the IEEE 802.16a standard defines three different PHY specifications:

- ***WirelessMAN-SCa*:** A single-carrier (SC) modulation technology
- ***WirelessMAN-OFDM*:** A 256-carrier orthogonal frequency division multiplexing (OFDM) modulation technology. A TDMA (Time Division Multiple Access)-based mechanism is used for multiple access. This air interface is mandatory for license-exempt bands.
- ***WirelessMAN-OFDMA*:** A 2048-carrier OFDM modulation technology. Multiple access is provided by assigning a subset of carriers to an individual receiver, thus referred to as OFD multiple access (OFDMA).

Each of these air interfaces can be used with the MAC layer to provide a reliable end-to-end link. Although all three air interfaces are designed for an NLOS environment, the two OFDM-based technologies are more suitable for NLOS operation owing to the implementation simplicity of the signal equalization. In addition, the 256-carrier ***WirelessMAN-OFDM*** compared to the 2048-carrier ***WirelessMAN-OFDMA*** is more favored by the vendor community because of a lower peak-to-average ratio, faster fast Fourier transform (FFT) calculation, and less stringent requirements for frequency synchronization. The MAC layer in 2–11 GHz is designed to address interference, fast fading, and different use of the air link. To work with PHY specifications designed for these spectra, additional MAC features such as automatic repeat request (ARQ) and support for an adaptive antenna system (AAS) are introduced. In addition to the PMP architecture, IEEE 802.16a also introduces and defines the key operation procedures for the mesh networking architecture.

All the aforementioned versions designed for fixed BWA were incorporated into the finalized standard, IEEE 802.16d-2004 [5]. Because the actual applications may require the mobility capability, the IEEE 802.16e was approved as an amendment in December 2005 to support mobile BWA systems [9]. The standard includes the components supporting PMP and mesh architectures, seamless handover operation, and location management schemes. The system works in 2–6 GHz licensed bands and may achieve data rates up to 15 Mbps in 5 MHz channel bandwidth.

In this chapter, we focus on the IEEE 802.16 wireless MAN operating in millimeter-wave bands, which provides a major broadband access solution to link home and business to core telecommunication networks worldwide. The organization of the chapter is as follows. Section 7.2 briefly introduces the IEEE 802.16 wireless MAN working in the PMP architecture. In Section 7.3, the ***WirelessMAN-SC*** PHY specified for millimeter-wave bands is addressed. Section 7.4 describes the basics of the IEEE 802.16 MAC. Section 7.5 discusses the quality-of-service (QoS) support issues. The security mechanisms and their enhancements are described in Section 7.6. Finally, the conclusions are drawn in Section 7.7.

7.2 IEEE 802.16 Wireless MAN Architecture

The IEEE 802.16 wireless MAN architecture, as shown in Figure 7.1, consists of two kinds of stations: base stations (BSs) and stationary subscriber stations (SSs). In millimeter-wave bands, the system is organized in PMP network topology. The BS is responsible for a specific area cell, and all communications in the cell are regulated by the BS; i.e., there is no peer-to-peer communication between the SSs. The BS provides the connectivity, management, and control of the SSs. Each of the SSs can deliver voice and data using common interfaces, such as telephony services, Ethernet, video, and other services with different QoS requirements [3,10].

The communication path between the SS and BS is bidirectional: downlink (DL) from BS to SS and uplink (UL) from SS to BS, multiplexed with either Time Division Duplex (TDD) or Frequency Division Duplex (FDD). The DL channel is a broadcast channel and the BS transmits to the SSs using time division multiplexing (TDM). The time in the UL channel is slotted, which is shared by the SSs through TDMA.

Figure 7.2 illustrates the layered protocol architecture for the IEEE 802.16 standard [4,5]. The MAC layer comprises three sublayers: the service-specific convergence sublayer (CS), the MAC common part sublayer (MAC CPS), and the privacy sublayer (PS). The CS is defined to be the interface to higher layers and provides mapping function from the transport layer diverse

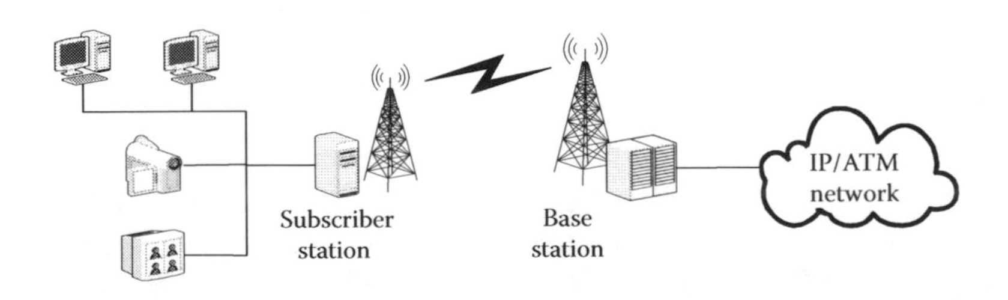

Figure 7.1 IEEE 802.16 wireless MAN architecture.

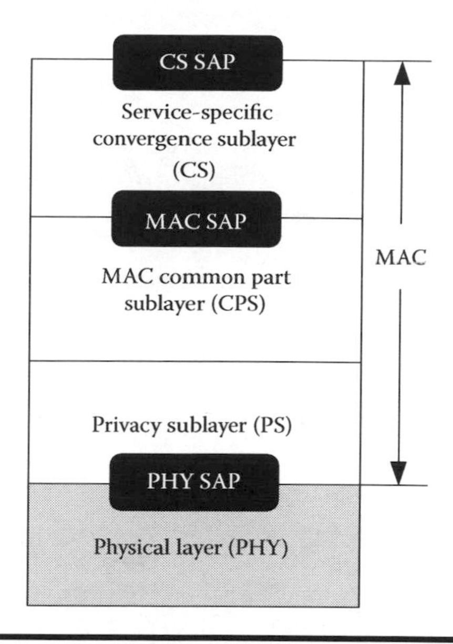

Figure 7.2 IEEE 802.16 layered protocol architecture.

traffics to the flexible MAC. The external network data is received through the CS service access point (SAP). The MAC CPS provides the core MAC function, including access control, collision resolution, control or data scheduling, and bandwidth request and allocation. The MAC SAP is used to receive the MAC service data units (SDUs) from the CS. The PS ensures secure connection establishment and provides network access authentication, secure key exchange, and data encryption. PHY specifies the frequency band, the modulation scheme, error-correction techniques, synchronization between transmitter and receiver, data rate, and the TDM structure. Data, PHY control, and statistics are transmitted between the MAC CPS and the PHY through PHY SAP.

7.3 WirelessMAN-SC PHY

Systems operating at millimeter wave bands require LOS and are prone to deep fades from rain and wet snow. For such harsh channel conditions, the single-carrier modulation was easily selected and the air interface was designed as ***WirelessMAN-SC*** [11]. In this section, we describe the details of the ***WirelessMAN-SC*** PHY specification.

7.3.1 Overview

The ***WirelessMAN-SC*** PHY specification is designed with a high degree of flexibility to provide the service provider the ability to optimize system deployment [5]. Both TDD and FDD modes are supported. The specified channel bandwidths are 20 or 25 MHz for typical U.S. allocation and 28 MHz for typical European allocation. A burst single-carrier modulation with adaptive burst profiling is used in which transmission parameters, including the modulation and coding schemes, may be adjusted individually to each SS on a frame-by-frame basis. The standard defines three different modulation schemes. For UL, the quadrature phase shift keying (QPSK) is mandatory and 16-state quadrature amplitude modulation (16-QAM) and 64-state QAM (64-QAM) are optional. In the DL, the BS supports QPSK and 16-QAM whereas 64-QAM is optional. In addition to these modulation schemes, the standard also defines various forward error correction (FEC) schemes, including the Reed–Solomon (RS) codes, RS concatenated with inner block convolution codes (BCCs), RS with parity check, and block turbo codes (BTCs). The FEC options are paired with difference modulation schemes to form burst profiles of varying robustness and efficiency.

The system supports frame-based transmission with frame duration 0.5 ms, 1 ms, or 2 ms. Each frame includes a DL subframe and a UL subframe. The frame is divided into physical slots for the purpose of bandwidth allocation and identification of PHY transitions. The number of physical slots in a frame is a function of the symbol rate. In the TDD mode, the UL subframe follows the DL subframe in the same carrier frequency. In the FDD mode, the UL and DL subframes come at the same time but in different frequencies.

Between the PHY and MAC layers, there is a transmission convergence (TC) sublayer for DL and UL. This sublayer segments various-length MAC protocol data units (PDUs) into fixed-length FEC blocks of data. The format of a TC PDU is shown in Figure 7.3. The TC PDU starts with a pointer byte indicating the beginning of the first MAC PDU to start in the packet or the beginning of any stuff byte that precedes the next MAC PDU. This format allows for resynchronizing to the next MAC PDU if the previous FEC block had irrecoverable errors.

7.3.2 Frame Structures

Figure 7.4 illustrates the frame structure for ***WirelessMAN-SC*** in TDD mode. The Tx/Rx transition gap (TTG) is used to separate the DL and

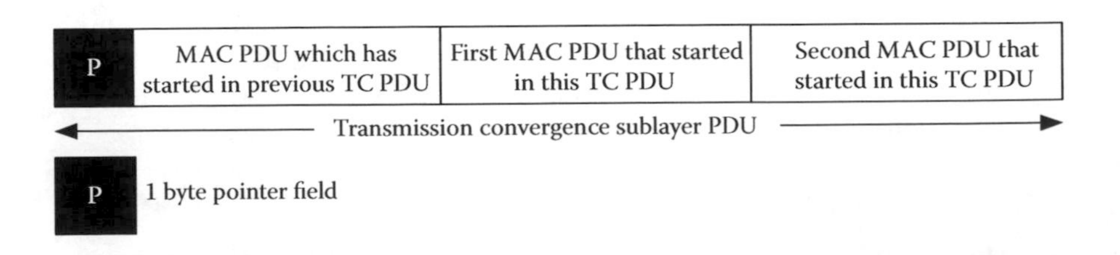

Figure 7.3 Transmission convergence sublayer PDU format.

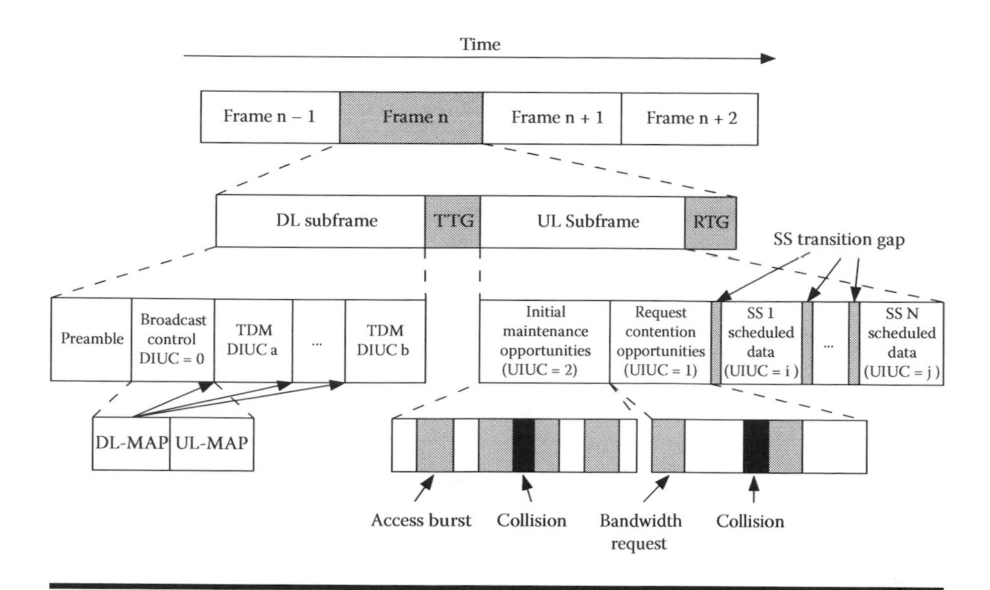

Figure 7.4 Frame structure for wirelessMAN-SC PHY in TDD mode.

UL subframes, which allows time for BS to switch from transmit to receive mode and SSs to switch from receive to transmit mode. In the same manner, the Rx/Tx transition gap (RTG) is used to separate the UL subframe and the DL subframe in the next TDD frame. This gap allows time for the BS to switch from receive to transmit mode and the SSs from transmit to receive mode.

The DL subframe starts with a frame start preamble modulated by QPSK, which is used for frame synchronization and control. The amplitude of the preamble depends on the DL power adjustment rule. Following the preamble is the frame control section, which carries control information for all SSs. The frame control section contains a DL-MAP message for the current DL subframe as well as a UL-MAP message for the associated UL subframe. The DL and UL channel descriptors (DCD and UCD) may be included in the section. DL-MAP defines the access strategy to the DL channel and UL-MAP specifies the access scheme to the UL channel. DCD and UCD are MAC messages used to describe the PHY characteristics of a DL channel and a UL channel, respectively. Following the frame control section, the DL subframe contains a TDM portion, which carries the data. The DL data is always FEC encoded and transmitted to each SS using a negotiated burst profile specified by the downlink interval usage code (DIUC) in the DL-MAP. The data is transmitted in order of decreasing burst profile robustness, thus the SSs can receive their data before being presented with a burst profile that may cause them to lose synchronization.

In the UL subframe, as shown in Figure 7.4, three classes of bursts may be transmitted by the SS to the BS: (1) those that are transmitted in contention opportunities reserved for initial ranging; (2) those that are transmitted in contention opportunities defined by Request Intervals reserved for response to multicast and broadcast polls; and (3) those that are transmitted in intervals defined by data grant information elements (IEs) specifically allocated to individual SSs. Any of the three burst classes may be present in a given frame. The order and quantity of their occurrence are indicated by the UL-MAP in the frame control section of the DL subframe. The uplink burst profiles for the bandwidth allocation are specified by the uplink interval usage code (UIUC) in the UL-MAP. The bandwidth allocated for initial ranging and request contention opportunities could be grouped together with the uplink burst profile specified for initial raging intervals (UIUC = 2) and request intervals (UIUC = 1), respectively. The remaining transmission slots are grouped by the SSs, which are separated by SS transition gaps (SSTGs). In a scheduled bandwidth, an SS transmits with the burst profile specified by the BS. The SSTG is used to ramp down the previous burst and allow the BS to synchronize with a new SS.

In FDD mode, the DL and UL channels are on separate frequencies, allowing the system to simultaneously support full-duplex SSs and half-duplex SSs. The full-duplex SSs can transmit and receive simultaneously whereas the half-duplex SSs can only transmit or receive at one time.

The UL subframe in FDD is similar to that in TDD mode. The structure of the DL subframe in FDD mode is illustrated in Figure 7.5. Like the DL subframe in TDD mode, the DL subframe in FDD mode starts with a frame start preamble followed by a frame control section and a TDM portion organized into bursts transmitted in decreasing order of burst profile robustness. The TDM portion continues with a TDMA portion that includes an extra preamble at the start of each new burst profile. This allows the half-duplex SSs to be better supported in the system because the half-duplex SSs may need to transmit earlier in the frame than they receive, thus losing the synchronization with the DL. The TDMA preamble helps the SSs to regain synchronization. Bursts in the TDMA portion do not need to be ordered by the burst profile robustness. The DL-MAP in the FDD frame control section contains a map of both the TDM and TDMA bursts.

7.3.3 Adaptive Burst Profiling

To achieve various trade-offs of data rate and robustness, the system can adaptively select an appropriate PHY modulation and FEC scheme based on the channel and interference conditions [12].

The allowed modulation constellations in ***WirelessMAN-SC*** PHY are QPSK, 16-QAM, and 64-QAM [5]. In the DL, QPSK and 16-QAM are mandatory and 64-QAM is optional. In the UL, QPSK is required and 16-QAM and

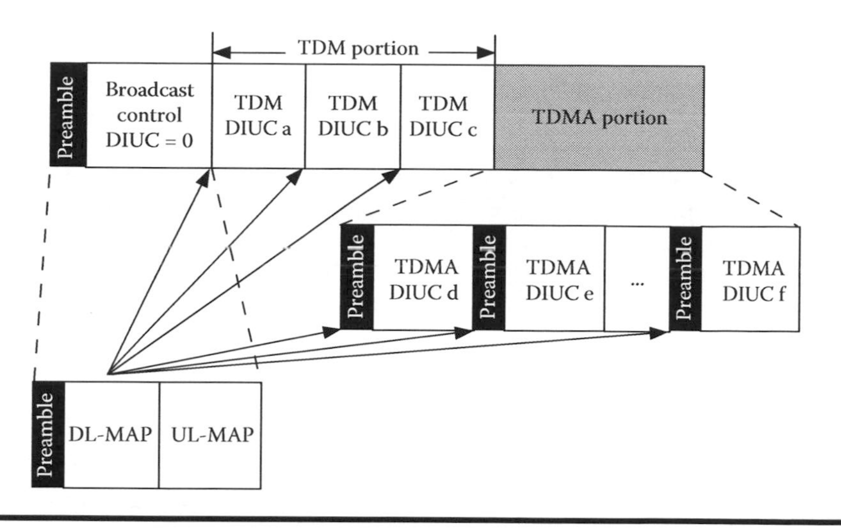

Figure 7.5 The structure of the DL subframe in FDD mode.

64-QAM arc optional. Thc modulation type defines the number of bits per symbol. In 16-QAM, each symbol can represent four bits instead of two bits per symbol with QPSK. For 64-QAM, it can send eight bits per symbol and thus achieve higher throughputs or better spectral efficiencies. However, it must be noted that to use a modulation technique such as 64-QAM, better signal-to-noise ratios (SNRs) are needed to overcome any interference and maintain a certain bit error ratio (BER). In ***WirelessMAN-SC*** PHY, the modulation constellation can be adaptively selected per SS based on the channel quality. If the link condition is good, a more complex modulation scheme can be used to maximize throughput while maintaining a reliable data transfer. If the link degrades over time, the system can change to a less complex constellation to keep the data transfer reliable.

Table 7.1 shows the four selectable FEC code types in ***WirelessMAN-SC*** PHY. In the standard, the implementation and use of code types 1 and 2 are mandatory but code types 3 and 4 are optional. Code type 1 uses an outer RS block code generated from GF(256) without an inner code. It is useful either for a large data block or when high coding rate is required.

Table 7.1 FEC Code Types

Code Type	*Outer Code*	*Inner Code*
1	RS over GF(256)	None
2	RS over GF(256)	(24,16) BCC
3	RS over GF(256)	(9,8) Parity check code
4	BTC	-

Code type 2 is an outer RS over GF(256) code concatenated with an inner BCC, which makes it efficient for low-to-moderate coding rates by providing the carrier-to-noise (C/N) enhancement. Code type 2 can be used only with QPSK modulation. Code type 3 uses an outer RS over GF(256) code concatenated with an inner parity check code for moderate-to-high coding rate with small-to-medium-size blocks. The parity code can be used for error correction. Code type 4 is a BTC, which requires significantly lower carrier-to-interference (C/I) level for reliable data transfer. It can improve the coverage and/or capacity of the system at the price of increased decoding latency and complexity.

Different PHY modulation and FEC schemes can be combined adaptively to form DL or UL burst profiles according to the direction of the traffic flow between a BS and an SS. The burst profiles in DL and UL are identified using the DIUC and UIUC specified in DL-MAP and UL-MAP, respectively. The parameters for each burst profile are included in the DCD message for DL and in the UCD message for UL. The DL burst profile includes the following parameters: modulation type, FEC code type, last codeword length, DIUC mandatory exit threshold, DIUC mandatory entry threshold, and preamble presence. If the FEC code type is 1, 2, or 3, the RS information bytes and RS parity bytes should also be included in the DL burst profile. Moreover, for FEC code type 2, the BCC code type needs to be specified. If the FEC code type is 4, the DL burst profile should also include the BTC row and column code types and BTC interleaving type. In the DL, 13 different burst profiles can be specified using DIUC with values from 0 to 12. The DL burst profile 1 (DIUC = 0) shall be stored in the SS instead of including it in the DCD message. The parameters included in the UL burst profile are modulation type, FEC code type, last codeword length, preamble length and randomizer seed. The extra parameters to be included for different FEC code types are the same as those of the DL burst profile. To fully utilize the adaptive burst profiling, one of the key issues is the link adaptation algorithms [12].

7.4 IEEE 802.16 MAC Layer Basics

7.4.1 Overview

In millimeter-wave bands, the IEEE 802.16 MAC layer is designed for PMP broadband wireless access applications [5,13]. It is similar to the MAC layer in the Data Over Cable Service Interface Specifications (DOCSIS) standard, which is a common protocol in cable-based networks [11,14]. The MAC protocol addresses the need of very high data rate applications with a variety of QoS requirements, both in DL and UL. The access and bandwidth algorithms are designed to accommodate hundreds of terminals per channel, which allows each terminal to be shared by multiple end users with quite different bandwidth requirements for services such as the legacy TDN (tactical data network) voice and data, the Internet Protocol (IP), and the packetized Voice-over-IP (VoIP). The IEEE 802.16 MAC layer protocol is designed to be flexible and efficient to support different demands. In addition to an analogue of the classic asynchronous transfer mode (ATM) service categories, it also provides new categories, such as guaranteed frame rate (GFR).

The IEEE 802.16 MAC layer consists of three sublayers: the CS sublayer, the MAC CPS sublayer, and the privacy sublayer. Because the IEEE 802.16 MAC protocol must support both ATM and packet-based protocols, the CS sublayers are used to map the transport-layer–specific traffic to the MAC that is flexible enough to carry any traffic type efficiently. The MAC CPS sublayer is independent of the transport mechanism, and it is responsible for fragmentizing and segmenting the MAC SDUs into MAC PDUs, bandwidth allocation, performing QoS control over transmission, and scheduling of the MAC PDUs [6]. In addition to the fundamental tasks of bandwidth allocation and data transportation, the MAC layer also includes a privacy sublayer to provide authentication of network access and connection establishment, key exchange, and encryption for data privacy.

7.4.2 CS Sublayer

The CS sublayers in the IEEE 802.16 MAC layer serve as an interface to higher layers by performing the following functions: (1) accept higher-layer PDUs from the higher layer; (2) classify the higher-layer PDUs; (3) process the higher-layer PDUs based on the classification if required; (4) deliver CS PDUs to the appropriate MAC SAP; and (5) receive CS PDUs from the peer entity. The IEEE 802.16 standard provides two CS sublayers: the ATM CS sublayer and the packet CS sublayer [5]. The ATM CS sublayer is specified for ATM services whereas the packet CS sublayer is defined for mapping packet services such as IPv4, IPv6, Ethernet, and virtual local area network (VLAN) [11].

7.4.3 MAC CPS Sublayer

The MAC CPS sublayer provides the core functionality of the IEEE 802.16 MAC layer. In PMP mode, the 802.16 wireless link operates with a central BS and a sectorized antenna to handle multiple independent sectors simultaneously. On the DL, the BS broadcasts data to SSs that are TDM multiplexed. On the UL, the SSs share the link on a demand basis using the Demand Assignment Multiplex Access (DAMA) TDMA technique.

The 802.16 MAC protocol is connection oriented. All data communications are in the context of a connection to be mapped to the services provided to SSs and associated with varying QoS levels. A connection includes the mapping between peer convergence processes utilizing the MAC and a service flow. The service flow specifies the QoS parameters for the PDUs exchanged on the connection. Service flows provide a mechanism for DL and UL QoS management, especially for the bandwidth allocation process. Connections are identified by 16-bit connection identifiers (CIDs). Once established, connections may need active maintenance or may be terminated when a customer's service contract changes.

In the IEEE 802.16 standard, each SS is assigned a universal 48-bit MAC address to serve as a unique equipment identifier. Once an SS enters the network, three pairs of management connections are established that reflect three different QoS levels for management traffic between the SS and the BS. The basic connection is used for exchanging short, time-critical MAC management messages between the SS and the BS. The primary management connection is used to exchange longer, more delay-tolerant MAC management messages like those used for authentication and connection setup. Finally, the secondary management connection is used for transferring delay-tolerant, standards-based messages such as Dynamic Host Configuration Protocol (DHCP), Trivial File Transfer Protocol (TFTP), Simple Network Management Protocol (SNMP) and so forth.

7.4.4 MAC PDU

The MAC PDU is the basic data unit in the 802.16 MAC layer. For ***WirelessMAN-SC*** PHY used for millimeter-wave bands, each MAC PDU starts with a fixed-length MAC header followed by a variable-length payload and an optional cyclic redundancy check (CRC). The standard defines two MAC header formats: the generic MAC header and the bandwidth request header. A single-bit header type (HT) field is used to distinguish these two header formats. It is set to 0 for the generic header and to 1 for the bandwidth request header.

The format of a MAC PDU with the generic MAC header is shown in Figure 7.6. The header starts with the HT and encryption control (EC) fields. The type field indicates the subheaders and special payload types present in the message payload. The 1-bit CRC indicator (CI) field is set to 0 if no

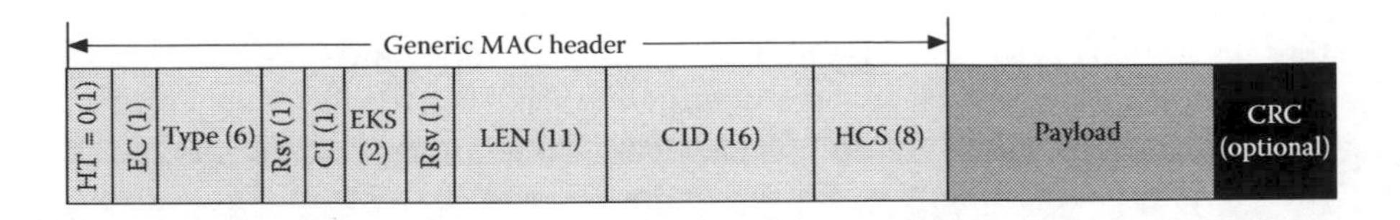

Figure 7.6 Format of a MAC PDU with generic MAC header.

HT = 0(1)	EC (1)	Type (6)	BR (19)	CID (16)	HCS (8)

Figure 7.7 Format of a MAC PDU with the bandwidth request header.

CRC is included. The encryption key sequence (EKS) field is the index of the traffic encryption key (TEK) and the initialization vector used to encrypt the payload. It is meaningful only when the EC field is 1. The 11-bit length (LEN) field defines the length in bytes of the MAC PDU followed by a 16-bit CID field. Finally, the 8-bit header checker sequence (HCS) field is used to detect errors in the header. The payload can be MAC management messages or CS sublayer data.

In the MAC PDU, three types of MAC subheaders may be present. The fragmentation subheader includes information indicating the presence and orientation of any fragments of SDUs in the payload. The 2-byte grant management subheader is used by the SS to convey bandwidth management needs to its BS. The packing subheader indicates the packing of multiple SDUs into a single MAC PDU. The fragmentation and grant management subheaders are pre-PDU subheaders that can be inserted in MAC PDUs immediately following the generic MAC header, as indicated in the type field. The packing subheader is the only pre-SDU subheader that may be inserted before each MAC SDU, if so indicated by the type field.

Figure 7.7 illustrates the format of a MAC PDU with the bandwidth request header. It contains the bandwidth request header alone without the payload. The length of the header is 6 bytes. The EC field is always set to 0 indicating no encryption. The 3-bit type field indicates the bandwidth request header type, where 000 is for incremental and 001 is for aggregate. The 19-bit bandwidth request (BR) field defines the number of bytes of UL bandwidth requested by the SS. Following the BR field, the CID field indicates for which connection the UL bandwidth is requested.

In the 802.16 MAC layer, multiple MAC PDUs can be concatenated into a single transmission in either the UL or the DL. An example of concatenated UL burst transmission is shown in Figure 7.8. Each MAC PDU has a unique CID such that the receiving MAC entity can send the MAC SDU to the correct instance of the MAC SAP after the MAC SDU is reassembled from one or more received MAC PDUs. MAC management messages PDU, user data PDU, and bandwidth request PDU can be concatenated into the burst transmission.

To use the available bandwidth efficiently, fragmentation and packing can be employed for the incoming MAC SDUs from corresponding CS

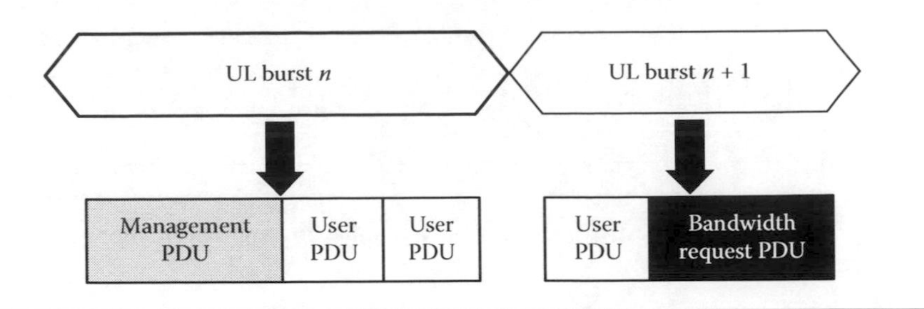

Figure 7.8 Concatenation of MAC PDUs.

sublayers [5,11]. A MAC SDU can be divided into one or more MAC PDU fragments. Multiple MAC SDUs can be packed into a single MAC PDU. Fragmentation and packing can be used together to achieve better bandwidth utilization. They can be initialized by a BS for a DL connection and an SS for a UL connection.

7.4.5 Radio Link Control

To fully utilize the adaptive burst profiling in ***WirelessMAN-SC*** PHY, the 802.16 standard defines an equally advanced radio link control (RLC) [11,12]. The RLC has the capability to transit from one PHY scheme to another. The UL or DL burst profiles are identified using the DIUC and the UIUC. The RLC can switch between different burst profiles on a per-frame or per-SS basis. The BS and SSs continuously negotiate the DL and UL burst profiles to maintain the link quality between them.

The carrier-to-interference-and-noise ratio, CINR $= C/(N + I)$, is recommended by the standard for SS initializing a burst profile change, where C is the received signal power, N is the noise floor, and I is the sum of the interference power [12]. In the DL, each SS continuously monitors the CINR and compares the average value with the preset threshold levels, as shown in Figure 7.9. If the received CINR exceeds or falls below the allowed operating region, the SS requests a change in the DL burst profile by using a range request (RNG-REQ) message with the DIUC of the desired burst profile. Based on the channel conditions, the SS can switch to a more robust or less robust burst profile. It should be noted that the standard defines only the framework to be used for link adaptation. The switching thresholds for CINR and the adaptation algorithm are not specified. A simple link adaptation algorithm has been proposed [12] using the framework defined in the standard. It has also been shown [12] how to determine the optimal CINR switching thresholds for different burst profiles.

In the UL, the BS is in control and monitors the UL signal quality directly, thus it is simple to change the UL burst profile for an SS. The BS specifies

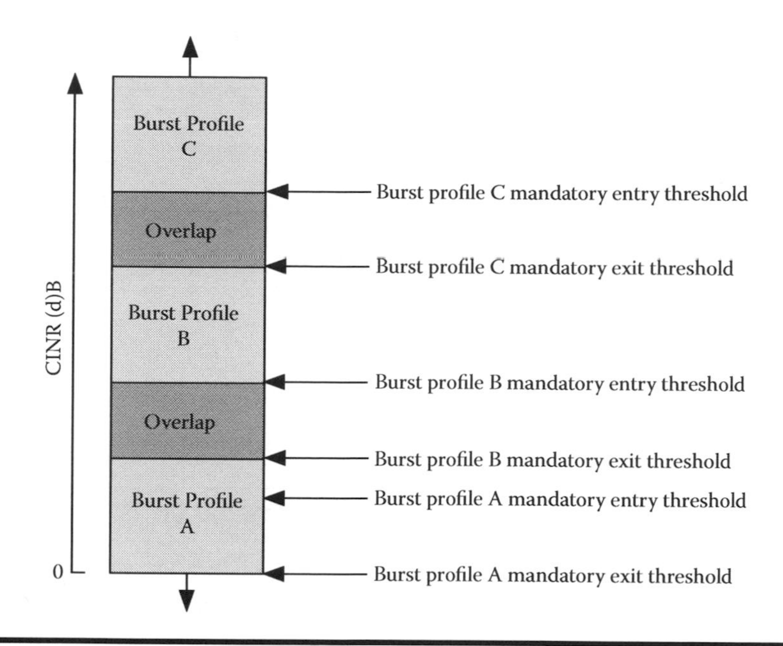

Figure 7.9 Threshold levels for burst profile usage.

the UIUC associated with the desired profile in UL-MAP whenever granting bandwidth to the SS in a frame. Because the SS will always receive either both the UIUC and the grant or neither of them, there is no need for an acknowledgment, and no mismatch of the UL burst profile will happen between the BS and the SS.

In millimeter-wave bands, the 802.16 standard does not support any form of ARQ schemes. For a system operating in lossy channels, it is more effective to sacrifice some bandwidth and increase the robustness of PHY than to use the ARQ support [15].

7.4.6 Network Entry and Initialization

In order to operate in the 802.16 network, a new SS must enter and register itself to the network. The network entry process of the new SS has several stages, including DL channel synchronization, initial ranging, capabilities negotiation, authorization message exchange, registration, and IP connectivity. The network entry state machine will reset if it fails to move from one stage to the next. Once the network entry process is completed, the SS can create one or more service flows to send data to the BS. The network entry process of a new SS is shown in Figure 7.10 and the details of each stage are described below.

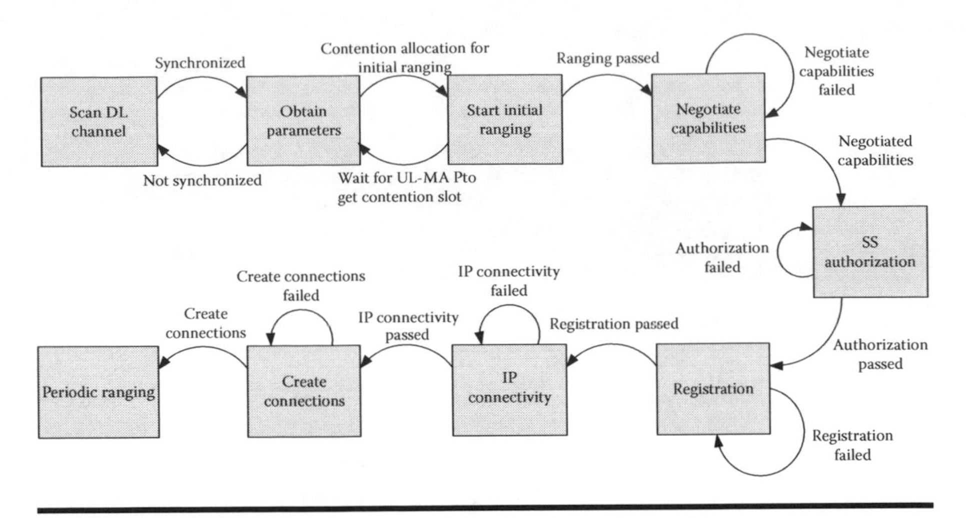

Figure 7.10 Network entry process of a new SS.

- *DL network synchronization*: Upon initialization, an SS scans its defined frequency list to find an operating channel. Usually the SS is programmed to register with a specific BS with a given set of operational parameters. If the SS finds a DL channel, it will try to synchronize to the DL channel by detecting the periodic frame preambles. Once the PHY is synchronized, the SS will look for DCD and UCD messages to obtain information on UL and DL parameters such as modulation and FEC schemes.

  Initial ranging: When an SS has synchronized with the DL channel and learned what parameters to use for its initial ranging transmission, it will scan the UL-MAP message to look for initial ranging opportunities. A truncated exponential backoff algorithm is used by the SS to find an initial ranging slot for sending a ranging request MAC message. The request message is sent using the minimum transmission power. If the SS does not receive a ranging response, it will send the request again using higher transmission power in the subsequent frame. Upon receiving the ranging request message from the SS successfully, the BS will return a ranging response message indicating either power level and timing offset corrections or ranging success. If the response indicates corrections, the SS performs fine-tuning of the parameters and sends another ranging request. If the response indicates ranging success, the SS joins the normal data traffic in the UL.

  Capabilities negotiation: After completion of initial ranging, the SS sends a capability request message to inform the BS of its PHY capabilities, including the supported modulation level, coding schemes and rates, and duplexing methods. The BS will send back a response to accept or deny the use of the capability reported by the SS.

SS authorization: After capability negotiation, the BS and the SS perform authorization and key exchange. The details of the SS authorization procedure are described in Section 7.6.1.2.

Registration: The SS needs to register with the network after successful authorization. To register with the BS, the SS sends a registration request message to the BS. The BS then sends back a registration response message. The registration process establishes the secondary management connection of the SS and determines capabilities related to connection setup and MAC operation. It also determines the IP version used on the secondary management connection.

IP connectivity: After registration, the SS invokes the DHCP to get the IP address and other parameters to establish IP connectivity. The BS and SS use the time of day protocol to maintain the current date and time. After DHCP is successful, the SS requests a configuration file from the DHCP server using TFP. The configuration file is the standard interface for providing vendor-specific configuration information.

Connection creation: After the transfer of the configuration file and completion of registration, transport connections are created. IEEE 802.16 uses the concept of service flows to define the transport of packets. Each admitted or active service flow is mapped to a MAC connection with a 16-bit unique CID. The service flows in IEEE 802.16 are generally preprovisioned. For preprovisioned service flows, the BS initiates the connection creation process by sending a dynamic service flow addition request message to the SS. The SS then sends a response to confirm the creation of the connection. For nonpreprovisioned service flows, the SS initiates the connection creation process by sending a dynamic service flow addition request message to the BS. The BS then sends back a response for confirmation.

7.5 QoS Support

7.5.1 Overview

IEEE 802.16 can support multiple communication services (data, voice, video) with different QoS requirements [10,16]. The data transmission on the DL is relatively simple because the BS manages the transmission. The BS broadcasts the data packets to all SSs and each SS picks up only its own packets. On the UL, the standard uses the DAMA-TDMA technique. DAMA is a capacity assignment technique to adapt to varying demands from multiple stations. TDMA divides time on a channel into a number of slots for allocating to SSs. The BS determines the number of time slots to be assigned to an SS for transmitting in a UL subframe through the BS UL scheduling module. This transmission information for the SS is contained in the UL-MAP at the beginning of each frame broadcasted by the BS. The UL-MAP includes an IE, which indicates the transmission opportunities; i.e., which time slots the SS can transmit during the UL subframe [10]. The BS UL scheduling module determines the IEs using the bandwidth request PDUs received from SSs. There are two ways to send the bandwidth request: (1) unicast (contention free): The SS is polled individually by the BS to send a bandwidth request message for bandwidth allocation. (2) contention based: If there is not enough bandwidth to individually poll inactive SSs, the contention-based bandwidth request is used by multicasting or broadcasting to a group of SSs, which have to contend for the opportunity to send bandwidth requests [13].

7.5.2 UL Scheduling Service Classes

IEEE 802.16 is a completely connection-oriented protocol. Each connection in the UL direction is mapped to a scheduling service. Each scheduling service is associated with a set of QoS parameters. In IEEE 802.16, four types of scheduling services with different QoS parameters are defined:

- *Unsolicited grant service (UGS)*: This type of service is designed to support constant-bit-rate (CBR) or CBR-like traffic (e.g., T1/E1, VoIP without silence suppression). Since these applications require constant bandwidth allocation, bandwidth requests are not required and there are no contention opportunities.
- *Real-time polling service (rtPS)*: This type of service supports real-time variable-bit-rate (VBR)-like service flows that are transmitted at fixed intervals, such as MPEG video or VoIP with silence suppression. These applications have specific bandwidth and delay requirements. The received packets with delays higher than required are considered useless.
- *Non-real-time polling service (nrtPS)*: This type of service is designed for delay-tolerant flows consisting of variable-size data packets with a minimum data rate requirement, such as FTP.
- *Best effort (BE) service*: This type of service is used to support flows without QoS guarantee, such as HTTP. It may be handled in an available space after the bandwidth has been allocated to the previous three service flows.

7.5.3 QoS Architecture for UL Scheduling

With the four kinds of the scheduling services, UL packet scheduling (UPS) runs in the BS to control all packet transmissions on the UL. The QoS

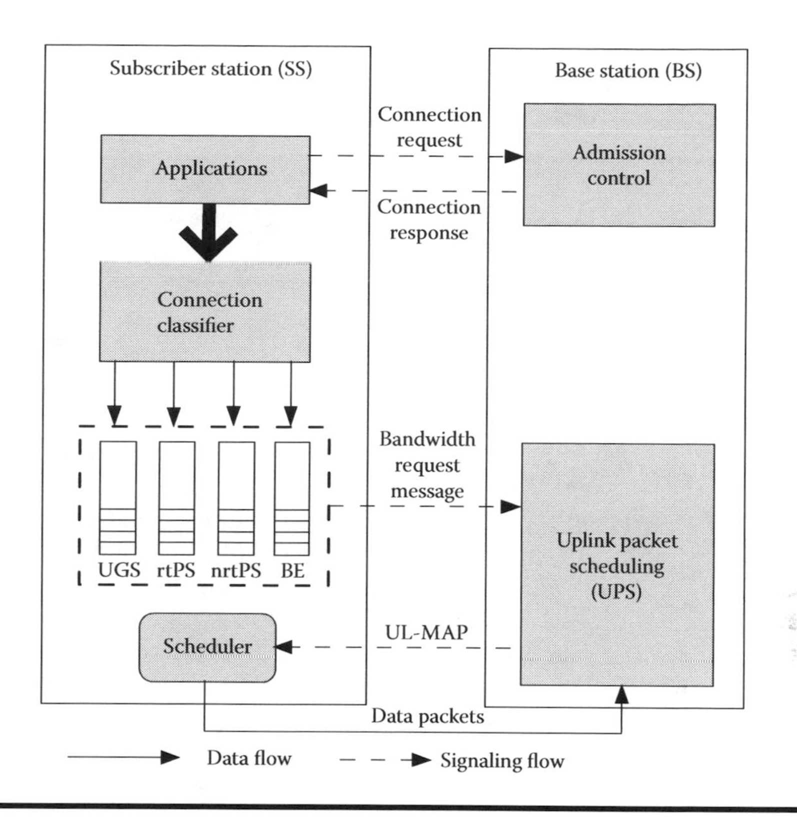

Figure 7.11 QoS architecture for IEEE 802.16 UL scheduling.

architecture for UL scheduling is shown in Figure 7.11 [16]. Because the 802.16 MAC protocol is connection oriented, an application in the SS first establishes the connection with the BS and each connection on the UL is mapped to a scheduling service flow (UGS, rtPS, nrtPS, or BE). A unique CID will be assigned to each connection. In millimeter-wave bands, the bandwidth is granted per SS. Thus the bandwidth granted to the connections for an SS is aggregated into a single grant to the SS. The connection signaling, including the connection request and response, is defined in the standard but the admission control process is not specified. In the SS, the connection classifier differentiates the packets from the application layer based on CID and puts them into an appropriate queue. Then the scheduler in the SS retrieves the packets from the queues and transmits them using the time slots granted by the UL-MAP sent by the BS. The UPS module in the BS determines the UL-MAP based on the bandwidth request message sent by the SS to report the current queue size per connection.

The SS has numerous ways to request bandwidth from the BS by combining the deterministic unicast polling with the responsiveness of contention-based requests and the efficiency of unsolicited bandwidth [11].

For UGS connection, SSs do not need to request bandwidth because the BS will grant it unsolicited. It is not allowed to use random access opportunities for bandwidth requests. To reduce the bandwidth requirements of individual polling, any SS with an active UGS connection can use the poll-me (PM) bit in the MAC grant management subheader to let the BS know it needs to be polled individually. The BS may choose to poll SSs that have UGS connections only when they have the PM bit set.

The more convenient way for bandwidth request is to send a bandwidth request message as a stand-alone bandwidth request MAC header. The bandwidth request message can be transmitted during any UL allocation except the initial ranging interval. Another closely related method for bandwidth request is piggyback request, where the grant management subheader is used to piggyback a request for additional bandwidth for the same connection within a MAC PDU. The requests may be incremental or aggregate, as indicated in the type field in the bandwidth request header. The BS will add the requested bandwidth to its current perception of the bandwidth needs of the connection after receiving an incremental bandwidth request message. For an aggregate bandwidth request message, the BS replaces the current perception of the bandwidth need with the requested quantity of the bandwidth. The piggyback request should always be incremental because there is no type field in the piggyback request message.

If there is no sufficient bandwidth to poll individual SSs, the BS may poll the SSs in multicast groups or send a broadcast poll by allocating a request interval to certain CIDs reserved for multicast groups and broadcast messages.

In summary, the existing QoS architecture in the IEEE 802.16 standard defines the signaling mechanism for information exchange between the BS and the SSs, such as the connection setup, bandwidth request, and UL-MAP [16]. It also specifies the UL scheduling for UGS service flow. However, the scheduling for rtPS, nrtPS, and BE service flows and the admission control and traffic policing processes are not defined, which is left up to the manufacturers [5,16].

7.5.4 QoS Enhancements

Because the IEEE 802.16 standard does not specify the admission control, traffic policing, and QoS-based packet scheduling algorithm, an appropriate architecture with those mechanisms is essential to support multiple types of services with different QoS requirements.

Several new architectures have been proposed recently to enhance the QoS support in IEEE 802.16 wireless MAN [3,10,16,20–24]. A new QoS architecture has been proposed [3], including a traffic classifier, and the SS's UL scheduler, and the BS's UL and DL schedulers. Traffic shaping and traffic policing are implemented as two modules in the BS for QoS control of the connections. A multiclass priority fair queuing (MPFQ) scheduling is used in the implementation of the SS's UL scheduler. In MPFQ, each service category has a corresponding priority class and its own scheduling policy. For service flows with high priority, such as UGS and rtPS, a wireless fair queuing (WFQ) policy is used to provide a lower delay bound. For nrtPS service flow with middle priority, weighted round robin (WRR) scheduling is used because there is no tight delay requirement. Finally, the first-in-first-out (FIFO) scheduling is used for low-priority BE flow. Wongthavarawat and Ganz [16, 20] proposed a new architecture design that includes an admission control module, a traffic policing module, and a UL scheduling algorithm. The admission control is designed to ensure that the existing sessions' QoS will not be degraded and the new session will be provided QoS support. The traffic policing module with the token bucket mechanism monitors violations of QoS contracts of the admitted connections. During UL scheduling, fixed bandwidth is allocated to UGS connections based on their requirements. The earliest deadline first (EDF) policy is used for rtPS connections such that packets with the earliest deadline will be scheduled first. For nrtPS connections, the WFQ is used for scheduling according to the ratio between the connection's average data rate and the total nrtPS connections' average data rate. The remaining bandwidth is then equally allocated to each BE connection. Cho et al. [10] propose an architecture with UL scheduling and admission control modules in the BS and an additional traffic management module in the SS. The architecture aims to provide QoS support to real-time traffic with high priority while maintaining throughput performance at an acceptable level for low-priority traffic. Alavi et al. [22] have designed an architecture based on the request and grant, polling, and bandwidth allocation mechanisms addressed in the 802.16 standard. The combination of token-bucket and leaky-bucket algorithms is applied for traffic shaping and policing. The WFQ algorithm is used in the UL and DL schedulers. A new traffic scheduler, referred to as the frame registry tree scheduler (FRTS), has been proposed to provide differentiated treatments to data connections based on their QoS requirements [23]. The scheduler uses a tree structure in order to prepare timeframe creation and reduce processing needs at the beginning of each timeframe. In order to allow more packets to be transmitted and increase throughput, the algorithm schedules each packet at the last timeframe before its deadline. The scheduler also manages to avoid fragmentation of transmissions to and from the same SS or of the same modulation. Niyato and Hossain [24] designed adaptive UL bandwidth allocation and rate control mechanisms in an SS for polling service. The designed bandwidth allocation and rate control mechanisms use the queue status information to guarantee the desired QoS performance for polling service.

Supporting different applications and services with various QoS requirements in IEEE 802.16 wireless MAN is very challenging. A promising technique to address the challenge is the cross-layer design. A cross-layer scheduling algorithm is proposed [25] with a priority-based scheduler designed at the MAC layer for multiple connections with diverse QoS requirements, where an adaptive modulation and coding scheme is employed at the PHY layer for each connection. The connections are scheduled based on a priority function (PRF) defined for each connection. The PRF is updated dynamically according to the wireless channel quality, QoS satisfaction, and service priority across layers. The connection with highest priority is scheduled first each time. The proposed scheduler offers prescribed delay and rate guarantees for real-time and non-real-time traffic as well as uses the wireless bandwidth efficiently. In addition, it enjoys flexibility, scalability, and low implementation complexity. A new integrated architecture is proposed [26–27] in the IP layer and the MAC layer to provide cross-layer QoS control for the IEEE 802.16 wireless MAN. The proposed architecture implements a cross-layer traffic-based mechanism in a comprehensive way. A mapping rule and a fast signaling mechanism for providing the integrated service (IntServ) and the differentiated service (DiffServ) are given. The aim of the architecture is to guarantee different level QoS, prioritize the traffic classes, conduct multigranularity traffic grooming efficiently, adjust resource allocation dynamically, and share resources fairly.

7.6 Security Management

The security mechanism in IEEE 802.16 is defined as a privacy sublayer within the MAC layer [5,17]. In this section, we introduce the security mechanism of IEEE 802.16 in PMP mode and discuss the enhancements.

7.6.1 IEEE 802.16 Security Architecture

The security architecture of IEEE 802.16 consists of two main components: (1) Encapsulation protocol for encrypting the packet data. It defines a series of data encryption and authentication algorithms, such as the RSA public key encryption algorithm, the data encryption standard (DES) algorithm to encrypt the MAC PDU payloads, and so forth. It also gives the rules for applying those algorithms. (2) Privacy and key management protocol (PKM) for secure distribution of keying data.

7.6.1.1 Security Association

The security association (SA) is a set of security information that a BS and one or more of its SSs share in order to maintain secure communications [5]. There are two SA types defined in the 802.16 standard: data SA and authorization SA.

The data SA, which the standard explicitly defines, consists of a 16-bit SA identifier, the supported crypto algorithms, two TEKs to encrypt data, a

TEK lifetime, a 64-bit initialization vector for each TEK, and an indication of the type of data SA. There are three types of data SA defined in the standard: primary, static, and dynamic. Primary SAs are established during the SS initialization process. Static SAs are configured within the BS. Dynamic SAs are established and eliminated as needed for dynamic transport connections. The static and dynamic SAs can be shared by multiple SSs.

The authorization SA, on the other hand, is never explicitly defined in the standard. The authorization SA contains an X.509 certificate to identify the SS, a 160-bit authorization key (AK), lifetime of AK, a key encryption key (KEK) used by the BS to encrypt the TEKs, a DL hash function-based message authentication code (HMAC) key, a UL HMAC key, and a list of authorized data SAs. An authorization SA is shared between an SS and the associated BS.

7.6.1.2 SS Authorization

In the PMP architecture, SSs cannot directly communicate with each other. Traffic occurs only between the BS and SSs. Thus, the BS is in charge of the authorization and authentication of SSs.

When an SS logs into a network, the BS has to authorize the connection so that the SS may have a security association with the BS. Each SS has a unique X.509 digital certificate issued by the manufacturer, which includes the SS's public key and MAC address. During the initial authorization exchange, the SS sends an authentication information (AI) message that contains its X.509 certificate to the BS. Then the SS sends an authorization request message to the BS immediately after the AI message to request an AK. The BS will verify the SS's certificate. If it is valid, the BS generates an AK, encrypts it with the SS's public key using the RSA algorithm, and sends it back to the SS in an authorization reply message. The SS can get the AK by decrypting it using its public key. The authorization reply message includes the following information: (1) an AK encrypted with the SS's public key; (2) a 4-bit sequence number used to distinguish between successive generation of AKs; (3) a key lifetime; and (4) the security association identities (SAIDs). After the initial authorization, the SS has to periodically reauthorize with the BS to refresh its AK because of the limited key lifetime. The procedure of reauthorization is the same as for authorization except that the AI message is not sent by the SS. The AK exchange procedure is shown in Figure 7.12.

7.6.1.3 Key Management

After achieving authorization, the SS begins to request keying materials from the BS. The SS will initiate a separate TEK state machine for each SAID in the authorization reply message. The TEK is used for managing the keys to encrypt the actual data traffic. During the TEK exchange, the SS

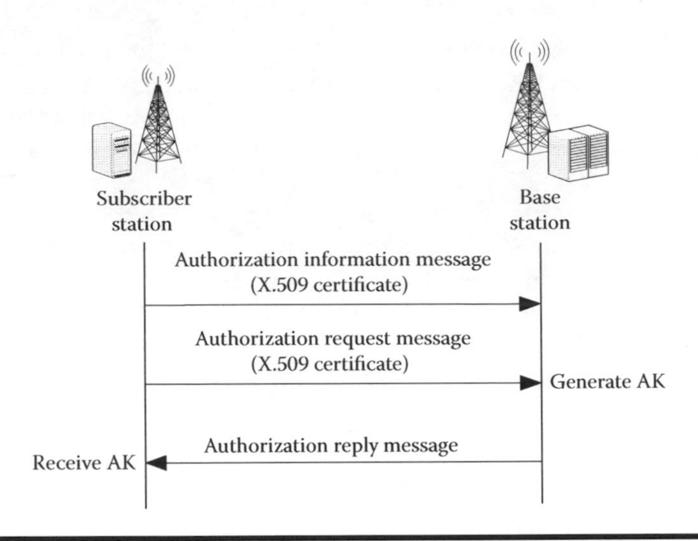

Figure 7.12 AK exchange procedure.

sends a key request message to the BS first. After receiving the request, the BS replies with a key reply message, which contains the BS's active keying material for the specified SAID. The TEK exchange procedure is shown in Figure 7.13.

The first message shown in Figure 7.13 is a rekey message, which is optional. It is sent from the BS only when the BS wants to rekey a data SA or create a new SA. The BS will choose a SAID from a SAIDList accessible by the SS. SeqNo is the AK's sequence number provided by the BS to the SS in the authorization process. The SS uses this number to determine which DL authentication key, HMAC_KEY_D, is used for message authentication. In message 2, the BS will use this number to determine which UL authentication key, HMAC_KEY_U, is used. The DL and UL authentication keys, HMAC_KEY_D and HMAC_KEY_U, are derived from the AK. By computing the value of digest HMAC(1), the SS can detect message corruption or forgery.

The SS uses message 2 (key request message) to request keying materials from the BS. If the SS does not receive message 1 before the current key expires, the SS will send a key request message when the current key is about to expire. The SS chooses the SAID from the SAIDList. HMAC(2) is the digest of message 2 computed by using HMAC_KEY_U, which allows the BS to authenticate the message.

Message 3 (key replay message) is replied from the BS to the SS after the BS receives message 2. The BS always keeps two active sets of keying materials for each SAID. The OldTEK is the set of keying material for the currently used TEK and the NewTEK is the set of keying materials to be used after the current TEK expires. The keying materials contain the TEK encrypted

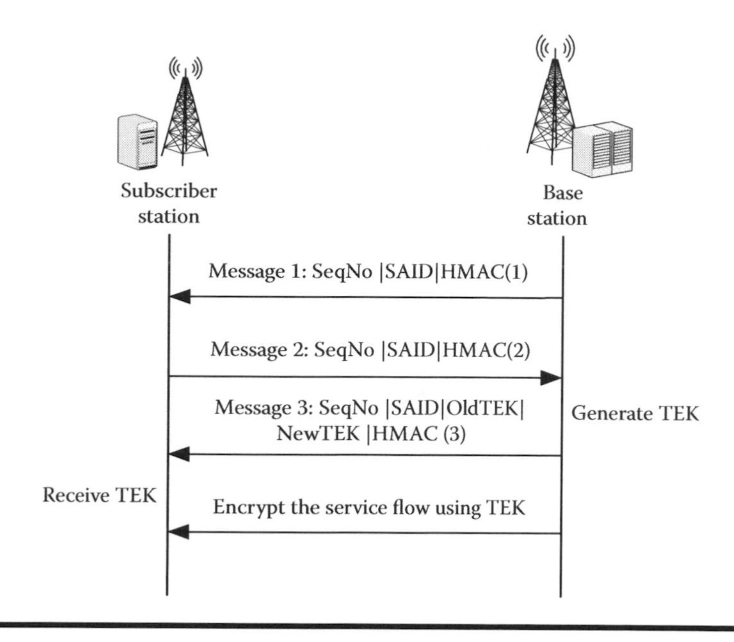

Figure 7.13 TEK exchange procedure.

by a KEK derived from AK using triple data encryption standard (3-DES) algorithm. HMAC(3) is the digest of message 3 computed from HMAC_KEY_D to ensure that the message the SS received has not been modified.

7.6.1.4 Cryptographic Suites

The cryptographic suites supported in IEEE 802.16 include: (1) data encryption algorithms: it supports no data encryption, encryption with DES in cipher block chaining (CBC) mode, or encryption with the U.S. Advanced Encryption Standard (AES) in CCM (counter with CBC-MAC) mode; and (2) TEK encryption algorithms: the TEK can be encrypted using DES, RSA, or AES algorithms.

7.6.2 Security Enhancements

To avoid the design mistakes of IEEE 802.11 [28–29] and improve network security, the IEEE 802.16 working group incorporated an existing standard, DOCSIS, into IEEE 802.16 [4,8]. Because DOCSIS is a wired technology and IEEE 802.16 is wireless, the security mechanism implemented in 802.16 is proved to have many problems in authentication and authorization procedures, key negotiation, and key management [17,30].

In IEEE 802.16, the security threats apply to both PHY and MAC layers [31]. Because the security mechanism operates at the MAC layer, there is no protection for attacks in the PHY level [17]. For MAC security support, there are several weaknesses [17,30]: (1) Because there is no BS certificate in authentication, the PKM protocol is open to forgery attack. (2) There is a serious AK-related problem in the authorization protocol. All key negotiation and data encryption key generation relies on AK's secret. However, there is no requirement in the standard for AK generation. (3) There is a TEK sequence space problem in the key management protocol. The TEK generated from AK has just a 2-bit identifier space, which is not enough during the AK lifetime. (4) The key distribution scheme offers no TEK freshness assurance.

The new IEEE 802.16e standard tries to address some of those weaknesses [9]. The Extensible Authentication Protocol (EAP) is adopted to authenticate devices on the network, which allows for nearly any imaginable authentication mechanism to be used, including those supporting mutual authentication [17]. EAP also works as the core part of the new IEEE 802.11i security standard [32].

A wireless key management infrastructure (WKMI) has been proposed [30] to improve the security of IEEE 802.16 WMAN. WKMI is a key management hierarchy infrastructure based on IEEE 802.11i. EAP frames under the TLS protocol are used for the authentication and key negotiation. In the message exchange of the infrastructure, EAP-TLS is combined with some modified information to improve the robustness of the system. Xu et al. [31] analyze the security flaws and illustrate possible attacks in the privacy and key management protocols of IEEE 802.16. They propose revised protocols by including the time stamps and extra signatures to prevent the attacks.

7.7 Conclusion

Fixed BWA systems have been considered as a promising approach for high-speed access networks. IEEE 802.16 provides a standard for the development and deployment of such systems. In this chapter, we presented a comprehensive survey of millimeter-wave–based IEEE 802.16 wireless MAN. The details of PHY and MAC layers in millimeter-wave bands were introduced. We also discussed the QoS support and security management issues.

References

[1] International Telecommunication Union, "ITU Internet report: Birth of broadband," Sept. 2003.

[2] C. Cicconetti, L. Lenzini, and E. Mingozzi, "Quality of service support in IEEE 802.16 networks," *IEEE Network,* pp. 50–55, March 2006.

[3] G. Chu, D. Wang, and S. Mei, "A QoS architecture for the MAC protocol of IEEE 802.16 BWA system," *IEEE Communications,* Circuits and Systems and West Sino Expositions, 2002.

[4] IEEE Std 802.16-2001, "IEEE standard for local and metropolitan area networks—Part 16: Air interface for fixed broadband wireless access systems," Apr. 2002.

[5] IEEE Std 802.16-2004 (Revision of IEEE Std 802.16-2001), "IEEE standard for local and metropolitan area networks—Part 16: Air interface for fixed broadband wireless access systems," 2004.

[6] A. Ghosh, D. R. Wolter, J. G. Andrews, and R. Chen, "Broadband wireless access with WiMax/802.16: Current performance benchmarks and future potential," *IEEE Communications Magazine*, pp. 129–136, Feb. 2005.

[7] Wireless MAN Working Group, http://WirelessMAN.org

[8] IEEE Std 802.16a, "IEEE standard for local and metropolitan area networks—Part 16: air interface for fixed broadband wireless access systems—Medium Access Control modifications and additional physical layer specifications for 2–11 GHz," 2003.

[9] IEEE Std 802.16e, "IEEE standard for local and metropolitan area networks—Part 16: air interface for fixed and mobile broadband wireless access systems, amendment 2: physical and medium access control layers for combined fixed and mobile operation in licensed bands and corrigendum 1," 2006.

[10] D.-H. Cho, J.-H. Song, M.-S. Kim, and K.-J. Han, "Performance analysis of the IEEE 802.16 wireless metropolitan area network," Proc. of the First International Conference on Distributed Frameworks for Multimedia Applications (DFMA'05), 2005.

[11] C. Eklund, R. B. Marks, K. L. Stanwood, and S. Wang, "IEEE standard 802.16: A technical review of the WirelessMANTM air interface for broadband wireless access," *IEEE Communications Magazine*, pp. 98–107, June 2002.

[12] S. Ramachandram, C. W. Bostian, and S. F. Midkiff, "A link adaptation algorithm for IEEE 802.16," Proc. of the IEEE Wireless Communications and Networking Conference, vol. 3, pp. 1466–1471, 2005.

[13] G. Nair, J. Chou, T. Madejski, K. Perycz, D. Putzolu, and J. Sydir, "IEEE 802.16 medium access control and service provisioning," *Intel Technology Journal*, vol. 8, no. 3, pp. 213–228, 2004.

[14] IEEE 802.16.1mc-00/01, "Media Access Control Protocol based on DOCSIS 1.1," 1999.

[15] P. Mahonen, T. Saarinen, and Z. Shelby, "Wireless Internet over LMDS: Architecture and experimental implementation," *IEEE Communications Magazine*, vol. 39, no. 5, pp. 126–132, May 2001.

[16] K. Wongthavarawat and A. Ganz, "Packet scheduling for QoS support in IEEE 802.16 broadband wireless access system," *International Journal of Communication Systems*, vol. 16, pp. 81–96, 2003.

[17] D. Johnson and J. Walker, "Overview of IEEE 802.16 security," *IEEE Security and Privacy*, vol. 2, no. 3, pp. 40–48, May 2004.

[18] I. Koffman and V. Roman, "Broadband wireless access solutions based on OFDM access in IEEE 802.16," *IEEE Communications Magazine*, pp. 96–103, 2002.

[19] S. Choi, G.-Ho Hwang, T. Kwon, A.-Ri Lim, and D.-Ho Cho, "Fast handover scheme for real-time downlink services in IEEE 802.16e BWA system," Proc. of IEEE 61st Vehicular Technology Conference, May–June 2005, pp. 2028–2032.

[20] K. Wongthavaranwat and A. Ganz, "IEEE 802.16 based last mile broadband wireless military networks with quality of service support," Proc. of IEEE MILCOM 2003, 2003, pp. 779–784.

[21] H. Wang, W. Li, and D. P. Agrawal, "Dynamic admission control and QoS for 802.16 wireless MAN," Proc. of 2005 Wireless Telecommunication Symposium, 2005, pp. 60–66.

[22] H. Alavi, M. Mojdeh, and N. Yazdani, "A quality of service architecture for IEEE 802.16," Proc. of Asian and Pacific Conference on Communications, Oct. 2005, pp. 249–253.

[23] S. A. Xergias, N. Passas, and L. Merakos, "Flexible resource allocation in IEEE 802.16 wireless metropolitan area networks," Proc. of the 14th IEEE Workshop on Local and Metropolitan Area Networks (LANMAN2005), Sept. 2005, pp. 1–6.

[24] D. Niyato and E. Hossain, "Queue-aware uplink bandwidth allocation and rate control for polling service in IEEE 802.16 broadband wireless networks," *IEEE Transactions on Mobile Computing*, vol. 5, no. 6, pp. 668–679, June 2006.

[25] Q. Liu, X. Wang, and G. B. Giannakis, "A cross-layer scheduling algorithm with QoS support in wireless networks," *IEEE Transactions on Vehicular Technology*, vol. 55, no. 3, pp. 839–847, May 2006.

[26] J. Chen, W. Jiao, and Q. Guo, "Providing integrated QoS control for IEEE 802.16 broadband wireless access systems," Proc. of IEEE 62nd Vehicular Technology Conference, Sept. 25-28, 2005, pp. 1254–1258.

[27] J. Chen, W. Jiao, and Q. Guo, "An integrated QoS control architecture for IEEE 802.16 broadband wireless access systems," Proc. of IEEE Globecom'05, Nov. 28–Dec. 2, 2005, pp. 3330–3335.

[28] W. A. Arbaugh, N. Shankar, and Y. C. Wan, "Your 802.11 network has no cloth," Mar. 2001; www.cs.umd.edu/waa/wireless.pdf

[29] N. Borisov, I. Goldberg, and D. Wagner, "Intercepting mobile communications: The insecurity of 802.11," Feb. 2001; www.Issac.cs.berkeley.edu/wep-faq.html

[30] F. Yang, H.-B. Zhou, L. Zhang, and J. Feng, "An improved security scheme in WAMN based on IEEE standard 802.16," Proc. of International Conference on Wireless Communications, Networking and Mobile Computing, 2005, pp. 1145–1148.

[31] M. Barbeau, "WiMAX/802.16 threat analysis," Proc. of 1st ACM Workshop on QoS and Security for Wireless and Mobile Networks (Q2SWinet'05), Montreal, Canada, 2005, pp. 8–15.

[32] J.-C. Chen, M.-C. Jiang, and Y.-W. Liu, "Wireless LAN security and IEEE 802.11i," *IEEE Wireless Communications Magazine*, pp. 27–36, Feb. 2005.

Chapter 8

Millimeter-Wave Dedicated Short-Range Communications (DSRC): Standard, Application, and Experiment Study

X. James Dong, Wenbing Zhang, Pravin Varaiya, and Jim Misener

Contents

This chapter talks about the ongoing IEEE 802.11p standard that enables intelligent transportation systems (ITS) applications between vehicles and roadside access equipment. As a variant of the indoor IEEE 802.11a PHY (physical layer) standard, IEEE 802.11p is designed to work over the multipath fading wireless channels. Extensive simulations have been performed to study the performance of IEEE 802.11a/p over multipath fading wireless channels. To demonstrate the potential deployment of IEEE 802.11p, a typical application, traffic probe application, has been implemented and is reported in this chapter. The system architecture, implementation, and experiments are presented.

8.1 Introduction

IEEE 802.11p is the incipient standard for dedicated short-range communications (DSRC) radio [9]. DSRC has been adopted as the radio service in the 5.850–5.925 GHz spectrum (5.9 GHz band) to provide the critical communication link for ITS. The recent national interest in vehicle infrastructure integration (VII) in the United States promises a revolution of ITS services given that DSRC implementation of wireless technologies can indeed be widespread. A VII project will lead to an infrastructure that continuously monitors the operational status of the transportation system, based on the information gathered by sensors in vehicles and devices installed in roadside stations. Many applications can only be implemented with the new infrastructure. Some of these applications include electronic toll collection, intelligent speed adaptation, and travel information collection. These VII-based applications will help to alleviate traffic congestion and achieve public safety goals such as collision avoidance.

This chapter describes an ongoing project between the California Department of Transportation (Caltrans), California Partners for Advanced Transit and Highways (PATH), and DaimlerChrysler Research and Technology North America (DCRTNA). The joint project demonstrates two potential VII services: one in traffic data probes and another with safety, using real cars and on Caltrans roadways. The chapter is organized as follows. First, we'll briefly introduce the IEEE 802.11p standard in Sections 8.2 and 8.3. Then the traffic probe application and our project goals will be described in Section 8.4. Sections 8.5 and 8.6 will present the overall system architecture, the main functionality of each software entity, and the implementation. The experiment will be introduced in Section 8.7. Finally, we'll summarize our lessons from the project and our future plan.

8.2 Overview of the Emerging DSRC Standard: IEEE 802.11p

Most high-speed packet wireless networks are not designed for mobile users. For example, the IEEE 802.11a standard explicitly states that it is proposed for indoor applications with low mobility [6]. However, the ongoing evolution of wireless devices and service requirements make inevitable the development of new high-speed packet wireless standards for mobile deployment. By far, IEEE 802.11p and IEEE 802.16e are the two major upcoming standards for mobile packet wireless networks. IEEE 802.11p is revised from IEEE 802.11a, whereas IEEE 802.16e extends IEEE 802.16a [11,19,20]. To combat multipath delay spread, OFDM (orthogonal frequency division multiplexing) is adopted in these standards due to its inherent multipath resistance [1]. In this section, we'll compare the PHY layer design of IEEE 802.11a and IEEE 802.11p.

8.2.1 IEEE 802.11a

Wireless local area networks (wireless LANs or WiFi) were developed in the 1990s as an extension of Ethernet, the dominant technology that enables today's Internet. Wireless LANs aim to transmit data and operate local networks without the constraints of wires and associated infrastructure normally required by Ethernet. Originally, the IEEE 802.11 standard specified the MAC (Media Access Control) layer and PHY (physical) layer in the 2.4-GHz band with data rates of 1 and 2 Mbps using either direct sequence spread spectrum (DSSS) or frequency hopping spread spectrum (FHSS) [5]. In 1999, the IEEE defined two high-rate extensions: 802.11b, which is based on DSSS and CCK (complementary code keying) with data rates up to 11 Mbps in the 2.4-GHz band, and 802.11a, which is based on OFDM technology with data rates up to 54 Mbps in the 5-GHz band [6–7]. In 2003, the 802.11g standard, which extends the 802.11b PHY layer to support data rates up to 54 Mbps in the 2.4-GHz band, was finalized [8]. The IEEE 802.11 standard and its several important supplements (IEEE 802.11a/b/g) provided a basis for interoperability of different products and triggered the explosive growth of the wireless LAN market. IEEE 802.11n, a new amendment to the 802.11 standard, is on the way [18]. IEEE 802.11n promises a higher data rate in excess of 100 Mbps, and longer operating distance than IEEE 802.11a/b/g wireless LANs using MIMO (multiple input multiple output) technology, and prestandard 11n products are already widely available.

The PHY layer of the 802.11a is divided into two entities: the Physical Layer Convergence Protocol (PLCP) and the Physical Medium Dependent (PMD) sublayers. The PLCP sublayer maps MPDUs (MAC protocol data units) from the MAC layer into a frame format suitable for the PMD layer and delivers incoming frames from the PMD layer to the MAC layer. The PMD layer deals with actual transmission and reception over the air medium [5].

The PHY frame format for 802.11a is shown in Figure 8.1. The PLCP preamble field has ten repetitions of a short training symbol, a guide interval, and two repetitions of a long training symbol. In the receiver, the short training symbols are used for automatic gain control (AGC; to prevent signals from saturating the output of the A/D converter) convergence, timing acquisition, and coarse frequency acquisition, while the long training symbols are used for channel estimation and fine frequency acquisition. Here the frequency domain channel estimation based on two long training symbols T_1 and T_2 assumes that the wireless channel remains the same till the end of packet transmission.

Binary input data from MAC layer is collected and coded with forward error correction (FEC) schemes such as convolutional encoder. Then the coded bits are punctuated to achieve high rates and interleaved to combat bursty bit errors. Afterwards, based on the selected baseband modulation scheme and its number of bits per symbol, interleaved binary bits are grouped and mapped to corresponding constellation points or data symbols. These symbols are serial complex numbers and are divided into

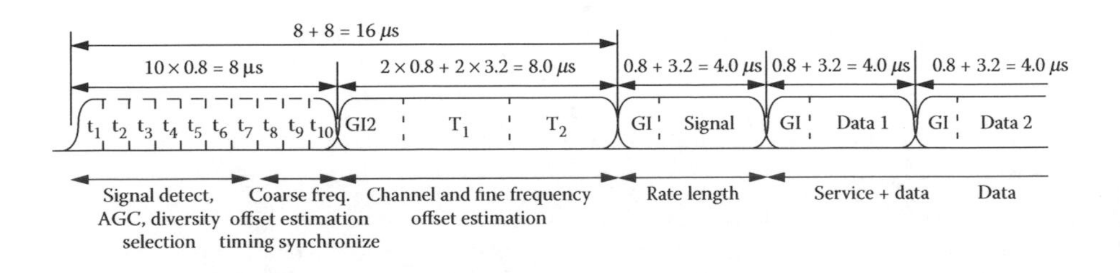

Figure 8.1 IEEE 802.11a PHY frame format [6].

groups of 48 symbols. Each such group is associated with one OFDM symbol. Assuming that the active subcarriers are sequentially numbered from −26 to 26, in each group, the baseband symbols are mapped into 48 OFDM subcarriers numbered −26 to −22, −20 to −8, −6 to −1, 1 to 6, 8 to 20, and 22 to 26. At this point, pilot symbols with a known modulation scheme may be inserted into four subcarriers, −21, −7, 7, and 21, with a known pattern. In addition, zero symbols for unloaded subcarriers are also inserted, so that the number of subcarriers is a power of 2, which is required by the IFFT (inverse fast Fourier transform)/FFT operation (64 in 802.11a) [6].

The OFDM modulation in the frequency domain, IFFT operation, is performed on the parallel symbols to generate parallel symbols. In OFDM, the symbols are split and transmitted over a large number of subcarriers and modulated at a low rate. The subcarriers are made orthogonal to each other by appropriately choosing the frequency spacing between them. For example, the frequency of the modulating sinusoid in each subcarrier is an integer multiple of a base frequency, $1/T_s$, where T_s is the symbol period. These parallel symbols help to address the intersymbol interference (ISI) problem. For a single-carrier wireless system, its symbol rate, R_s, is inversely proportional to the symbol period, T_s. Higher R_s means smaller T_s ($T_s = 1/R_s$), which might cause serious ISI problems in multipath channels with delay spread larger than T_s. In OFDM systems with data rate R_s, because data is split to N subcarriers, the symbol rate in each subcarrier is R_s/N and the symbol period in each subcarrier is $N \cdot T_s$. Therefore, in multipath channels, OFDM systems are more robust to ISI than single-carrier systems with equivalent data rate.

These parallel symbols are then serialized and inserted with the cyclic prefix to form an OFDM baseband symbol for one OFDM symbol period in the time domain. The data portion of an OFDM symbol (as seen from Figure 8.1) comes from the time domain signals. A cyclic prefix is introduced as guard interval (GI) by prepending to the IFFT waveform a circular extension of itself and truncating the resulting periodic waveform to a single OFDM symbol length. Due to the cyclic prefix, the transmitted time domain signal becomes periodic, as long as the length of the cyclic prefix is larger than the delay spread of the multipath channel. Therefore, in the time domain, the effect of the multipath channel becomes a circular convolution with the channel impulse response function. In the frequency domain, the effect of the multipath channel is then a point-wise multiplication of the constellation symbols and the channel frequency response function. Periodicity of an OFDM symbol also ensures that subcarriers after FFT are orthogonal to each other. Thus the intercarrier interference (ICI) may be reduced [1,4].

The OFDM symbols are appended one after another, starting after the PLCP header, until the length is reached. The PLCP header field has information about the transmission rate and the length of the payload, a parity bit, and six 0 tail bits. The rate field conveys information about the type of modulation and coding rate used in the rest of the packet. The length field specifies the number of bytes in the physical layer service data unit (PSDU).

The operating frequencies of 802.11a in the United States fall into the National Information Structure (U-NII) bands: 5.15–5.25 GHz, 5.25–5.35 GHz, and 5.725–5.825 GHz. Within this spectrum, there are twelve 20-MHz channels, and each band has different output power limits. The complex baseband time domain waveform will be upconverted to a radio frequency according to the center frequency of the desired channel and transmitted.

8.2.2 IEEE 802.11p

Recently, the upcoming IEEE 802.11p has been endorsed by the ASTM as the platform for the PHY and MAC layers of DSRC. Providing wireless communications between vehicles and the roadside, and between vehicles, DSRC enables a whole new class of applications that enhance the safety and productivity of the transportation system.

As a variant of IEEE 802.11a, 802.11p follows many design features of 802.11a, such as frame structure, training sequences, scrambler, convolutional coding, interleaving, modulation schemes, pilot subcarriers, IFFT/FFT size, cyclic prefix, and pulse shaping [19–20]. IEEE 802.11a is basically designed for low-mobility indoor applications, where the wireless channel is assumed to be stationary for the frame duration. Therefore, all system parameters are chosen to achieve best performance in indoor propagation environments. However, IEEE 802.11p is proposed for high-mobility outdoor environments. Thus, it has to deal with the impairments brought by high mobility. Some of these challenging requirements for PHY layer design include:

- The transceiver has to combat increased multipath delay spread.
- The transceiver has to combat increased Doppler spread, which means the multipath channel varies more rapidly. It is no longer a realistic assumption that the channel estimation acquired at the beginning of the transmission will be valid till the end of the transmission.
- Longer communication range.

IEEE 802.11p works in the 5.850–5.925 GHz ITS radio service (ITS-RS) band, which accommodates seven channels in a total spectrum of 75 MHz. In 802.11p, each mandatory channel operates with 10-MHz bandwidth, rather than the 20-MHz bandwidth used in 802.11a. Accordingly, rates of 802.11p are half those of 802.11a and the symbol period of 802.11p is twice the symbol period of 802.11a. The advantage of a longer symbol period is that the longer cyclic prefix of each OFDM symbol enables 802.11p to combat possibly larger delay spread introduced by outdoor channel environments. In addition, the transmission power limits designated by 802.11p are different from the power limits of 802.11a. The maximum antenna input power for some DSRC mandatory channels is 28.8 dBm (750 mW), which enables longer range.

8.3 Performance of IEEE 802.11a/p in Multipath Fading Wireless Channels

The convenience brought by mobility support in an urban environment comes with the harsh and challenging time-varying channel condition. In addition to the line-of-sight (LOS) or direct-path component (if there is one), the radio signal emitted by the transmitter is reflected by many environment objects, such as buildings, trees, and cars, to reach the receiver over many different paths. Since the wavelength of modern high-speed packet wireless networks is very short (e.g., only about 5 mm at a carrier frequency of 5.9 GHz), the small distance difference between paths followed by the same signal means a huge phase difference between received radio waves. These radio waves at the receiver may add (constructively) or cancel (destructively) each other. Due to the relative movement between receiver and transmitter in the communications system or the random movement of environment reflecting objects, the travel paths of signals for mobile applications always change, and accordingly the constructive/destructive effect on signals keeps changing too. As a result, the wireless channel for mobile users, called the multipath fading channel, is much more unpredictable than a static channel [3].

Although the multipath fading channel is unpredictable, it shows some statistical characteristics that can be modeled in terms of several important parameters. Among them, delay spread and Doppler spread are two fundamental parameters of a mobile multipath channel. Doppler spread indicates how fast the channel changes, whereas delay spread shows the time dispersion of signal arrival along different paths.

We study how these two important parameters of multipath fading channels affect the performance of high-speed packet wireless networks. Many techniques may be used to perform the evaluation. Experiments using real radio, hardware, and software are practical. However, since the behavior of mobile multipath channels strongly depends on environment, such as terrain, foliage, buildings, and moving objects such as cars, it will be hard to reproduce experiment results. Therefore, a simulation-based approach is preferred to study the effect of time-varying channel conditions because of its great advantages such as repeatability, controllability, and low cost.

Extensive simulations are reported in this section investigating how Doppler spread and delay spread affect the performance of several popular high-speed packet wireless networks, including IEEE 802.11a and IEEE 802.11b. Of them, IEEE 802.11p is similar to IEEE 802.11a and has similar performance behavior.

8.3.1 Multipath Fading Channel Model

A baseband multipath fading channel is usually modeled with a multiplicative fading component and an additive noise component [3]. Two typical models for the multiplicative component are the Rayleigh and Rician distributions.

Assuming that the received signal is the sum of signals with different phases caused by different paths, the amplitude of the received signal, r, can be modeled as a random variable with a Rayleigh distribution, whose probability density distribution, $f(r)$, is given by

$$f(r) = \frac{r}{\sigma^2} \exp\left(-\frac{r^2}{2\sigma^2}\right), \quad r \geq 0,$$

where $2\sigma^2$ is the predetection mean power of the received signal [12].

Although Rayleigh fading is often a good approximation of realistic channel conditions, it is considered by many to be a worst-case scenario of signal fading. If a wireless receiver works in a Rayleigh fading channel, then it is likely to work in other types of channels.

In simulations, as shown in Figure 8.2, the Rayleigh fading channel is modeled as

$$r(t) = h(t) \cdot s(t) + n(t),$$

where $r(t)$ is the received signal, $n(t)$ is the noise, $s(t)$ is the transmitted signal, and $h(t)$ refers to a multiplicative distortion of the transmitted signal, $s(t)$. The Rayleigh channel simulator generates the fading envelope coefficients, $h(t)$, using a statistically accurate and computationally efficient approach.

Rayleigh fading with a strong LOS is called Rician fading with probability density distribution, $f(r)$,

$$f(r) = \frac{r}{\sigma^2} \exp\left(-\frac{r^2 + K^2}{2\sigma^2}\right) I_0\left(\frac{rK}{\sigma^2}\right),$$

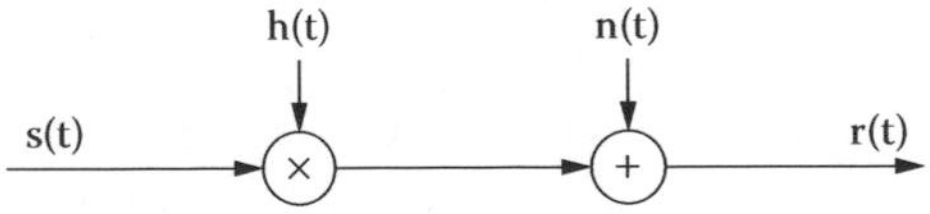

Figure 8.2 Multiplicative and additive model for fading channels.

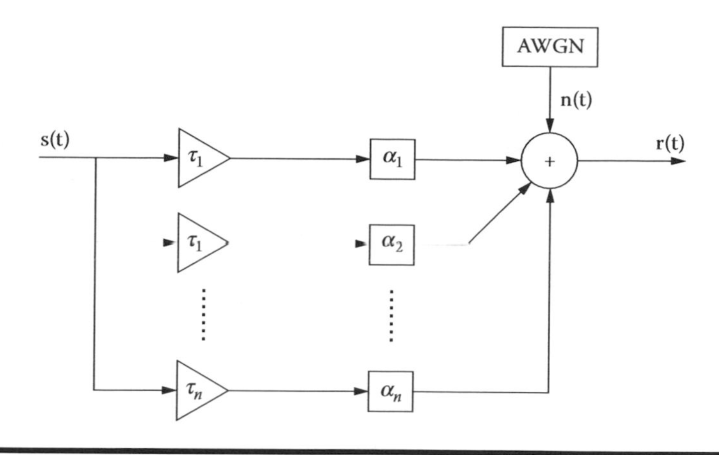

Figure 8.3 Tapped delay line channel model.

where K is the coefficient that indicates how strong the LOS component is compared to the rest of the received signal. I_0 is the zero-order modified Bessel function [3, 12–13].

Both Rayleigh fading and Rician fading are in the standard Simulink block library. Jakes' model is used to generate the Rayleigh fading coefficients in the block.

A general model, the tapped delay line channel model [3,13], is used to model a time-varying multipath fading channel with L multipath signal components, as illustrated in Figure 8.3. The channel model consists of a tapped line with different delays. The tap coefficients, denoted by $\alpha_i(t)$, are usually modeled as complex-valued Rayleigh fading coefficients or Rician fading coefficients that are uncorrelated with each other. The tap delay, τ_i, corresponds to the amount of time dispersion in the multipath fading channel. In this model, Doppler spread is used to generate the tap coefficients and delay spread is used to configure tap delays.

8.3.2 IEEE 802.11a and IEEE 802.11b Simulators

The two simulators used for this chapter are modified from the existing IEEE 802.11a PHY Simulink model and IEEE 802.11b PHY Simulink model on the MATLAB Central exchange Web site [14]. To make a fair performance comparison, the two simulators share the same code of multipath fading channel model.

As shown in Figure 8.4, the IEEE 802.11a digital baseband transceiver can be used to perform a complete end-to-end simulation. The baseband simulator assumes that the underlying passband subsystem works perfectly without any ICI. Inside the frequency domain equalizer, the first four

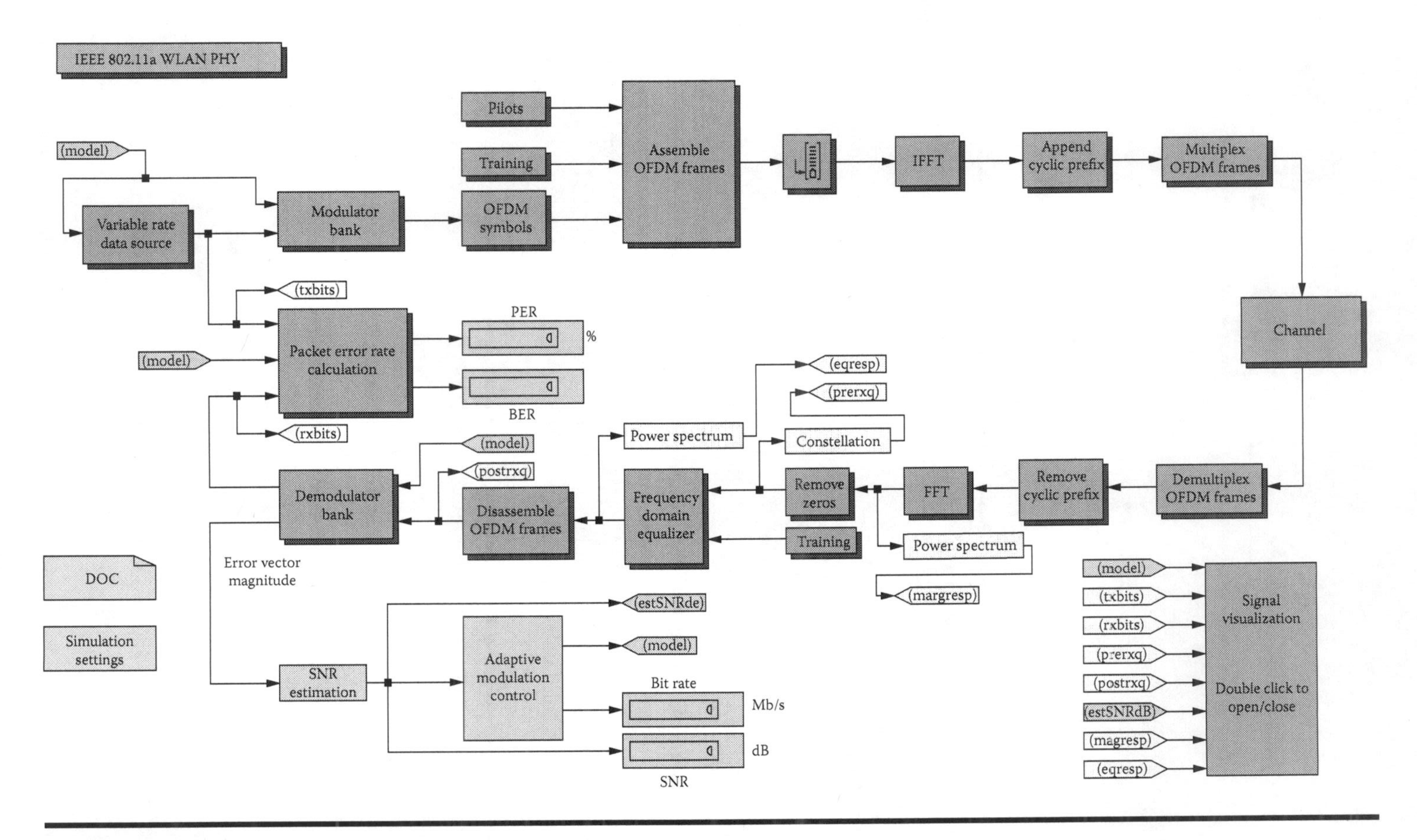

Figure 8.4 IEEE 802.11a PHY simulator.

training symbols in each of these 52 subcarriers are extracted and divided by the corresponding four known training symbols to calculate the channel estimate, which is used to correct the received data symbols within the same OFDM block. Each subcarrier has its own channel estimator. The adaptive modulation block is disabled in our simulations. The simulator works in (BPSK, 1/2) mode, e.g., at a data rate of 6 Mbps.

Figure 8.5 shows the DSSS simulator for an IEEE 802.11b digital baseband transceiver. The simulator follows the IEEE 802.11b standard [7]. It supports 1 Mbp, 2 Mbps, 5.5 Mbps, and 11 Mbps rates. The basic components includes DBPSK (differential binary phase shift keying) and DQPSK (differential quadrative phase shift keying) modulation, Barker code spreading, and CCK. However, only the 1 Mbps rate (e.g., DBPSK and Barker code spreading) is used in our simulations. Perfect synchronization is assumed by the simulator.

We now present key numerical results to highlight the effects of delay spread and Doppler spread. Simulations have been performed under various mobile multipath channel conditions (via different Doppler spread and delay spread). Each simulation run lasts 50 s in the simulated system time. In our simulations, PHY traffic is assumed to be in saturation state; i.e., the PHY layer buffer is never empty.

8.3.3 Effect of Doppler Spread

To concentrate on the effect of Doppler spread, only one tap of multipath fading channel is used in our simulations, as shown in Figure 8.6. The SNR (signal-to-noise ratio) for all simulations is set to 10 dB. As the controllable parameter, the Doppler spread in the Rayleigh multipath fading block or Rician multipath fading block is changed for each simulation run.

From Figures 8.7 and 8.8, it is observed that both 802.11a and 802.11b perform worse when the mobile multipath channel changes rapidly (the speed and Doppler spread becomes larger). However, the performance of 802.11a degrades much more than 802.11b. The fundamental reason is that 802.11b has a much higher symbol rate or shorter symbol period than 802.11a.

In the 802.11a standard, channel estimation and fine frequency offset estimation are done via training symbols at the beginning of a PHY frame. The channel estimates obtained during this period are used to compensate for multipath effects for the entire frame. The implicit assumption is that the channel will remain stationary for the duration of the OFDM frame. Although the assumption may be valid for low-mobility channel conditions, it doesn't hold for mobile multipath channels. The fast-varying channels are not equalized adequately based on the available training symbol placement scheme. In addition, Doppler spread may translate to carrier frequency

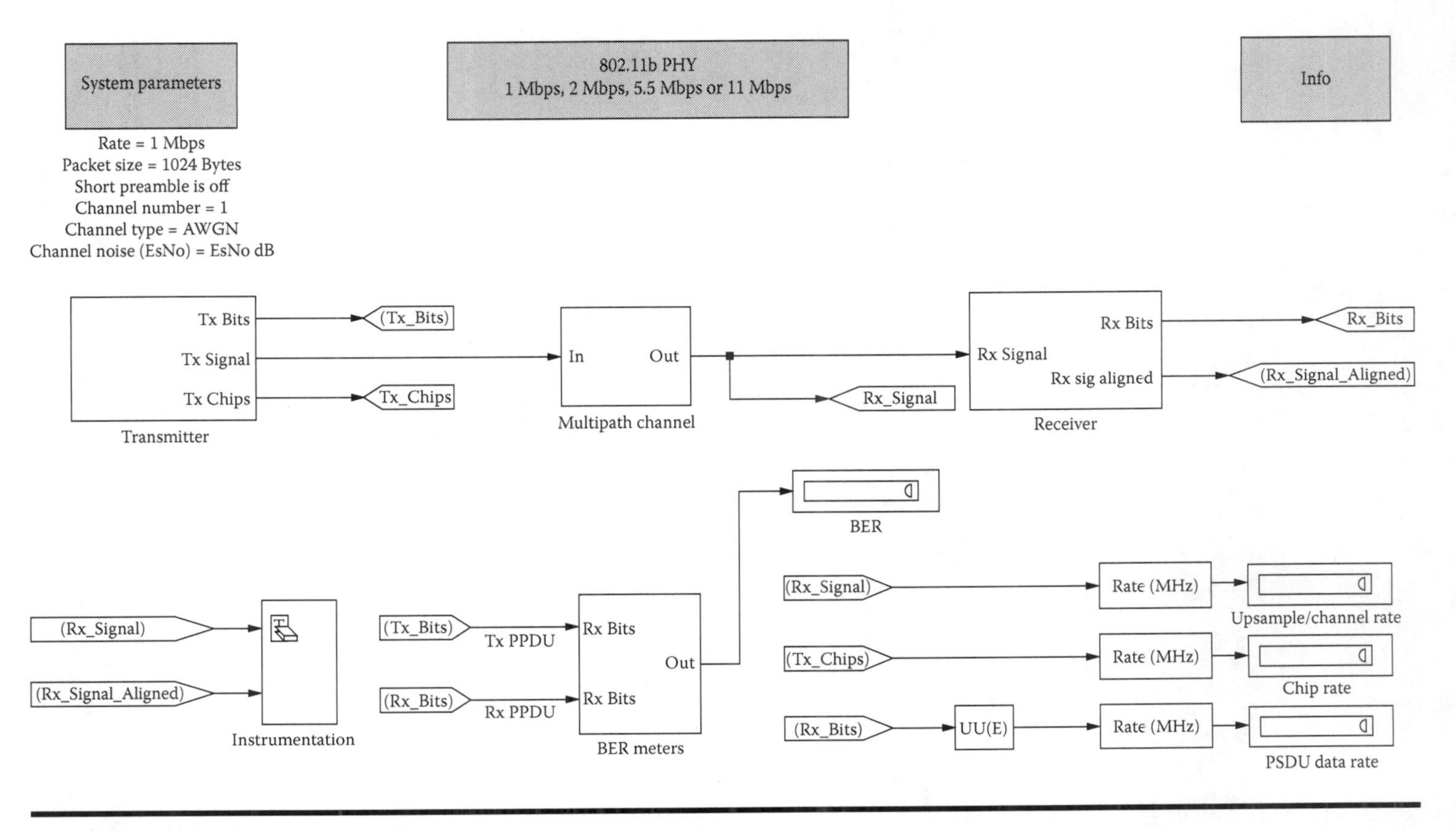

Figure 8.5 IEEE 802.11b PHY simulator.

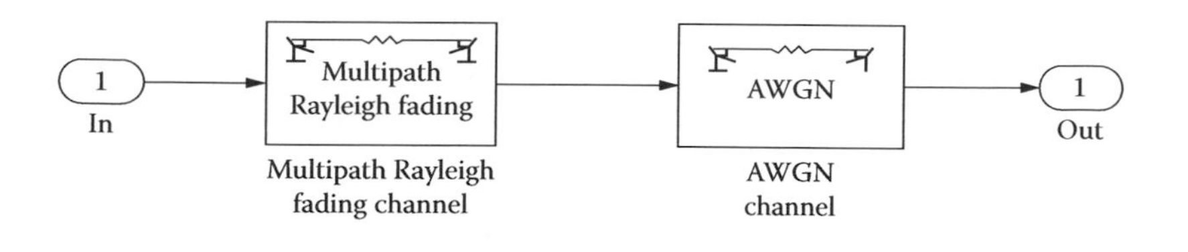

Figure 8.6 Channel blocks for the effect of Doppler spread.

offset, which causes subcarriers of OFDM to lose their orthogonality relative to each other. (The perfect synchronization assumption in this chapter produces more optimistic simulation results.) For 802.11a in mobile multipath channels, an equalizer is certainly required to adapt and compensate for the fast-changing channel. To combat distortion introduced by large Doppler spread, OFDM-based wireless systems need more advanced channel estimation techniques than what is currently proposed for 802.11a. Indeed, 802.16e does propose a better pilot symbol placement scheme so that efficient channel estimation can be implemented [11].

Because 802.11b has a high symbol rate, well above the usual Doppler spread (less than 5000 Hz), the channel is considered to be fairly constant during each symbol transmission. In addition, 802.11b's DBPSK modulation scheme enhances its intrinsic ability to handle a larger Doppler spread than 802.11a.

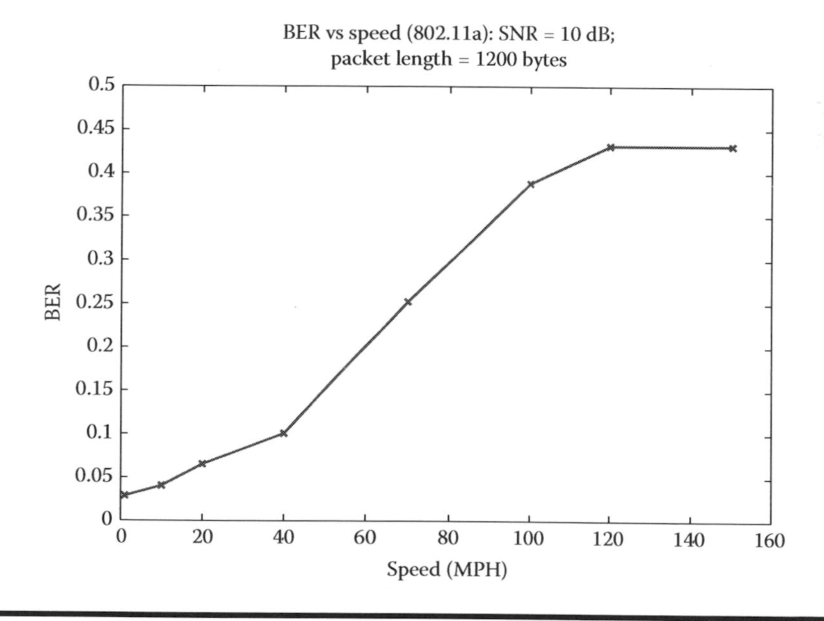

Figure 8.7 Effect of Doppler spread on IEEE 802.11a.

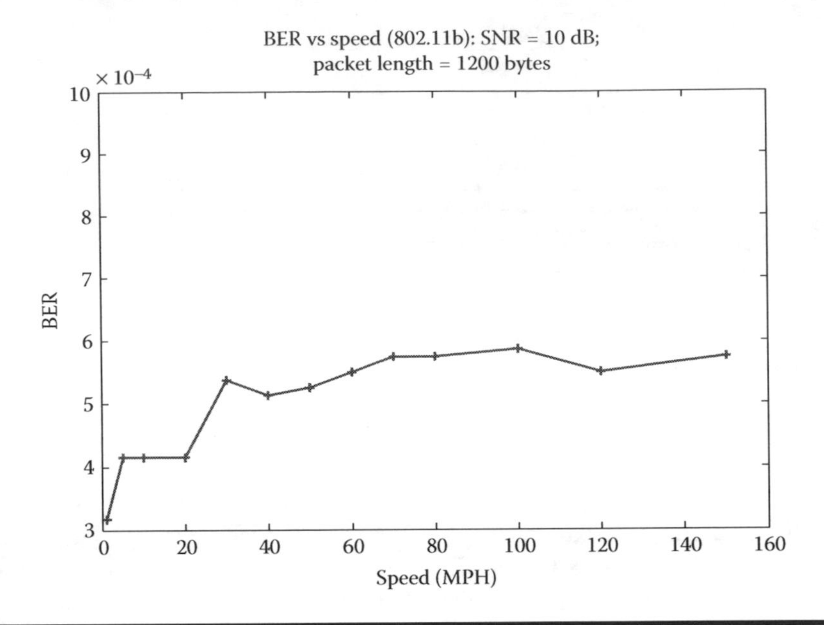

Figure 8.8 Effect of Doppler spread on IEEE 802.11b.

8.3.4 Effect of Delay Spread

Delay spread (or equivalently coherent bandwidth) describes the time-dispersive nature of multipath fading channels. But it does not provide information about the time-varying nature (caused by relative motion of transmitter and receiver). There are two ways to define delay spread: maximum excess delay and root mean square (RMS) delay.

Maximum excess delay is defined as the overall time span from the earliest arrival to the latest arrival. As the simplest way to define delay spread, maximum excess delay does not exhibit the relative amplitudes of multipath components (or intensity-delay profiles), which will strongly affect the system performance. Thus, a better measure of delay spread is the RMS delay spread, which is given mathematically by

$$\tau_{rms} = \sqrt{\overline{\tau^2} - (\bar{\tau})^2},$$

where, given L propagation paths,

$$\overline{\tau^n} = \frac{\sum_{i=1}^{L} \tau_i^n |\alpha_i|^2}{\sum_{i=1}^{L} |\alpha_i|^2}, \qquad n = 1, 2.$$

The RMS delay spread reveals the distribution of arriving signal power along different paths, while maximum excess delay spread tells only the

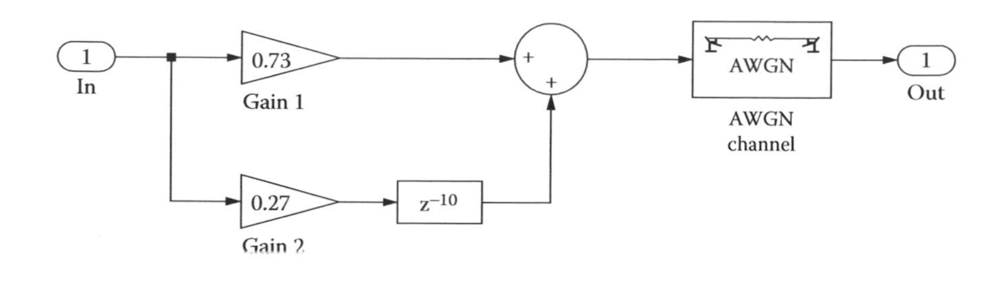

Figure 8.9 Channel blocks for the effect of delay spread.

time difference between the first arrival signal and the last arrival signal. However, to simplify the parameter configuration in our simulations, we use maximum excess delay as the metric for delay spread.

As shown in Figure 8.9, a two-tap multipath channel model is used in the simulations. There is no fading for either of these two signal components. Adding fading for each component will lead to worse performance. Here we are trying to isolate the effect of delay spread from Doppler spread. During the simulations, the maximum excess delay has been gradually increased by setting the parameter of delay block in one of two signal components. The ratio of signal power on each tap is 0.73:0.27.

BER (bit error rate) versus maximum excess delay for 802.11a and 802.11b are plotted in Figures 8.10 and 8.11, respectively. It is easy to see

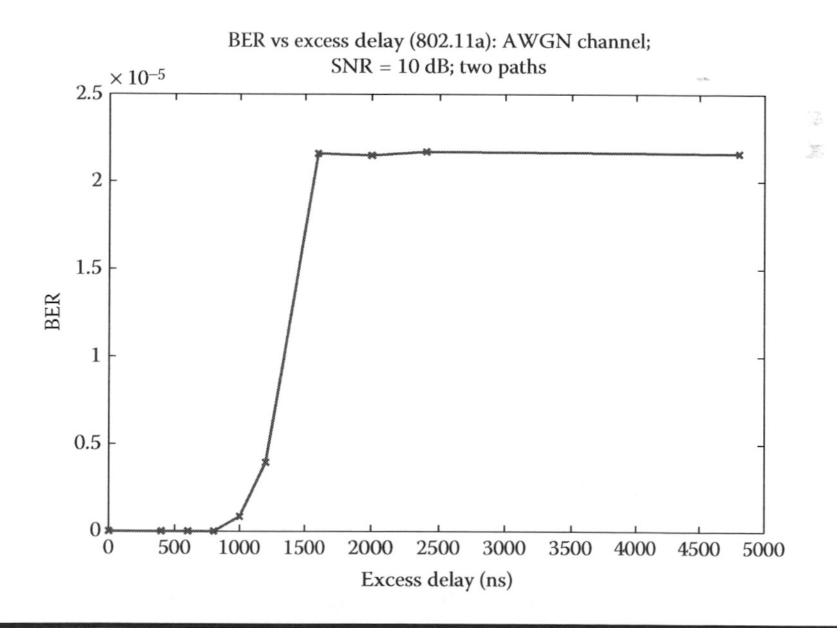

Figure 8.10 Effect of delay spread on IEEE 802.11a.

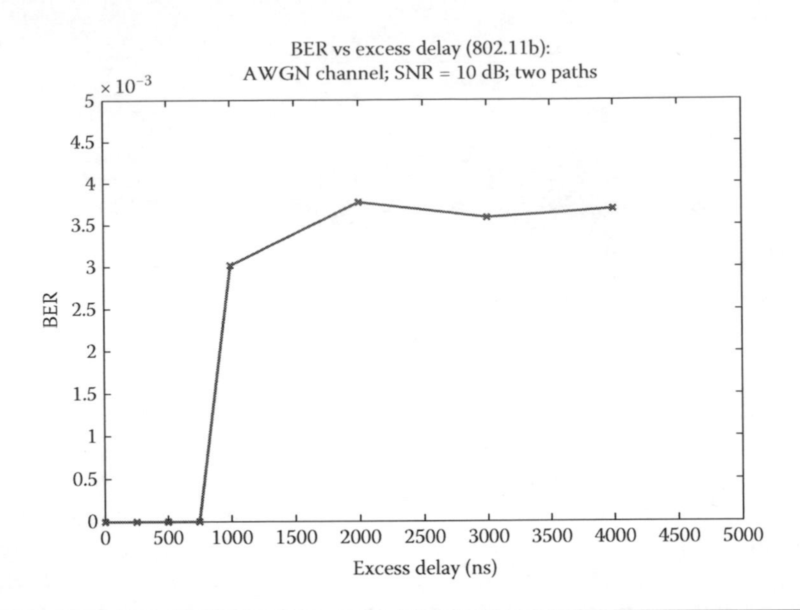

Figure 8.11 Effect of delay spread on IEEE 802.11b.

that performance of both 802.11a and 802.11b is degraded, as the delay spread increases. These figures show that 802.11a tolerates a larger delay spread (around 1.6 ms) than 802.11b (around 1 ms), although 802.11a (6 Mbps) has a higher data rate than 802.11b (1 Mbps). In addition, even with the same delay spread, 802.11a achieves better BER performance than 802.11b. For example, with delay spread around 2 ms, the BER scale of 802.11a is 10^{-5}, but the BER scale of 802.11b is only 10^{-3}. We have to emphasize that maximum excess delay has been used as delay spread instead of RMS delay spread. Otherwise, it is easy to get confused by simulation results reported elsewhere.

IEEE 802.11a's relative immunity to delay spread is due to a combination of the slower symbol rate (longer symbol period) and placement of significant guard time (cyclic prefix) around each symbol, which provide protection against ISI. To combat very large delay spread in an outdoor urban environment, the upcoming 802.11p standard enhances the robustness of 802.11a by reducing the symbol rate by half. By contrast, 802.11b is very sensitive to larger delay spread because of its higher symbol rate or short symbol period. Reception of 802.11b in a multipath channel can be substantially improved by techniques such as the RAKE receiver principle and equalization [15]. However, they come with complexity and cost.

One more thing worthy to point out is that the introduction of additional taps (paths) or fading effect on each path will degrade the system performance further. What we see in Figures 8.10 and 8.11 are the best multipath scenarios.

8.4 Introduction to Traffic Probe Application

Travel time and traffic speed between predetermined freeway locations provide important information to daily commuters, highway administrators, and public safety personnel. A number of techniques, such as magnetic loop detectors, video monitors, and radar detection, have been used to measure traffic speed and travel time. The incipient DSRC technology provides a new way to estimate travel time and traffic speed.

The Freeway Performance Measurement System (PeMS) has been developed jointly by the University of California, Berkeley, CalTrans, California PATH, and Berkeley Transportation Systems, Inc. PeMS has real-time and historical data for many of California's freeways in its database. More important, its built-in software modules help PeMS users conduct freeway operational analysis, planning, and research. Details of the PeMS are available [16, 17].

The traffic application software developed for the EVII project feeds GPS (global positioning system) data records from the traffic probe vehicles into the PeMS, produces the vehicle travel time between two predetermined freeway locations in a format that is compatible with that provided by the metropolitan transportation commission's (MTC) toll tag readers, and allows users to access the travel time via Web browsers such as Internet Explorer and Firefox over Internet.

8.5 System Architecture

8.5.1 Software Components

There are mainly four components in the system: the OBU sender, the RSU collector, the EVII-PeMS adaptor, and the PeMS traffic probe application, as shown in Figure 8.12.

The EVII probe data onboard unit (OBU) developed by DCRTNA is composed of three functional components: data (CAN and GPS) interface devices, an onboard computer, and a data-transmitting device (DSRC radio), as illustrated in the following schematics.

The EVII onboard system includes software for traffic probe data collection. In addition, it has DSRC listening and transmission components, as shown in Figure 8.13. The onboard computer gets GPS data from the computer's serial interface and calculates the distance traveled by the vehicle based on the GPS position data. It recorded a GPS point about every 50 m. The onboard computer also monitors CAN activities on both the high-speed and low-speed CAN buses through a USB interface and selectively records CAN parameters that are important for traffic probe applications, such as vehicle windshield wiper activities.

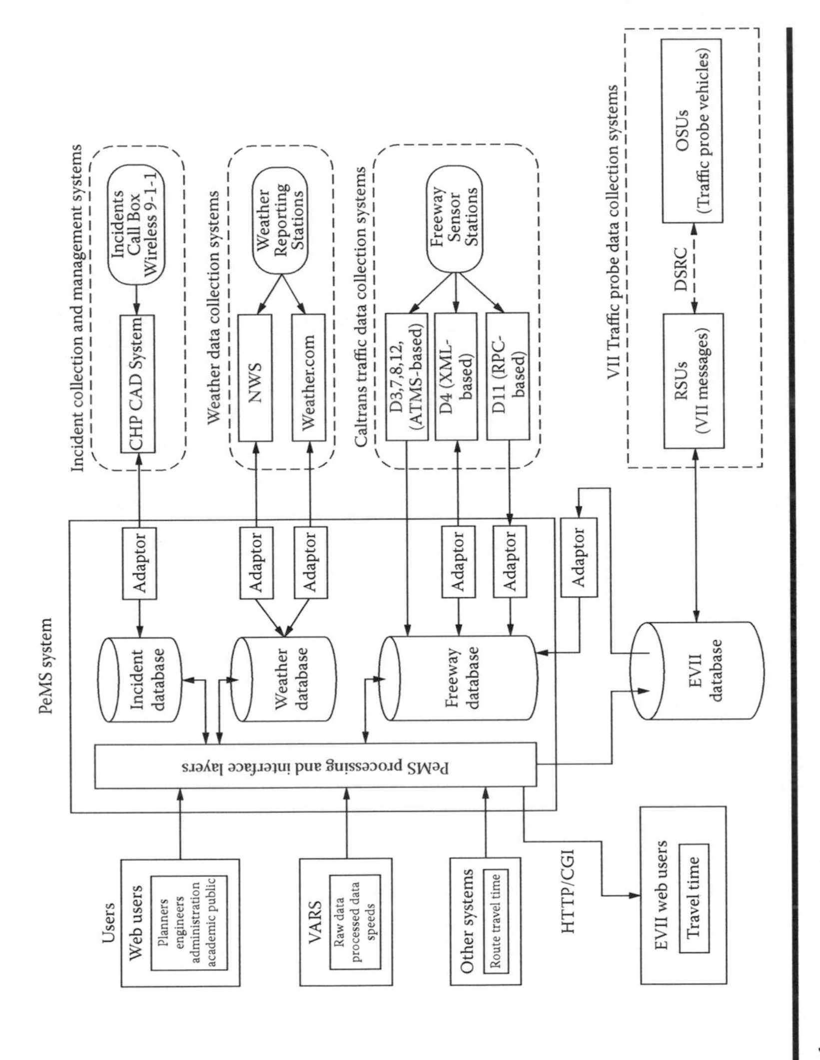

Figure 8.12 Software components.

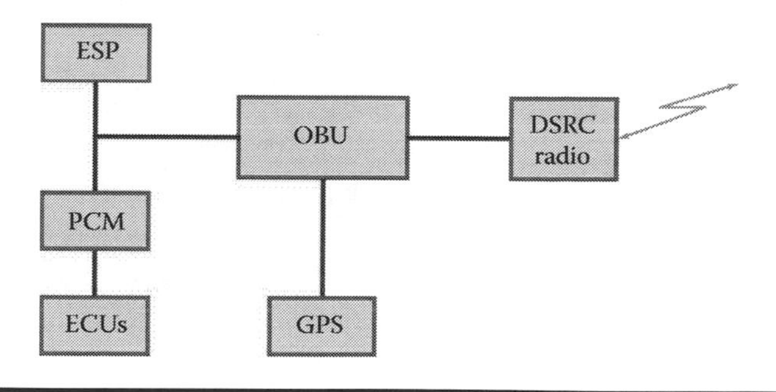

Figure 8.13 The onboard system.

The OBU sends broadcast UDP (User Datagram Protocol) packets over the DSRC wireless link to an RSU (roadside unit) along a highway. The payload portion of those UDP packets is defined in the VII message set, as shown in Figure 8.14. The VII probe data message uses binary encoding and presents the probe data in snapshots. In the EVII project, each GPS data point (a combined NMEA, GGA, and RMC) was encoded in a snapshot. In EVII, the number of vehicle device status fields was set to zero for probe data applications.

The RSU collector knows the PeMS server and is responsible for collecting all traffic probe messages from OBUs, aggregating them, and forwarding them to the PeMS server reliably via the GPRS (general pocket radio service) backhaul radio and Internet.

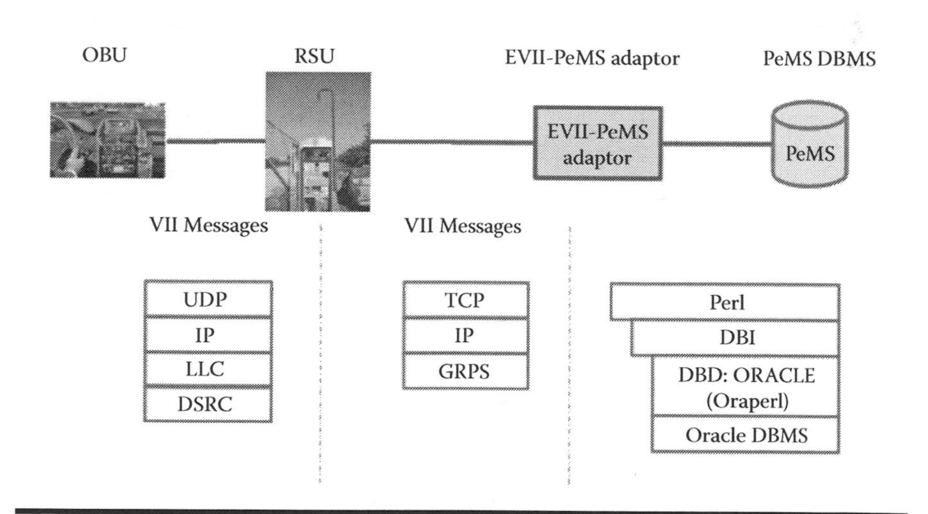

Figure 8.14 Protocol stack.

The EVII-PeMS adaptor collects the traffic probe data records from RSU, converts all GPS data into FDM (frequency division multiplexing) data, and stores them in the PeMS database. In addition, the EVII-PeMS adaptor estimates the traffic speed and travel time based on the received traffic probe data records and updates the PeMS database.

The PeMS traffic probe application is a CGI (Common Gateway Interface) program that accepts Web users' queries and returns intuitive graphical travel time and traffic speed information to users.

8.5.2 Protocol Stack

The protocol stack between different software entities is shown in Figure 8.14. The UDP-based protocol between OBUs and RSUs helps to increase the throughput over the DSRC wireless link and combat mobility inherent with OBUs. Reliable TCP (Transmission Control Protocol) is used between RSUs and the EVII-PeMS adaptor to avoid packet loss. However, since the GPRS backhaul is slow and not stable, we put more intelligence in nodes closer to the PeMS server and developed special protocols between RSU and the EVII-PeMS adaptor.

8.6 Implementation of EVII Prototype System

8.6.1 Hardware Equipment

A traffic probe vehicle with a 5.9-GHz DSRC transceiver (OBU) and a GPS is used to regularly generate probe records, as shown in Figure 8.15. Based on its memory and link capacity, the OBU also stores probe records temporarily until a connection with an RSU is detected.

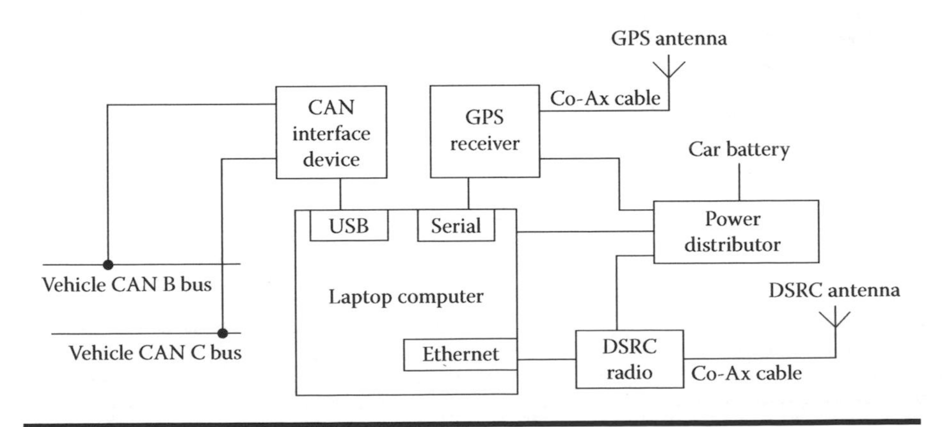

Figure 8.15 The on-board hardware.

An RSU is a computer system equipped with a 5.9-GHz DSRC transceiver, an interface to the backbone networks (GPRS modem), application-oriented intelligent software, and enough storage capacity. In our demo, the RSU receives traffic probe data records from the OBU via DSRC radio link. Then it stores, aggregates, and forwards the probe records to the backbone networks via GPRS backhaul radio, which is provided by Cingular.

The PeMS server is the destination of traffic probe data records from OBUs. It is fully accessible from the Internet. Actually, it also stores other data from the Caltrans WAN (wide area network), which is connected to road equipment used to collect traffic information. Because Cingular GPRS enables Internet connection, probe records from OBUs are forwarded by many routers to reach the PeMS server.

Any computer with an Internet connection can be used to read the traffic speed and travel time information via standard Web browsers. The overview of equipment elements can be found in Figure 8.16.

8.6.2 Software Functionality

8.6.2.1 OBU

Traffic Probe Application using DSRC links is different from other applications using always-on communication channels, such as a GPRS modem. The connectivity is sporadic along the route, and this makes the buffering

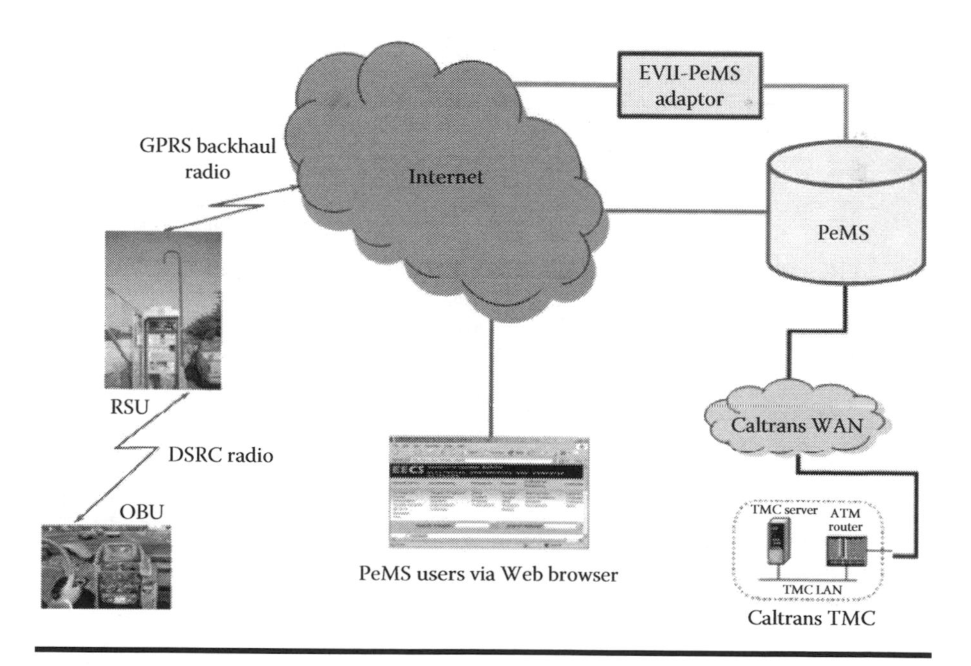

Figure 8.16 Equipment overview.

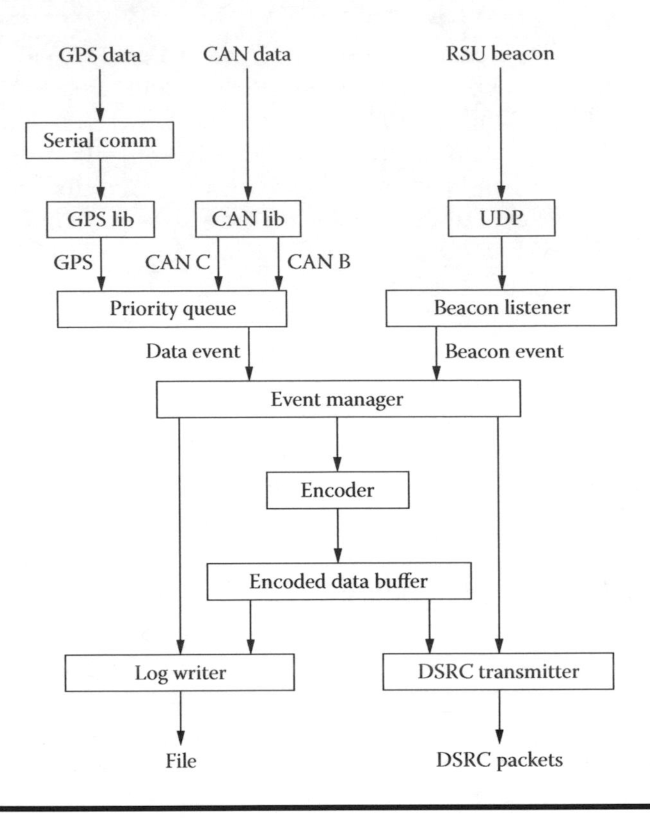

Figure 8.17 The control and data flow of the on-board software.

of the probe data necessary. Due to the limitations of on-board computer memory, a ring buffer is used to just keep the most recent traffic probe data if the car is driven for a long distance before encountering any DSRC RSUs. The control and data flow of the on-board software logic is depicted in Figure 8.17.

There are several possible ways of transmitting a DSRC probe message to RSUs. For example, one scenario could be an OBU that periodically broadcasts a probe message at a certain time interval. Although this is relatively easy to implement, it is not a decent implementation because the vehicle OBU is doing DSRC transmission in vain for most of the time. For this reason, in the EVII project we used an RSU DSRC beacon to announce the presence of an RSU. An OBU starts to transmit only if a beacon message is received by an RSU. Here is what happened during a drive-by scenario:

1. Caltrans RSU broadcasts "I Am Here" beacon message at 5 Hz.
2. DaimlerChrysler vehicle starts data collection and listens for RSU beacon.

3. DaimlerChrysler OBU records a traffic data point every 50 min and converts it into the VII California message format.
4. OBU gets DSRC beacon, starts sending all messages in computer buffer via DSRC.
5. OBU repeats, sending multiple times to enhance the chance of getting through.
6. RSU gets probe messages, forwards them to PeMS database.
7. OBU leaves RSU range, clears message buffer, and starts recording new data.

8.6.2.2 RSU Collector

The main function of the RSU collector is to parse the broadcast UDP packets from OBUs, classify them, and then send required data to the PeMS adaptor via UDP packets. Upon the arrival of a UDP packet, the main procedures followed by the RSU collector include:

1. Parse the message based on predefined message formats. What we need to transfer includes traffic safety messages and traffic probing messages.
2. Extract useful information to the EVII-PeMS adaptor and cache all recent probe records so that the number of records is large enough to write into a file with predetermined size.
3. Since the live time of a connection between an OBU and an RSU is short, the arrival of traffic probe data records is bursty. After the RSU collector receives the first data record from a new connection, it can detect the completion of transmission for this connection. Once the data records from the current OBU are ready, the RSU collector initializes the TCP transmission to the EVII-PeMS adaptor.
4. After the TCP transmission to EVII-PeMS adaptor is finished, go to step 1.

8.6.2.3 EVII-PeMS Adaptor

The main function of the EVII-PeMS adaptor is to parse the traffic probe data records from RSUs, classify them, and then store required data in the corresponding tables in the PeMS database. Upon the arrival of a file from RSUs, the main procedures followed by the EVII-PeMS adaptor include:

1. Check the receiving file. Keep checking if there is no new record.
2. If new records are detected, wait for a stabilizing period so that recent transmission from RSUs will be fully completed.
3. Either store new traffic probe records immediately to the PeMS database via embedded SQL (structured query language) commands or cache them and store them to the PeMS database periodically

(e.g., every 30 s). Note that the traffic safety data should be stored in an EVIL_TRAFFIC_SAFETY table and the traffic probing data should be stored in an EVIL_TRAFFIC_PROBE table.

4. Convert all traffic probe records with GPS coordinates into records with FDM coordinates and store them in the database table.
5. Estimate the traffic speed and travel time for predetermined freeway segments and store them in the database table. Go to step 1.

8.6.2.4 *Traffic Probe Application*

The entry point of the traffic probe application is a Web link: http://pemscs.eecs.berkeley.edu/xjdong/traffic_probing.html. Users need to use the GUI to select a source freeway location and a destination freeway location by clicking the two points in the map or finding the locations in the list. Then the two locations are validated and sent to the PeMS server. The Perl CGI program developed for the traffic probe application in the PeMS server parses the query commands, generates a new page that contains the estimated traffic speed and travel time, and sends the new Web page back to the user.

8.6.2.5 *GPS-FDM Convertor*

Most GPS receivers can report local information based on latitude, longitude, and altitude, as shown in Figure 8.18. These GPS coordinates are

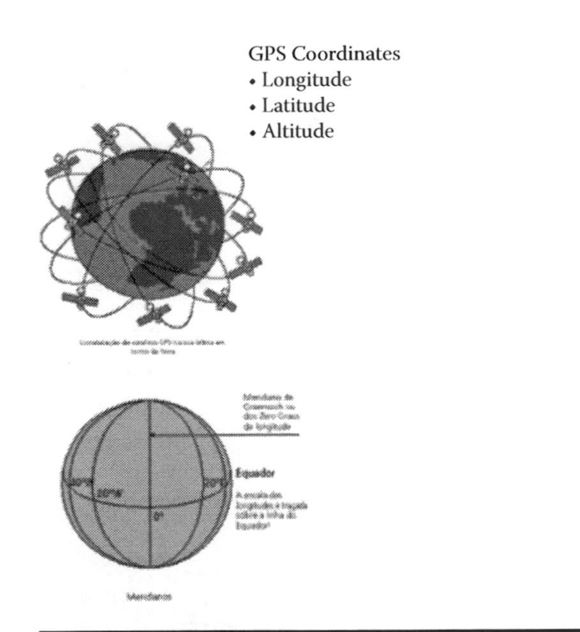

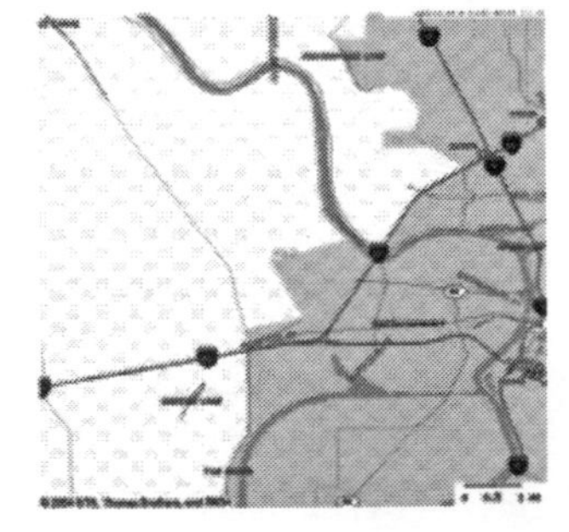

Figure 8.18 GPS coordinates and FDM coordinates.

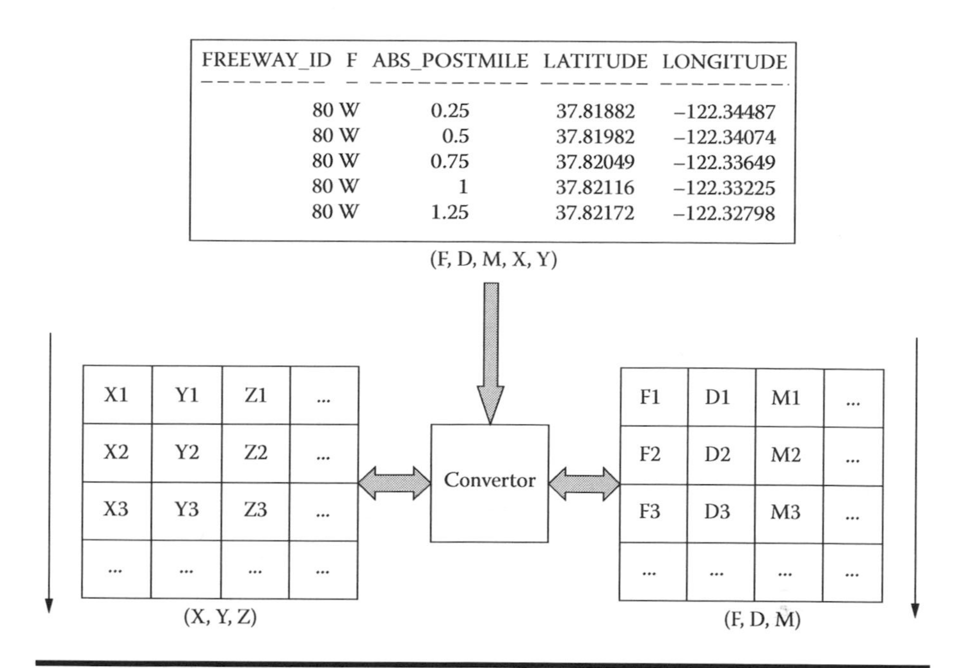

FREEWAY_ID	F	ABS_POSTMILE	LATITUDE	LONGITUDE
80	W	0.25	37.81882	−122.34487
80	W	0.5	37.81982	−122.34074
80	W	0.75	37.82049	−122.33649
80	W	1	37.82116	−122.33225
80	W	1.25	37.82172	−122.32798

Figure 8.19 GPS-FDM convertor.

recorded in angular units of degrees, minutes, and seconds. GPS receivers may display coordinates in degrees with minutes to four decimal places. In addition, GPS gives the UTC time information instead of local time. However, MTC tag data and PeMS data records have different formats. They give local time and location information based on freeway, direction, and postmile (so called FDM coordinates). We designed special algorithms and developed the software to convert GPS records into FDM records. The conversion algorithm is based on a freeway map with both FDM coordinates and segments with lat-long coordinates in the PeMS system, as shown in Figure 8.19.

8.7 Experiment and Performance Measurement

DaimlerChrysler RTNA provided the probe vehicle and the program to generate and transmit traffic probe data records. The RSU provided by California PATH is installed in the Berkeley Highway Lab (BHL). The probe vehicle with the DSRC OBU travels from US101@Dumbarton (north of Stanford University) to the BHL mainly along freeway I-880, as seen from the map in Figure 8.20.

The real-time probe data collection over DSRC was successfully demonstrated for Caltrans in October 2005. In the demo, the probe vehicle

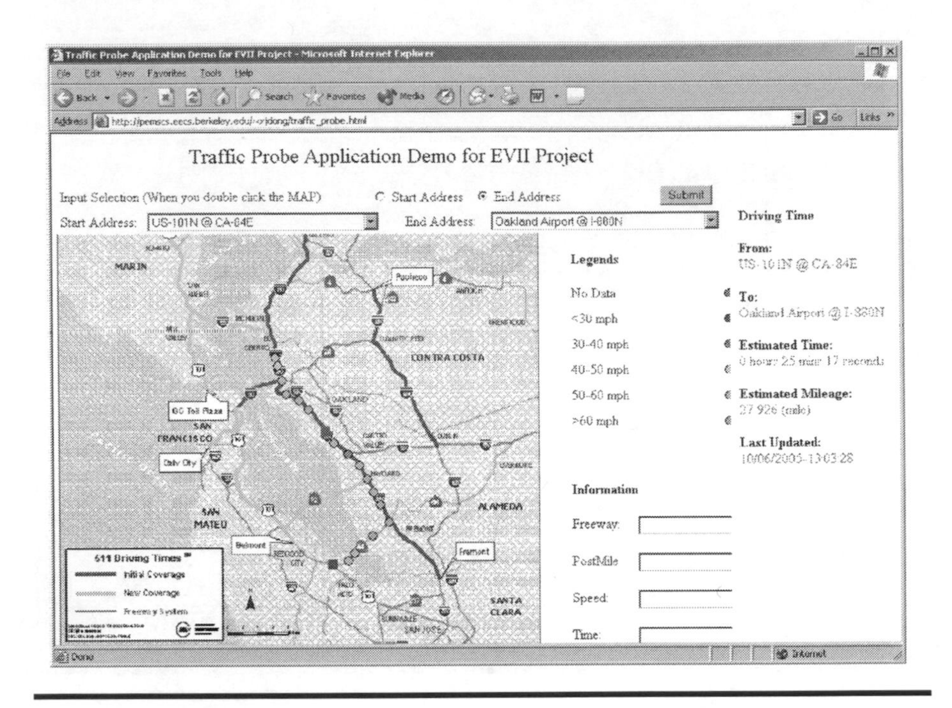

Figure 8.20 User interface.

generates a probe record every 100 m. Once an RSU is detected, the probe vehicle sends all records to the RSU. At the RSU along Interstate 80 in Berkeley, onboard traffic data was successfully relayed to the RSU through DSRC. The RSU sends all probe data to the PeMS server. PeMS stores, interprets, and analyzes probe data. The traffic speed and travel time information can be browsed by any Internet users, as shown in Figure 8.20.

During the EVII project, we tested different DSRC UDP packet sizes used for transmission and different time intervals between two consecutive DSRC packets. Since the packet loss rate heavily depends on the traffic and the type of vehicles nearby, we will not provide an accurate performance measurement report. From our experiments we found that packet length of 320 bytes is a good choice in terms of packet loss rate and data throughput in the EVII drive-by settings. Generally speaking, with our test settings, as packet length increases, the average packet loss rate will be higher. In addition, if a big truck blocks the LOS direction, the performance degrades dramatically.

8.8 Conclusion

This chapter describes an applied research project that presages an operational test and a deployment. It explores and resolves key engineering issues associated with the point deployment of these services in a realistic setting. As such, it will pave the way for VII by in effect jump-starting technological work. The demonstration of the individual vehicle collection of data for traffic and safety applications shows that indeed this data can be provided and received for operational purposes by the roadway management infrastructure with reasonable performance over IEEE 802.11p wireless links.

Acknowledgment

We would like to thank Susan Dickey, Joel VanderWerf, Jeff Ko, and Kang Li from California PATH for their nice assistance and helpful discussion.

References

[1] A. R. S. Bahai and B. R. Saltzberg, *Multi-Carrier Digital Communications—Theory and Applications of OFDM*, Kluwer Academic/Plenum, London, 1999.
[2] T. Cooklev, *IEEE Wireless Communication Standards*, IEEE Press, New York, 2004.
[3] Andrea Goldsmith, *Wireless Communications*, Cambridge University Press, 2005.
[4] J. Heiskala, and J. Terry, *OFDM Wireless LANS: A Theoretical and Practical Guide*, Sams Publishing, Indianapolis, IN, 2001.
[5] IEEE 802.11 Standard Part II: Wireless LAN Medium Access Control (MAC) and Physical Layer (PHY) Specifications, August 1999.
[6] IEEE 802.11a (Supplement to IEEE 802.11 Standard Part II): High-Speed Physical Layer Extension in the 5 GHz Band, September 1999.
[7] IEEE 802.11b (Supplement to IEEE 802.11 Standard Part II): High-Speed Physical Layer Extension in the 2.4 GHz Band, September 1999.
[8] IEEE 802.11g (Supplement to IEEE 802.11 Standard Part II): Further Higher Data Rate Extension in the 2.4 GHz Band, 2003.
[9] IEEE 802.11p draft standard, August 2005.
[10] IEEE 802.16, IEEE Standard for Local and Metropolitan Area Networks–Part 16: Air Interface for Fixed Broadband Wireless Access Systems, October 2004.
[11] IEEE 802.16e (Supplement to IEEE 802.16 Standard Part 16): Physical and Medium Access Control Layers for Combined Fixed and Mobile Operation in Licensed Bands and Corrigendum 1, 2006.
[12] W. C. Jakes, Jr., *Microwave Mobile Communications*, John Wiley & Sons, New York, 1974.
[13] William C. Y. Lee, *Mobile Communications Engineering*, McGraw-Hill, Columbus, OH, 1998.
[14] http://www.mathworks.com/matlabcentral/
[15] K. Pahlavan and P. Krishnamurthy, *Principles of Wireless Networks—A Unified Approach*, Prentice Hall, Upper Saddle River, NJ, 2002.

[16] T. Choe, A. Skabardonis, and P. Varaiya, Freeway Performance Measurement System (PeMS): An Operational Analysis Tool, TRB 81st Annual Meeting, January 2002.
[17] http://pemscs.eecs.berkeley.edu
[18] Yang Xiao, IEEE 802.11n: Enhancements for Higher Throughput in Wireless LANs, *IEEE Wireless Communications*, vol. 12, no. 6, pp. 82–91, 2005.
[19] J. Yin, T. ElBatt, G. Yeung, B. Ryu, S. Habermas, H. Krishnan, and T. Talty, Performance Evaluation of Safety Applications over DSRC Vehicular Ad Hoc Networks, Proceedings of the First ACM Workshop on Vehicular ad hoc Networks, October 1, 2004, Philadelphia, PA.
[20] J. Zhu and S. Roy, MAC for Dedicated Short Range Communications in Intelligent Transport System, *IEEE Communications Magazine*, vol. 41, no. 12, pp. 60–67, 2003.

Chapter 9

Interference in Millimeter-Wave Wireless MAN Cellular Configurations

Pantel D. M. Arapoglou, Athanasios D. Panagopoulos, and Panayotis G. Cottis

Contents

9.1 Introduction

Wireless metropolitan area networks (WMAN) operating in the millimeter-wave range are usually referred to in the literature under the broad term fixed broadband wireless access (BWA), which encompasses a plethora of fixed wireless systems. Among others, the term fixed BWA includes LMDS (local multipoint distribution service) networks [1,2] operating in licensed frequency bands above 20 GHz. Specifically, the International Telecommunication Union (ITU) in Europe and the Federal Communications Commission (FCC) in the United States have allocated spectrum in the range 24.5–26.5 GHz/40.5–43.5 GHz and 28 GHz/38 GHz, respectively, for implementing LMDS. Furthermore, the air interface specified in recent technical WMAN standards issued by the IEEE (802.16 WirelessMAN [3]) and ETSI (BRAN HIPERACCESS [4]), covers frequencies from 10 GHz to 66 GHz to provide access at high data rates (up to approximately 100 Mbps). These two standards constitute the WiMax* forum, aiming at the convergence and interoperability of the two technologies.

The main characteristic of an LMDS system, as implied by its acronym, is the distribution of service to local areas (cells) in a point-to-multipoint (PMP) fashion, from a hub or base station (BS) to fixed customers or subscriber stations (SSs) equipped with the necessary customer premises equipment (CPE). It is well known that any wireless system employing a cellular architecture is interference sensitive due to frequency reuse in multiple cells. The present chapter is particularly targeted at investigating intercell interference issues related to fixed BWA networks in the 10–66 GHz frequency range, where interference exhibits distinct features in terms of system architecture and propagation characteristics (see Section 9.2) in comparison to conventional wireless networks (e.g., mobile communications). Whenever the source of interference employs the same carrier frequency, it is referred to as cochannel interference (CCI), whereas for interfering sources employing spectrally adjacent carrier frequencies or the same carrier frequency with an orthogonal polarization, the term adjacent channel interference (ACI) is used. The cumulative effect of CCI and ACI imposes a major limitation on the capacity and scalability of fixed BWA systems [5], which, in turn, affects the degree of future penetration of millimeter-wave WMAN in the competitive market of access technologies.

* Worldwide Interoperability for Microwave Access.

To quantify performance in the presence of interference, throughout this chapter, the signal-to-interference ratio (SIR) is adopted. The statistical behavior of the SIR is studied assuming fading conditions for both the desired and the interfering signals. In this context, we review the relevant fading mechanisms proposed in the literature to best approximate the LMDS propagation environment with respect to the degree of path clearance characterizing a specific radiopath. Emphasis is put on line-of-sight (LOS) and nearly LOS propagation conditions. The main objective of the chapter is the analytical determination of the interference outage probability

$$OP_I = \Pr\left\{\frac{S}{I} < \gamma_{th}\right\} \tag{9.1}$$

where S is the desired signal power received at the victim station, I is the sum of the CCI and ACI contributions, and γ_{th} is the SIR threshold value; that is, the interference fade margin. OP_I represents a critical quality-of-service (QoS) specification in cellular systems because it dominates the determination of the total outage characteristics.

Apart from issues related to system architecture (sectoring, frequency reuse) and propagation (path clearance, type of fading), intercell interference depends on the multiple access scheme employed. Section 9.3 presents the SIR statistical analysis for the conventional multifrequency time division multiple access (MF-TDMA) case, which is an integral part of all recent standards [3,4]. The presentation of the state-of-the-art interference models is separated into LOS (Section 9.3.2) and nearly LOS (Section 9.3.3). Then, we go into the corresponding three-dimensional problem including the effects of elevation and building height.

Although TDMA is the current scheme employed in LMDS, direct-sequence code division multiple access (DS-CDMA), especially popular in third-generation mobile communication networks, has attracted the interest of a number of researchers. After discussing the basic aspects of this multiple access scheme implementation for fixed BWA, Section 9.4 compares the CDMA and TDMA alternatives based on a few simplified assumptions and deduces several interesting conclusions. The final section of the chapter extends the simple deterministic SIR model for the CDMA version of fixed wireless networks to a comprehensive statistical model taking into account fading conditions.

9.2 Cellular WMAN Architectures

9.2.1 Single-Cell Planning

Early microwave PMP systems consisted of a collection of narrow beam antennas, each serving a small number of subscribers. Hubs of this type

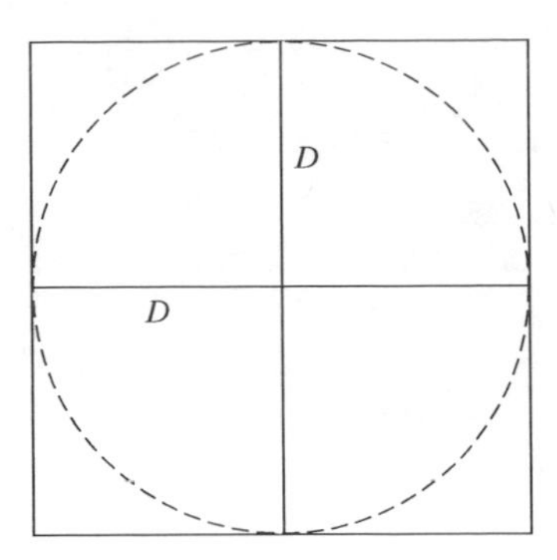

Figure 9.1 Square cell architecture for LMDS. D **denotes the cell radius.**

are more accurately characterized as multiple point-to-point links instead of PMP BSs. For a true PMP hub, where antennas are intended to serve multiple SSs in a specific cell area, several factors such as coverage, spectral efficiency, and infrastructure cost must be balanced [6]. Although cells in mobile communication systems are commonly modeled as hexagonal, in the majority of the relevant literature the cell coverage for LMDS is approximated by a square divided into four 90° sectors, as shown in Figure 9.1, with the cell radius, D, ranging from 1 to 5 km. The number of sectors per hub can be further increased (sectoring) to achieve a higher capacity; however, this will result in higher equipment cost.

Another factor limiting the practical number of sectors per BS is antenna performance. For frequencies above 20 GHz, the current trend for hub sector antennas is the use of horn radiators with an azimuth pattern similar to the one shown in Figure 9.2 for horizontal and vertical polarization. Candidate locations to install the BS antenna are high structures, such as tall building rooftops or existing radio cellular towers. On the other hand, the fixed position of customers implies the use of very directional SS antennas, usually parabolic dishes with half-power beamwidth (HPBW) ranging from 2° to 5°. ETSI specifies the electrical requirements for linearly polarized fixed SS* antennas in the 11–60 GHz band. In Figure 9.3, the antenna pattern specification relative to the maximum gain† for Class 1 terminal stations at frequencies 24–30 GHz [7] and 40.5–43.5 GHz [8] is presented. Subscriber antennas, together with the outdoor and indoor units, form the CPE and are usually installed at residences and small- to medium-size enterprises.

Due to the harsh transmission medium in the frequency range of interest (discussed in Section 9.2.1), a major concern when dimensioning an LMDS cell is the existence of LOS between the BS and the subscribers.

* Subscriber stations are referred to as terminal stations according to ETSI terminology.
† In general, ranging from 30 to 40 dBi.

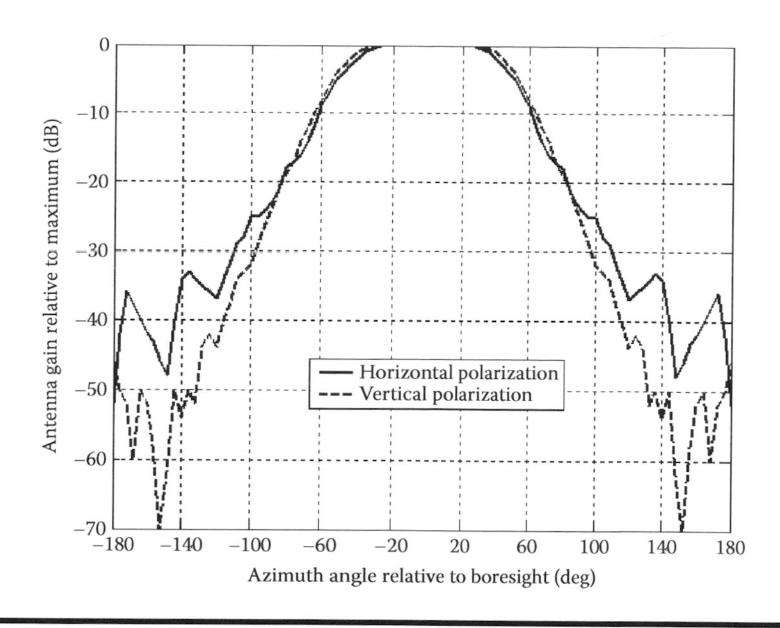

Figure 9.2 Typical antenna radiation pattern in the azimuth plane for an LMDS sector. Both horizontal and vertical antenna polarizations are shown.

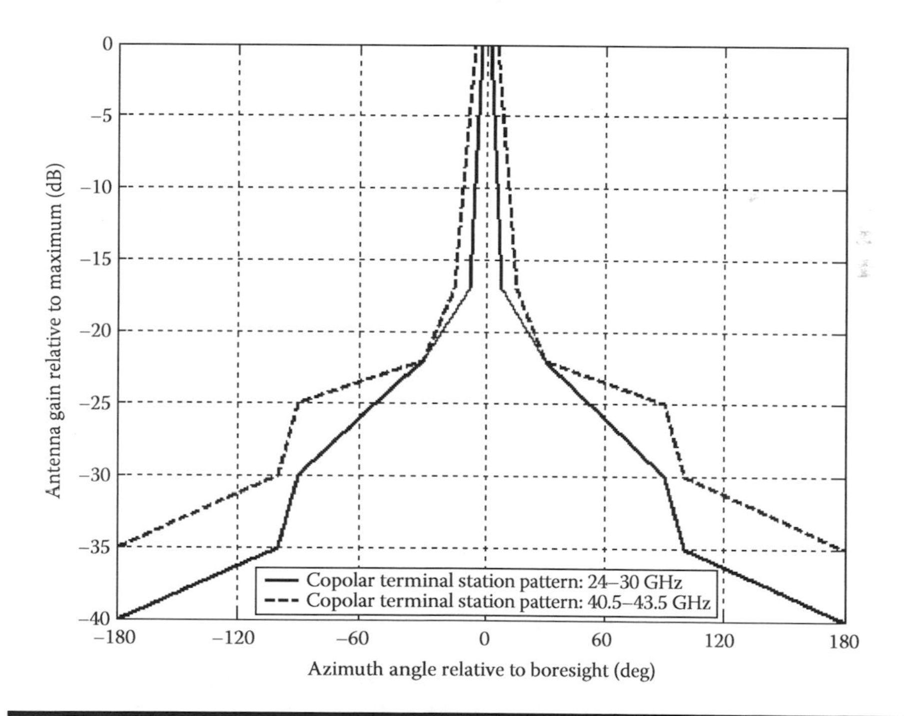

Figure 9.3 ETSI specification of the Class 1 terminal antenna pattern for subscribers of fixed BWA.

Since LOS is not always possible, within the frame of the European project CABSINET (Cellular Access to Broadband Service and INtEractive Television), a two-layered cell architecture was proposed, dividing the cell area into macrocells operating under LOS in the 42-GHz band and microcells operating under non-LOS in the 5.8-GHz band (see Figure 9.4). The physical (PHY) [9] and Media Access Control (MAC) [10] layer details concerning this architecture have been discussed elsewhere. Two-layered architectures seem to be the trend of future fixed BWA networks, since they naturally lend themselves to the air interface design of the IEEE 802.16 and ETSI BRAN standards, which comprise two corresponding physical layer interfaces [11]: one for the 10–66 GHz band, where LOS is deemed a practical necessity, and one for the 2–11 GHz range, where even portable users having no LOS to the intermediate microbase station may be accommodated.

9.2.2 Frequency Reuse

The available operational bandwidth in multicell LMDS networks may reach 2 GHz depending on the choice of frequency bands. Nevertheless, in an effort to maximize data throughput, aggressive frequency reuse topologies are adopted in LMDS by reusing in every cell the entire available bandwidth or splitting it into two or, at most, four bands. As a consequence, different levels of CCI are created. This suggests a trade-off between throughput and intercell interference during the system planning procedure.

To alleviate the aggravating effect of this reuse policy, polarization interleaving (i.e., alternating polarization in adjacent cell sectors) is widely used, leading to a frequency reuse optimization by means of sectoring and polarization [12]. As an example, in the sector topologies of Figure 9.5, which are only a few out of many possible alternatives, horizontal and vertical polarization are deployed in an alternating pattern to maximize isolation between bordering sectors, while doubling channel utilization. In Figure 9.5 a and b, the total bandwidth is partitioned into two bands (frequency reuse of 2) and four bands (frequency reuse of 1) per cell, while in Figure 9.5 c and d, the whole available bandwidth is reused in every sector (frequency reuse of 4). Figure 9.5c and d of their sector antenna positions; that is, in the periphery (corners) and the center of the cell, respectively. While the use of orthogonal polarizations offers additional bandwidth and CCI improvement, stations must be supplied with a sufficient cross-polar discrimination (XPD) to avoid ACI effects due to cross-polarization, especially during adverse propagation conditions. Finally, instead of frequency segmentation, sectoring in the case of DS-CDMA LMDS networks can also be carried out in the code domain or in the frequency, code, and polarization domains combined. Such CDMA architectures have been investigated [13,14] and will be reviewed in more detail in Section 9.4.

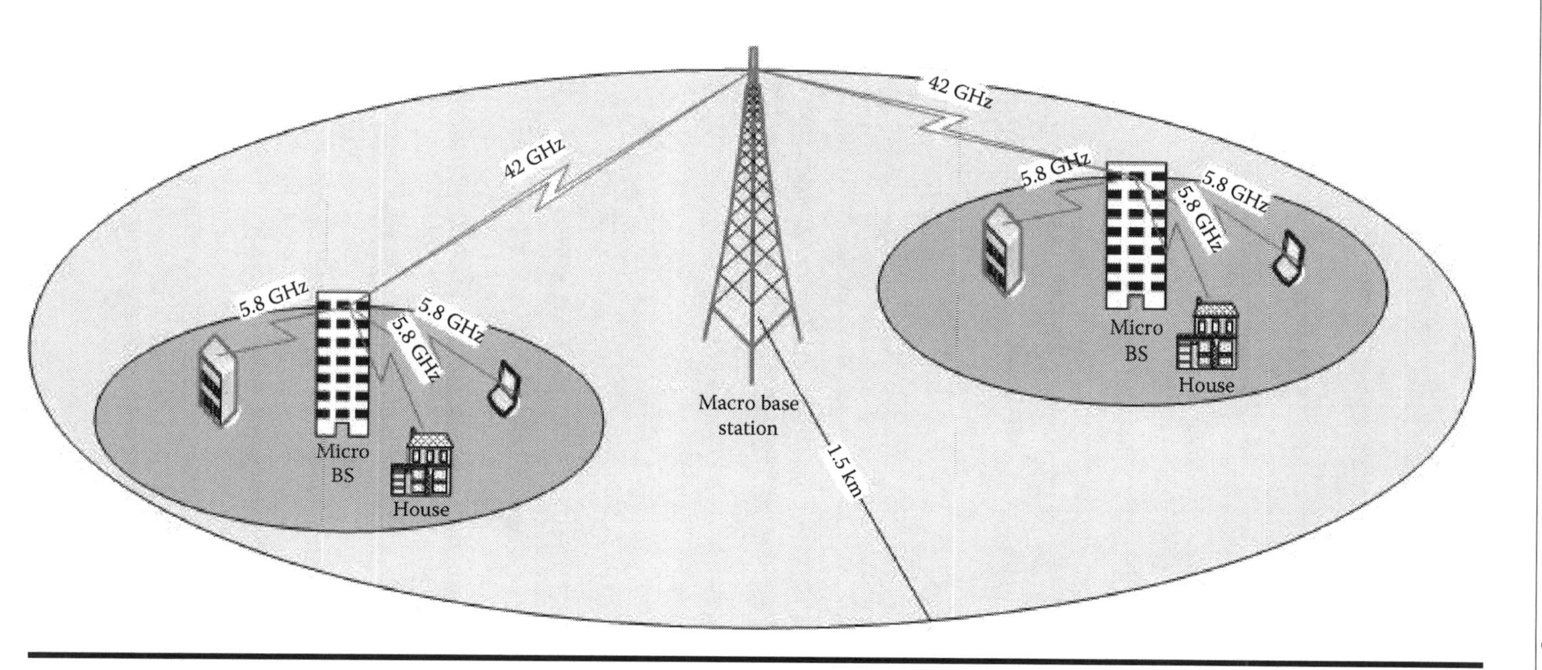

Figure 9.4 Two-layered LMDS cell architecture in the frame of the CABSINET project to accommodate both LOS (macrocells) and non-LOS (microcells) customers.

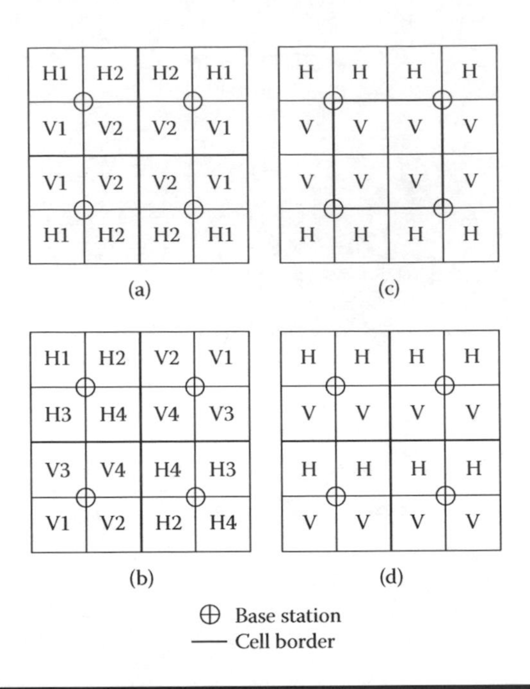

Figure 9.5 Rectangular cellular patterns in LMDS employing different frequency and polarization sectoring: (a) frequency reuse of 2; (b) frequency reuse of 4; (c) frequency reuse of 1 (sector antennas at the corners of the cell); (d) frequency reuse of 1 (sector antennas at the center of the cell).

9.2.3 Propagation Issues

As already mentioned, the first concern when designing millimeter-wave WMAN destined to operate above 20 GHz is to achieve clearance of at least 60 percent of the first Fresnel zone for all BS–SS links; i.e., achieve LOS, avoiding buildings and vegetation/foliage. Otherwise, as reported in a number of measurement campaigns (see [15–18] and references therein), obstructing structures will result in severe power loss due to scattering and diffraction. To this end, the efficient design of LMDS requires environmental information including topographic maps, building data, and morphology. The various categories of the relevant environmental databases necessary for the proper design of fixed BWA systems are thoroughly described elsewhere [19].

Once path clearance is ensured, the effects of short-range fading must be taken into consideration. In millimeter-wave frequencies, short-range fading exhibits distinct characteristics compared to conventional wireless networks for the following reasons:

- The fixed position of customers, which reduces the dynamic properties of the channel.
- The extremely high frequency bands employed.
- The installation of antennas on rooftops high above the reflecting ground or the street level, thus increasing the reflection angle of incidence.
- The narrow HPBW of the SS antenna, which strongly suppresses multipath components. As a result, multipath fading becomes less significant and its effect to link availability may be neglected. Instead, the dominant fading mechanism above 20 GHz is attenuation from precipitation, particularly from rain fading,* which falls under the class of slow-varying flat fading.

In the following, apart from cases where the LOS condition is satisfied, intercell interference in cases when the BS–SS link is partially obstructed will be investigated, a situation referred to as nearly LOS. Such cases arise when, for example, the radiopath is close to the vertical edge or side of a building or due to the movement of tree leaves under windy conditions and result in excess loss. To cover similar situations, modeling procedures that include the effect of shadow fading are presented in Section 9.3.

Furthermore, the benefit from polarization interleaving is related to ACI, which depends on the degree of signal depolarization. Frequency reuse becomes more effective as the XPD, defined as the ratio of the copolarized to the cross-polarized received power, becomes higher. Although rain fading is considered the primary contributor to signal attenuation, depolarization due to vegetation results generally in lower XPD values than depolarization due to rain [15], especially in rural areas. Concluding, the ITU-R Recommendation P.1410 [20] provides general guidance for system engineers on propagation data and requirements concerning fixed BWA networks and classifies all the relevant atmospheric phenomena with reference to a number of prediction methods included in ITU-R Recommendation P.530 [21] for point-to-point terrestrial LOS links. For instance, Figure 9.6 illustrates the implementation of the rain attenuation prediction method (time percentage during which a specified attenuation value is exceeded) for a 5-km millimeter-wave terrestrial link operating under the annual rain rate characteristics of Paris, France.

* Note, however, that observations of multipath during rain but not under clear weather have been reported [17].

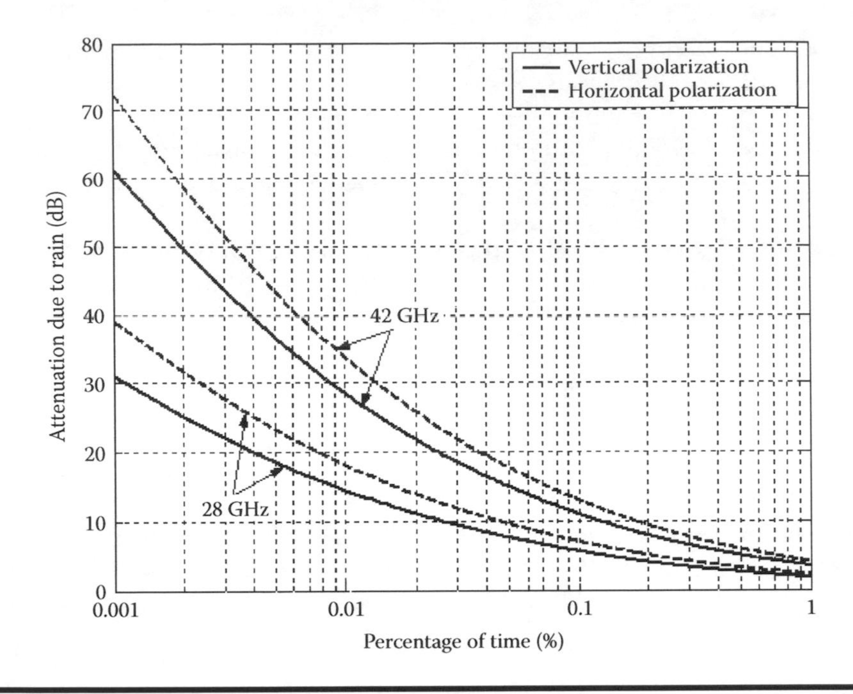

Figure 9.6 Illustration of the ITU-R P.530 rain attenuation prediction method for a 5-km terrestrial link located in Paris, France.

9.3 Interference Scenarios in TDMA-Based LMDS

9.3.1 General Considerations and Interference Scenarios

The current air interface for WiMax-based millimeter-wave WMAN specifies a single-carrier modulation for 10–66 GHz, where, due to its PMP nature, in the downstream, the BS transmits a time division multiplexed (TDM) signal, while, in the upstream, access from multiple SSs to the BS is achieved by employing TDMA. For the lower NLOS 2–11 GHz band, orthogonal frequency division multiplexing/multiple access (OFDM/OFDMA) is selected. As to duplexing, 802.16 WirelessMAN and BRAN HIPERACCESS support both FDD (frequency division duplex) and TDD (time division duplex). Optionally, the possibility of dynamic TDD exists, where in each frame the time allocated between upstream and downstream transmissions is configurable so that it matches the asymmetric nature of multimedia traffic. However, the dynamic assignment of downstream and upstream time in neighboring cells gives rise to significant CCI, when signals with opposite transmission directions coincide. An analytical approach of CCI in dynamic TDD fixed wireless applications together with some resource allocation algorithms for its reduction are presented elsewhere [22].

Let us now consider the dual-frequency, dual-polarization cell architecture of Figure 9.5a, reproduced in Figure 9.7a and b to demonstrate typical CCI scenarios for the downstream and upstream. It should be noted that this specific topology was identified by the CRABS project [23] as the optimum combination of spectral efficiency and implementation cost. For the downstream, it is observed that, for any SS, there are three dominant interfering BSs employing the same frequency and polarization that are responsible for CCI. For example, apart from receiving the desired signal from BS_{11}, SS also receives interfering signals from the H1 sectors of BS_{13}, BS_{31}, or BS_{33}, depending on which of the three is visible to the SS. The first two interfering BSs are called parallel interferers and their distance from BS_{11} is $4D$, whereas the latter is called diagonal interferer at a distance $4\sqrt{2}D$ from BS_{11} [24]. No interference is considered here between downstream and upstream transmissions, assuming that either FDD or static TDD is used.

In the same way, ACI originates from cell sectors employing the same frequency as SS but being orthogonally polarized; that is, sectors V1 of the configuration that are visible to SS. There are three possible interfering BSs; namely, sectors V1 of BS_{21}, BS_{23}, or BS_{13}. The interference originating from BS_{43}, which is also within the HPBW angle of SS but not shown in the figure, has been neglected.

The following equation can be used for the evaluation of the unfaded downstream SIR experienced by a typical SS:

$$\left(\frac{S}{I}\right)_u = \frac{EIRP_{BS,D}\, G_{R,SS}(0)\,(1/d_D)^m}{\sum_{k=1}^{K} EIRP_{BS,I}^{(k)}\, G_{R,SS}^{(k)}(\phi_k)\,(1/d_I^{(k)})^m\, 1/X^{(k)}} \tag{9.2}$$

Where the subscripts D and I correspond to the parameters of the desired and interfering station, respectively. Elaborating on Equation (9.2), it is assumed that there are K sources of interference. $EIRP_{BS,i}^{(k)}$, $i = D$ or I, is the effective isotropic radiated power (EIRP) from the relevant BS; $d_i^{(k)}$, $i = D$ or I, is the BS–SS path length for either the desired or an interfering BS (IBS); and m is the propagation exponent. Furthermore, $G_{R,SS}(\phi_k)$ is the directive gain of the subscriber antenna in the direction of the IBS relative to its maximum, $G_{R,SS}(0)$ (see Section 9.2.3). Finally, $X^{(k)}$ is equal to 1 when the cochannel unfaded SIR is evaluated, and equal to $XPD^{(k)}$ when the adjacent channel SIR (from orthogonal polarized sources) is evaluated. Note that, in the above, the power from spectrally adjacent carriers that is intercepted by the receiver filter of the victim station is not accounted for, since adequate channel spacing is assumed.

On the other hand, for the evaluation of the upstream intercell interference, BS_{33} is used as a reference (see Figure 9.7b). The SS located in sector H1 of this cell is the subscriber transmitting the desired signal, while users in sectors H1 of BS_{31}, BS_{11}, and BS_{13} are possible interfering SS (ISS). An expression similar to Equation (9.2) holds also for the upstream unfaded SIR.

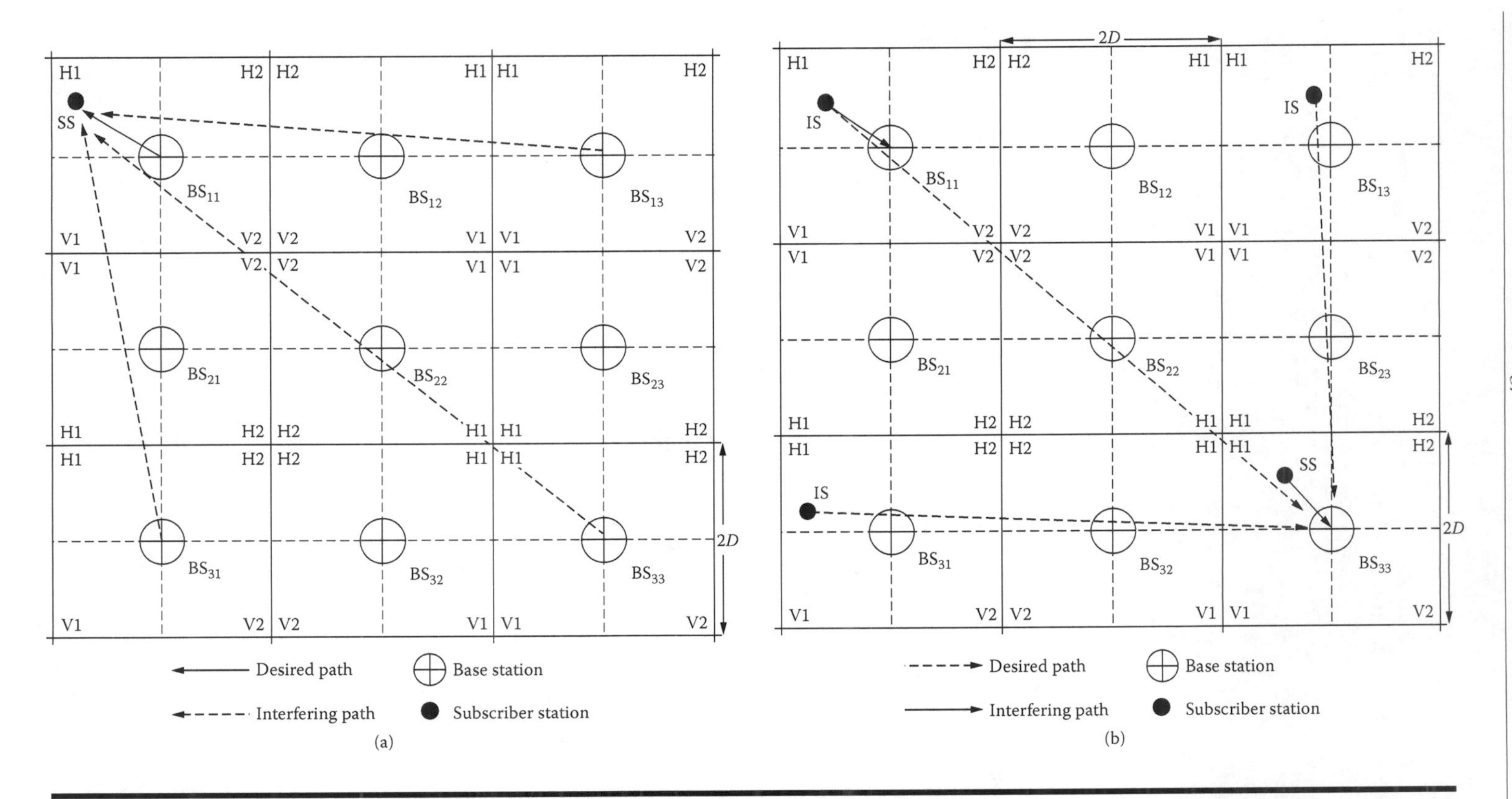

Figure 9.7 Intercell interference scenarios for a dual-frequency, dual-polarization sectoring scheme. (a) Downstream case. (b) Upstream case.

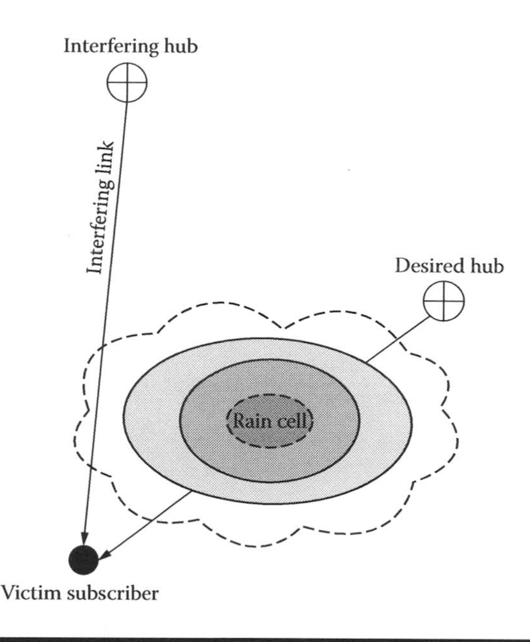

Figure 9.8 Two converging links affected to a different degree by rain attenuation from an overlying rain cell. The darker shaded volumes of the rain cell correspond to higher rain rate values.

9.3.2 LOS Analysis and Correlated Rain Fading

Fading in LOS microwave links operating above 20 GHz is mainly attributed to precipitation events. This propagation impairment has been studied in depth for both terrestrial and satellite systems by a number of researchers [25] who have proposed many empirical and physical prediction methods. However, only a fraction of these research works have addressed the effect of rain fading on the SIR performance of LMDS cellular networks [26–28].

Rain attenuation, which originates from hydrometeor scattering and absorption of radiowaves propagating through the rain medium, exhibits spatial inhomogeneity due to the structure and movement of rain cells*. The spatial–temporal dynamics of this atmospheric phenomenon result in correlation of the attenuation along different links in the same area, a fact that increases the interference whenever the rain fading is higher along the desired path than along the interfering one. This is shown in Figure 9.8, where darker shaded volumes of the rain cell correspond to higher rain rate levels. To assess the effect of correlated rain fading on the SIR statistics,

* Compact areas of high rainfall with a diameter of a few kilometers surrounded by wider areas of stratified rain.

Panagopoulos et al. in [28] proposed an analytical methodology based on the convective rain cell physical model [29]. It relies on a spatial correlation coefficient of specific rain attenuation for two points separated by a distance, d, expressed by

$$\rho_o(d) = \begin{cases} \frac{g}{\sqrt{g^2+d^2}} & d \leq D_r \\ \frac{g}{\sqrt{g^2+D_r^2}} & d > D_r \end{cases} \tag{9.3}$$

where D_r denotes the diameter of a rain cell and g is the spatial factor of the rainfall medium ($g \in [0.75, 3]$).

The model also makes use of the fact that the random variable representing rain attenuation, A[dB], is successfully approximated by the lognormal distribution with probability density function

$$p(A) = \frac{1}{\sqrt{2\pi}\,\sigma_A\, A} \exp\left\{ -\frac{(\ln A - m_A)^2}{2\sigma_A^2} \right\} \tag{9.4}$$

The statistical parameters, (m_A, σ_A), depend drastically on the rainfall characteristics of the particular geographical region,* the path lengths, the frequency of operation, and the polarization tilt angle.

Adjusting expression (9.2) of the previous section to the LOS case, the propagation exponent m becomes equal to 2 and, due to the narrow HPBW of the subscriber, only one cochannel plus one adjacent channel IBS are taken into account. Then, introducing rain attenuation over both the desired and the interfering paths, the following expression comes up for the total SIR (in decibels):

$$\frac{S}{I_{CCI} + I_{ACI}}[\text{dB}] = A_{CCI} + A_{ACI} - A_D - 10\log\left[10^{\frac{A_{CCI}-(S/I_{ACI})_u}{10}} + 10^{\frac{A_{ACI}-(S/I_{CCI})_u}{10}}\right] \tag{9.5}$$

where A_D, A_{CCI}, and A_{ACI} are the rain attenuation random variables along the desired link and the links related to the two interfering BSs, respectively. The terms $(S/I_{CCI})_u$, and $(S/I_{ACI})_u$ correspond to expressions analogous to Equation (9.2) in decibels.

Implementing the methodology proposed by Panagopoulos et al. [28] for the interference scenario depicted in Figure 9.7a—assuming a frequency of operation $f = 42$ GHz, a cell radius $D = 3$ Km, and $XPD = 16$ dB—one obtains the downstream SIR sector distribution of Figure 9.9. More specifically, the figure consists of SIR isolines corresponding to every possible customer position within the sector, under a certain specification for

* In case no rainfall rate measurements are available for the specific area, the ITU-R Recommendation P.837 [30] provides such statistics for every possible geographical location.

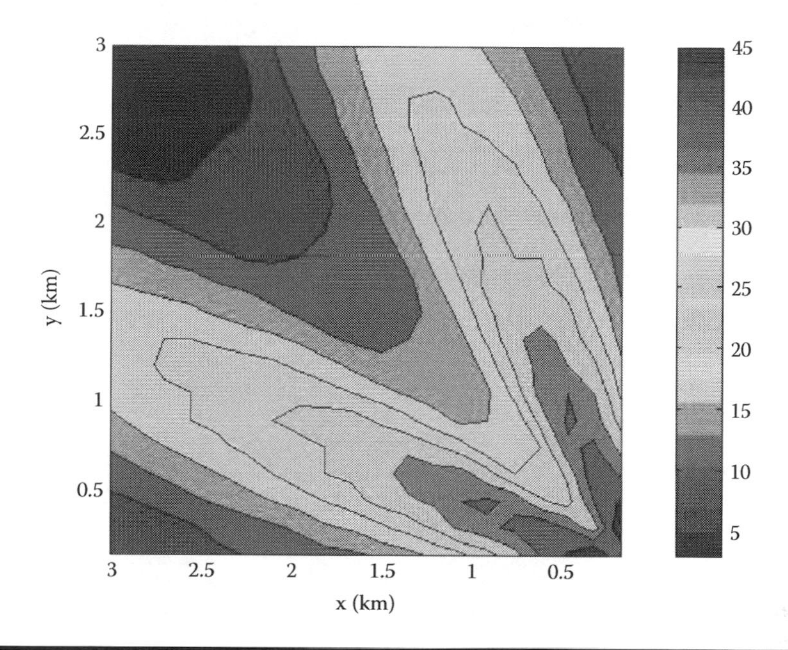

Figure 9.9 Downstream SIR sector distribution for an LMDS system operating in Athens, Greece, under the specification $OP_I = 0.001\%$. (Frequency of operation 42 GHz. Cell radius 3 km.)

the interference outage probability ($OP_I = 0.001\%$). As far as rain rate is concerned, climatic data from Athens, Greece, has been assumed. From Figure 9.9, where the BS is located at the lower right corner of the sector, it becomes evident that worst-case positions are those at the other corners, since, then, the SS antenna is aligned to the IBS.

For the upstream, similar conclusions may be drawn as to the worst sector position. However, during upstream transmission, fixed BWA subscribers are usually equipped with automatic power control (APC) modules to counterbalance fading and, thereby, preserve QoS. Such an APC scheme was analyzed elsewhere [28], where the transmitted power from an SS, $P_{T,SS}(d, A)$, was adjusted statically according to the distance from its desired BS and dynamically for rain attenuation levels up to the power control margin, M_{APC}, according to the expression

$$P_{T,SS}(d, A) = \begin{cases} P_{T,SS}^{min} \times (d/d_{\min})^2 & A[\text{dB}] < 0.5 \\ P_{T,SS}^{min} \times (d/d_{\min})^2 \times 10^{A/10} & 0.5 \leq A[\text{dB}] \leq M_{APC} \\ P_{T,SS}^{min} \times (d/d_{\min})^2 \times 10^{M_{APC}/10} & A[\text{dB}] > M_{APC} \end{cases} \tag{9.6}$$

where $P_{T,SS}^{\min}$[W] is the minimum output power from the closest subscriber to the BS at a distance $d_{\min}$ under clear sky conditions. Fading depths larger

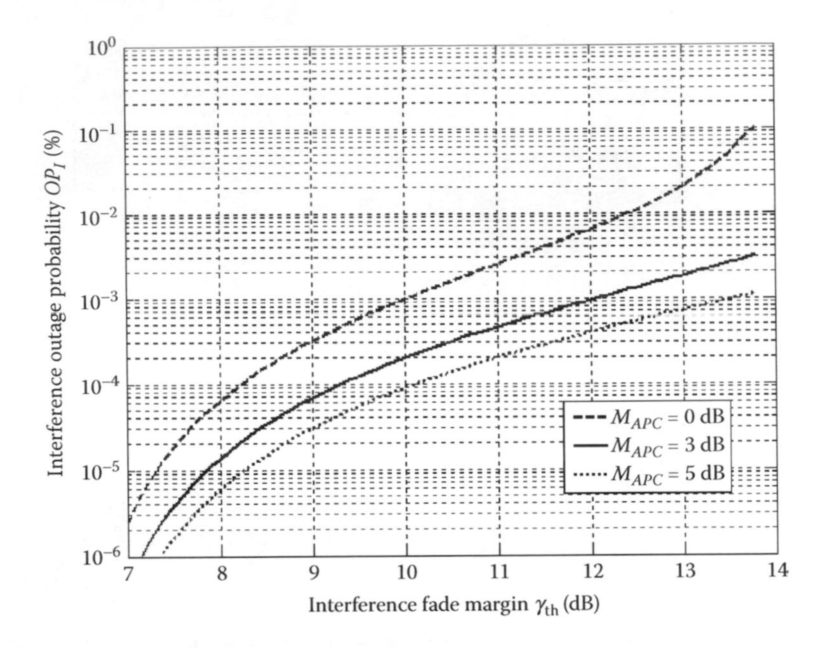

Figure 9.10 Interference outage probability, OP_I versus fade margin, γ_{th}, for rain attenuation levels exceeding the power control capability of the desired user, while not exceeding that of the interfering users (worst-case scenario).

than 0.5 dB signify the existence of rain attenuation. Note that in practical cases, unlike Equation (9.6), the control loop increases the output power step-wise; for example, in increments of 1 dB. The impact of this single-cell–oriented APC scheme on CCI is investigated further in Figure 9.10: The OP_I distribution for the corner position is drawn when rain fading along the desired path exceeds the power control limit, while attenuation over the interfering paths does not.

Up to a certain extent of rain fading, APC is an appropriate countermeasure for the upstream. It is limited, however, by the dynamic range of the amplifier, which must operate in its linear region to avoid the sidelobe restoration that appears near saturation. For the downstream, a more effective fade mitigation technique is cell site diversity, a technique that exploits the spatial characteristics of rain by linking each user with more than one BS. This technique allows each SS to choose the least attenuated signal. Empirical approaches based on measured data or theoretical models based on the understanding of the physical mechanisms responsible for rain attenuation have been reported [31–36]. Nevertheless, the majority of these works focus on the improvement of the downstream signal-to-noise ratio (SNR) resulting from cell site diversity protection. The relationship between cell site diversity and the CCI problem has been recently investigated [37].

9.3.3 Nearly LOS Analysis and Shadow Modeling

Nearly LOS propagation through partially obstructed paths is characterized by shadow fading. This type of fading, denoted by ξ, is represented by a zero-mean lognormal random variable, or, if expressed in decibels, by a Gaussian random variable with probability density function

$$p(\xi) = \frac{1}{\sqrt{2\pi}\,\sigma_\xi} \exp\left\{-\frac{\xi^2}{2\sigma_\xi^2}\right\} \tag{9.7}$$

where the parameter σ_ξ is the shadow standard deviation.

LMDS cellular architectures subjected to shadowing have been investigated elsewhere [38,39], where detailed analytical procedures and simulations were applied to investigate the effect of a series of factors on the SIR: omnidirectional versus sectored BS antennas, SS antenna gain, power control, and macrodiversity.*

Compared to the LOS approach presented earlier, in the case of shadowing, multiple sources of interference must be taken into account for the calculation of the total interference:

$$I = 10\log\left[\sum_{k=1}^{K} 10^{I_k/10}\right] \tag{9.8}$$

where I_k[dB] represents the interfering power from station $k = 1, \ldots, K$ affected by shadowing. Hence, to statistically model the SIR, it is necessary to define a new random variable, I, as the sum of a finite number of lognormally distributed random variables. Although there is no exact expression known concerning the probability distribution of this sum, various authors have derived several approximations. All of them treat the sum of lognormal random variables as another lognormal random variable. A method matching the first two moments of the approximating distribution has been developed by Fenton [40], whereas the Schwartz–Yeh method [41] uses nesting and recursion by splitting the K random variables into groups of two and then imposing the exact expression for the first two moments.

Another difference with the LOS model presented in Section 9.3.2 is the value of the propagation exponent, m, which, in the presence of shadowing, is higher than 2. The effect of the propagation exponent, m, on the average SIR and the outage probability, OP_I, has been studied [39], based on the Schwartz–Yeh approximation method. The relevant results are presented in Figures 9.11 and 9.12, respectively, for the worst-performing

* Macrodiversity refers to the possibility of accessing more than one transmitter in order to combat shadowing, similar to the cell site diversity technique for the mitigation of rain fading.

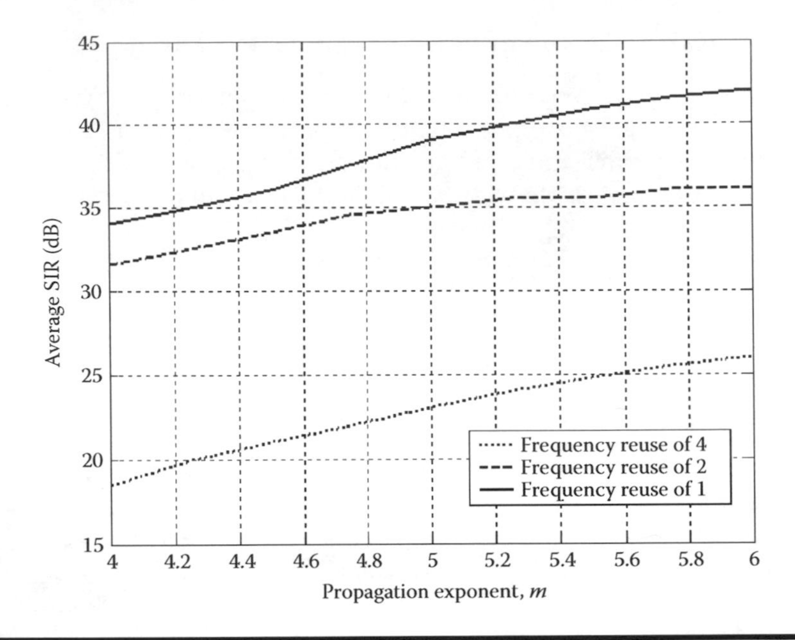

Figure 9.11 Average SIR versus propagation exponent, *m*, for various nearly LOS cell architectures.

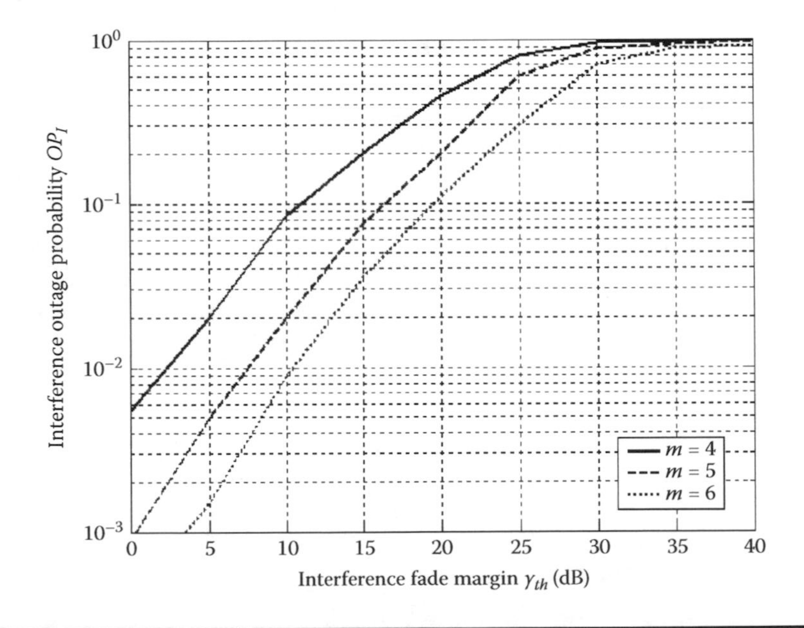

Figure 9.12 Interference outage probability, OP_I, versus fade margin, γ_{th}, in a shadowed propagation environment for different values of propagation exponent, *m*.

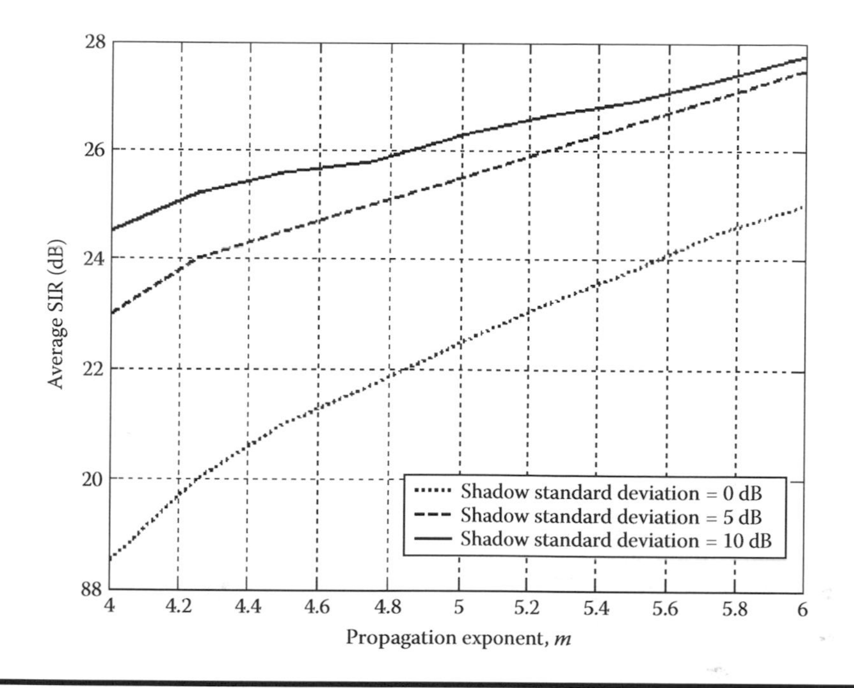

Figure 9.13 Average SIR versus propagation exponent, *m*, for different values of the shadow standard deviation, σ_ξ.

subscriber. Figure 9.11 was drawn in relation to the various topologies shown in Figure 9.5, revealing an average SIR improvement of more than 15 dB for the architecture reusing frequency by a factor of 1 compared to that reusing frequency by a factor of 4.

Although for mobile applications σ_ξ is well documented, ranging from 5 to 12 dB in macrocells or from 4 to 13 dB in microcells, and exhibiting a slight increase with frequency (0.8 dB higher at 1800 MHz than at 900 MHz) [42], for fixed LMDS systems operating in millimeter waves, relevant measurements are limited. Referring again to the configuration with frequency reuse 4, in Figures 9.13 and 9.14 results are presented concerning the average SIR and the OP_I for three values of the shadow standard deviation ($\sigma_\xi = 0$ dB, 5 dB, and 10 dB) [40]. $\sigma_\xi = 0$ dB means that the signal and interfering powers are deterministic and that no shadowing occurs.

9.3.4 Three-Dimensional Analysis: The Effect of Elevation and Building Height

So far, the interference has been considered in two dimensions exclusively in the azimuth plane. This is a common practice in order to provide an interference analysis of general purpose. However, for real-life LMDS networks

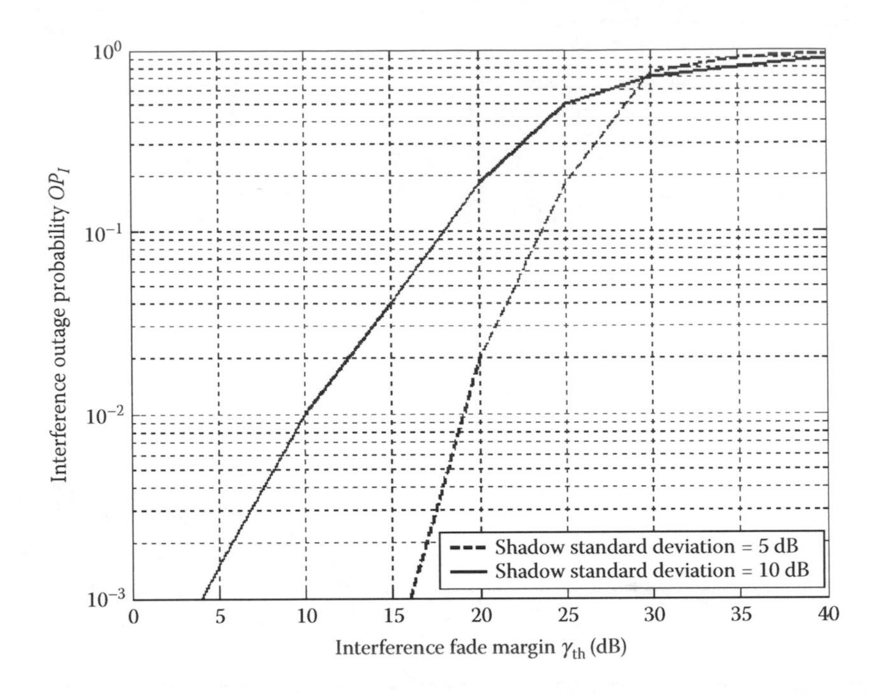

Figure 9.14 Interference outage probability, OP_I, versus fade margin, γ_{th}, in a shadowed propagation environment for different values of the shadow standard deviation, σ_ξ.

located in urban areas, where the requirement for LOS imposes that BS antennas should be placed as high as possible, it is imperative to take into account elevation and building heights. Along these lines, Bauer et al. [43] present a three-dimensional interference study for LMDS networks based on the urban database of Munich, Germany.

To incorporate the spatial filtering effect due to the HPBW of the SS antenna with respect to elevation, its directive gain must be modified to $G_{R,SS}(\phi_k, \theta_k)$ to form a stereo angle pattern. The existence of high buildings, apart from blocking LOS coverage within the cells,* also has a favorable impact on the CCI. This is especially true when LOS exists, since then, each customer receives a large part of the total interference from a dominant source (base station of an adjacent cell). In this situation, CCI could be completely eliminated by placing the customer antenna significantly below its serving BS, so that the IBS lies outside the SS HPBW angle, ensuring, however, LOS to the desired BS. Obviously, when the SIR is determined in the two-dimensional space, such cases cannot be accounted for.

* The ITU-R Recommendation P.1410 [20] provides advice on area coverage calculations, including a number of prediction methods.

Concluding, from the SIR study in three dimensions, the following useful design rules come up [43]:

- For a fixed height of the desired BS, increasing the IBS height leads to an increase in the interference experienced by a customer.
- For a fixed height of the IBS, increasing the desired BS height leads to a decrease of the percentage of covered customers experiencing interference.
- Increasing the height of all BSs leads to an increase in covered customers facing interference problems.
- Since every desired BS also acts as an IBS for its surrounding cells, an efficient design would be to place all BS antennas at nearly the same height.

9.4 An Alternative Access Solution: CDMA-Based LMDS Networks

9.4.1 General Considerations and Interference Scenarios

In the last decade or so, spread-spectrum techniques and particularly DS-CDMA–based systems have experienced a large growth due to their incorporation in modern mobile communication networks [44]. Third-generation cellular systems, such as cdma2000 and UMTS* wideband-CDMA belong to this category. On the other hand, standardization institutes have not yet mandated the application of DS-CDMA for fixed BWA networks, with the exception of the UMTS terrestrial radio access (UTRA) [45] and the wireless LAN standard IEEE 802.11b at 2.4 GHz [46].

Furthermore, as stated in previous sections, the recent evolution of millimeter-wave fixed BWA technologies based on WiMax standards relied at large on MF-TDMA, while DS-CDMA received little attention due to practical reasons. Although a significant portion of the spectrum is available for LMDS applications, the high data rates make a sufficient processing gain (or spreading factor) questionable. In addition, as pointed out in Section 9.2.3, multipath in millimeter-wave LOS paths is insignificant. Therefore, constructive use of the resolved multipath signals through a Rake receiver is not possible.

Despite the above reservations, some pioneering researchers [47,48] have carried out preliminary investigations on CDMA-LMDS, concluding that, at least for the upstream, DS-CDMA offers a higher immunity to intercell interference compared to conventional TDMA. Furthermore, various sectoring approaches for application in the CDMA version of LMDS

* UMTS: Universal Mobile Telecommunications System.

have been proposed [13–14, 47]. Once more, the well-studied techniques of bandwidth segmentation and polarization alternation were considered. DS-CDMA offers an additional domain to discriminate between neighboring sectors of a cell; namely, different spreading sequences. In particular, it has been found preferable [13] to carry out sectoring exclusively in the code domain (single-frequency scheme with frequency reuse of 4) to ensure the highest possible spreading factor. This results in the cellular architecture of Figure 9.15, where the numbering of sectors does not designate different frequency channels, but different spreading codes.

Spreading sequences designed for DS-CDMA are differentiated according to their cross-correlation properties into orthogonal, quasi-orthogonal, and pseudo-noise sequences. Orthogonal code sequences have zero cross-correlation provided they are synchronized in time; quasi-orthogonal and pseudo-noise code sequences have nonzero, though small, cross-correlation values. From the above alternatives, both Sari [48] and Arapoglou et al. [49] chose orthogonal spreading sequences to execute the sectoring function. This is done to avoid intracell interference between users belonging to the same cell. In LMDS, the little effect of multipath, albeit canceling the possibility of attaining any diversity gain through a Rake receiver, helps preserve the orthogonality of the codes at the receiver.

To take advantage of the mature technology of second-generation DS-CDMA cellular communication networks, both Sari [48] and Arapoglou et al. [49] proposed a code assignment scheme similar to the one specified in IS-95 [50]. For the spreading operation, an orthogonal Walsh–Hadamard (WH) sequence of length $4N$ is assigned to every subscriber, so that each fully loaded sector can accommodate up to N subscribers. The resulting signal is then multiplied with a long scrambling pseudo-noise sequence, with a different time-offset for every BS to differentiate between BSs. In each cell, sectoring is carried out by separating the set of WH codes into four disjoint subsets; that is, each rectangular cell consists of four sectors (numbered 1, 2, 3, and 4 in Figure 9.15).

As seen from Figure 9.15a, in the downstream direction, the BSs in adjacent cells are all possible cochannel interferers. However, for a specific SS position in the sector, the set of interfering BS is limited by the narrow HPBW of the receiving antenna. Preserving the notation of the previous sections, a general expression for the unfaded downstream SIR is

$$\left(\frac{S}{I}\right)_u^{down} = \frac{EIRP_{BS,D}\, G_{R,SS}(0)\,(1/d_D)^m}{(1-\alpha)(N-1)EIRP_{BS,D}\, G_{R,SS}(0)\,(1/d_D)^m + N\sum_{k=1}^{K} EIRP_{BS,I}^{(k)}\, G_{R,SS}^{(k)}(\phi_k)\,(1/d_I^{(k)})^m} \times \frac{W}{R} \tag{9.9}$$

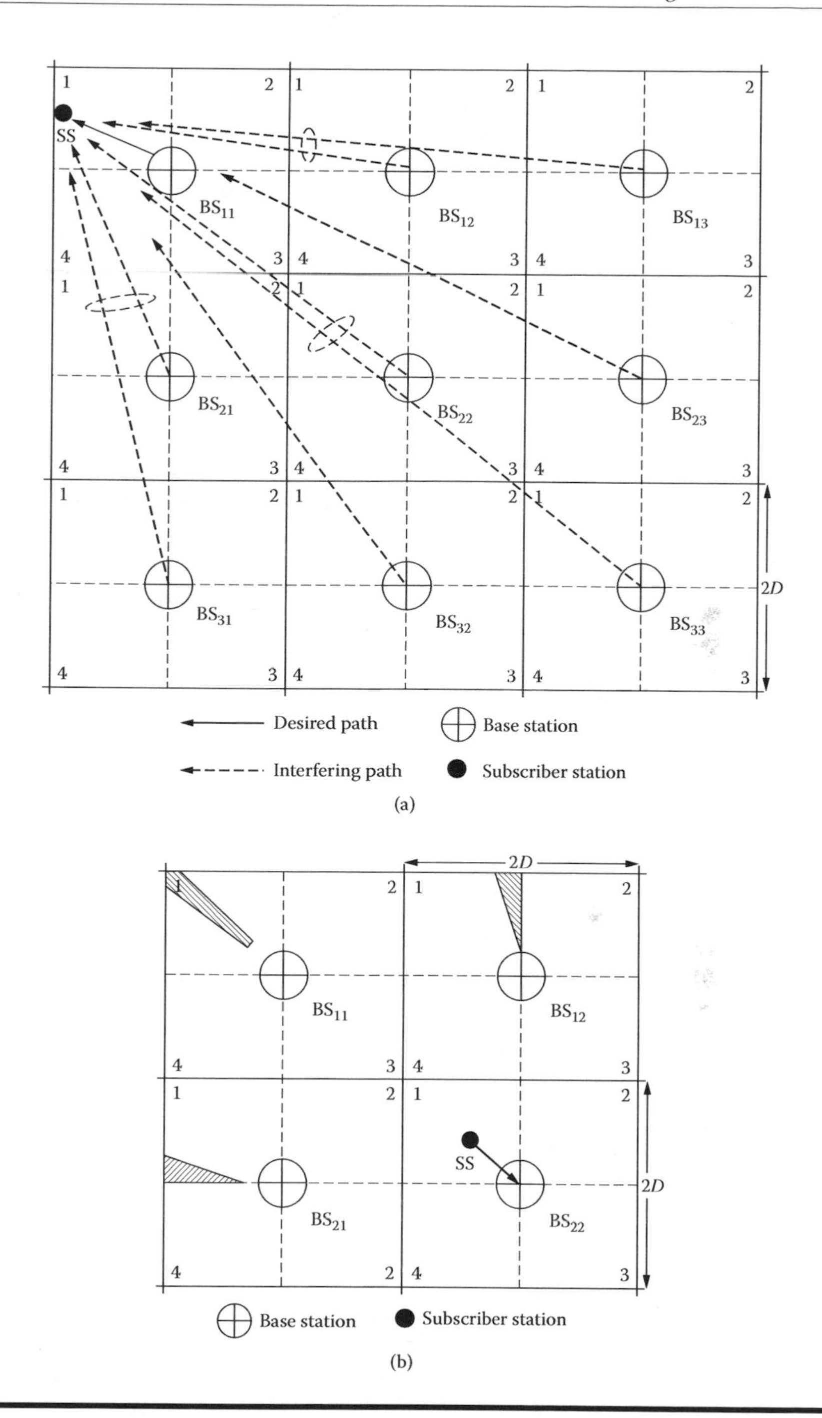

Figure 9.15 Intercell interference scenarios for a single-frequency CDMA-based LMDS network. Sectoring is carried out by partitioning the set of orthogonal spreading sequences between the sectors. (a) Downstream case. (b) Upstream case.

Where the first term in the denominator stands for the total power from the home hub to the rest $(N - 1)$ of the nonintended users in the sector (intracell interference), whereas the second one comes from interfering BS k, $k = 1, \ldots, K$, located in a neighbor cell providing service to N active users (intercell interference). α is the orthogonality factor between the spreading sequences. Adopting WH codes and taking into account the lack of multipath in LMDS systems, $\alpha = 1$, and, therefore, the first term of the denominator is eliminated. Furthermore, W is the chip rate of the spreading sequence (WH in this case) and R the information data rate. The fraction W/R corresponds to the spreading factor and reflects the system capability to reject interference.

In Figure 9.15b, where BS_{22} serves as the reference hub, interference scenarios in the upstream direction are examined. Once more, it is revealed that any subscriber in an adjacent cell is a possible ISS for the victim BS. But strong interference originates only from regions of the sector for which the antenna of the ISS is aligned both to its desired and the interfered BS (shaded regions in Figure 9.16). The corresponding unfaded upstream SIR relationship may now be written as

$$\left(\frac{S}{I}\right)_u^{up} = \frac{P_{T,SS}\, G_{T,SS}(0)\,(1/d_D)^m}{(1-\alpha)(1-j)\sum_{n=1}^{N-1} P_{T,SS}^{(n)}\, G_{T,SS}^{(n)}(0)\,(1/d_D^{(n)})^m + N\sum_{k=1}^{K}\sum_{n=1}^{N} P_{T,SS}^{(k,n)}\, G_{T,SS}^{(k,n)}(\phi_{k,n})\,(1/d_I^{(k,n)})^m} \times \frac{W}{R} \times q \tag{9.10}$$

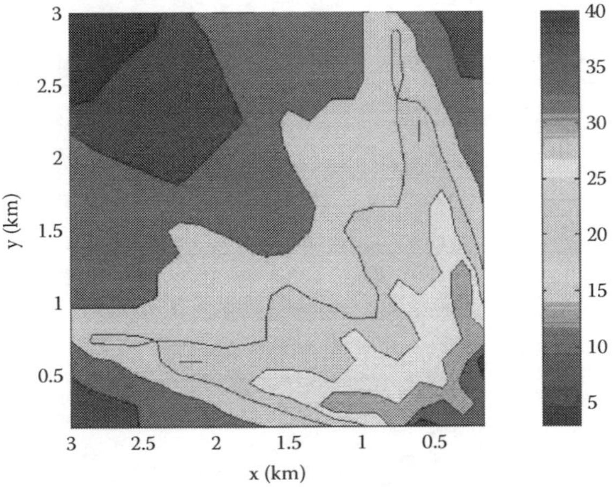

Figure 9.16 Downstream SIR sector distribution for a CDMA-based LMDS system operating in Athens, Greece, under the specification $OP_I = 0.001\%$ (frequency of operation, 42 GHz; cell radius, 3 km).

where $P_{T,SS}$ and $G_{T,SS}$ represent the output power and antenna gain, respectively, of the transmitting SS. The pair (k, n) refers to user n, $n = 1, \ldots, N$, residing in the interfering cell k, $k = 1, \ldots, K$. Equation (9.10) also contains some unknown coefficients: First, q is a general factor ranging from 0 to 1 to capture the effect of imperfect power control for upstream transmission. Since user positions in LMDS are fixed and the fading phenomena are slowly varying processes, q may be set equal to 1. Second, the parameter j, ranging also from 0 to 1, represents the improvement in terms of intracell interference when multiuser detection is employed. More on multiuser detection as a multiple access interference (MAI) reduction technique applied specifically to CDMA-LMDS can be found elsewhere [51].

9.4.2 Comparison between TDMA- and CDMA-Based LMDS in Terms of Intercell Interference

Sari [48] carried out a simple comparison concerning CCI for the two LMDS multiple access alternatives investigated in the present chapter. The signal attenuation is assumed proportional to the square of the distance ($m = 2$) and all stations involved have the same individual information rate. To ensure a fair comparison of the two multiple access schemes, this is based on a cell configuration consisting of four sectors. In TDMA, sectors are separated by splitting the total bandwidth, $4N$, into four orthogonal frequency bands, whereas in CDMA this is achieved by using a set of $4N$ orthogonal codes in every cell.* For simplicity, alternating polarization and XPD effects are not taken into account in either type of sectoring. In both cases, each cell sector can accommodate up to N subscribers. Furthermore, a common mean $EIRP_{BS}$ for all BSs is assumed, whereas the SIR is calculated taking into consideration only CCI to and from first-tier subscribers† that are aligned with the source of interference, that is, they transmit under maximum antenna gain. Hence, this approach to the problem is worst-case oriented and the level of the SIR experienced depends exclusively on the related BS–SS distances, d.

* A number of $4N$ orthogonal codes require sequences of length $4N$. Consequently, the spreading factor is also $4N$, because the length of orthogonal spreading sequences is equal to the spreading factor.

† In TDMA, due to the frequency reuse scheme adopted, the first tier of interference refers to interfering stations two cells away, whereas in CDMA this distance is reduced to one cell.

For the configuration described, the TDMA downstream SIR for an SS interfered by a single IBS can be computed after simplifying Equation (9.2):

$$\left(\frac{S}{I}\right)_{\text{TDMA}} = \left(\frac{d_I}{d_D}\right)^2 \tag{9.11}$$

If an SS located at the corner of the sector along the diagonal (worst sector point) is considered, then $d_D = \sqrt{2}D$ and $d_I = 5\sqrt{2}D$ (corresponding to an interfering BS two cells away) and the SIR turns out to be 14 dB. For the upstream of a millimeter-wave fixed BWA system, APC is usually employed. Here, a simplified version of Equation (9.6) is adopted, ignoring fading effects, according to which the transmitted power from an SS is proportional to the length of the BS–SS link. After straightforward algebra, it is concluded that, in TDMA-based LMDS, the worst-case SIR for the downstream and the upstream directions is the same and equal to 14 dB.

For the CDMA downstream SIR, one can write Equation (9.9) as

$$\left(\frac{S}{I}\right)^{down}_{\text{CDMA}} = \frac{4N}{N}\left(\frac{d_I}{d_D}\right)^2 \tag{9.12}$$

where $W/R = 4N$, $\alpha = 1$, and interference comes from the IBS one cell away transmitting toward N active subscribers within its own sector (fully loaded). For exactly the same geometrical arrangement of the desired SS, $d_D = \sqrt{2}D$, $d_I = 3\sqrt{2}D$, and Equation (9.12) yields 15.5 dB. In the case when the IBS located two cells away ($d_I = 5\sqrt{2}D$) is also incorporated in the derivation of the SIR, its value reduces to 14.2 dB, which is still slightly higher than in TDMA. CDMA upstream calculations are more involved because CCI is produced by the percentage of sector users in the shaded areas of Figure 9.15b (strong interference regions of the cell). According to Sari [48], a realistic assumption for this number is 25 percent, whereas a geometrical test performed by Arapoglou et al. [49] suggested 13 percent for users in the three interfering sectors (horizontal, vertical, diagonal). In the present worst-case approach, it is assumed that 50 percent of the interferers generate strong interference. Substituting this into (9.10) (neglecting joint user detection and APC irregularities), an SIR value of 18.5 dB is determined.

Concluding, the downstream SIR performances of TDMA- and CDMA-based LMDS are comparable, whereas CDMA exhibits a higher immunity to interference in the upstream. To support this interesting result, which provides valuable insight as far as multiple access in millimeter-wave fixed BWA is concerned, further study in the frame of a more realistic operational environment and perhaps measured data is needed. Nevertheless, the potential capacity improvement offered by CDMA-LMDS cannot be neglected. Finally, for reasons of completeness, we cite Bose [52], who reports a similar capacity improvement in fixed BWA networks achieved by adopting trellis-coded modulation (TCM), showing many similarities with the CDMA case investigated here.

9.4.3 Correlated Rain Fading

The CDMA counterpart of the statistical model outlined in Section 9.3.2 for the prediction of the interference outage OP_I under rain fading conditions was presented by Arapoglou et al. [49]. Equations (9.9) and (9.10) were applied to calculate clear-sky SIR in the downstream and upstream links, respectively, superimposing the convective rain cells model to account for the correlation between multiple converging LOS paths undergoing rain attenuation. It should be noted that the analysis under discussion is of global applicability, because it provides a methodology relating the statistical parameters of interference, $\{\gamma_{th}, OP_I\}$, to the ITU-R global rain rate database [30]. Nevertheless, the most reliable way to proceed is to design fixed wireless systems based on local propagation measurements, the availability of which, however, is rather limited for most geographical regions.

To illustrate the implementation of the statistical model, isoline curves have been plotted in Figure 9.16 for the CDMA version of the hypothetical LMDS system presented in Figure 9.9. The downstream SIR values shown correspond to the specification $OP_I = 0.001$ percent and, as expected,

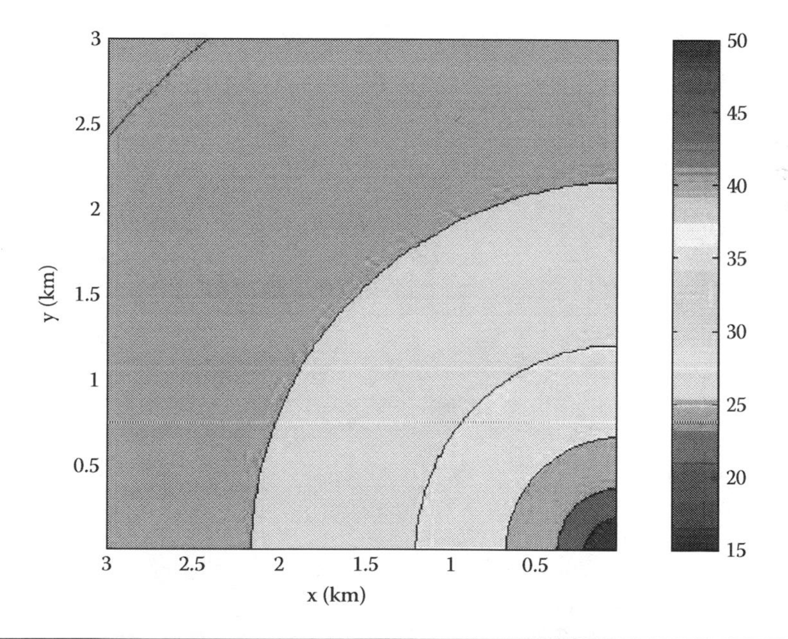

Figure 9.17 Upstream SIR sector distribution for a CDMA-based LMDS system operating in Athens, Greece, under the specification $OP_I = 0.001$ percent.

strong interference within the cell sector is in the form of wedges. In contrast to the wedge-form SIR distribution of the downstream, upstream CCI consists of SIR isolines that are concentric quartercycles centered at the reference station BS_{22} (see Figure 9.17). This demonstrates that, for a given interference outage probability, specific sector areas (or, equivalently, a percentage of customers in the sector) suffer from SIR values below the clear-sky minimum. Thus, it is imperative to incorporate fading in the interference analysis for a reliable design of fixed wireless networks. On the other hand, when APC is employed, the upstream SIR becomes independent of the position of the SS and, for a specific OP_I, there is a balancing effect on the SIR over the sector area.

9.5 Conclusions

The present chapter intended to present the state of the art concerning SIR modeling in millimeter-wave WMAN cellular configurations operating above 20 GHz. Such a study is motivated by the fact that fixed BWA networks exhibit distinct characteristics in comparison to more conventional wireless networks in terms of system architecture and propagation conditions. The main elements of an LMDS system influencing intercell interference were described in Section 9.2 and can be summarized by the following points:

- The fixed position of subscriber stations.
- The narrow HPBW of the SS antenna.
- The aggressive frequency reuse policy within each cell.
- The LOS requirement for all hub–subscriber links.
- The fact that the dominant fading mechanism is rain attenuation instead of multipath.

Apart from the above points, intercell interference in LMDS networks is also strongly related to the multiple access scheme adopted, according to which the presentation of the material is separated into two major parts: the conventional MF-TDMA case from Section 9.3 (included in recent fixed BWA technical standards) and the alternative of employing DS-CDMA discussed in Section 9.4. Section 9.3.2 presented the SIR statistical analysis under correlated rain fading for cleared (LOS) paths. On the other hand, the appropriate modeling procedures under shadowing when nearly LOS conditions prevail were discussed in Section 9.3.3. Finally, in Section 9.3.4, some useful design rules for the corresponding three-dimensional interference problem were outlined.

After stating the general assumptions and interference scenarios for the less common DS-CDMA–based LMDS system, a qualitative comparison between the two multiple access schemes was provided in Section 9.4.1. The interesting conclusion comes up that DS-CDMA exhibits a higher immunity to interference. Then, similar to the TDMA case, the corresponding statistical analysis under rain fading is presented.

References

[1] H. Sari, "Broadband radio access to homes and businesses: MMDS and LMDS," *Computer Networks*, vol. 31, pp. 379–393, 1999.

[2] A. Nordbotten, "LMDS systems and their application," *IEEE Commun. Mag.*, vol. 38, no. 6, pp. 150–154, June 2000.

[3] IEEE 802.16, "IEEE Standard for Local and Metropolitan Area Networks—Part 16: Air Interface for Fixed Broadband Wireless Access Systems," April 2002.

[4] ETSI TS 101 999 V1.1.1, "Broadband Radio Access Networks (BRAN); HIPERACCESS; PHY protocol specification," 2002.

[5] B. Fong, N. Ansari, A.C.M. Fong, G.Y. Hong, and P.B. Rapajic, "On the scalability of fixed broadband wireless access network deployment," *IEEE Radio Commun.*, vol. 1, no. 3, pp. 12–18, September 2004.

[6] P.A. Gray, "Optimal hub deployment for 28GHz LMDS system," in *Proc. IEEE Wireless Commun. Conf. 97*, pp. 18–22, August 1997.

[7] ETSI EN 301 215-2 V1.2.1, "Fixed Radio Systems; Point to Multipoint Antennas; Antennas for point-to-multipoint fixed radio systems in the 11 GHz to 60 GHz band; Part 2: 24 GHz to 30 GHz," Final Draft, European Standard.

[8] ETSI EN 301 215-3 V1.1.1, "Fixed Radio Systems; Point to Multipoint Antennas; Antennas for point-to-multipoint fixed radio systems in the 11 GHz to 60 GHz band; Part 3: Multipoint Multimedia Wireless System in the 40.5 to 43.5GHz," Final Draft, European Standard.

[9] P. Mähönen, A. Jamin, T. Saarinen, Z. Shelby, L. Muñoz, and T. Sukuvaara, "Two-layer LMDS architecture: DAVIC-based approach and analysis," *Wireless Commun. Mobile Comput.*, vol. 2, no. 4, pp. 319–338, 2002.

[10] T. Martin, J. You, and A. Marshall, "Analysis of a multiple service MAC layer for two-way two-layer LMDS networks," in *Proc. 4th Europ. Conf. Multimedia Applic., Services Techn.*, pp. 169–180, 1999.

[11] C. Eklund, R.B. Marks, K.L. Stanwood, and S. Wang, "IEEE Standard 802.16: A technical overview of the WirelessMAN air interface for broadband wireless access," *IEEE Commun. Mag.*, vol. 40, no. 6, pp. 98–107, June 2002.

[12] V.I. Roman, "Frequency reuse and system deployment in local multipoint distribution service," *IEEE Pers. Commun.*, vol. 6, no. 6, pp. 20–27, December 1999.

[13] C. Novák, A. Tikk, and J. Bitó, "Code sectoring methods in CDMA-based broadband point-to-multipoint networks," *IEEE Microwave Wireless Compon. Lett.*, vol. 13, no. 8, pp. 320–322, August 2003.

[14] M.-K. Tsay and F.-T. Wang, "Cellular architecture on CDMA based LMDS," in *Third Int. Symp. CSNDSP*, Staffordshire, UK, July 2002.

[15] P.B. Papazian, G.A. Hufford, R.J. Achatz, and R. Hoffman, "Study of the local multipoint distribution service radio channel," *IEEE Trans. Broadcast.*, vol. 43, no. 2, pp. 175–184, June 1997.

[16] A. Paulsen and A. Seville, "Attenuation and distortion of millimeter radio waves propagation through vegetation," in *Millennium Conf. Antennas Propagat.*, vol. AP2000, 2000.

[17] H. Xu, T.S. Rappaport, R.J. Boyle, and J.H. Schaffner, "Measurements and models for 38GHz point-to-multipoint radiowave propagation," *IEEE J. Select. Areas Commun.*, vol. 18, no.3, pp. 310–321, March 2000.

[18] P. Soma, L.C. Ong, S. Sun, and M.Y.W. Chia, "Propagation measurements and modeling of LMDS radio channel in Singapore," *IEEE Trans. Vehic. Technol.*, vol. 52, no. 3, pp. 595–606, May 2003.

[19] H.R. Anderson, *Fixed Broadband Wireless System Design*, John Wiley & Sons, New York Chapter 5, 2003.

[20] ITU-R, "Propagation data and prediction methods for the design of terrestrial broadband millimetric radio access systems operating in a frequency range of about 20–50 GHz," *Propagation in Non-Ionized Media*, Rec. P.1410-3, Geneva, 2005.

[21] ITU-R, "Propagation data and prediction methods required for the design of terrestrial line-of-sight systems," *Propagation in Non-Ionized Media*, Rec. P.530–11, Geneva, 2005.

[22] W. Jeong and M. Kavehrad, "Cochannel interference reduction in dynamic-TDD fixed wireless applications, using time slot allocation algorithms," *IEEE Trans. Commun.*, vol. 50, no. 10, pp. 1627–1636, October 2002.

[23] ACTS Project 215, "Cellular Radio Access for Broadband Services (CRABS)," February 1999.

[24] R. Bose, G. Bauer, and R. Jacoby, "Two-dimensional line of sight interference analysis of LMDS networks for the downlink and uplink," *IEEE Trans. Antennas Propagat.*, vol. 52, no. 9, pp. 2464–2473, September 2004.

[25] R.K. Crane, *Electromagnetic Wave Propagation Through Rain*, Wiley, New York, 1996.

[26] A. Paraboni, G. Masini, and A. Elia, "The effects of precipitation on microwave LMDS networks—Performance analysis using a physical raincell model," *IEEE J. Select. Areas Commun.*, vol. 20, no. 3, pp. 615–619, April 2002.

[27] C.-Y. Chu and K.S. Chen, "Effects of rain fading on the efficiency of the Ka-band LMDS system in the Taiwan area," *IEEE Trans. Vehic. Techn.*, vol. 54, no. 1, pp. 9–19, January 2005.

[28] A.D. Panagopoulos, P.-D.M. Arapoglou, J.D. Kanellopoulos, and P.G. Cottis, "Intercell radio interference studies in broadband wireless access networks," *IEEE Trans. Vehic. Technol.*, vol. 56, No. 1, pp. 3–12, 2007.

[29] A.D. Panagopoulos and J.D. Kanellopoulos, "Statistics of differential rain attenuation on converging terrestrial propagation paths," *IEEE Trans. Antennas Propagat.*, vol. 51, no. 9, pp. 2514–2517, September 2003.

[30] ITU-R, "Characteristics of precipitation for propagation modeling," *Propagation in Non-Ionized Media*, Rec. P.837–3, Geneva, 2001.

[31] I.S. Usman, M.J. Willis, and R.J. Watson, "Route diversity analysis and modelling for millimetre wave point to multi-point systems," in *1st International Workshop COST 280*, PM3032, July 2002.

[32] A.D. Panagopoulos and J.D. Kanellopoulos, "Cell-site diversity performance of millimeter-wave fixed cellular systems operating at frequencies

above 20GHz," *IEEE Antennas Wireless Propagat. Lett.*, vol. 1, pp. 183–185, 2002.

[33] G. Hendrantoro, R.J.C. Bultitude, and D.D. Falconer, "Use of cell-site diversity in millimeter-wave fixed cellular systems to combat the effects of rain attenuation," *IEEE J. Selec. Areas Commun.*, vol. 20, no. 3, pp. 602–614, April 2002.

[34] C. Sinka and J. Bitó, "Site diversity against rain fading in LMDS systems," *IEEE Microwave Wireless Comp. Lett.*, vol. 13, no. 8, pp. 317–319, August 2003.

[35] P.-D.M. Arapoglou, E. Kartsakli, G.E. Chatzarakis, and P.G. Cottis, "Cell-site diversity performance of LMDS systems operating in heavy rain climatic regions," *Int. J. Infrared Millimeter Waves*, vol. 25, no. 9, pp. 1345–1359, September 2004.

[36] A.D. Panagopoulos, P.-D.M. Arapoglou, G.E. Chatzarakis, J.D. Kanellopoulos, and P.G. Cottis, "LMDS diversity systems: A new performance model incorporating stratified rain," *IEEE Commun. Lett.*, vol. 9, no. 2, February 2005.

[37] K.P. Liolis, A.D. Panagopoulos, and P.G. Cottis, "Use of cell-site diversity to mitigate co-channel interference in 10-66GHz broadband fixed wireless access networks," in *IEEE Radio and Wireless Symp.*, pp. 283–286, 2006.

[38] S.Q. Gong and D. Falconer, "Cochannel interference in cellular fixed broadband access systems with directional antennas," *Personal Wireless Commun.*, vol. 10, no. 1, pp. 103–117, June 1999.

[39] S. Farahvash and M. Kavehrad, "Co-channel interference assessment for line-of-sight and nearly line-of-sight millimeter waves cellular LMDS architecture," *Int. J. Wireless Inform. Networks*, vol. 7, no. 4, pp. 197–210, 2000.

[40] L.F. Fenton, "The sum of log-normal probability distributions in scatter transmission systems," *IRE Trans. Commun.*, vol. 8, pp. 57–67, March 1960.

[41] S. Schwartz and Y.S. Yeh, "On the distribution function and moments of power sums with log-normal components," *Bell System Tech. J.*, vol. 61, pp. 1441–1462, September 1982.

[42] G.L. Stüber, *Principles of Mobile Communication*, 2nd Edition, Kluwer Academic Publishers, Norwell, MA, Chapters 2–3, 2002.

[43] G. Bauer, R. Bose, and R. Jakoby, "Three-dimensional interference investigations for LMDS networks using an urban database," *IEEE Trans. Antennas Propagat.*, vol. 53, no. 8, pp. 2464–2470, August 2005.

[44] R. Prasad and T. Ojanperä, "An overview of CDMA evolution toward wideband CDMA," *IEEE Commun. Surveys & Tutorials*, vol. 1, no. 1, Fourth Quarter 1998.

[45] M. Haardt, A. Klein, R. Koehn, S. Oestreich, M. Purat, V. Sommer, and T. Ulrich, "The TD-CDMA based UTRA TDD mode," *IEEE J. Selec. Areas Commun.*, vol. 18, no. 8, pp. 1375–1385, August 2000.

[46] IEEE 802.11 Working Group Web site: http://ieee802.org/11

[47] H. Halbauer, P. Jaenecke, and H. Sari, "An analysis of Code-Division Multiple Access for LMDS networks," in *7th Europ. Conf. Fixed Radio Systems Network.*, pp. 173–180, Dresden, September 2000.

[48] H. Sari, "A multimode CDMA with reduced intercell interference for broadband wireless networks," *IEEE J. Select. Areas Commun.*, vol. 19, no. 7, pp. 1316–1323, July 2001.

[49] P.-D.M. Arapoglou, A.D. Panagopoulos, J.D. Kanellopoulos, and P.G. Cottis, "Intercell radio interference studies in CDMA-based LMDS networks," *IEEE Trans. Antennas Propagat.*, vol. 53, no. 8, pp. 2471–2479, August 2005.

[50] EIA/TIA IS-95B, "Mobile station—base station compatibility standard for dual-mode wideband spread spectrum cellular system," 1997.

[51] C. Novák, D. Tóth, and J. Bitó, "Uplink interference analysis of LMDS networks applying CDMA with interference cancellation," in *IEEE 7th Symp. Spread Spectrum Techn. Applic.*, vol. 1, pp. 288–292, 2002.

[52] R. Bose, "Improving capacity in LMDS networks using trellis coded modulation," *EURASIP J. Wireless Commun. Networking*, Issue 2, pp. 365–373, November 2004.

Chapter 10

Millimeter-Wave Radar: Principles and Applications

Felix Yanovsky

Contents

This chapter provides a radar perspective for millimeter wave propagation and scattering. It considers radar principles and the features of subsystems and components of millimeter-wave radars. Rather complicated notions of radar theory are stated in a very simple manner as a kind of overview. Then, application of millimeter-wave radar for intelligent transportation systems

(ITS) is analyzed in detail as the main section of the chapter. The role of radar sensors in ITS is shown, and frequency allocation for ITS radars is analyzed. Traffic surveillance radar, which is an element of ITS infrastructure, is described. Then automotive radars are categorized into long range, medium range, and short range, and each category is considered. On this basis different functional applications of automotive radar are briefly described. Radar-based communications for both car-to-car and the entire ITS system are considered. The necessity and possibility of data exchange using WLAN (wireless local area network) is indicated. Millimeter-wave radars are employed in a wide range of commercial, military, and scientific applications for remote sensing, safety, and measurements. That is why other important millimeter-wave radar applications are considered. Finally, the conclusion is that millimeter-wave radar is a universal instrument for numerous promising applications and a device quite suitable for network use and data exchange via wireless networks.

10.1 Introduction

Millimeter waves were once considered unfit for practical use in radar. One of the main reasons was the absence of suitable means of generation, reception, channelization, and transmission of electromagnetic waves of millimeter range. Moreover, the laws of millimeter-wave propagation in the nonhomogeneous atmosphere were not studied enough.

Nowadays, creation of modern and prospective millimeter-wave radar systems is based on the research of propagation and scattering features in the millimeter-wave range as well as on the development of methods and means for millimeter-wave generation and reception.

Step by step both theory and practice discover new advantages of millimeter waves, and millimeter-wave radars become more and more applicable in different fields. Today millimeter waves are increasingly used in car radar, cloud radar, radar, and radiometry for concealed weapon detection (CWD), high-speed wireless access, ultra-high-speed wireless local area networks (WLAN), and other means of communications including radar-based communication systems.

High antenna gain at rather small aperture and the possibility to use wideband and very short waveforms facilitate improved resolution and accuracy of radar measurements, which is an obvious advantage of millimeter-wave radars, as well as huge information content and information rate with application of radar-based communication data systems. Moreover, as we now know, millimeter waves are characterized by stable propagation properties in unfavorable environments and improved noise immunity.

The millimeter-wave spectrum has become the focus of attention in recent years also because the lower frequency bands are filling up very quickly. New wideband applications, such as WLAN and radar, require large bandwidths, which are readily available at millimeter-wave frequencies.

However, significant attenuation, signal depolarization, amplitude, and phase changes are attached to millimeter-wave propagation, and atmospheric attenuation tends to be higher if the frequency increases, and it also depends on weather conditions. The selection of a particular frequency band for new radar systems depends therefore on a number of factors, including (1) the propagation environment, (2) the frequency bands available for a particular service, (3) backscattering properties of targets, and (4) the availability of appropriate technology.

Because of rather large absorption in the atmosphere, millimeter waves are used mainly in short-range radar systems.

Theory and applications of millimeter-wave radar are described elsewhere in books [1–3], book chapters [4–6], and survey papers [7]. Problems and advantages of millimeter-wave radar are discussed at various international conferences and symposia [8–12]. This is a field of rapid development and, from time to time, it is necessary to make a short stop to survey what has already been done in order to have a possibility to further advance and face the future with confidence. Thus, a consideration of the trends in development and application of millimeter-wave radar systems and their various functions, which have received significant development during recent years, is of doubtless interest.

In this chapter, the state of the art for millimeter-wave radar design and application is presented. First, a brief overview of millimeter-wave propagation and scattering is provided. This is followed by basic radar design principles. On the basis of millimeter-wave radar possibilities and real practical needs, different millimeter-wave radar systems can be created, particularly for intelligent transportation systems weather information systems, remote sensing, and other applications. These items are used as the core of the chapter.

10.2 Propagation and Scattering of Millimeter-Length Waves

Understanding millimeter-wave interaction with molecules of atmospheric gases, hydrometeors, turbulent inhomogeneity of air, and also the estimation of influence of vertical stratification of the atmosphere and reflections from underlying terrain on characteristics of received signals are rather important in radar applications.

Now the problem of millimeter-wave propagation is appreciably investigated. Results of research and theoretical calculations of scattering and molecular absorption in hydrometeors coincide quite well. The features of millimeter-wave behavior in the atmosphere that are the most significant for radar applications and the features of backscattering that underlie active radar will be considered in this section.

10.2.1 Molecular Absorption

The theoretical description of gaseous absorption is well established and a number of models have been developed to calculate the transmission and attenuation through the Earth's atmosphere. Millimeter-wave attenuation in atmospheric gases can be assessed, relying primarily on data on temperature, pressure, and humidity, all of which are normally available.

The theory of molecular absorption in the atmosphere is quite complicated. Therefore, normally researchers use numerous approximations when calculating molecular absorption spectra of gaseous components of the atmosphere. This influences the accuracy of the final result. Perhaps the most comprehensive theory of millimeter-wave gaseous absorption was developed by Kalmykov and Titov [13] and Zagorin et al. [14]. As a result of theoretic and experimental work, the method of the memory functions was developed for the description on the molecular level of the complex refractive index of polar gases for arbitrary symmetric molecules. According to Zagorin et al. [14], the developed model of molecular gaseous absorption describes sufficiently the phenomena under study in a broad range of pressures. In Figure 10.1, adapted from the work of Zagorin et al. [14], the absorption spectra of oxygen in dry atmosphere are shown at 29.7°C for a frequency band of 54–66 GHz. Solid curves represent the calculations in

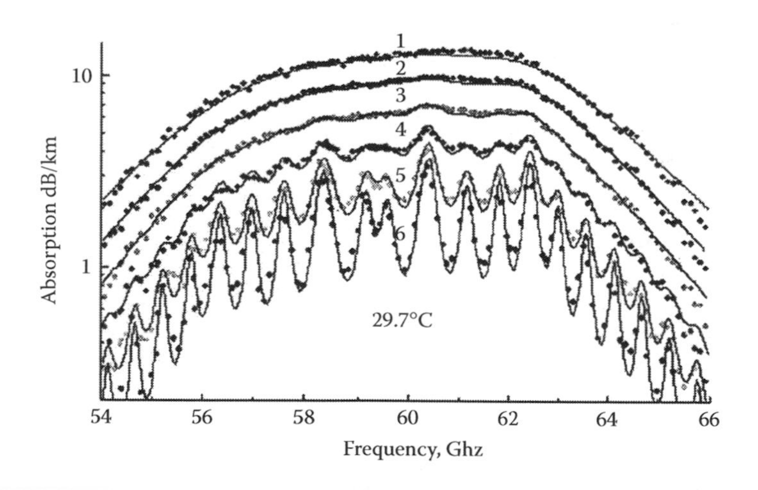

Figure 10.1 Absorption of oxygen at 54–66 GHz band in dry atmosphere as a function of frequency at 29.7°C and different pressure (height). The six curves correspond to six heights:(1) 0; (2) 3 km; (3) 6 km; (4) 9 km; (5) 12 km; (6) 15 km. Dots are experimental results. From Zagorin et al. [14].

the framework of the model [13]; dots are experimental data. Curves 1–6 correspond to different pressures (in kilopascals) and therefore to different heights (in kilometers): 1, 101.3 kPa (0 km); 2, 70.11 kPa (3 km); 3, 47.19 kPa (6 km); 4, 30.81 kPa (9 km); 5, 19.49 kPa (12 km); 6, 12.1 kPa (15 km).

Such calculations can be performed at different frequency bands for different gaseous components, particularly for water vapor, and also for the complex influence of both oxygen and vapor.

In the area of submillimeter wavelengths, the absorption is made by molecules of water vapor, carbonic gas, and oxygen. Air temperature decreases in the troposphere with increased height. That is why water content is also rather sharply reduced with height. Therefore, the infrared area of a spectrum is substantially accessible to monitoring from balloons and high-altitude planes. In this area of a spectrum, along with gaseous absorption, the self-radiation of the atmosphere is also essential, which is especially important for research of background radiation of the universe.

Summarizing, one can say that absorption of millimeter waves increases on average if frequency rises; however, there are pronounced resonance peaks of absorption in the atmosphere. They are caused by the presence of oxygen and water vapor. These phenomena are typical, for example, at the frequencies of 22.2 GHz (vapor), 60 GHz (oxygen), 118.8 GHz (oxygen), and 180 GHz (vapor). Under the condition of moderate air humidity (about 7.5 g/m^3 at the Earth's surface) the complete millimeter-wave attenuation at the separate parts of the spectrum runs up to 200 dB and even more. Between such stable absorption bands, the window regions exist. Average wavelengths and typical absorption factors of window regions are presented in Table 10.1.

The window regions are rather wide; therefore, frequencies around the mean values indicated in Table 10.1 normally may be chosen as standard values for different applications. Special practical interest is traditionally typical for window frequencies of about 35, 94, 140, and 220 GHz.

10.2.2 Attenuation in Hydrometeors

In addition to molecular absorption in air, attenuation during propagation is caused by particles in the atmosphere, especially condensed water vapor particles (hydrometeors) in the form of fog, cloud, rain, snow, and hail. The

Table 10.1 Absorption of Millimeter and Submillimeter Waves in Window Regions

Wavelength, mm	8.6	3.5	2.4	1.4	0.85	0.72	0.6	0.46	0.36	0.02
Frequency, GHz	35	86	125	214	353	417	500	652	833	1500
Absorption, dB/km	0.07	0.42	0.45	1.0	6	14	35	37	45	5

attenuation is specified by two mechanisms: (1) absorption of the incident radiation energy in a volume of a hydrometeor (e.g., a raindrop) and (2) diffraction scattering of the incident radiation by a hydrometeor into the ambient space.

When the radiation wavelength is much greater than the size of the hydrometeors, as in the case of microwave radar, the absorption cross-section is at least an order of magnitude greater than the scattering cross-section, and scattering can thus be neglected. However, this simplification is not valid in the millimeter region, because hydrometeors are comparable in size with the sounding wavelength. As a result, scattering occurs, leading to significant attenuation [15].

Rain, clouds, and fog are the most prevalent forms of hydrometeors encountered in the atmosphere, and rain, in particular, plays a dominant role in determining the availability and reliability of radar and communications systems operating at millimeter waves. Assuming spherical drops, in particular for clouds and light rains, the attenuation can be calculated using classical Mie scattering theory [16] using the known drop size distribution, terminal velocities of drops, and the complex refractive index of water. For nonspherical drops, particularly in the case of strong rain, more complex approaches are available that take into account polarimetric properties [17,18]. However, such approaches are nontrivial. At the same time very simple approximations have been developed in terms of power-law relationships between attenuation, γ, and rainfall rates, R, such as $\gamma = aR^b$ [19–20]. In this simple model, particular attention has been given to the data on the microstructure of rain used to calculate specific attenuation and to the applicability of such calculations to real rainfall situations.

As has been noted elsewhere [15], only a forward scattering brings the contribution to attenuation, and for rain, which is a very rare medium of randomly situated drops, especially in relation to millimeter-wave wavelength, this forward scattering is always coherent.

The contribution of scattering in the attenuation of millimeter waves is essentially different from centimeter waves. Attenuation of millimeter waves is substantially determined by scattering radiation, and the albedo of millimeter-wave single scattering practically does not depend on wavelength, rain rate, drop size distribution, and thermodynamic temperature of water in droplets. Note for comparison that in microwaves (centimeter waves) albedo is a decreasing power function of wavelength and also depends on rain rate, drop size distribution, and temperature of droplets.

The most important property of millimeter-wave propagation in all directions different from the direction of propagation (different from forward scattering) is the fact that for raindrops distributed randomly in space, the scattering on different droplets can be considered independent. This is because raindrops are distant enough from each other; that is, they are in the far-field region regarding the adjacent scatterers. Thus, each droplet

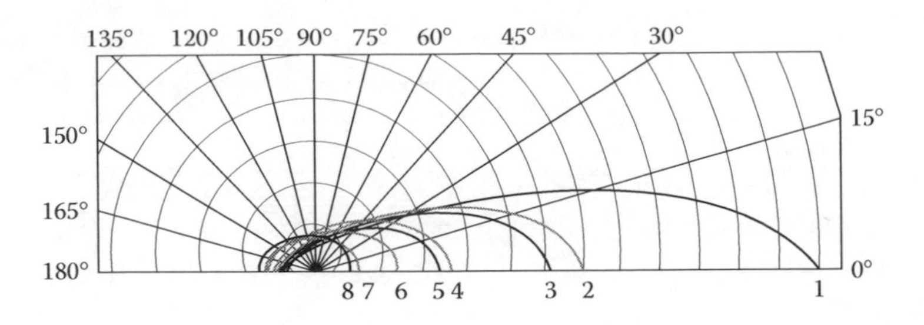

Figure 10.2 Normalized scattering indicatrices of the rain element. (1) $\lambda = 1.4$ mm, R = 12.5 mm/h; (2) $\lambda = 1.4$ mm, R = 1.56 mm/h; (3) $\lambda = 2.2$ mm, R = 12.5 mm/h; (4) $\lambda = 2.2$ mm, R = 1.56 mm/h; (5) $\lambda = 3.3$ mm, R = 12.5 mm/h; (6) $\lambda = 3.3$ mm, R = 1.56 mm/h; (7) $\lambda = 8.6$ mm, R = 12.5 mm/h; (8) $\lambda = 8.6$ mm, R = 1.56 mm/h. From Zagorin et al. [14]

interacts with electromagnetic waves in such a way as if other droplets do not exist. This allows the assumption that radiation scattered by the aggregate of different raindrops is incoherent scattering (but not forward scattering!).

Distribution of radiation scattered by rain in different directions is described by normalized scattering indicatrix. According to Zagorin et al. [14] in Figure 10.2, scattering indicatrices are shown for radio waves with wavelength λ = 1.4, 2.2, 3.3, and 8.6 mm in rains with rain rate R = 1.56 and 12.5 mm/h. Calculations were done on the basis of Mie theory [16]. Despite multilobed scattering indicatrix of separate particles (number of lobes $|m|\ x$, with $|m|$ as the modulus of complex refractive index particle material, and $x = 2\pi a/\lambda$, with a the particle radius), the shape of a scattering indicatrix of polydisperse medium, like rain, is very smoothed. Nevertheless, the effect connected with Mie scattering is very well expressed in the millimeter-wave band, in contrast to centimeter waves where indicatrix is close to the Rayleigh case.

This effect consists of two things: (1) the main lobe is considerably prolonged in the direction of wave propagation, and (2) the degree of oblongness depends on the wavelength, rain rate, and drop size distribution function. The intensity of scattered radiation in the direction of wave propagation increases with lessening wavelength and increasing rain rate. One can see from Figure 10.2 that the scattering indicatrix of a rain element at $\lambda = 8.6$ mm is almost symmetric and close to the Rayleigh case, whereas if λ decreases, the forelobe becomes elongated, and anisotropy of scattering is expressed brighter if rain is more intensive. The last fact is clear because increasing rain rate corresponds to an increase in the number of large drops.

Normalized scattering indicatrices of millimeter waves can be approximated by two-parameter expression [17] with sufficient accuracy to analytically estimate the level of mutual interferences between radio-electronic systems due to scattering radio waves in rain.

In the strict sense, falling raindrops are not spherical, orientation of their axes of symmetry has the preferable direction, and sizes are commensurable with wavelength. Therefore, polarization effects appear brightly at millimeter-wave propagation in rain. This phenomenon is useful for remote sensing.

Frozen precipitation, in the form of snow and hail, produces, on average, less influence on millimeter-wave propagation. In fact, snow, hail, and ice have much smaller dielectric constants than water, and frozen hydrometeors appreciably interact with electromagnetic radiation only when melting is taking place within the particles. Measurements of attenuation and backscatter by falling snow and rain at 96, 140, and 225 GHz can be found in Nemarich et al. [21].

10.2.3 Integrated Influence of Gaseous and Hydrometeor Attenuation

Perhaps the most comprehensive model for integrated attenuation calculations is the millimeter-wave Propagation Model (MPM) developed by Liebe [22]. This model was adopted by the International Telecommunication Union in recommendation [23]. It predicts propagation effects of loss and delay for the neutral atmosphere at frequencies up to 1000 GHz with contributions from dry air, water vapor, suspended water droplets (haze, fog, cloud), and rain. For clear air, the local line base (44 O_2 plus 30 H_2O lines) is complemented by an empirical water–vapor continuum. Input variables are barometric pressure, temperature, relative humidity, suspended water droplet concentration, and rainfall rate. The calculation example in Figure 10.3 is adapted from McMillan [7]. It shows the results of calculating the atmospheric attenuation over the range 40–1000 GHz for relative humidities of 50 and 100 percent and rainfall rates of 5 and 20 mm/h.

These calculations were done with a computer program [24] that calculates attenuation as a function of a variety of factors for a number of conditions such as rain and fog. This program has been shown to give results accurate to about 0.2 dB/km in the atmospheric window regions of interest in the range 0–1000 GHz.

The curve for 100 percent relative humidity includes attenuation due to 0.5 g/m^3 of condensed water vapor, corresponding to a fog that would give only 100 m visibility in the visible spectrum. Note that this thick fog has a considerably smaller effect on propagation than rainfall, especially at the lower frequencies, because attenuation due to fog results mainly from Rayleigh scattering in these bands. Rainfall is another matter, however. It does not limit optical visibility but produces strong attenuation in millimeter waves. Larger drops occur normally at higher rain rates. Since raindrops are

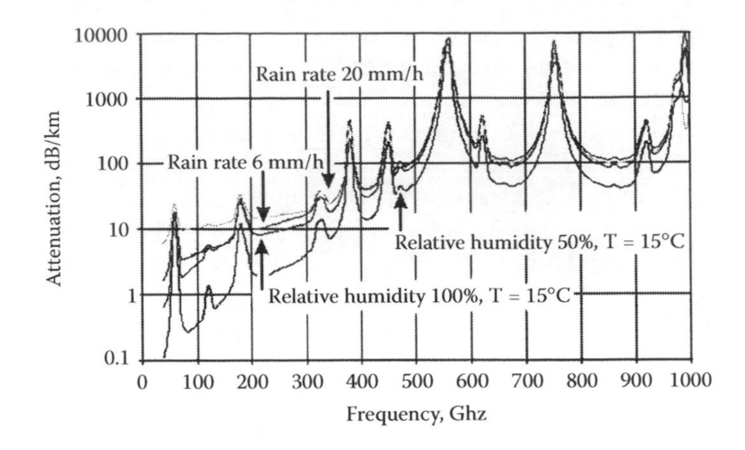

Figure 10.3 Atmospheric attenuation in the range 40–1000 GHz. From McMillan [7].

on the order of a few millimeters in diameter, strong attenuation due to Mie resonance scattering takes place.

10.2.4 Refraction

Refraction is a change in direction of propagating radio energy caused by a change in the refractive index or density of a medium.

The field of refraction coefficients undergoes strong variations in the surface layer of the atmosphere. Therefore, it is important to take into account the influence of weather conditions on the trajectory of millimeter waves propagating near the horizon.

Results of long-term measurements [14] have shown that the maximal value of an angular difference between apparent and true directions on a source of radiation is observed in autumn in anticyclonic weather at five angular minutes. From the same data it follows that maximal speed of change of this difference for the vertical component of angle of arrival is three angular minutes per hour. However, simulation under the worst conditions showed that essential trajectory curvature is possible on a surface radio path, which can result in errors and even loss of the target because of probable presence of air layers with super-refraction properties. The phenomenon of multipathing is also not incredible.

Simultaneous refraction measurements in vertical and horizontal planes have proved that horizontal refraction is at least two orders of magnitude less than vertical refraction. This result is in good coordination with known data on the structure of the troposphere. Note that the change of the angle of horizontal refraction was always less than the metering error (one angular second). Therefore, the results concerning vertical refraction are of interest.

On the biennial cycle of measurements in the middle latitudes [14], the distribution of vertical refraction angles appeared asymmetrical and has been precisely enough approximated by gamma distribution. The mean angle of refraction and root-mean-square angle made up accordingly 70 and 50 angular seconds. On average, the refraction at night is two to three times more than in the afternoon; in summer it is about two times more than in winter. The maximal magnitudes of refraction were observed at the moment of sunrise and sunset, which is associated with rising inversion layers of temperature and humidity at the surface layer in the morning and in the evening.

As a result of simultaneous measurements of refraction of millimeter waves and optical radiation, it was revealed that in summer, no steady correlation was observed between refraction angles at two frequency bands (correlation coefficient was 0.4), whereas in winter the correlation coefficient rose to 0.97. The explanation is rather simple: when air temperature is reduced, absolute air humidity also goes down and, hence, the distinction in parameters of refraction and beam trajectories of the millimeter-wave and optical radiation becomes less.

10.2.5 Underlying Terrain Irregularities

An interfacial area between air and earth is a weakly rough surface that provides, actually, a kind of mirror reflection in the microwave band. Such reflection is described by Fresnel's formulas [25]. In millimeter waves under the same conditions, wavelength and root-mean-square roughness are comparable, that is, surface irregularities play a significant role and the incoherent component of scattered reflection becomes dominant. Even in this case, however, an interference structure of the millimeter-wave field is apparent at a short distance and small grazing angle.

With increased distance, the zone of effective reflection, which is essential to formation of a fringe pattern, increases and includes both small-scale and large-scale irregularities, which can result in destruction of the interference structure. Besides, fluctuations of amplitude, phase, and a direction of propagation begin to appear that also deform the interference structure of a total field [26]. Complex phase of the fluctuating millimeter-wave field is defined by the superposition of three components: (1) direct waves; (2) the waves re-reflected by the underlying terrain irregularities; and (3) the waves scattered on turbulent inhomogeneities of the atmosphere, which will be described in the next section.

10.2.6 Turbulence

Atmospheric turbulence creates small-scale inhomogeneities in the refractive index, which are manifest in rapid fluctuations in the amplitude, phase, and angle of arrival of radio waves. This results in so-called scintillation phenomena, which can impact significantly on radar and communications systems and must be taken into account in system design. The scintillation effects are likely to be significant in the lower regions of the troposphere (the surface layer), where turbulent fluctuations produce mixing of air and are responsible for vertical transport processes, and in clouds, where turbulent eddies are the cause of mixing of air and hydrometeors.

The theory of wave propagation through a turbulent medium has been developed by Tatarskii [27]. Later, the further detailing to the microwave and millimeter-wave regions was done [14,28]. Using these theories, one can relate the general characteristics of a stationary scintillation event, such as the scintillation variance, Doppler spectrum width, and others, to the structure parameter, C_n^2, which is a measure of the turbulence-induced inhomogeneities in the refractive index, and also to other parameters such as the eddy dissipation rate. Such theories can be useful not only to estimate the turbulence effects to millimeter-wave propagation but also for deriving information about atmospheric turbulence from radar signals [29]. However, there is currently a paucity of reliable statistical data on atmospheric turbulence parameters, and empirical models have accordingly been developed that are based generally on surface parameters such as temperature and humidity, and which can hence be applied to the design of radar and communications systems.

10.2.7 Scattering and RCS

In Section 10.2.2, the phenomenon of scattering was considered as a source of wave attenuation. However, in the case of radar, the wave scattering should be additionally considered as a source of useful signal and also clutter. Features of this useful aspect of millimeter-wave scattering will be outlined now.

Considering a rain element as a radar target (for cloud radar, as an example), all directions of scattering, as shown in Figure 10.2, except of the wave line along the abscissa, represent this aspect of scattering (if bistatic radar is taken into account). The direction of 180° (back lobe) represents the backscattering, the only component of scattering that is important for monostatic radar. Hereinafter, the term radar means monostatic radar, when both transmitter and receiver are located in the same position.

Thus, the energy scattered back to the source of the wave (called backscattering) constitutes the radar echo of the object. The intensity of the echo is described explicitly by the radar cross-section (RCS) of the object (target). The units of the RCS are the equivalent area, usually expressed in square meters.

Millimeter-wave radar targets and clutter have been considered by Kulemin [3], who described reflections from land, sea, and precipitation, including land and sea backscattering for the small and extremely small grazing angles that are necessary for clutter rejection in radar systems. A summary of the interactions between radiated waveform and different targets and clutter affecting operation of radar in the millimeter-wave band is also presented.

General principles of target RCS description and measuring in the millimeter-wave band are the same as in other radar frequency bands [30]; however, small wavelengths require (1) more accurate theoretic considerations, such as calculations using the Mie approach instead of the Rayleigh approximation for raindrops, and (2) more difficult measurements because of the application of complicated millimeter-wave technology.

Comprehensive consideration of backscattering phenomena and the RCS for a wide variety of targets is beyond the scope of this book. Taking into account that millimeter-wave radar is actually a short-range sensor and some of its important applications concern the intelligent transportation systems (ITS) and automotive safety systems such as the forward collision avoidance assistance system (FCAAS), which has to be capable of detecting not only other vehicles but many other objects including pedestrians, let us consider the RCS of a pedestrian as an example of a millimeter-wave radar target. This is perhaps one of the most complicated targets because the RCS is difficult to measure. This consideration is completely based on the results of work by Yamada [31], who has researched the RCS of pedestrians for 76-GHz radar. The measurements were done with an FMCW (continuous-wave frequency-modulated) radar system from a distance of 5 m. Experimental data were obtained by rotating the pedestrian on the turntable and measuring the radio wave reflection intensity from all directions. The obtained scattering pattern is shown in Figure 10.4a. Figure 10.4b shows the results obtained for the moving average of the same data about an angle of 2.6 degrees (10 samples). The RCS is measured in decibels relative to a square meter, dBsm.

The average value of the RCS was found to be −8.1 dBsm. This is about 15–20 dB less than the reflection intensity of the rear of a vehicle. A spread of the RCS was more than 20 dB. Moreover, the results showed that the reflection intensity of the pedestrian's front and back is about 5 dB higher than the pedestrian's side. The radio wave reflection intensity depends on the pedestrian's aspect. The reflection intensity of a naked human body is almost the same as that of clothes that have a comparatively high radio wave reflection intensity, such as a cotton shirt.

In the case of the unwrinkled shirt, there is very little change in the reflection intensity, even when the shirt is swung to the right and left. For the wrinkled shirt, however, the reflection intensity changes considerably with the swing. Two or more scatter points (where the radio waves are reflected strongly) normally are created by wrinkling the clothes. This is why the RCS changes when the shirt moves. Perhaps this change in the

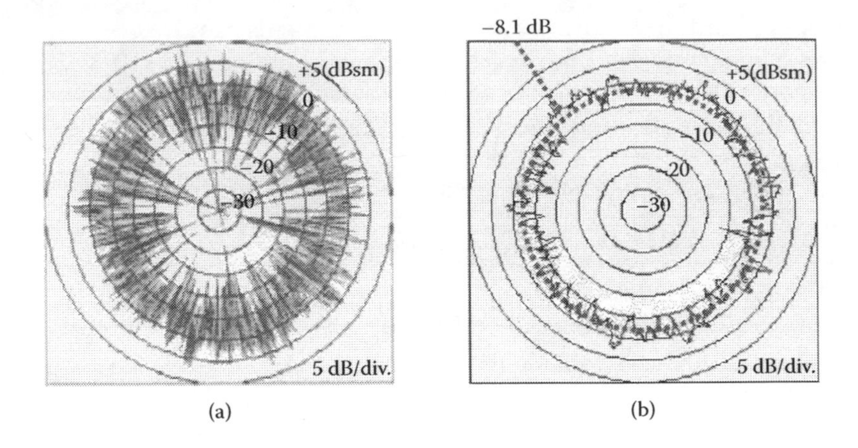

Figure 10.4 (a) Pedestrian scattering pattern raw data (left); (b) Averaged scattering pattern (right) [31].

RCS and the reflection intensity occurs because the waves reflected from these scattering points produce phase interference.

This example shows that millimeter-wave backscattering contains a lot of information about the target; it can be measured and used in many different applications. Even an insect may provide a measurable RCS in the millimeter-wave band.

Significant attenuation of millimeter wave can be considered a disadvantage of this frequency band; however, the same property can also be used for deriving information about the environment in remote sensing systems as well as for improving interference protection and electromagnetic compatibility. These aspects, radar principles, and applications will be considered later in this chapter.

10.3 Radar Design Principles

The basic principles of millimeter-wave radar are similar to those of microwave radar. The stronger wave attenuation discussed above can even be advantageous in certain radar and communications applications. For example, the transmission frequency can be selected such that the atmosphere imposes additional attenuation, thus minimizing unwanted interference from other cofrequency systems and improving the efficiency of spectrum usage. In fact, millimeter-wave radar systems have intermediate properties between microwave and infrared systems. They should be designed in such a way to take the best features from each. In this section we consider common principles of radar in relation to the features of millimeter-wave systems.

10.3.1 Radar Tasks

Assume that an observer is situated at a point O, and his task is to learn what object is located at another point, A. The observer (a radar) can radiate electromagnetic energy (waves) and concentrate it (with the help of an antenna) in a given direction. It is important to note the following. The main energy flow is spatially concentrated by the shape of the main beam; in spite of this, some energy (more or less) is, nevertheless, radiated in all directions without exception. The same, of course, concerns the direction of receiving reflected signals. The observer can have some a priori information about the object (radar target) and the environment. Under these conditions, the main tasks of radar are

- *Detection*: decision making regarding the presence or absence of a target with minimum allowable probabilities of erroneous decisions
- *Measurement*: a process to estimate the coordinates and parameters of motion of a target with minimum allowable errors
- *Resolution*: separate surveillance (detection and measurement) of an individual target in the presence of other targets
- *Recognition*: ascertainment that a resolved target belongs to the given class of targets
- *Target tracking*: the use of radar measurements for continued tracking of a given target

The numerous applications of radar can be reduced to this limited number of tasks.

10.3.2 Physical Processes

Let us assume a radar radiates a sounding signal (waveform), which after a time reaches a point where an object A (a target) is located. As a result of interaction with target A, the incident wave induces both electric and magnetic currents in the target, and they, in turn, generate electromagnetic waves propagating over all directions, including backscattering signal directed to point O. Reflected signal reaches point O and causes a corresponding signal in the form of a current or voltage. The delay time of the reflected signal contains range information. If the target is moving relative to the radar, the reflected signal gets a Doppler shift of frequency proportional to the radial velocity and inversely proportional to the radar wavelength, that is essentially higher in the millimeter-wave band than in longer wavelength bands. If the target has a nonspherical shape, the polarization of the reflected signal is different from the polarization of the radiated signal and also depends on the mutual orientation (in space) of a target and antenna beam.

It is clear that all information on the target can be obtained only by comparison of radiated and received signals. This information can be in the form of electric signals, and further interpretation and transformation into physical or geometrical parameters is an additional independent problem.

10.3.3 Sounding Waveforms

Various kinds of sounding waveforms can be used in millimeter-wave radar. The most obvious is division of all possible waveforms on continuous-wave (CW) and pulse sounding signals. A modulation of the sounding radiation is necessary at least to measure target range. Frequency (FM), phase code (PCM), amplitude (AM), and polarization modulations are possible. Different modulating functions can be used, including noise (random) function, but in the majority of cases the waveform is deterministic. Depending on the spectrum width, sounding signals are divided into narrow band (NB), wideband (WB), and ultrawideband (UWB).

Now we consider only a classical type of radar. In this case the sounding waveform is rather simple. It can be a CW FM wave normally with linear FM or a train of short pulses with pulse repetition time, which is much more than pulse duration. The spectrum width of such signals, B, most often is many times less than the carrier frequency of radiation, f_o, especially in millimeter-wave radar; that is, for the radar sounding signal, $B/f_o << 1$, excluding some special cases that will be considered in Sections 10.3.6 and 10.3.10. This inequation is a condition of the NB signal. Any NB signal $U(t)$ can be expressed as [32]

$$U(t) = A(t)\ \cos\,[2\pi f_0 t + \varphi(t)], \tag{10.1}$$

where $A(t)$ and $\varphi(t)$ are functions of time that are very slowly changing during the time $T = 2\pi/f_o$.

The theory of radar that was initially developed reasoning from such simple models as expression (10.1) now is very much advanced and is valid for a wide variety of waveforms (sounding signals), both NB and WB. Along with a law of time modulation being fulfilled in the transmitter, variations with time of orientation and parameters of antenna affect essentially the aggregate space–time modulation of radar signals. Selection of the sounding signal during radar design normally defines key features of the radar system, namely: resolution and measurement accuracy of range, radial velocity and angular data of targets; possibility to provide required energetic parameters at limited peak power of millimeter-wave generators; and interference immunity of the system.

Important properties of sounding signals are described with the help of ambiguous functions [32,33]. The most recent and advanced radar theory [33,34] uses the term "mismatch function" instead of ambiguous function.

This theory notes a difference between (1) matched and (2) optimal cases of signal resolution. Matched resolution takes place as a result of signal processing that is optimized for the background of uncorrelated stationary interference with known parameters; that is, without taking into account any other possible signals except predicted signal. In this case, improvement of resolution is provided only by the selection of the signal structure. Optimal (mismatched) processing is used for better reduction of strong unwanted signals, taking into account their features. Optimal (mismatched) processing can be nonadaptive, oriented on a predetermined typical situation, or adaptive, which means it can adjust to the unknown in advance of a specific situation.

10.3.4 Radar Signals and Information

It is natural to expect that a reflected signal looks like the sounding waveform. In the case of NB radar, it corresponds to model (10.1). If target A is a stationary target, only the amplitude and phase are changed while the moving target could change the frequency of the reflected signal relative to the radiated one. Note that other targets will be illuminated as well, specifically those that are located on the same distance as target A; reflected signals from such targets will simultaneously reach the point O where the radar is situated. Thus the reflected signal in point O is defined as a sum of all signals returned simultaneously and can be described also by expression (10.1).

In Section 10.2.6 we considered the RCS of a target without taking into account polarization properties. Now let us clarify in more detail what information can be obtained from radar returns. Suppose that neither interferences nor any effect of propagation medium exist, and confine our consideration to linear polarization. Under this condition, suppose at first that an electric field vector of incident wave E_i is horizontally oriented: horizontal component $(E_H)_i \neq 0$ and vertical component $(E_V)_i = 0$. For many targets the scattered waves have different polarization than the incident waves. This phenomenon is known as cross-polarization. So, in the general case, the electric field vector of a backscattered signal, E_b, has another spatial orientation. That means it has both $(E_H)_b \neq 0$ and $(E_V)_b \neq 0$ orthogonal components. Naturally, vector E_b is proportional to vector E_i, hence both $(E_H)_b$ and $(E_V)_b$ components of a backscattered vector are proportional to $(E_H)_i$:

$$(E_H)_b = S_{HH}(E_H)_i; \quad (10.2)$$

$$(E_V)_b = S_{VH}(E_H)_i, \quad (10.3)$$

where S_{HH} and S_{VH} are coefficients, which generally are modulus constituents of the scattering matrix $[S]$ composed of four numbers, S_{xy}, $(x = H; V, y = H; V)$.

Having recollected expression (10.1), one can see that each component of backscattered wave, in the general case, gets a phase shift relative to the incident (or radiated) wave. In the considered case, the orthogonal components of the electric field vector can be expressed by the following time representation:

$$(E_H)_b = S_{HH}A(t)\ \cos\,[2\pi f_o t + \varphi(t) + \psi_{HH}], \tag{10.4}$$

$$(E_V)_b = S_{HV}A(t)\ \cos\,[2\pi f_0 t + \varphi(t) + \psi_{HV}]. \tag{10.5}$$

In addition to modules S_{HH} and S_{HV}, phase shifts ψ_{HH} and ψ_{HV} are argument of complex constituents of the scattering matrix, [S]. It is seen from formulas (10.3) and (10.5) that in the case of a horizontally polarized incident wave, the backscattered wave is defined by four parameters, S_{HH}, S_{HV}, ψ_{HH}, and ψ_{HV}. Consideration of a vertically polarized incident wave will result in another four parameters, S_{VV}, S_{VH}, ψ_{VV}, and ψ_{VH}.

In general, an incident wave, radiated by radar, may have an arbitrary or any given polarization, which is characterized by two orthogonal polarization components, $(E_H)_i$ and $(E_V)_i$ in the case of a linear polarization basis. Thus complete description of the radar target can be done with the help of the indicated eight numbers, which constitute the scattering matrix mentioned above:

$$[S] = \begin{bmatrix} S_{HH} & S_{HV} \\ S_{VH} & S_{VV} \end{bmatrix}, \tag{10.6}$$

where s_{xy}, $x = H; V$, $y = H; V$, and quantities s_{xy} are in general complex with modulus S_{xy} and argument ψ_{xy}. The scattering matrix is individual for each object or a class of objects and provides target signature.

In reality it is difficult to measure absolute values of amplitudes and phases. That is why relative measurables are often used; for example, such values as S_{HH}/S_{VV}; S_{HY}/S_{VV}; ψ_{HH}-ψ_{YY}, and some others.

The backscattered RCS is related to the scattering matrix components by the following relation:

$$\begin{bmatrix} \sigma_{HH} & \sigma_{HV} \\ \sigma_{VH} & \sigma_{VV} \end{bmatrix} = 4\pi R^2 \begin{bmatrix} |S_{HH}|^2 |S_{HV}|^2 \\ |S_{VH}|^2 |S_{VV}|^2 \end{bmatrix}, \tag{10.7}$$

where σ_{xy}, $x = H; V$, $y = H; V$ is the RCS at given polarization properties on transmitting and receiving. Similar and equivalent equations can be obtained through circular polarization [35].

Polarization is mostly sensitive to the shape and orientation of a target or its components. Velocity of a target is reflected in corresponding Doppler frequency that gives additional information.

10.3.5 Spatial Resolution

Reflected signals are received by the radar antenna mainly within some solid angle, $\Delta\Omega$. Quantitative estimation of this solid angle can be done separately for two plane angles, $\Delta\theta$ and $\Delta\phi$, normally in the horizontal and vertical plane, correspondingly, and each of these two angles is defined by the ratio λ/d of wavelength, λ, to the linear size of the antenna, d, in the appropriate section. Thus, a reflected wave and finally a signal at the input of a radar is formed by the currents that are induced by the incident wave on the elementary area with linear sizes $R\Delta\theta$ and $R\Delta\phi$ within distance R. Obviously, for millimeter-wave radar it is easy to reach the λ/d ratio at least an order of magnitude better than for microwave radar.

Suppose we have a pulse radar that radiates a pulse with a rectangular envelope, and pulse duration is τ. In this case, all objects located along the same direction within distance $\Delta R = c\tau/d$ (c is speed of light) will be perceived as a single object. A typical pulse duration of microwave radar is of 1 μs, which gives $\Delta R = 150$ m. In order to improve range resolution, radar designers work to decrease τ. However, a decrease of pulse length leads to decreasing radiated energy and, finally, radar barrier in accordance with the radar equation [30,36], and this is one of the problems.

On the other hand, the shorter τ, the broader spectrum width B, and the theory of radar proves that, in general, a better range resolution can be achieved by broadening the spectrum of radiated waveform, which may be not only pulsed radiation but CW as well (see Section 10.3.3). Range (and also velocity) resolution potential of a waveform can be estimated with the help of the ambiguous function that was also considered in Section 10.3.3. Generally speaking, millimeter-wave radar has very high potentials to generate short (nanosecond) pulses and WB waveforms.

Finalizing the notion of spatial resolution, note that all objects within a parallelepiped of the size $\Delta R \times R\Delta\theta \times R\Delta\phi$, named resolution volume, will be perceived as a united single object. Actually, decreasing resolution volume is one of the core problems of radar design.

10.3.6 Pulse Compression and Synthetic Aperture

Now we know that a way to lessen the range size of a resolution volume consists in decreasing pulse duration or, more exactly, spectrum spreading. Special kinds of modulation are used within a pulse to generate WB signals ($\tau B >> 1$) without lessening pulse duration. Thus, radar designers may choose a pulse duration, τ, such that it provides the necessary radiating energy (and radar barrier) and then achieve the necessary range resolution by broadening the spectrum with the help of within-pulse modulation (mostly FM and PSM). Special processing of WB signals provides pulse compression at the output of the matched filter [32]. The compression ratio equals the time–bandwidth product, τB, and may be rather high (up to thousands). The theoretical limit of range resolution is defined by the wavelength.

Whereas lessening pulse duration and broadening spectrum width are related to engineering constraints, the problem of lessening tangential sizes of resolution volume ($R\Delta\theta$ and $R\Delta\phi$) is confronted with physical constraints because angles $\Delta\theta$ and $\Delta\phi$ are proportional to λ/d. Obviously, the first way to improve angle resolution is to decrease operational wavelength. Use of millimeter-wave band allows one to make the linear sizes of the resolution cell up to ten times smaller than in the case of microwave radar. The second way, an increase of antenna sizes, however, leads to the appearance of very large antenna construction, which causes technological difficulties apart from the fact that a large antenna is inconvenient, expensive, heavy, impossible to be applied on board, and so forth. Accuracy of antenna beamforming depends on relations between the phases of electric current in different points of the antenna, and in the millimeter-wave band, even very small distances, like a fraction of a millimeter, correspond to extremely big phase changes. Accuracy of adjustment is very critical in such antennas. That is why protection against thermal broadening, wind, and rain and avoidance of earth wave influence and other exposures are independent problems of high importance. Very big ground-based antennas are mostly unique and extremely expensive at both development and maintenance. In the case of airborne and spaceborne applications, antenna size is limited by the linear size of a platform, and to get a really high λ/d ratio is impossible. The tendency of azimuth resolution improvement (lessening angle $\Delta\theta$) resulted in the creation of a long-fuselage aircraft antenna. However, nowadays the much more productive idea of synthetic antenna aperture is implemented. Actually, any antenna produces the composition of signals obtained from different elements of the antenna surface by taking into account the correspondent phase incursion, caused by the features of antenna configuration. Synthetic aperture is created artificially by the following steps: (1) measurement of sequential values of amplitudes and phases of field strength in different points of the space as antenna elements; (2) memorization of these values; and (3) special composition of them. These sequential measurements are done in flight, which provides a brilliant possibility to create artificially an antenna, the size of which is defined by the distance between the first and the last inflight measurement. This means that such an antenna can be practically unlimited in length.

All theoretical and engineering difficulties, which accompanied the implementation of synthetic aperture, were successfully got over, and a kind of radar that uses such antenna is called synthetic aperture radar (SAR) [30].

As is easy to see, a special signal processing in the synthetic aperture antenna provides antenna beam compression that is similar to the pulse compression considered above. These two compression technologies improve spatial resolution dramatically.

Modern SAR can provide a λ/d ratio of up to thousands. Use of SAR has reduced the radar resolution cell so significantly that in some cases radar images become similar to photography.

10.3.7 Target Selection

Clutters created as a result of reflections from unwanted objects or background returns have a considerable influence on the quality of radar operations. There are different possibilities for selecting wanted targets on the background of clutter. Among the most efficient are polarization methods. Some targets do not transform polarization of incident waves; in matrix (10.6), they have $\sigma_{HH} = \sigma_{VV}$ and $\sigma_{HV} = \sigma_{VH} = 0$. Normally, they are smooth, convex, conducting bodies that have sizes and radii of curvature much more than wavelength. However, the majority of targets are polarization dependent. It is possible to find such polarization for radiated waves, which provides maximal ratio of returns from target and clutter within a single resolution volume. Radar contrast may reach up to 20 dB at polarization selection. A significant increase of the contrast opens the possibility to correlate scattering matrix with a target to solve the inverse problem and implement target recognition [37].

In the case of target movement, a reflected signal has a Doppler shift in the frequency relative to the radiated signal. Doppler shift is proportional to the ratio of the radial component of the target velocity to the wavelength. If among the scatterers inside the resolution volume only a wanted target is moving, it can be selected on the background of other scatterers, which are clutters, with the help of frequency (velocity) selection. Moving-target indication (MTI) is used in many modern types of radar [30].

10.3.8 Radar Detection

Radars often deal with very weak signals. Their intensity can be comparable with receiver noise, which is described by the same model as useful signal (10.1), where both amplitude and phase are random time functions. Noise acts all the time whereas signal may be either present or absent (a binary problem) [36]. A voltage at the input of a radar receiver can be represented as two summands:

$$U_{in}(t) = aU_S(t) + U_N(t), \tag{10.8}$$

where $U_S(t)$ corresponds to the wanted signal described by model (10.1), and $U_N(t)$ corresponds to all other sources including noise voltage and also is described by the same model (10.1) with different randomly changing values of amplitude and phase. Note that parameter $a = 1$ in the presence of a target or $a = 0$ when a target is absent.

A problem definition of radar detection can be formulated as follows way: determine a value of parameter a in Equation (10.8); if $a = 1$, a target is detected.

However a rigorous solution of Equation (10.8) is impossible. Thus it is fundamentally impossible to reply accurately to the question of whether a target is present in the resolution volume or it is absent. Such a reply can be only of a conjectural nature. Quantitative estimates of the appropriate conjectures are probabilities of their reliability. The reliability of a conclusion about a value for parameter a can be improved by increasing the observation period; that is, the interval of the function, $U_{in}(t)$, which is analyzed. A key distinction between functions $U_{in}(t)$ in two situations (a target is present and a target is absent) consists in the difference of statistical laws to which the stochastic functions $U_{in}(t)$ submit.

Let us try to understand those types of errors that inevitably arise during decision making on the presence or absence of a target. Assume a radar receiver produces some processing of input signal, $U_{in}(t)$ that resulted in a signal, $U_{out}(t)$, which is a function of an additive mixture of the wanted signal and the receiver noise:

$$U_{out}(t) = f[aU_S(t) + U_N(t)]. \tag{10.9}$$

Receiver noise, $U_N(t)$ is always present and acts as the input of a receiver while a signal, reflected from the target, $U_S(t)$, acts as a component of the mixture only if $a = 1$ (presence of the target in the resolution volume). A function, $f[*]$, symbolizes some procedure of signal processing. It should be chosen in such a way as to maximize the signal-to-noise ratio (SNR) at the output of the receiver. However, in any type of function, $f[*]$, in all cases the only suitable decision rule is a threshold rule: if $U_{out}(t)$ is more than some threshold value, U_0, the decision "target is present" should be made. Otherwise, the decision "target is absent" should be made.

This consideration clearly demonstrates that a threshold decision-making procedure is accompanied by two kinds of errors: (1) false alarm, when at the absence of a target in the resolution volume, the decision of detection is made; and (2) target skip, when at the presence of a target in the resolution volume, the decision of absence of a target is made. Corresponding probabilities of erroneous decisions are false alarm probability, F, and target skip probability, $(1 - D)$, respectively. Correct detection is a contrary event relative to a skip of the target. That is why probability of true detection is D.

Radars can be designed for different applications. However, in all cases, it is desirable to make erroneous decisions as rarely as possible. The use of a threshold decision rule limits influence to only one possibility: to change the value of the threshold, U_0.

An increase in the threshold naturally results in lessening the false alarm probability, F, but entails a decrease in the detection probability, D. On the contrary, lessening the threshold results in a reduction of the target skip probability (growth in D) but entails growth of the false alarm probability, F.

How might a radar designer determine the appropriate threshold level? Normally the level of false decision probabilities is defined by the user of the radar information on the basis of required radar functions. In turn, probabilities of false decisions constitute the basis for determination of the threshold. However, the fact that two independent probabilities define the reliability of radar detection leads to an unlimited number of possible criteria of detection quality. In radar theory and practice, the Neumann–Pearson criterion is generally accepted. According to this criterion, the false alarm probability is fixed at the acceptable level, F, and under this condition the detection probability, D, is estimated. Of course, we seek to have D as high as possible; it might be close to 1. However, how high can it be in reality? To answer this question, let us revert to formula (10.8), where signal processing is defined by the function $f[*]$. The value of D depends not only on the threshold value but also on the efficiency of the signal processing algorithm. Function $f[*]$ should be designed in such a way as to maximize the SNR at the input of a threshold device. For a given level, $F = F_o$, the maximum possible level of D will be reached. The problem of signal processing consists in the choice or synthesis of such a function, $f[*]$, at which in the framework of a given criterion the detection probability becomes maximum. Thus the Neumann–Pearson criterion of radar detection can be written as

$$F = F_o; D = D_{max}. \tag{10.10}$$

It is obvious that at a threshold rule of decision making, an algorithm $f[*]$ cannot be linear. In many cases, a function $f[*]$ is implemented in the matched filter, which is appropriated to the expected signal (radiated waveform). It gives at its output a value proportional to the energy of the reflected signal by the processing of input signal during its duration.

Development of signal processing algorithms is associated with knowledge of statistical information on wanted signals, background reflections, and receiver noise. This knowledge is used as a priori information for the synthesis of optimal detection algorithms [36,38]. The more complete a priori information, the more effective algorithm can be developed; that is, value D_{max} can be closer to the limiting value equals to unit.

However, in many cases real statistics of signals may differ from the models accepted during the synthesis. This may explain the essential decrease in algorithm efficiency though it was synthesized as optimal in accordance with a priori information. That is why special attention is focused on the development of robust nonparametric [30,33,39] and adaptive [33,40] algorithms for radar detection and recognition.

10.3.9 Radar Measurement

Measurement is a separate radar task. It is important to provide the required accuracy when measuring basic parameters and characteristics of reflected signals to allow determination of spatial target coordinates (range and angular position), velocity of target, and other target parameters. Whereas potential reliability of radar detection depends exclusively on signal energy, the accuracy of measurement depends not only on energy but also on the waveform.

The accuracy of direction finding depends on the antenna pattern when it is directed to the target. The best accuracy can be achieved using the monopulse technique. Monopulse radar splits the antenna beam into parts and compares the signal strength of the various parts. That means the comparison always takes place based on the reflection of a single pulse. This radically helps to avoid the problem that appears when radar signals change in amplitude for reasons that have nothing to do with beam alignment.

Classical radar theory rigorously proves that to provide high-accuracy measurements of both target range and target velocity, the waveform should be long-continued (accurate measurement of Doppler shift) and as wideband as possible (accurate measurement of time delay). This means the requirement of using WB waveforms with the time–bandwidth product $\tau B >> 1$ like that was in respect to radar resolution (Section 10.3.5). A sounding signal that satisfies this inequation can be called a complex signal in the sense that such a waveform is very different from the simplest radar waveform, which is a harmonic curve.

A complex signal is compressed as a result of processing in the radar receiver (Section 10.3.6), and the properties of complex signals, particularly FM and PCM pulses, noise-like waveforms, and others, can be estimated and compared among themselves with the help of the ambiguity function (Section 10.3.3).

Whereas radar detection was implemented by a threshold rule that was applied to voltage (10.8), the range measuring procedure consists of searching for a maximum of the same function, $U_{out}(t)$. Note that the best waveform with respect to a criterion of simultaneously measuring target range and velocity is the ideal noise signal because it has an ambiguity function, $U(t, f)$, that is akin to the delta function located in a point of the searched maximum. Coordinates of this maximum, $U(t_m, f_m)$, correspond to the target range, $R_m \sim t_m$, and target velocity, $v_m \sim f_m$.

Another condition of accurate measurements is a high enough SNR, which represents the energetic aspect of any radar measurement.

10.3.10 Nonclassical Types of Radar

Let us now consider some nonclassical kinds of radar that are especially important with respect to millimeter-wave applications.

UWB radar. In accordance with the FCC (the U.S. Federal Communications Commission) definition, a system is referred to as UWB if its fractional bandwidth, $B_F = (f_b - f_l)/f_c$, is more than 0.2 or total bandwidth $B > 0.5$ GHz; here f_b and f_l are the upper and lower spectrum components measured at -10 dB points, and f_c is the central frequency. Based on the second part of this definition, UWB radar systems with bandwidth $B > 0.5$ GHz but less than 20 percent of the center operating frequency, $B < 0.2f_c$, in the millimeter-wave band can be designed using classical radar theory and the traditional technology of millimeter-wave components, whereas UWB systems designed at $B < 0.2f_c$ have some specific differences and require a more novel approach. A comprehensive theory of UWB radar has still not been developed. However, significant features of UWB radar are known [41]. One of them is a change of the waveform during radiation by the antenna and also during propagation, at the time of scattering, and when receiving the backscattered signal. Therefore, traditional correlation processing or matched filter application can be senseless in the case of UWB radar. Some special approaches have been developed. Moreover, signal shape and some other parameters depend on the direction of radiating or receiving. These and many other peculiarities differ UWB radar theory from classical radar. Nevertheless, UWB radar offers a number of attractive advantages that make it promising and useful in many practical applications as a radar and sensing tool. UWB systems are particularly applicable as vehicular radar ground-penetrating radar (GPR), through-wall imaging sensors, medical imaging devices, and so forth.

Noise radar can be used in both NB and UWB versions [42,43]. This direction is quickly developing nowadays, provides nice resolution, solves the problem of ambiguous measurements, and is quite applicable for the millimeter-wave band.

Bistatic and multistatic radar, in which target radiation is performed from one position and scattered signals are received in other positions, is of great interest. Our vision—when the sun serves as the source of illumination and we perceive the light, which is scattered by entourage objects —is a good illustration of bistatic (multistatic) radar. The phenomenology of bistatic radar offers a significant benefit for different military and civilian systems, enabling separation of emitters and collectors, greatly increasing survivability. The theory of bistatic radar can be found in work by Mahafza [35]. Numerical simulations of the bistatic millimeter-wave radar return from a rocket-shaped object were performed by the Advanced Sensors Collaborative Technology Alliances [44] to identify the best bistatic configurations for detection of low-flying missile-looking objects. Polarimetric measurements of the bistatic scattered fields from a rough, dry soil surface were performed by Nashashibi and Ulaby [45] at 35 GHz over the entire upper hemisphere.

Secondary radar uses one radar station as interrogator, which illuminates an object, usually a vehicle, and stimulates another radar station, called a transponder, installed on the vehicle to reply. The transponder-reply contains special information, for example, about the parameters of the flight and the state of some on-board systems. Secondary radar of the L-band is widely used in civil and military aviation [46]. Millimeter-wave secondary radar systems are of interest for automotive ITS, which will be considered later.

Passive radar uses the electromagnetic radiation that is naturally produced by a target. As is well known, any lukewarm body emits radiation to a greater or lesser extent. At a body temperature of 300 to 350 K, the maximum intensity of this radiation falls in the infrared (IR) band. Passive radar that is based on detection of this radiation is called radiometry [47] and has been used successfully. The advantages of such IR radar consist of emission security and difficulties of jamming and target camouflage. Drawbacks are associated with the impossibility of straightforward selection of targets on range and with the strong influence of atmospheric conditions. The same principle is applied to millimeter-wave passive radar systems, which are almost free from these disadvantages in comparison with IR systems. Useful signal is essentially less in the millimeter-wave than in the IR band; nevertheless, this is not a problem for use of such millimeter-wave radiometric systems [48].

10.4 Radar Subsystems and Components

Millimeter-wave radars are designed for very different applications with different functionalities. That is why radar configuration can also be quite different. Nevertheless, any radar system has several major subsystems that perform standard functions. For example, a typical pulse radar system always contains a synchronizer, a transmitter, a duplexer, a receiver, and an indicator. The main features of millimeter-wave radar are related to radio frequency (RF) subsystems and components.

The intermediate position of the millimeter-wave band between microwaves with corresponding waveguide technology and infrared waves with typical optical methods allows application of both approaches for the calculation and development of RF components at millimeter waves.

Theoretical, physical, and engineering fundamentals of millimeter-wave components have been described [49]. Here we just briefly indicate the main possibilities for building RF components of different functions.

10.4.1 Transmitters

The choice of a reliable, high-powered millimeter-wave-transmitting device usually is a serious problem. This is due to the necessity to use transmitters that should provide both the rather high transmitting power needed to achieve a high radar sensitivity and spatial resolution simultaneously (see Section 10.3). Transmitters for millimeter-wave radars can be designed on the basis of both electrovacuum and semiconductor generators. A series of efficient millimeter-wave magnetrons were designed in the mid-1960s [50]. Among them, magnetrons were designed with champion power (e.g., pulse power of 100 kW at a wavelength of 4 mm).

The transmitter problem has been solved using spatial-harmonic magnetrons with cold secondary-emission cathodes [51]. In comparison with classical magnetrons, such magnetrons can operate effectively in almost an entire millimeter-wave band. Moreover, they have smaller dimensions and weight, higher peak and averaged output power, and larger lifetimes while maintaining other well-known advantages of magnetron tubes. A disadvantage of magnetrons is noncoherence of the train of pulses, which may be very important in some applications.

Some other vacuum tubes, such as klystrons, backward-wave tubes, and gyrotrons, are also suitable for coherent radar. Modern gyrotrons can be designed for power outputs of 22 kW CW at 2 mm and 210 kW pulsed at 2.4 mm [7]. An array of gyrotrons was developed to build a megawatt radar operating in the Ka band [52]. However, in many important applications such a huge power is not necessary, fortunately. For many experimental radars, more common power is about 1–2 kW. Such tubes, like the extended interaction amplifier klystron [53], can be used in real coherent radars as amplifiers in the output stage of a transmitter. These millimeter-wave devices are positioned between the very-high-power tubes and solid-state devices.

A good alternative to vacuum tubes is the solid-state generator. Mostly Gunn oscillators [54] and impact avalanche transit-time (IMPATT) diodes [55] are used. The high electron mobility transistor (HEMT) amplifier is an important new development for fully solid-state millimeter-wave radars. Such devices are suitable for high-bandwidth, medium-power amplifiers; experimental UWB radars can nowadays be designed with their help, as described elsewhere [56,57].

A set of power amplifier modules containing InP (indium phosphide substrates) HEMT monolithic millimeter-wave integrated circuit (MMIC) chips were designed for oscillator sources in the 90–130 GHz band [58]. The modules feature 20–45 mW of output power, to date the highest power from solid-state HEMT MMIC modules above 110 GHz.

Solid-state devices suffer from the same decrease in size of the frequency-determining elements as do vacuum tubes; thus, solid-state device operation mostly is limited to about 230 GHz [7]. Nevertheless, solid-state oscillators of higher frequencies can be definitely designed [59].

Kasatkin and Chayka [60] describe methods of analysis and design of millimeter-wave transmitters with semiconductor diodes and transistors. Physical principles of different millimeter-wave diodes and transistors, optimization methods of frequency-stabilized and wideband self-oscillating systems, as well as synchronized and frequency-multiplying systems are considered in this book. The power of generated oscillations of different active devices can be summarized in branched and hybrid electrodynamic systems, summation resonators and waveguides, and quasi-optical spatial-developed systems [60].

Millimeter- and submillimeter-wave lasers can be selected as a separate class of millimeter-wave generators. Methods of millimeter-wave femtosecond waveforms generation based on two-stream free electron lasers [61] are very promising.

10.4.2 Antennas

A wide variety of antennas can be effectively used in millimeter-wave radar, including lens antennas, Cassegrain antennas, patch antennas, microstrip antennas, and slot antennas.

Different configurations of planar antennas and antenna arrays have been developed. As an example, a millimeter-wave planar antenna array of 64 elements is described elsewhere [62].

The state-of-the-art performances of the single-layer waveguide arrays have been considered [63]. Various types of antenna input ports are being designed for the compact interface to millimeter-wave RF circuits. Low side-lobe design as well as a beam scan/switch capability have also been developed. The latest millimeter-wave wireless systems with these antennas are directed toward applications such as fixed wireless access (FWA), LAN, automotive radars and road monitors in ITS, and so on.

New developments of modeling techniques and technologies for multi-frequency antennas, conformal arrays, and smart beamforming are reported [64]. Some new applications require multifunction antennas with multiband capability even beyond millimeter waves.

As indicated in McMillan [7], at frequencies above about 100 GHz, and at lower frequencies for many applications, metal waveguides become unacceptably lossy because of skin effect losses and our inability to make these waveguides with the precision required for low-loss operation. In many cases, these problems can be solved using techniques developed for the visible and IR portions of the spectrum. This approach to the propagation and handling of millimeter-wave and tetrahertz radiation has been called quasi-optics.

Quasi-optic components of the bidirectional amplification array for transmit/receive front ends have been surveyed [65]; advantages of their application in wireless communications and radar are discussed and a millimeter-wave 22-element multilayer lens array using solid-state integrated circuits is presented. Finally, a review of antenna technology for millimeter-wave automotive sensor applications has also been presented [66].

10.4.3 Receivers

Both direct-detection receivers and heterodyne detectors are used in millimeter-wave radar. The earliest detectors of millimeter-wave and terahertz radiation were simple point–contact diodes. This technique is still used for the higher frequencies today. It has been refined in recent years by the use of Schottky barrier structures. These diodes have operating frequencies extending well into the terahertz range. Another detector suitable for the submillimeter-wave range can be built on the metal oxide–metal (MOM) diode. An interesting variant of the point–contact mixer is the back-to-back diode configuration in which the fundamental frequency and all odd harmonics are canceled, resulting in a mixer that operates at twice the fundamental, and in some cases at four times the fundamental [67]. This configuration extends the frequency range to 320 GHz and higher. In beam-led Schottky barrier diode detectors and mixers, the diodes are fabricated by the same techniques used to make integrated circuits, and for this reason, they can be included in these circuits. More details can be found in McMillan's overview [7].

10.4.4 Integrated Circuits Technology

Monolithic integrated millimeter-wave circuits have emerged as an attractive option in the field of millimeter-wave communications and millimeter-wave sensorics and radar. The combination of active devices with passive planar structures, including antenna elements, allows single-chip realizations of complete millimeter-wave front ends. The state-of-the-art silicon- and SiGe-based MMICs have been reviewed [68]. Compact heterojunction field-effect transistor (HJFET) monolithic integrated-circuit switches will contribute to the low-cost and high-performance millimeter-wave radar and communications systems [69]. The millimeter-wave radar possibilities have been greatly aided by recent large-scale government and industry investment in millimeter-wave integrated circuit development in such programs as MMIC and others.

10.4.5 Other Components

Among advanced radar component technologies, micro-electro-mechanical system (MEMS) devices [70] are of particular interest. They will be important for many applications, including millimeter-wave phased arrays for terahertz radiometric systems and millimeter-wave SAR.

The MMIC program has developed monolithic chips, such as voltage-controlled oscillators, driver amplifiers, power amplifiers, doublers, mixers, switches, and monolithic transceivers. This chapter does not focus on some important components, such as the low-noise amplifiers, power amplifiers, frequency multipliers, feed horns, power dividers, slot couplers, matched hybrid tees, directional couplers, PIN switches, and other integral parts of a radar system. Nevertheless, all of them have been developed, are produced by many firms, and are suitable for different millimeter-wave radar applications, a part of which will be considered in the next three sections.

10.5 Intelligent Transportation System Applications

10.5.1 Role of Radar Sensors in ITS Structure

The ITS programs (Intelligent Transportation Systems in the United States and Intelligent Transport Systems in European Union countries) were started in the beginning of the 1990s. The U.S. Department of Transportation's ITS program [71] is based on the fundamental principle of *intelligent vehicles* and *intelligent infrastructure* and the creation of an intelligent transportation system through integration within and between these two components (Figure 10.5). This principle is shared in other parts of the world. In the European Union, ERTICO (Intelligent Transport Systems Europe) [72] is a public/private partnership working to facilitate the safe, secure, clean, efficient, and comfortable mobility of people and goods in Europe through the widespread deployment of ITS. Other countries all over the world, particularly Canada, China, India, South Africa, Brazil, East European countries, and Russia are also interested and participate in ITS research, development, and deployment.

The **intelligent infrastructure** segment consists of several major components. Among them are those using radar systems as sensors: arterial

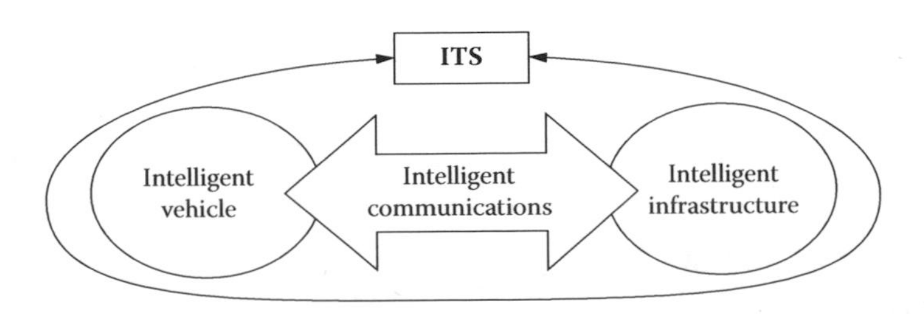

Figure 10.5 General structure of ITS.

management, including surveillance, traffic control, lane management, and enforcement, particularly speed and stop/yield enforcement based also on radar; freeway management with traffic surveillance systems using detectors and video equipment to support the most advanced freeway management applications; incident management, including surveillance and detection; emergency management with early warning systems; and **road weather management** based on surveillance, monitoring, and prediction.

Even more radar-based components are associated with the **intelligent vehicles** segment, which consists of collision avoidance systems, collision notification systems, and driver assistance systems. Let us consider them in more detail.

Collision avoidance systems can implement at least seven functions with the help of radar:

1. Intersection collision warning systems, designed to detect and warn drivers of approaching traffic at high-speed intersections.
2. Obstacle detection systems, which use vehicle-mounted sensors to detect obstructions, such as other vehicles, road debris, or animals, in a vehicle's path and alert the driver.
3. Lane change warning systems, deployed to alert bus and truck drivers of vehicles, or obstructions, in adjacent lanes when the driver prepares to change lanes.
4. Lane departure warning systems, which warn drivers that their vehicle is drifting out of the lane.
5. Road departure warning systems, designed to detect and alert drivers to potentially unsafe lane-keeping practices and to keep drowsy drivers from running off the road.
6. Forward collision warning systems, which use radar detection to avert vehicle collisions. These systems typically use in-vehicle displays or audible alerts to warn drivers of unsafe following distances. If a driver does not properly apply the brakes in a critical situation, some systems automatically assume control and apply the brakes in an attempt to avoid a collision.
7. Rear-impact warning systems, which use radar detection to prevent accidents. A warning sign is activated on the rear of the vehicle to warn tailgating drivers of impending danger.

Collision notification systems supply public/private call centers with crash location information; they use in-vehicle crash sensors, satellite (GPS, GLONASS, GALILEO) technology, and wireless communication systems.

Driver assistance systems (DAS) include navigation/route guidance and driver communication systems, which use GNSS (global navigation satellite system) and wireless communication technologies, drowsy driver warning alert, precision docking, roll stability control, and other systems that do not directly use radar information. However, some important components are actually based on radar sensorics. Among them:

1. Object detection systems that warn the driver of an object (front, side, or back) that is in the path or adjacent to the path of the vehicle.
2. Cruise control and adaptive cruise control (ACC), which is intelligent cruise control, speed control, and distance control to adjust speed in order to maintain a proper distance between vehicles in the same lane.
3. Vision enhancement, which improves visibility for driving conditions involving reduced sight distance due to night driving, inadequate lighting, fog, drifting snow, or other inclement weather conditions [73].
4. Lane-keeping assistance systems, which are related to ACC and radar sensors [74].

Thus radar technology and sensorics play an important role in the development and deployment of ITS and automotive advanced electronics. Because of the requirements to provide high accuracy, nice resolution, and small equipment size, the overwhelming majority of automotive radar sensors are designed at the millimeter-wave band. Millimeter-wave radar sensors today are the heart of ITS. Sensing moving targets and complex obstacles using active millimeter-wave radars is the most important component in automotive collision avoidance and intelligent cruise control systems.

An overview of the application of automotive radars was provided by Mende and Rohling [75]. They noted that in the last few years numerous new DAS applications have been under evaluation or development. First automotive radar sensors were introduced in passenger cars as a comfort feature. Now they are designed also or even first of all to improve safety. New functions of millimeter-wave radar appear very quickly, and no list can be complete because different manufacturers sometimes use different names as well as a slightly different understanding for desired functions of the equipment. Each manufacturer has its own predicted sequence for the introduction of those functions. It depends on what the customer research and marketing studies have identified so far as functions a customer is willing to pay for.

In accordance with the findings of Mende and Rohling [75], both current and promising applications of automotive radar fall into four categories, as follows:

1. Applications related to sensor and display to improve comfort:

   - Parking aid: Invisibly mounted distributed sensors behind the bumpers.

- Blind spot surveillance: The zones beside a vehicle are covered by radar sensors; a warning is displayed when the driver is about to change lanes when the (radar) field of view (FOV) is occupied.

2. Vehicle control applications to improve both **comfort and control:**

   - ACC: Longitudinal vehicle control at a constant speed with additional distance control loop.
   - ACC plus: Improves the handling of cut-in situations with a wider FOV at medium range.
   - ACC plus stop and go: Improves/allows the vehicle control function in an urban environment, with complete coverage of the full vehicle width.

3. Restraint systems applications to improve safety:

   - Closing velocity sensing: The main technical challenge in this application is to decide whether a crash will happen and to measure the impact position and speed before it happens to adaptively adjust thresholds/performance of restraint systems (which are not fired by the radar system).
   - Precrash firing for nonreversible restraints: See above. Nonreversible restraint systems (like airbags) are directly fired by the sensor system. This can be done even before the crash happens, with crash position and severity selective. This function is of most importance for side crashes, to gain a few life-saving milliseconds to fire before the crash happens.

4. Collision-related applications to improve both safety and control:

   - Collision mitigation: See restraint systems–related functions. The sensor system detects unavoidable collisions and applies full brake power (by overruling the driver).
   - Collision avoidance: In future function, the vehicle would automatically take maneuvers to avoid a collision and calculate an alternative path, overruling the driver's steering commands.

Key problems with each function implementation have also been discussed [75]. The typical key questions are cost, number of sensors, frequency approval problems, mounting position, sensor fusion, low false alarm rate, liability issues, and so forth. Regarding the sequence of new functions introduction, the authors [75] mentioned that official organizations are already evaluating the feasibility of certain functions and a forced introduction is under discussion.

10.5.2 Frequency Allocation

Intelligent infrastructure's surveillance function can be implemented with radar sensors. Such radars designed for roadside traffic data collection and monitoring are limited by FCC regulations to operating frequency bands near 10.5, 24.0, and 34.0 GHz. Advanced imaging techniques for traffic surveillance and hazard detection may use also 94-GHz mid-millimeter-wave radar and 35-GHz long-millimeter-wave radar [74].

With regard to automotive radars, the feasibility of millimeter-wave radar technology has already been demonstrated in the neighboring 76–77 GHz band for long-range radar (LRR) ACC systems [76]. Typically, short-range radar (SRR) requires high-range resolution. As was shown in Section 10.3, a wideband signal provides high resolution. That is why UWB waveforms are frequently used in these applications. If cost-effective 24-GHz band operation is desired, legal restrictions varying from country to country have to be taken into account. On a worldwide basis, only a few parts of bandwidth are free to use. In the United States, the FCC has allowed the use of UWB radar sensors since 2002 [77]. In Europe, the European Commission approved the decision on allocation of the 24-GHz frequency band (from 21.625 to 26.625 GHz) for automotive SRR temporarily, until 30 June 2013. Included is the task to work toward an early introduction of equipment operating in the 79-GHz band by means of a research and development program. From mid-2013, new cars have to be equipped with SRR sensors operating in the 79-GHz frequency range. This frequency band was designated for the use of automotive SRR starting 19 March 2004 [78], and the following regulations are fixed:

- The 79-GHz frequency range (77–81 GHz) is designated for SRR equipment on a noninterference and nonprotected basis, with a maximum mean power density of −3 dBm/MHz EIRP (effective isotropic radiated power) associated with a peak limit of 55 dBm EIRP.
- The maximum mean power density outside a vehicle resulting from the operation of one SRR equipment shall not exceed −9 dBm/MHz EIRP.
- The 79-GHz frequency range (77–81 GHz) should be made available as soon as possible and not later than January 2005.

The European approach of a temporary use of 24 GHz with a transition to 79 GHz is called a packaged solution, to make an early contribution to the enhancement of road safety possible and to allow time for the development of the 79-GHz technology, which is not yet mature for SRR sensors. The current state and prospect of frequency band usage in automotive radar can be generalized, as presented in Table 10.2.

Table 10.2 Automotive Radar Frequency Bands

Frequency Band(s)	*Country, Organization*	*Functions*	*Note*
76–77 GHz	European Telecommunications Standards Institute (ETSI)	Automotive LRR, also road surveillance radar	From 1998 and 1992
21.625–26.625 GHz	European Commission	Automotive SRR	Until 2013
46.7–46.9 GHz, 76–77 GHz	United States, FCC	Vehicle-mounted field disturbance sensors, including vehicle radar systems	
76–77 GHz	European Conference of Postal and Telecommunications Administrations	Vehicular radar systems	
60–61 GHz, 76–77 GHz	International Telecommunication Union	Transport information and control systems	From 2000
60–61 GHz, 76–77 GHz	Japan, Ministry of Post and Telecommunication (MPT)	Transport information and control systems	
60–61 GHz, 76–77 GHz	Asia-Pacific Telecommunications Standardization Program (ASTAP)	Low-power short-range vehicle radar equipment	
76–77 GHz	Australian Communications Authority	ITS automotive radar component	From 2001
77–81 GHz	European Commission	Automotive SRR	From 2004

Hence, currently operating bands are 76-GHz NB, 77–81-GHz UWB, 47-GHz NB (U.S.), 24-GHz NB (200-MHz bandwidth), and 24-GHz UWB (3-GHz bandwidth). Additionally, baseband impulse UWB radars with bandwidth $B < 10$ GHz can be used, and 152 GHz is suggested for future research [79].

10.5.3 Traffic Surveillance Radar

Road transport and traffic telematics (RTTT) radars are used as sensors in surveillance systems. Traffic surveillance technologies play an essential role in incident detection, traffic management, and travel time collection. There are two basic types of traffic surveillance systems: road based and vehicle based [80]. Road-based detection systems can be intrusive or nonintrusive. Traditional intrusive sensors include inductive loops, magnetometers, and other devices that are installed directly on the pavement surface or in the road surface.

Comparatively new, nonintrusive aboveground sensors can be mounted above the lane of traffic they are monitoring or on the side of a roadway where they can view multiple lanes of traffic at angles perpendicular to or at an oblique angle to the flow direction. The technologies currently used in aboveground sensors are video image processing, millimeter-wave radar, laser radar, passive infrared, ultrasonic, passive acoustic array, and combinations of sensor technologies. Like the subsurface sensors, the aboveground sensors measure vehicle count, presence, and passage. However, they can additionally provide vehicle speed, vehicle classification, and multiple-lane, multiple-detection zone coverage [81].

Roadside-mounted millimeter-wave radar transmits energy toward an area of the roadway from an overhead antenna. The beamwidth of millimeter-wave radar can be rather narrow (Section 10.3.5); that is, an area in which the millimeter-wave radar energy is transmitted in short distance can be designed in accordance with the lane width. When a vehicle passes through the antenna beam, a portion of the transmitted energy is reflected back toward the antenna. The energy then enters a receiver where the detection is made, and vehicle data, such as volume, speed, occupancy, and length, are calculated by signal processing. Two types of radar sensors are used in roadside applications: continuous-wave (CW) nonmodulated Doppler radar and CW frequency modulated (FMCW) radar. The traffic data they receive is dependent on the shape of the transmitted waveform. A generalized diagram of a CW radar sensor is shown in Figure 10.6.

The CW nonmodulated sensor transmits a signal that is constant in frequency with respect to time. According to the Doppler principle, the motion of a vehicle in the detection zone causes a Doppler shift, which is extracted from the beat frequency at the output of the mixer and can be used to detect moving vehicles and determine their speed. Such CW Doppler radar is

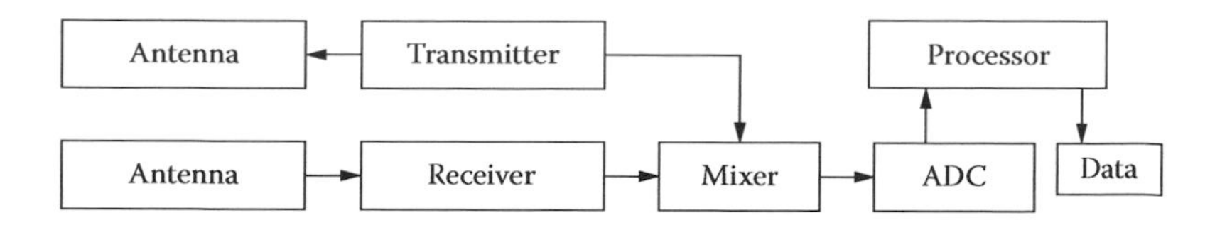

Figure 10.6 A CW radar generalized diagram.

also used to implement the speed enforcement function. CW systems measure the instantaneous range rate and maintain continuous contact with the target. That is why they can be used in tracking by speed systems. However, CW Doppler sensors that do not incorporate an auxiliary range measuring capability cannot detect motionless vehicles. Range measuring capability can be implemented in FMCW radar. The principle of FMCW radar has been described in many books [30]. The simplest way to modulate the wave is to linearly increase the frequency during one half period of modulation, T_M, and then decrease it back during the second half period. It is illustrated in Figure 10.7 (upper graph), where the solid line is the frequency of the transmitted waveform and the dashed line is the frequency of the received signal reflected from an immovable object. The lower graph in Figure 10.7 shows beat frequency, which depends on delay time and, thus, on the range of the object. When a moving car is the target, the beat frequency depends also on car speed, as illustrated in Figure 10.8, where beat frequency during the positive (up) and negative (down) portions of the slope, are denoted, respectively, as f_u (Doppler shift is added) and f_d (Doppler shift is subtracted) [35].

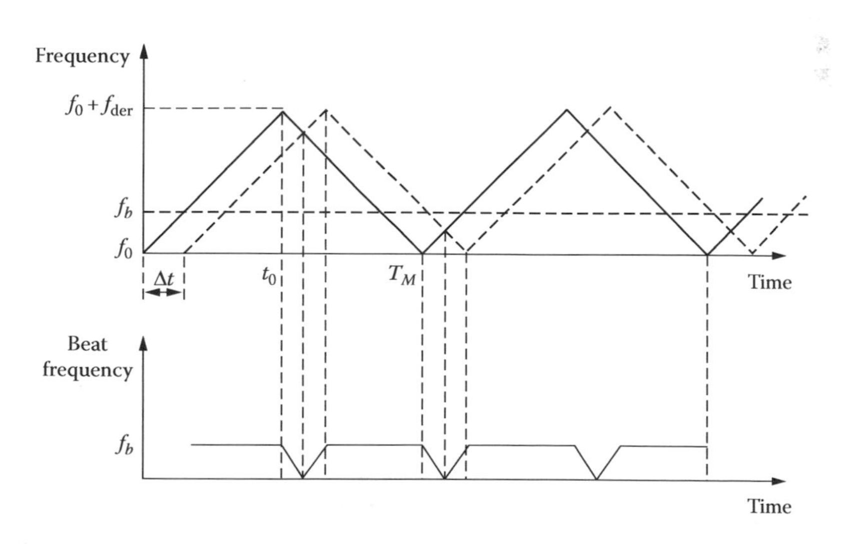

Figure 10.7 Frequency–time dependence and beat signal for a motionless target.

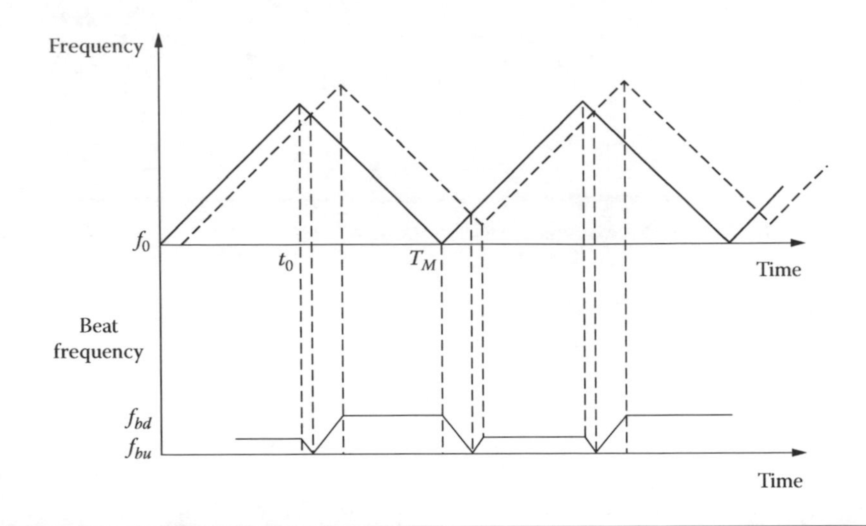

Figure 10.8 Frequency–time dependence and beat signal for a moving target.

Both range information and velocity information may be extracted from such a beat signal by signal processing in each range. As was shown in Sections 10.3.5 and 10.3.6, radar range resolution is defined by the spectrum width of the radiated waveform. In the case of FMCW radar, the spectrum width, B, directly depends on the frequency deviation, f_{dev}, and the potential range resolution is approximately $\Delta R = c/2f_{dev}$.

One can easily calculate that, for NB 24-GHz radar with $B = 200$ MHz, the potential range resolution, ΔR, is about 75 cm. It can be much higher in the case of, say, 76-GHz millimeter-wave radar. A range bin size is normally close to ΔR and less than the antenna beam footprint. Thus the forward-looking surveillance FMCW radar can measure vehicle speed in a single lane using a range-binning technique that divides the FOV in the direction of vehicle travel into range bins, as shown in Figure 10.9.

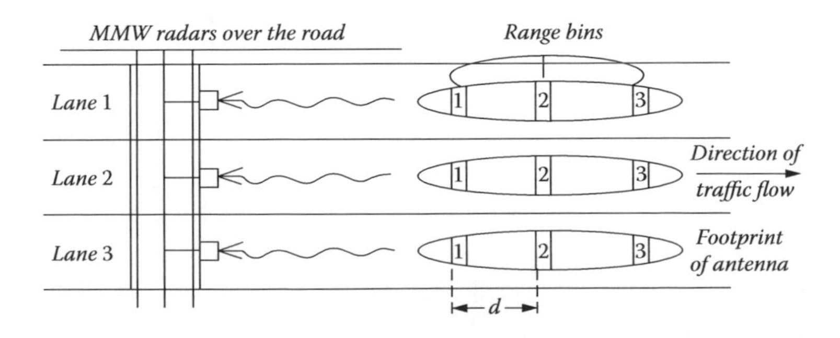

Figure 10.9 Range-binned footprints of radar sensors in traffic lanes [81].

A range bin technique allows the reflected signal to be partitioned and identified from smaller regions on the roadway. Vehicle speed, V, is calculated as $V = d/\Delta t$, where d is the known distance between leading edges of two range bins and Δt is the difference between points of time when a vehicle passes through these range bins.

Another promising development of millimeter-wave radar for intelligent infrastructure has been reported [82] where a radar sensor for an advanced cruise-assist highway system is proposed. This radar uses WB pseudo-noise–modulated signal in millimeter-wave radar. The radar prototype is built at the central frequency of 76.5 GHz. As was shown in Section 10.3, WB and noise signal provide better potential performances in comparison with traditional NB, in particular, an FMCW signal.

Along with road-based surveillance, recent advances in vehicle sensors and detection algorithms open the opportunity to implement or enhance vehicle-based surveillance systems. Vehicle-based traffic surveillance systems [80] involve probe vehicles equipped with tracking devices, such as transponders, that allow the vehicles to be tracked by a central computer facility. Such systems are rather promising. They can be used to detect incidents, provide rich data on travel times, and estimate flows and origin–destination patterns. These vehicle-based technologies are built on secondary radar principles close to those used in air traffic control radar beacon systems.

Another purpose of vehicle transponders is for automatic vehicle identification (AVI) or radio frequency identification (RFID). Such transponders [83] are essential to any electronic toll collection function; AVI tags have become increasingly sophisticated and intelligent.

10.5.4 Automotive Radar

A simple but suitable parameter to technically distinguish radars is the maximum range. Along with traditional distinguishing SRR and LRR, the medium-range automotive radar (MRR) has been introduced [75,84]. Therefore, we distinguish the following radar categories: LRR, with a maximum range of 150 m; MRR, with a maximum range of 40 m; and SRR, with a maximum range of 15 m. Their applications for the automotive functions listed in Section 10.5.1 are given in Table 10.3, where FSK means frequency shift keying. The content of this table is completely taken from Mende and Rohling [75], however, it is supplemented in the last and the next-to-last columns by additional suitable frequency bands and radar principles. It should be mentioned that possible functions are not limited by those listed here. For example, pedestrian detection, overtake support, rear collision warning, lane keeping, and other functions are quite expedient and possible technically to be implemented.

Table 10.3 Automotive Applications of Millimeter-Wave Radar

Function	Requirements			Suitable Sensors Category	Suitable Radar Principle	Suitable Carrier Frequency
	Range	Velocity	FOV			
1. Parking aid	0.2–5 m,	0 to ±30 km/h	Full vehicle width	2–4xSRR per bumper	UWB, pulsed	24 GHz (21.625–26.625), 77–81 GHz
2. Blind spot surveillance	0.5–10 m/ 0.5–40m	Reasonable velocity interval	Two lanes beside vehicle	1–2xSRR or 1–2xMRR per side	FMCW/FSK/pulsed, UWB	24 GHz, 77–81 GHz
3. ACC	1–150 m	Reasonable velocity interval	Three lanes in front of vehicle in 65 m	1xLRR	FMCW/FSK/pulsed	76–77 GHz
4. ACC plus	1–150 m/ 0.5–40 m	Reasonable velocity interval	Three lanes in front of vehicle in 20 m	1xLRR/1xMRR	FMCW/FSK/pulsed	76–77 GHz, 24 GHz, 77–81 GHz
5. ACC plus stop & go	0.5–150 m/0.5–40 m	Reasonable velocity interval	Three lanes in front of vehicle in 10 m Full vehicle width in 0.5 m	1xLRR/1xMRR	FMCW/FSK/pulsed	76–77 GHz, 24 GHz, 77–81 GHz
6. Closing velocity sensing	0.5–10 m/0.5–30m	Any velocity	About 45°	1xSRR/1xMRR	FMCW/FSK, UWB	24 GHz, 77–81 GHz
7. Precrash reversible restraints	0.5–10 m/ 0.5–30 m	Any velocity	Full vehicle width in 0.5 m	2xSRR/2xMRR	FMCW/FSK, UWB	24 GHz, 77–81 GHz
8. Precrash nonreversible restraints	0.5–10 m/ 0.5–30 m	Any velocity	Full vehicle width in 0.5 m	2xSRR/2xMRR	FMCW/FSK, UWB	24 GHz, 77–81 GHz
9. Collision mitigation	0.5–150 m/ 0.5–40 m	Any velocity	Three lanes in front of vehicle in 10 m Full vehicle width in 0.5 m	1xLRR/2xMRR	FMCW/FSK	76–77 GHz, 24 GHz, 77–81 GHz
10. Collision avoidance	0.5–150 m/ 0.5–40 m	Any velocity	Three lanes in front of vehicle in 10 m Full vehicle width in 0.5 m	1xLRR/2xMRR	FMCW/FSK	77 GHz, 24 GHz, 77–81 GHz

[a] An ultrasonic may be considered as an alternative to the first function. Functions 2 to 5 can also be implemented by laser radar; however, MMW radar is preferable because it is practically an all-weather automotive sensor. Functions 6 to 10 do not have any suitable alternatives.

[b] Adapted from Mende and Rohling [75].

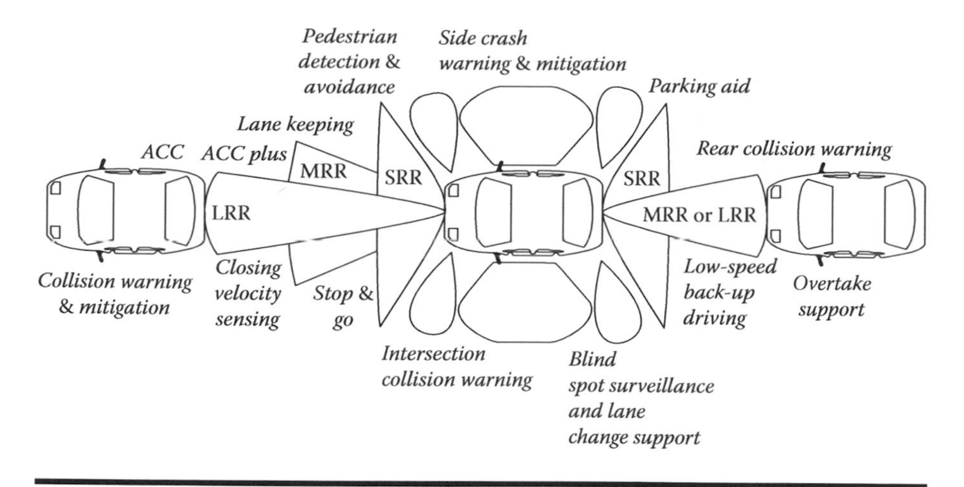

Figure 10.10 Layout of antenna patterns along the perimeter of the car to implement different functions.

Certainly, each radar sensor may be used as a multifunctional device, which can reasonably reduce the cost. An example of layout of schematic antenna patterns to realize a number of automotive radar functions is given in Figure 10.10.

10.5.5 Long-Range Radar

Conventionally, LRR delivers a target list with distance, angular position, relative speed, and reflectivity of significant objects within typical cycle times of 50 ms.

The size of a target can be smaller or bigger than the spatial resolution of the radar. Naturally, high-resolution radar is preferable because vehicles appear as laterally and longitudinally extended objects, which opens an additional possibility of target recognition. Normally, LRR maximum range is 150 m (ACC application) and MRR can be 1–1.5 m. LRR is typically implemented at 76.5 GHz center frequency (1 GHz bandwidth available), which generally corresponds to ETSI and FCC specifications. LRRs of 24 GHz have also been developed [85] and are in use. LRR has a limited FOV. Coverage in azimuth is typically 12° (up to 20°) and 6° in elevation. Range resolution is 1–5 m in different models, with a speed interval of 250–500 km/h. In some cases the maximum range can be increased to 200 m, but the higher the range, the bigger the problem of target-lane association. LRR can be applied as a single unit.

LRR can be built on different radar principles, such as FMCW radar, CW radar with FSK, or pulse Doppler radar. Noise modulation can also be considered.

An FSK CW LLR is described by Takezaki et al. [86]. This radar unit employs an MMIC for the high-frequency module that transmits and receives 76-GHz band radio waves. The MMIC chipset consists of four chips: a voltage-controlled oscillator, a power amplifier, and two receivers. These chips are integrated in a monolithic RF module combined with an integrated flat antenna. The RF module and signal processing board are integrated in the same housing to achieve a small, lightweight radar unit. This radar unit employs two-frequency CW radar modulation, in which an FSK signal is transmitted and the relative speed of objects is measured from the Doppler frequency and the distance to the objects is measured from the phase of the reflected signals. For azimuth angle measurement, the monopulse technique (Section 10.3.9) is used, in which the signals reflected from the vehicle are received by two antennas and the direction of the vehicle is determined from the amplitude ratio of those two received signals.

A pulse Doppler LRR is discussed in Schneider and Wenger [87]. This radar is designed to provide imaging of vehicle underbodies and detection of hidden objects via reflections from the road surface. That requires higher sensitivity with respect to a conventional ACC radar sensor. An increased target dynamic of 50 dB was assumed. Together with the dynamic due to propagation attenuation in the 5–150 m range of 60 dB, an overall requirement for the system dynamic results, which is difficult to be handled by a CW system. Hence pulsed waveform was chosen for this imaging radar system design because of its robustness against saturation.

At a center frequency of 76.5 GHz, the FOV in azimuth is ±10°, the azimuth beam width is 1.0°, the elevation beam width is 5.0°, the real range resolution is 1 m, and the velocity resolution is 1 km/h.

The central component of the system is an embedded PC performing system control and radar image composition. Its main periphery is formed by timing and control electronics, an IF processor module for converting the IF signal from the receiver module into baseband I/Q (inphase and quadrature) signals and digitization, a digital signal processor performing the Doppler FFTs (Fast Fourier Transforms) and other preprocessing, and a number of common devices.

A ferrite circulator for the transmit and receive duplex is mounted together with a receive low-noise amplifier directly at the antenna and connects to the transmit and receive modules. The transmit signal is generated in a phase-locked dielectric resonator oscillator (DRO) at 19.125 GHz, pulse gated in a two-stage PIN modulator, and multiplied to 76.5 GHz in an active quadrupler with an output power of about 16 dBm. The minimum pulse length is 3.5 ns. The key component in the transmit module is a biased, balanced fundamental mixer. It converts to an IF signal of 3 GHz, which is led to the base system on a high power level in order to minimize interference in the long transmission line. The system is connected to the radar control PC via Ethernet LAN.

A 76–77 GHz pulse Doppler radar module for LRR with a three-beam antenna is described by Gresham et al. [88].

An FMCW LRR is the most typical in automotive applications. The FMCW principle was considered in Section 10.5.3. As a way to keep the hardware and architecture relatively simple, the peak power down, and at the same time achieve robust performance, an FMCW approach is frequently used as opposed to a more sophisticated pulse Doppler implementation [54]. A radar of this type has a relatively low peak power output and can meet the range and velocity resolution requirements if a sufficiently large frequency deviation is used. Such radar sensors are often developed as multibeam antenna systems with sequential lobing technique [75] or with monopulse technique [85].

10.5.6 Medium-Range Radar

This category of radar introduced by Mende and Rohling [75] can fill the gap between SRR and LRR regarding maximum range. They cover a range from 0.5 m to 70 m and a speed interval of –500 km/h (closing) to +250 km/h (opening) [84]. The minimum range is usually less than 0.6–1 m. They can be implemented as a radar network of distributed MRRs and a central processing unit. But single-sensor operation is also possible and is preferred in many cases for cost reasons. Measuring range, angle, and relative radial velocity in multitarget scenarios at one time, they are capable of handling highly dynamic situations. For direct angular measurement, either sequential lobing or monopulse techniques may be applied.

As an example, the universal medium-range radar (UMRR) [84] can be considered. The UMRR frequency band is 24 GHz. The UMRR has multimode capability and may be switched between UWB pulse operational mode and FMCW NB mode within one measurement cycle. In NB FMCW mode, UMRR performances correspond to typical MRR requirements: min. range, 0.75 m; max. range, 60 m; velocity interval, –69.4 ± 69.4 m/s. UMRR uses a combination of FSK and LFM (linear frequency modulation) waveform design principles [89]. It also has a flexible bandwidth by operating in modes of NB operation and a 1–500 MHz maximum modulation bandwidth. Such a universal and flexible device can be used in different configurations, improving the functionality of automotive radar.

10.5.7 Short-Range Radar

The requirements of SRR are short range but wide FOV and high accuracy. Typically, such radar should detect targets in the range 0.2–30 m, with an azimuth FOV of about 100°, and a maximum velocity of more than 240 kmh. SRR sensors were first developed in the 24-GHz band. More recent development in Europe has been fulfilled in the 79-GHz band. In many cases, SSRs are designed as UWB pulse radars ($B = 3$ GHz). In this case, target velocity is estimated from the range rate. However, FMCW technique is also applicable.

It is difficult and expensive to design quick scanning in the FOV to measure azimuth position of targets. Another way could be in principle using very narrow antenna beams per sensor and multiple sensors around the car. However, the cost of such a system grows by the number of sensors and by the number of antenna beams per sensor.

That is why the multistatic principle for automotive radar networks was developed [90]. That was a brilliant idea to provide accurate angle measurements with a number of cheap radar sensors, which have no angle resolution at all but can measure target range and target velocity with high accuracy. The simplest technique consists in using a linear array of several sensors, spaced along the bumper, each having up to 60° beamwidth. Each sensor measures the range to the target with high accuracy, and the ranges are different for each sensor in the general case. From these measured results it is possible to estimate an angular position of the target. The real situation may be complicated because of multiple and extended targets in the FOV. Position reconstruction of a large target based on triangulation is not accurate enough because the different sensors detect different parts of the same target. Moreover, the range-velocity estimation should be provided. The multilateration procedure, which is used to solve this problem, has been described in detail [91,92]. It is a real technical challenge to handle the large number of detections in the data association and tracking as well as in the range, velocity, and angle parameter estimation.

Let us consider the signal processing procedure of four FMCW radar sensors as described by Rohling and Fölster [91]. For data acquisition, a waveform is used that consists of four individual chirp signals (Figure 10.11). The waveform parameters are center frequency, 77 GHz; number of chirps, 4; single chirp duration, 2 ms; first sweep bandwidth, 1 GHz; second sweep bandwidth, 500 MHz.

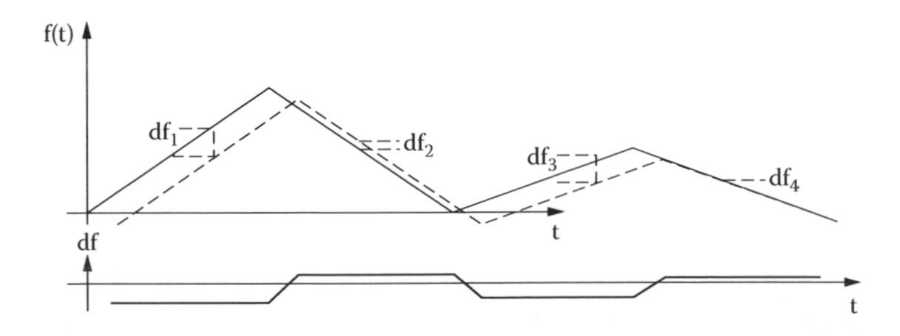

Figure 10.11 Four-chirp FMCW waveform and the receive signal for a single-target situation.

Four individual chirps provide a sufficient redundancy in multitarget or extended target situations to suppress ghost targets in the range–velocity processing. For each individual chirp signal, the beat frequencies, $df_1, \ldots df_{41}$, shown in Figure 10.11, will be estimated at each individual radar sensor by applying an FFT, respectively. A single-target detection with range r_s and velocity v_S leads to a deterministic beat frequency for each chirp signal of the waveform. This beat frequency is related to target range and velocity by the linear equation $f_{CS} = a_C r_S + b_C v_S$, where parameters a_C and b_C depend on chirp duration, bandwidth, and carrier frequency. Based on the four beat frequencies measured by a single sensor, the point target range and velocity can be derived simultaneously by an intersection process. In this case and in a single-point target situation, the four measured frequencies are transformed into target range and velocity, unambiguously. But in multiple or even extended target situations, this range velocity calculation could lead to some ghost targets.

Each sensor of the radar network has an individual position behind the front bumper. Therefore, each sensor will calculate individual values for target range and velocity based on the four measured beat frequencies inside the FMCW waveform. Thus, a set of linear equations can be derived that describes the relationship between 16 measured beat frequencies and sensor-specific target range and velocity parameter (two physical quantities, r and v, multiplied by four chirps and four sensors).

In multi-target situations the association of the detected beat frequencies at the FFT output to different targets is not trivial. Therefore, a data association process has to be performed using the redundancy given by the four chirp signals inside a single waveform. The radar network signal processing described by Rohling and Fölster [91] calculates the azimuth angle (or target position in Cartesian coordinates) of each target in multiple or even extended target situations based on the precise range measurement of each radar sensor. Furthermore, the tracking procedure is part of the network processing. In different configurations of the system, two to six SRRs may be used in such a network. The features of sensor design and its application to automotive problems are considered in Rickett and Manor [93].

10.5.8 Adaptive Cruise Control System

Unlike conventional cruise control, ACC can automatically adjust speed to maintain a proper distance between vehicles in the same lane. This is achieved through a forward-looking millimeter-wave LRR sensor, digital signal processor, and longitudinal controller. Jaguar and Mercedes first introduced 76-GHz automotive radar to provide ACC in 1999, operating throttle and brakes to maintain headway. Jaguar used FMCW LRR with a mechanical scanned antenna, and Mercedes applied pulse Doppler LRR with a quasi-optic antenna [79].

The latest innovations in automotive radar systems are based on the positive experiences with the LRR-based ACC function.

10.5.9 Road Departure Warning System

In 2001 the project of the road departure warning system was tested in field operational research. The project defines and evaluates a system that warns drivers when they are about to drift off the road and crash into an obstacle, as well as when they are traveling too fast for an upcoming curve. Technologies include a vision- and radar-based lateral drift warning system and a map-based curve speed warning system [83]. The system is based on the information about the current situation obtained by two forward-looking LRR and two side-looking SRRs. The technology also includes a camera-based lane detection and map-based curve speed warning system using GPS data. More details see UMTRI [94]. Forward- and side-looking radar systems detect the presence of any obstacles on the shoulder (such as parked cars) or the roadside (such as poles or guardrails).

Another approach [95] is based on lane position sensing using a special selective stripe detected by radar. The radar sensor measures lateral position by sensing backscattered energy from a frequency-selective surface constructed as lane striping and mounted in the center of the lane.

10.5.10 Blind Spot Monitoring and Lane Change Control System

A lane change assistant should help to increase safety when drivers change lanes. Such a system needs two SRRs looking at the blind spots; that is, laterad and a little back to the adjacent lanes. An example of a lane change assistance system is described by Valco Raytheon [96]. It consists of two 24-GHz radar sensors joined to a control box and two LED warning indicators mounted in the side rear-view mirrors. The sensors continuously monitor the presence, direction, and velocity of vehicles in the roadway lanes adjacent to the protected vehicle. The radar image sensors create a digital picture of the traveling environment. The digital information is processed by a central control module. When vehicles, motorcyclists, or bicyclists move into a blind spot of the protected automobile, the central control box alerts the driver by lighting the warning indicators in the side rear-view mirrors. The system range extends to 40 m, with a 150° broad FOV. The radar is a multibeam system operating in a narrow bandwidth. Using several beams to recognize objects in the blind spot allows for high accuracy in determining the position and distance of the object as well as its relative speed.

The second example [83] is the Visteon system, which is a close-in blind spot monitor, covering a detection range of 6 m.

Another example of a blind spot detection sensor is a UMRR [89] optimized for short-range detection, which implements detection algorithms, tracking, object-to-lane mapping, and a warning algorithm (illuminating warning LEDs). This blind spot monitoring system's range is 0.3–8 m. A sensor operates in a speed interval from 7 to 250 km/h. For the sensor fusion case, it is reasonable to combine the two blind spot sensors with a third medium-range (70 m) or long-range (120 m) sensor. Additional applications of a blind spot detector may be (1) warn to open door when object approaches, and (2) side/rear precrash or presafe applications.

10.5.11 Obstacle Detection

LRR, MRR, and SRR can be used as sensors for obstacle detection. Object classification can also be performed on the basis of radar data. Probabilities of true and erroneous classification using a millimeter-wave radar system are presented in Kruse et al. [97].

10.5.12 Radar-Based Communications

Air traffic control surveillance systems widely use secondary radar technology to get more precise and comprehensive information and data exchange. A similar technology adapted for motor transport needs may be established in the framework of ITS. It can be considered as radar-based communications. This kind of data exchange can be used for both the vehicle–vehicle and vehicle–infrastructure communications arena.

Two aspects of radar-based communications in millimeter waves related to intelligent infrastructure were considered in Section 10.5.3. Now we add some other possibilities that may be important for an intelligent vehicle system.

One concept is communication using a read–write RF tag. This idea is similar to the radio frequency ID that has become more and more important nowadays. An RF tag can be activated when illuminated by standard vehicle 77-GHz radar. The return signal carries additional coded information. In this case the automotive millimeter-wave radar receives not only reflections but digital data. After signal detection, useful information is extracted and used to improve safety. For example, an RF tag placed on a bridge footing in a complex highway environment can provide key data as to the size of this potential obstacle, thus enabling the vehicle system to better interpret the situation. Every vehicle may also be provided with an individual RF tag, creating a very robust and cost-effective secondary radar system. For example, stationary and moving targets can be distinctly identified by the data encrypted into the tag, which actually is a kind of transponder.

Such an interrogator–transponder system can be applied to establish car-to-car communications based on the secondary radar principle.

Another possibility relates to direct-use millimeter-wave radar for communications. A car-to-car communications operation together with millimeter-wave radar systems can combine the data from different vehicles, allowing observation of the complete car environment. An improvement can be achieved if one car has access to the sensor information of the preceding and following cars. The potential use of the millimeter-wave radar spectrum for communication needs is analyzed by Winkler et al. [98]. The developed system consists of a pulse radar and a communication transceiver with 4 Mb/s. The main lobe of the pulse radar waveform spectrum occupies a bandwidth of 250 MHz. The frequency band of the transceiver, which is small compared to the pulse spectrum, is placed at the first side-lobe of the pulse spectrum. The pulse radar is designed to allow simultaneous sensing over a range of nearly 40 m, and the communication link has a maximum range of 200 m. The feasibility of the concept and its superior dynamic range has been shown [98] in case PN-code is used for ranging and data transmission. A configurable RISC processor, consisting of FPGA (field-programmable gate array) logic resources, is connected to the radar and the transceiver. It sets operation parameters and addresses a LAN chip for data exchange between the system and a PC over Ethernet.

More details on wireless LAN application to car-to-car communications are provided by Zlocki and Zambou [99].

10.5.13 Radar-Based Automatic Road Transportation System

The availability of high-resolution millimeter-wave radar sensors, secondary radar systems, and radar-based communications along with modern technologies creates more possibilities to improve ITS. Kramer [100] presents an idea of a fully automatic road transportation system and proposes the system's architecture. The concept of this hypothetical system includes a network of short-range, high-resolution millimeter-wave radar sensors that monitors traffic space; a network of sensors, some in the vehicles, that registers road conditions and traction parameters; and a computer network that receives information from the two sensor systems and the road users and generates guidance commands.

The system could use special navigation signals in addition to the radar signals, similarly to GPS. The in-vehicle precision navigation system may transmit the vehicle's position, velocities, turn rates, and so on to the central guidance system. This would substantially reduce the tasks for the fixed installation radars, which would then only detect and observe noncooperative road users and random obstacles. Noncooperative road users and pedestrians might also carry responders that communicate with the central guidance system. Because of the low signal powers involved, the responders could even be integrated in wristwatches.

During the conversion to an exclusively automatic road transportation system, vehicles might have transceivers that communicate with the central guidance system. The reception of guidance information would allow for participation in optimum traffic management. The realization of the automatic transportation system should start with a small prototype to successively refine the definition of sensor requirements and the algorithms for the central guidance system.

10.5.14 Automatic Driving

Future vehicles will fuse data from the millimeter-wave radar sensors (up to 12 per vehicle) with vision, navigation, communications, and vehicle dynamics to provide safety and convenience. The dataware level of such a vehicle approaches to what is enough to dramatically unload the driver.

Automatic driving systems are actually under development; for example, AutoDrive [101]. AutoDrive is a system for automatic driving of cars and commercial vehicles on highways. It uses GPS positioning, radars, intervehicle signaling, and wireless Internet to aid driving, which is mainly those electronic facilities considered above. AutoDrive is a completely automated highway driving system. By 2030, all major interstates in the United States are expected to support AutoDrive lanes [101] with completely automated car and commercial vehicle driving. Any car updated for AutoDrive will be able to enter these lanes and enjoy speed limits of around 130 mph.

Such a system will use GPS to accurately locate the vehicle's position on the map and then steer the car according to the path specified by the map. Intervehicle signaling (IVS) will be used to keep track of neighboring vehicles, detecting speed changes and responding to lane change requests from other vehicles. The radar feed will be used to identify objects in the vicinity of the car. The main objective of the radar in AutoDrive will be to detect nonvehicle objects that might be present on the highway. The precision radar will detect objects that are more than 3 cm in height.

Map updates can be obtained over a wireless Internet connection. Most map updates are obtained at the time of trip start. The maps are extremely detailed and store information about the different lanes with an accuracy of ±1 cm.

Weather feed is used to adjust driving parameters according to the current weather conditions. Construction and road closure detection is implemented using IVS. An IVS transmitter is placed in a closed lane. Vehicles receive the IVS signal from the transmitter and decide to change lanes. Fully automated driving can be considered as an ultimate level of ITS development.

10.6 Remote Sensing Applications

10.6.1 Cloud Radar

10.6.1.1 Introductory Notes

Applications of millimeter-wave radar to study clouds and precipitation are known from the 1950s, but early in the development of millimeter-wave meteorological observations they were limited to backscatter power measurements mainly because of hardware problems. As early as 1966 weather radar MRL-1 (Oblako) was produced and then widely used for weather observations in the Soviet Union. It was two-channel incoherent radar with wavelength $\lambda = 32$ mm in one of the channels and $\lambda = 8$ mm in the second one. However, the weather forecasters underestimated the millimeter-wave channel at that time. It did not seem to be a highly effective instrument for practical needs due to little range in comparison with the 3-cm channel and lack of developed observation techniques and data interpretation. This situation changed completely during the last two decades. Millimeter-wave Doppler radars are considered now as an actual instrument for permanent monitoring of the atmosphere and investigation of clouds and precipitation.

As distinct from ITS applications, weather radars are comparatively long-range sensors. That is why a majority of millimeter-wave weather radars are designed to operate in the relatively transparent window regions, mainly about 34–36 and 93–95 GHz. Nevertheless, water vapor substantially attenuates millimeter waves in the troposphere, which is a disadvantage, of course, but can also be used to derive some useful information about clouds and precipitation. Scatterers include nonprecipitating cloud water and ice-phase precipitation, insects, and seeds.

Weather objects are volume distributed targets. High spatial resolution of millimeter-wave radar turns it into a rather fine instrument for cloud structure investigation.

Nowadays a number of scientific and industrial institutions all over the world perform research in the field of remote sensing of troposphere with millimeter-wave radars, design and produce new radars, and develop new techniques for cloud observation and algorithms for signal and data processing.

This is a separate, important, and very wide area of knowledge. Here we will just touch on some features of millimeter-wave radar applications in the field of meteorology.

10.6.1.2 Features, Methods, and Advantages

The phenomenon of electromagnetic wave scattering on hydrometeors forms the basis for weather observations and measurements with the help of millimeter-wave radar. As was considered in Section 10.2, in case size r

of the hydrometeor is small comparatively to wavelength λ (Rayleigh case), an RCS (and reflected power) is proportional to r^6/λ^4. This is valid in the millimeter-wave band for cloud droplets but, in the case of larger raindrops, especially in the short-wave part of the millimeter-wave band (95 GHz), resonance phenomena can play an essential role. A high sensitivity of millimeter-wave weather radar results from this RCS proportionality to $1/\lambda^4$.

Dual-wavelength radar (microwaves plus millimeter waves) allows measuring attenuation of radar signal and improving estimation of rain rate or cloud water content. Inasmuch as in the millimeter waves band λ may be comparable with r, the scattering law appreciably differs from the Rayleigh case, and by knowing frequency-dependent wave attenuation (see Section 10.2) one can estimate raindrop size distribution from measured data at different wavelengths. An example of retrieving cloud and rain parameters from dual-wavelength measurements at the Ka-band and W-band is provided by Bezvesilniy and Vavriv [102].

Real hydrometeors in many cases are not spherical. For example, raindrops are normally oblate, and they are more oblate if they are larger; cloud droplets are mostly spherical; and ice particles are complicated in shape. The polarization properties of a radar signal at scattering on nonspherical hydrometeors depend on the shapes of the scatterers and the mutual orientation of scatterers and radar beam. That is why application of polarimetric measurements opens up possibilities to recognize hydrometeor shapes and hence their aggregative state.

Hydrometeor movement cases Doppler shift. This results in change of Doppler spectrum of the returns from a radar resolution volume. Measurements of spectrum parameters give possibility to study microstructure and dynamic processes in the atmosphere.

The physical basis for detecting different weather phenomena and measuring meteorological variables is a relationship of echo-signal parameters with the peculiarities of microstructure and the dynamics of scatterers (droplets, ice crystals, snowflakes, etc.) within a resolution volume or an aggregate of resolution volumes. In operational observations and research works, the following types of radars are used:

- Incoherent: conventional pulse radar
- Coherent: pulse-Doppler radar, FMCW radar, or CW with other modulation formats
- Polarimetric: application of polarization diversity on the basis of a conventional pulse radar
- Doppler-polarimetric: coherent radar with application of polarization diversity for measurements

Incoherent radar measures just the power parameters of reflected signals that are related with weather object reflectivity.

Coherent radar measures parameters of Doppler spectra. Theory fundamentals describing relations between detected signal and features of scatterer movement as well as the sounding techniques by ground-based weather radar were developed already in the 1960s. At that time it was important that Doppler spectrum width could be estimated by means of special signal processing with incoherent radar for the narrower Doppler frequency dynamic range.

The instruments for remote sensing of the atmosphere that were developed on the basis of incoherent radar and pulse Doppler radar are widely used in radar meteorology and operational weather observation for detection and measuring of winds, storms, turbulence, and related phenomena [103].

Polarimetric radars measure relative polarimetric parameters of reflected signal, such as differential reflectivity (a ratio of reflectivity values at the alternatively polarized radar signals; for example, at horizontally and vertically polarized electromagnetic waves), linear depolarization ratio (a ratio of a cross-polarized component of reflected signal by a copolarized one), correlation between reflections at orthogonal polarizations, and specific differential phase (a difference between propagation constants for horizontally and vertically polarized electromagnetic waves per unit of range).

Doppler-polarimetric radars simultaneously use spectral and polarimetric characteristics [104]. It can be not only a simple combination of Doppler and polarimetric parameters but also new integrated characteristics, such as a curve of the spectral differential reflectivity and its slope parameter, differential Doppler velocity, and others [105]. Fully polarimetric Doppler radars are able to enhance dramatically information possibilities of the remote sensing of the atmosphere.

Application of sophisticated Doppler-polarimetric techniques in high-resolution millimeter-wave radars is made possible by the great advances in digital signal processing technology and the significant increase in the role of software in modern radar system design, which has resulted in single-chip processors and special-purpose IC that can simultaneously execute algorithms to compute all necessary quantities in real time. The development of solid-state millimeter-wave componentry, high-power oscillators and amplifiers (Section 10.4) hastened the evolution of reliable and stable radars, which have become extremely useful instruments for ground-based, airborne, and even spaceborne remote sensing of clouds and precipitation. The 35-GHz frequency band is considered now as the most suitable band for performing observation with the help of ground-based millimeter-wave radar, whereas the 95-GHz band has obvious advantages for developing airborne and spaceborne tools for meteorological observations though it is not a rule. The usefulness of millimeter-wave radar does not mean that weather radars of other frequency bands (S, C, and X bands) are not important anymore. They have their own advantages. However, millimeter-wave radar systems with high sensitivity and resolution are required to research even rather weak and thin cloud layers.

10.6.1.3 Examples of Systems and Applications

Built using modern reliable technology, millimeter-wave radar systems are capable for long-term, unattainable operation and real-time, accurate measurements of various cloud and precipitation characteristics. Such radars should include a reliable system of permanent calibration, a possibility of remote control, diagnostics, and access to radar data via a network. The millimeter-wave cloud radars developed and produced at the Institute of Radio Astronomy in the Ukraine satisfy these requirements to a great extent [51]. They use the spatial-harmonic magnetron with the cold secondary-emission cathodes described in Section 10.4. The application of the magnetron in weather radar has predetermined the development of a coherent on-receive technique for the implementation of Doppler spectrum measurements. This technique proposes memorizing in some way the values of the phase of the RF pulses emitted by a transmitter and comparing these values with those measured by a receiver. The recent advances in microprocessor technique and digital signal processing have enabled [51] the memorization and comparison procedures in digital form. This has resulted in accurate measurements of Doppler spectrum and its moments with performance similar to those offered by truly coherent radar systems.

Pulsed Doppler radar from the University of Massachusetts, Amherst, is described by Bluestein and Pazmany [106]. It is a 3-mm-wavelength mobile Doppler radar system mounted in a Ford 350 Crewcab pickup truck. A 1.2-m Cassegrain dish antenna has a half-power beamwidth of 0.18° and provides azimuthal resolution of just under 10 m in the 3-km range. The radial resolution is 15 m at all ranges. The radar transmitter is designed with a multiplier chain. It has an operating frequency of 95.04 GHz. It was successfully used for tornado observations [106] and for fine-scale observations of dynamic phenomena via a pseudo-multiple-Doppler radar processing technique to decompose radial velocity vectors into the individual components of motion [107].

A low-power solid-state W-band FMCW radar system for airborne measurements of clouds and precipitation is presented by Mead et al. [108]. A millimeter-wave I/Q homodyne detector is used in this radar instead of a simple mixer.

Different compact millimeter-wave radar for airborne studies of clouds and precipitation, this time a pulsed solid-state radar with fully coherent transmitter and receiver is described by Bambha et al. [109]. Another example is the Wyoming cloud radar (WCR), which is an observational system of 95 GHz for the study of cloud structure and composition. It is installed on the Wyoming KingAir airplane. WCR provides high-resolution measurements of reflectivity, velocity, and polarization fields. Depending on the antenna configuration used, the scanned plane from the KingAir can be vertical or horizontal, and with two antennas, dual-Doppler analysis is possible [110].

Spaceborne applications of millimeter-wave radar were discussed previously in some papers, especially the use of satellite-based 95-GHz radars for measuring the vertical distribution of clouds [104]. Now millimeter-wave weather radar has been used already for cloud observation from space. CloudSat was launched April 28, 2006. The main CloudSat instrument is a 94-GHz cloud profiling radar (CPR) [111]. CloudSat uses this advanced radar to slice through clouds to see their vertical structure, providing a completely new observational capability from space. Earlier satellites could image only the uppermost layers of clouds. Thanks to millimeter-wave CPR, CloudSat is among the first satellites to study clouds on a global basis. It looks at their structure, composition, and effects. CloudSat measurements have applications in air quality, weather models, water management, aviation safety, and disaster management. However, low-power solid-state scanning millimeter-wave radar systems cannot provide the necessary sensitivity to detect low-reflectivity cloud particles, especially at longer ranges from space. That is why new high-power electronic scanning millimeter-wave radar is proposed by Remote Sensing Solutions, Inc. [112] on NASA demand. It is expected that this system will lead to future generations of large-aperture spaceborne electronic scanning radars.

Another millimeter-wave application to atmospheric remote sensing is a kind of passive radar—millimeter-wave radiometry. This can be realized from both a ground-based platform and spaceborne apparatus.

Taking into account different scattering mechanisms, various combinations of instruments may give promising results at tropospheric observations. An example is the 4-D-Cloud Project, which makes intensive use of dual-wavelength (35/95 GHz) radar together with radiometry, namely the 22-frequency radiometer MICCY [113].

Another good example of different observation tools with joint application is the international CloudNet project, which aims to provide a systematic evaluation of clouds in forecast and climate models by comparing the model output with continuous ground-based observations of the vertical profiles of cloud properties. CloudNET is a research project supported by the European Commission [114]. Three experimental sites called cloud observing stations (COSs) have been arranged in Chilbolton (UK), Sirta (France), and Cabauw (The Netherlands) where different instruments from a number of institutions of European countries were collected. Every COS is equipped with millimeter-wave radars and radiometers (94-GHz Doppler cloud radar, 35-GHz cloud radar, 24/37-GHz radiometer, 22-channel MICCY radiometer) and also microwave radars of different frequency bands, lidars, and other tools including standard meteorological instruments, rain gauges, and disdrometers. This project brings significant results in different aspects of the remote sensing of troposphere and its applications [115].

In this chapter it is impossible even to list all known projects or the corresponding institutions that are doing research in the field of millimeter-wave weather radar in different countries. This overview of millimeter-wave meteorological radar does not provide comprehensive and deep knowledge on the topic, but it is quite sufficient in order to understand that this is one of the promising applications of millimeter-wave technology that are nowadays under quick development.

It is worthy also to note that millimeter-wave weather radars may be useful for observations of other objects in the atmosphere besides meteorological objects and phenomena [116]. Among them, blowing dust, birds, and insect echoes [117].

The only thing that should be added is the importance of weather radar networking. Millimeter-wave radar is really a modern instrument that is quite suitable for network application, data transmission, and data exchange. The radars must have network capabilities enabled for remote radar control and data receiving via any network supporting the TCP/IP protocol, including the Internet. Remote control and diagnostics of the radar operation from any network computer are needed as well. Any authorized user should have access to radar data and images via local Ethernet in real time. The Internet access should provide viewing of both current information and previously stored data. Wireless PAN, LAN, and MAN are applicable in appropriate cases.

10.6.2 Remote Sensing of the Terrain

Radar sensors operating in millimeter waves can be installed on board aircraft or spacecraft. They play an important role for both aircraft navigation (altimeters) and surface investigation (SAR, radiometer, scatterometer, etc.). An overview of spaceborne remote sensing instruments was provided by Glackin [118]. Millimeter-wave radar for remote sensing of the Earth falls into the general classes of passive and active sensors. Passive sensors, called radiometers, collect and detect natural radiation, whereas active sensors emit radiation and measure the returning signals.

Passive millimeter-wave sensors include imaging radiometers, synthetic-aperture radiometers, and submillimeter-wave radiometers. Active millimeter-wave sensors include classical radars, SAR, altimeters, and scatterometers.

Passive imagers and sounders generally operate at frequencies ranging from 6 to 200 GHz. Submillimeter-wave radiometers have recently been used for measuring cloud ice content.

The problem of high spatial resolution in radiometry can be overcome through a technique of aperture synthesis, as was explained in Section 10.3

for active radars. In the passive version this concept was previously used in radio astronomy, where the operation of a large solid dish antenna is simulated by using only a sparse aperture or thinned-array antenna. In such an antenna, only part of the aperture physically exists and the remainder is synthesized by correlating the individual antenna elements. This kind of aperture synthesis differs a little from a similar technique that has a long and successful history in airborne and spaceborne SAR.

Active sensors can be broadly divided into real-aperture radars and SARs. Some are interferometric, meaning that they exploit the signals that are seen from two somewhat different locations, which is a powerful means of elevation measurement. This can be done using two antennas separated by a rigid boom, or using a single antenna on a moving spacecraft that acquires data at two slightly different times, or using similar antennas on two separate spacecraft. Modern SARs frequently also use the radar polarimetric principle, taking into account the polarization properties of the surface. The most powerful SARs apply both principles: polarimetry and interferometry.

Real-aperture radars can be further categorized as scanning (imaging) radars, altimeters, and scatterometers. (Atmospheric radars were considered above.) A prototype of new high-power electronic scanning millimeter-wave radar was recently designed [112]. A scanning millimeter-wave radar is a critical tool for improving the remote sensing of the Earth and other bodies in our solar system.

Altimeters measure surface topography, and radar altimeters are typically used to measure the surface topography of the ocean (which is not as uniform as one might think). They operate using time-of-flight measurements and typically use two or more frequencies to compensate for ionospheric and atmospheric delays. Altimeters have been flying since the days of Skylab in 1973. Aperture synthesis and interferometric techniques can also be employed in altimeters, depending on the application. A millimeter-wave altimeter using an FMCW system in the radio navigation frequency band can be installed on a helicopter.

Scatterometers are a form of instrument that uses radar backscattering from the Earth's surface. The most prevalent application is for the measurement of sea surface wind speed and direction. This type of instrument first flew on Seasat in 1978. A special class of scatterometer called delta-k radar can measure ocean surface currents and the ocean wave spectrum using two or more closely spaced frequencies.

SAR also flew for the first time on Seasat. These radars sometimes transmit in one polarization (horizontal or vertical) and receive in one or the other. A fully polarimetric SAR employs all four possible send/receive combinations. SARs are powerful and flexible instruments that have a wide range of applications, such as monitoring sea ice, oil spills, soil moisture, snow, vegetation, and forest cover.

The specific aspects of millimeter-wave SAR application are related at least to three advantages in comparison with longer wavelength SAR systems:

1) Compactness of the instrument
2) Potential not only to reach extremely high angular (lateral) resolution due to the SAR principle but also really high-range resolution because of a possible broader spectrum width
3) Ease of signal processing due to short aperture length necessary for good lateral resolution and consequently low importance of imaging errors because of high robustness against the instabilities of the airborne carrier platform

All these potential advantages of millimeter-wave SAR are characteristic for experimental radar MEMPHIS [119], which has two front ends: one at 35 GHz and the other at 94 GHz. The radar waveform is a combination of a stepped frequency waveform and an FM chirp. For the high-resolution mode, the frequency is stepped up from pulse to pulse over a bandwidth of 800 MHz in steps of 100 MHz to gain a range resolution of about 19 cm. Hägelen [120] studied the detection of moving vehicles and the determination of their ground speed with the application of experimental radar MEMPHIS in monopulse mode (Section 10.3.9). The results show that monopulse processing is an adequate technique to process millimeter-wave SAR data in the case of a ground MTI.

The indicated features of millimeter-wave SAR open a lot of interesting and useful applications of such radars. SAR systems operated from small airborne platforms like motor gliders or even remotely piloted air vehicles have to take advantage of millimeter-wave systems because only this frequency region offers short aperture times, which guarantee ease of processing and a less stringed demand on platform stability [120].

10.7 Imaging Systems for Security and Safety Applications

The capability of millimeter waves to pass through fog, clouds, drizzle, dry snow, smoke, and other substances makes the millimeter-wave imaging systems the most efficient instrument to resolve a number of problems that cannot be solved with the help of infrared and visible imaging systems. In particular, object detection and recognition for provision of security and safety is a very important and promising field of millimeter-wave radar applications. In this brief review we consider two groups of such applications: (1) the devices for concealed weapon detection (CWD) and (2) aviation safety applications, including foreign object detection (FOD) of airfield.

10.7.1 Miniature Radar and Radiometric Systems for CWD Applications

CWD at a safe distance is a new field of great importance in which millimeter-wave radar can be effectively applied. Normally, CWD systems include instruments, devices, equipment, and technologies to detect the weapons most commonly concealed on human bodies, but also in containers or vehicles. Millimeter-wave passive systems, which are also called radiometers, can be applied for imaging different scenes, objects, and structures.

All bodies emit, absorb, reflect, and transmit millimeter-wave energy, and the quantity of energy that is emitted, absorbed, reflected, and penetrated depends on the material, shape, temperature, and surface condition of the body and also on the frequency. This actually is the basis for passive detection of a concealed weapon and other objects. Millimeter waves penetrate well through clothing; the human body basically absorbs millimeter waves (which is equivalent to thermal self-radiation); and weapon material reflects ambient radiation. For example, indoor radiometric sensors will see the contrast between radiation of a human body of 309 K and room temperature of 297 K. At the same time, metal components reflect almost all radiation.

A tutorial overview of development in imaging sensors and CWD processing is done by Chen et al. [121]. In the first passive millimeter-wave imaging sensors the millimeter-wave data were obtained by means of scans using a single detector that took up to 90 min to generate one image [121].

The single-channel radiometric CWD imaging system was designed in the Kiev Research Center "Iceberg" [122]. It had a heterodyne detector of 90–94 GHz and a parabolic antenna of 300 mm diameter, which formed the beamwidth of 0.6°. The beam could scan mechanically in the horizontal plane with a shift in the vertical plane after each scan. The computer control allows a change in the scan rate, the size, and the number of angular pixels. As shown by Denisov et al. [122], a human body passive image of 100 × 150 pixels, obtained indoors, was formed at the screen of the display during 1 min. It demonstrated that the system could quite clearly detect a plastic handgun hidden in the clothing. In order to provide system operation in realtime, the authors proposed a multichannel imaging system with a reception sensor array providing simultaneous reception of signals from different parts of an object under investigation. In particular, the 16-channel system [122] provides an image during 1 s.

Recent advances in millimeter-wave sensor technology have led to video-rate (30 frames/s) millimeter-wave cameras [121]. One such camera is a 94-GHz radiometric pupil-plane imaging system that employs frequency scanning to achieve vertical resolution and uses an array of 32 individual waveguide antennas for horizontal resolution [123,124].

Another approach was considered by Essen et al. [125]. They emphasized the development of a demonstrator for CWD applications and an imaging system for medium-range applications, up to 200 m. The short-range demonstrator is a scanning system operating alternatively at 35 GHz or 94 GHz to detect hidden materials such as explosives, guns, or knives beneath the clothing. The demonstrator uses a focal plane array approach using four channels in azimuth, while mechanical scanning is used for the elevation. The medium-range demonstrator employs a single-channel radiometer on a pedestal for elevation over azimuth scanning. To improve the image quality, methods have been implemented using a Lorentzian algorithm with Wiener filtering.

Millimeter-wave passive sensors can be combined with other devices to increase the quality of imaging and the reliability of weapon detection. Currie et al. [126] considered the application of infrared and millimeter-wave sensors to solve CWD, through-the-wall surveillance, and wide-area surveillance under poor lighting conditions. The operation of different sensors, in particular, infrared cameras, millimeter-wave passive and active cameras, and millimeter-wave real-aperture and holographic radars, have been described. All of these sensors form images, but the images are of varying quality. That is why methods using multiple sensors to improve performance [126] are expedient.

An example of active millimeter-wave radar for CWD is described by Chang and Johnson [127]. FMCW radar is designed for unobtrusive detection of concealed weapons on persons or in abandoned bags. The developed 94-GHz radar system provides image scanning and is suitable for portable operation and remote viewing of radar data. This system includes a fast image-scanning antenna that allows for the acquisition of medium-resolution 3D millimeter-wave images of stationary targets with frame times on the order of 1 s. It allows CWD on the background of the body and environmental clutter such as nearby furniture or other people. The 94-GHz radar-emitted power of approximately 1 mW is considered low and poses no health concerns for the operator or the targets [127]. The low-power operation is still sufficient to penetrate heavy clothing or material. In contrast to passive imaging systems, which depend on emission and reflection contrast between the weapon and the body and clothing, this active radar system operates the same way both indoors and outdoors.

Information about small radar units designed for using a robotic remote-controlled machine to inspect roadway objects and debris for concealed bombs has been published [128]. Such devices could be used to detect explosives at checkpoints for buildings and transportation hubs. Government officials are particularly interested in being able to spot suicide bombers and to monitor large areas for the introduction of explosives, without terrorists being aware that they are under surveillance [128]. This publication indicates that the goal achievement relies on a technology of active millimeter-wave radar, already used today in automotive collision-avoidance systems. Combined with a video camera and software, such a system is designed to detect explosives from afar [128].

A comparison of passive and active millimeter-wave radar sensors for CWD applications shows the following advantages and disadvantages of both technologies. The benefits of a passive radar sensor are (1) it is undetectable if one doesn't take into account a leakage of local oscillator, which can be used in some devices for heterodyne detection; (2) it is absolutely safe for people; (3) the image is similar to a photograph and easy to understand; and (4) the SNR depends less on range. On the other hand, the shortcomings of a passive method are (1) stringent requirements on receiver sensitivity; (2) greater dependence on external conditions, first of all, on ambient temperature; and (3) acquisition of fast, high-quality images requires application of quite expensive sensors such as the focal plane array.

In contrast, the positive properties of an active sensor are formed by its reduced requirements on receiver sensitivity, less dependence on external factors, and acceptability of scanning architectures that makes it cheaper. However, the active radar sensor has a number of disadvantages: (1) it is obviously detectable; (2) it causes problems with irradiating noncooperative targets (difficulty of covert surveillance); (3) sometimes it can be unsafe for both operators and objects; (4) harder image interpretation; and (5) the SNR is hardly dependent on range; that is, active radar has a limited range of detection.

A combination of different nature sensors, such as millimeter waves, infrared, and normal camera [129] with image fusion [130] gives the most promising results. Different works on CWD are supported by the U.S. Office of Justice Programs [131].

10.7.2 Safety Navigation Applications Including FOD of Airfield

As was shown above, the better sensitivity allows for increased scanning velocity due to the reduction of adequate integration time to provide good image quality. The diffraction-limited spatial resolution of a passive millimeter-wave imaging system is inversely proportional to the diameter of the quasi-optical antenna. Certain improvement of image quality can be achieved by the digital processing of the obtained image with help from mathematic methods. Thus the development of the practical millimeter-wave imaging system requires an increase in the number of receiving sensors in the focal plane array and use of the quasi-optical antenna with sufficiently big diameter. Gorishyake et al. [132] present a 32-channel 33–38-GHz imaging system that provides fast image with rather high image quality in a full angle of view of 90° in the horizontal plane in 3 s. Such an imaging system can be used to provide navigation on the ground or at sea, especially at short distances where surface radar clutter is high, for remote sensing in space and air investigations, for all-weather surveillance, and in many other commercial and special applications.

One of the important directions of millimeter-wave imaging system development is application for improvement of airport operations under the condition of increasing flight intensity. In accordance with Qinetiq [133], the Qinetiq Tarsier T1100 is the first radar designed for detection of debris on runways. It gives a real-time airfield picture. The system scans the runway for debris approximately every 60 s, typically giving an inspection between every aircraft movement. It is FMCW radar with a central frequency of 94.5 GHz, 600 MHz bandwidth, 100 mW transmit power, sawtooth modulation type, and a 2.56-ms sweep repetition interval. The system provides a 2-km detection range at location accuracy of 0.5 m in the range and 0.2° in azimuth. It fulfills a 180° scan during 72 s and supports simultaneous targets. When a piece of FOD is detected, a visual and audible alarm is raised and the FOD location is shown on the display. Until the FOD is removed, Tarsier will continue to highlight the FOD location on the runway. The system records all events in a log for future reference, as well as for process analysis and improvement [134].

Another new field for millimeter-wave radar applications is helicopter collision avoidance and piloting radar. Traditional radar instruments cannot be applied as an airborne system for autonomous flight guidance purposes due to lack of resolution. On the other hand, optical sensors such as infrared systems provide excellent resolution but are nearly blind in adverse weather conditions such as fog and rain. A new radar technology called ROSAR (synthetic aperture radar based on rotating antennas) promises to overcome the deficiencies of the traditional radar systems. In 1992 Eurocopter Deutschland and Daimler-Benz Aerospace started a research program to investigate the feasibility of a piloting radar based on ROSAR technology: HELIRADAR [135]. HELIRADAR has been designed to provide a video-like image with a resolution good enough to safely guide a helicopter pilot under poor visibility conditions to the target destination. To yield very high resolution a similar effect as for synthetic aperture radar systems can be achieved by means of a rotating antenna. This principle is especially well suited for helicopters, because it allows for a stationary carrier platform.

A multisensor autonomous approach landing capability (AALC) for air mobility platforms that increases aircrew situational awareness in low- and no-visibility conditions has been developed by BAE Systems [136]. The AALC system fuses millimeter-wave radar and an optical sensor—either infrared or low-light television—processing those inputs to render the best available image on a head-up, head-down, or helmet-mounted display. The system, drawing on this technology in the weather-penetrating capabilities of the 94-GHz imaging radar technology, is designed to permit aircraft landings in zero-ceiling/zero-visibility (0/0) conditions such as fog, dust, smoke, snow, and rain. It provides visual situational awareness of the runway environment, enhances obstacle avoidance, and minimizes pilot spatial disorientation caused by lack of visual perspective. AALC is fully autonomous, placing all sensors aboard the aircraft with no need for ground-based landing aids, significantly improving flexibility in mission planning and execution.

10.8 Conclusion

We have considered millimeter-wave radar principles in this chapter. We saw features and advantages of millimeter-wave radars, as well as some of their important applications. Numerous other possible applications of millimeter-wave radar cannot be considered even briefly in the framework of a chapter such as this. An example is a millimeter-wave tomograph [137], which allows retrieval of both images of slices in a 3D isosurface for objects buried under the plane surface of a medium. The instrument can find application in medicine, biophysics, nondestructive test analysis, and many other areas utilizing a novel detection technology that can provide plan and elevation views of probing spaces. Numerous millimeter-wave radar applications follow from the consideration of millimeter-wave properties, radar principles, and some examples considered brief. Some of the applications are briefly reviewed here:

- *Industrial applications*: Speed and range measurement for industrial uses; industrial depth measurement in hostile environments
- *Meteorological applications*: Severe weather studies and measurement, clear air turbulence detection, and wind field measurements
- *Aviation applications*: Aircraft collision warning and obstacle detection system for helicopters, airport airfield surveillance, runway visualization, and wind shear detection
- *Artificial vision*: Robotic vision, unmanned aerial vehicle, unmanned surface vehicle.
- *Vision and sensing* in adverse weather or environment
- *Harbor monitoring* and navigation guidance
- *Safety and security devices*: Passive imaging for security applications, radio-wave imaging, presence and motion sensors for automated systems, intrusion detection
- *Military applications*: Surveillance, air defense, sniper and artillery location-tracking, missile guidance and tracking, seekers, and finally passive detection of targets in millimeter waves

The tendency of millimeter-wave–band applications to solve various tasks has a steady character now. The opportunity for wide application of millimeter waves in the radar systems of different functions has opened.

This resulted from advances in the developing componentry and creating perfect engineering devices on this basis, as well as from the necessity of quality improvement of radar measurements and the transfer of large information content.

Millimeter-wave radars are employed in a wide range of commercial, military, and scientific applications for remote sensing, safety, and measurements. Millimeter-wave sensors are superior to microwave- and infrared-based sensors in most applications. Millimeter-wave radars offer better range resolution than lower frequency microwave radars and can penetrate fog, smoke, and other obscurants much better than infrared sensors. Millimeter-wave radars are applicable for integrated use and data fusion with other sensors.

Millimeter-wave radar is a universal instrument that can serve as an additional communication channel, for example, in car-to-car communications; it is quite suitable for network application, data transmission, and data exchange. Millimeter-wave radars have good network capabilities and enable remote control and data receiving via any network including the Internet and wireless PAN, LAN, and MAN.

References

[1] N. C. Currie and C. E. Brown, *Principles and Applications of Millimeter-Wave Radar*, Artech House, Inc., Norwood, MA, 1987.

[2] N. C. Currie, R. D. Hayes, and R. N. Trebits, *Millimeter-Wave Radar Clutter*, Artech House, Inc., Norwood, MA, 1992.

[3] G. P. Kulemin, *Millimeter-Wave Radar Targets and Clutter*, Artech House, Norwood, MA, 2003.

[4] E. K. Reedy and J. C. Wiltse, “Fundamentals of millimeter-wave (MMW) radar systems,” in *Aspects of Modern Radar*, by E Brookner, Ed., Artech House, Norwood, MA, 1998.

[5] E. K. Reedy and W. L. Cassaday, “Millimeter radar: Current assessment, future directions,” in *Millimeter and Microwave Engineering for Communications and Radar*, SPIE Press, Bellingham, WA, 1994.

[6] E. K. Reedy and G. W. Ewell, “Millimeter radar,” in *Infrared and Millimeter Waves*, K. Button and J. Wiltse, Eds. Academic Press, New York, 1981.

[7] R. W. McMillan, Terahertz imaging, millimeter-wave radar, NATO Advanced Study Institute: Advances in sensing with security applications, 17–30 July 2005, Il Ciocco, Italy, www.nato-asi.org/sensors2005/papers/mcmillan.pdf

[8] 3rd ESA Workshop on Millimeter Wave Technology and Applications, Espoo, Finland, 21–23 May 2003.

[9] “Physics and engineering of millimeter and submillimeter waves, 1998,” MSMW ’98. Third International Kharkov Symposium, 15–17 Sept. 1998.

[10] “Physics and engineering of millimeter and submillimeter waves, 2001,” MSMW 2001. Fourth International Kharkov Symposium, 4–9 June 2001.

[11] "Physics and engineering of microwaves, millimeter and submillimeter waves, 2004," MSMW 2004. Fifth International Kharkov Symposium, 21–26 June 2004.

[12] "25th International Conference on Infrared and Millimeter Waves," Beijing, China, 12–15 Sep. 2000.

[13] Yu. P. Kalmykov and S. V. Titov, "A semiclassical theory of dielectric relaxation and absorption in polar fluids: Memory function approach to the extended rotational diffusion models. Relaxation phenomena in condensed matter," W. T. Coffey, *in Advances in Chemical Physics*, Ed., Series Editors I. Prigogine and S. A. Rice, Wiley, New York, 1994, vol. 87, pp. 31–122.

[14] G. K. Zagorin, A. Yu. Zrazshevsky, Ye. V. Konkov, A. V. Sokolov, S. V. Titov, G. I. Khokhlov, and L. F. Chernaya, "Factors having an influence to millimeter wave propagation in the surface layer of atmosphere" [in Russian], *Journal of Radio-Electronics*, no. 8, 2001.

[15] H. C. van de Hulst, *Light Scattering by Small Particles*, J. Wiley, New York, 1957.

[16] D. Deirmendjian, *Electromagnetic Scattering on Spherical Polydispersions*, Elsevier, New York, 1969.

[17] F. J. Yanovsky, "Phenomenological models of Doppler-polarimetric microwave remote sensing of clouds and precipitation," *IEEE International Geoscience and Remote Sensing Symposium IGARSS-02*, Toronto, Canada, vol. 3, 2002, pp. 1905–1907.

[18] F. J. Yanovsky, C. M. H. Unal, H. W. J. Russchenberg, and L. P. Ligthart. "Doppler-polarimetric weather radar: Returns from wide spread precipitation," *Proceedings International Workshop on Microwaves, Radar and Remote Sensing MRRS 2005*, September 19–21, Kiev, Ukraine, pp. 139–146.

[19] D. Maggiori, "Computed transmission through rain in the 1–400 GHz frequency range for spherical and elliptical drops and any polarization," *Alta Frequenza*, vol. L.5, 1981, pp. 262–273.

[20] ITU-R Recommendation P.838-1, "Specific attenuation model for rain for use in prediction methods," *Radiocommunication Bureau, ITU*, Geneva, Volume 2000 P Series, 2001.

[21] J. Nemarich, R. J. Wellman, and J. Lacombe, "Backscatter and attenuation by falling snow and rain at 96, 140 and 225 GHz," *IEEE Trans. Geosci. and Remote Sens.*, 26, 1988, pp. 319–329.

[22] H. J. Liebe, "MPM—an Atmospheric millimeter-wave propagation model," *International Journal of Infrared and Millimeter Waves*, vol. 10, 1989, pp. 631–650.

[23] ITU-R Recommendation P.676-4, "Attenuation by atmospheric gases," *Radiocommunication Bureau, ITU*, Geneva, Volume 2000 P Series, 2001.

[24] H. J. Liebe and D. H. Layton, NTIA Report 87-224 National Telecommunications and Information Administration, Boulder, CO, 1987. www.its.bldrdoc.gov/pub/ntia-rpt/87-224/

[25] F. G. Bass and I. M. Fuks, *Wave Scattering from Statistically Rough Surfaces*, Pergamon Press, 1979.

[26] G. A. Andreev, "Millimeter-wave beams in ground-based telecommunication systems," *Journal of Communications Technology and Electronics*, vol. 46, no. 9, 2001, pp. 1022–1031.

[27] V. I. Tatarskii, "The effects of the turbulent atmosphere on wave propagation," Israel Program for Scientific Translations Ltd., Jerusalem, 1971. IPST Cat. No. 5319. (Available from U.S. Dept. of Commerce, UDC 551.510, ISBN 07065 0680 4, NTIS, Springfield, VA, USA.)

[28] A. Ishimaru, *Wave Propagation and Scattering in Random Media*, vol. 2, Academic Press, New York, 1978.

[29] R. J. Doviak and D. S. Zrnich, *Doppler Radar and Weather Observations*, Academic Press, New York, 1993.

[30] *Radar Handbook*, 2nd ed., M. I. Skolnik, Ed., McGraw-Hill, New York, 1990.

[31] N. Yamada, "Radar cross section for pedestrian in 76GHz band," *R&D Review of Toyota CRDL*, vol. 39 no. 4, pp. 46–51.

[32] C. E. Cook and M. Bernfeld, "Radar signals, in *An Introduction to Theory and Application*, Academic Press, New York, 1967.

[33] Ya. D. Shirman, *Computer Simulation of Aerial Target Radar Scattering, Recognition, Detection, and Tracking*, Artech House, Norwood, MA, 2002.

[34] Ya. D. Shirman and V. N. Manzhos, "Theory and technique of processing radar information in noise background" [in Russian], Radio i Svyaz', Moscow, 1981.

[35] B. R. Mahafza, *Radar Systems Analysis and Design Using Matlab*, CRC Press, Boca Raton, FL, 2000.

[36] D. K. Barton, *Radar System Analysis and Modeling*, Artech House, Norwood, MA, 2005.

[37] W. M. Boerner, "Polarization dependence in electromagnetic inverse problems," *IEEE Trans. AP*-29, March 1981, pp. 262–274.

[38] F. Le Chevalier, *Principles of Radar and Sonar Signal Processing*, Artech House, Norwood, MA, 2002.

[39] F. J. Yanovsky, R. B. Sinitsyn, and I. M. Braun, "Reliability of detection of radar signals from hailstones by using parametric and non-parametric algorithms," *Proc. 2nd European Radar Conference, EuMA, IEEE*, 2005, Paris, pp. 121–124.

[40] L. P. Ligthart, F. J. Yanovsky, and I. G. Prokopenko, "Adaptive algorithms for radar detection of turbulent zones in clouds and precipitation," *IEEE Transactions on Aerospace and Electronic Systems*, vol. 39, no. 1, Jan. 2003, pp. 357–367.

[41] *Introduction to Ultra-Wide Band Radar Systems*, J. D. Taylor, Ed., CRC Press, Boca Raton, FL, 1995.

[42] K. A. Lukin, "Millimeter wave noise radar applications: Theory and experiment," *Proc. 4th International Kharkov Symposium Physics and Engineering of Millimeter and Sub-Millimeter Waves*, Kharkov, Ukraine, 2001, pp. 68–73.

[43] X. Xu and R. M. Narayanan, "FOPEN SAR Imaging UWB Step-frequency and Random Noise Waveform," *IEEE Transactions on Aerospace and Electronic Systems*, vol. 37, no. 4, Oct. 2001, pp. 1287–1300.

[44] "2001 Annual Report of the Advanced Sensors Collaborative Technology Alliance." www.arl.army.mil/main/ResearchOpportunities/alliances/supporting_files/annual_reports/2001/as01ar.doc
[45] A. Y. Nashashibi and F. T. Ulaby, "Millimeter-wave bistatic radar: Clutter characterization and SAR imaging," *The IASTED Conference on Antennas, Radar, and Wave Propagation*, Banff, Alberta, Canada, 3–5, July 2006. www.astapress.com.Content_Of_Proceeding.aspx?ProceedingID=385
[46] M. Kayton and W. R. Fried, *Avionics Navigation Systems*, John Wiley & Sons, New York, 1997.
[47] R. McCluney, *Introduction to Radiometry and Photometry*, Artech House, Norwood, MA, 1994.
[48] "Microwave and millimeter wave radiometry." www.millitech.com/pdfs/Radiometer.pdf
[49] V. E. Lyubchenko, Ed., *Physics and Technology of Millimetre Wave Devices and Components*, Taylor & Francis, London, 2002.
[50] A. I. Nosich, Y. M. Poplavko, D. M. Vavriv, and F. J. Yanovsky, "Microwaves in Ukraine," *IEEE Microwave Magazine*, Dec. 2002, pp. 82–90.
[51] D. M. Vavriv, "Ukrainian Institute of Radio Astronomy High-Resolution Radars," *IEEE Aerospace and Electronic Systems Magazine*, vol. 20, no. 10, Oct. 2005, pp. 19–24.
[52] A. A. Tolkachev, B. A. Levitan, G. K. Solovjev, V. V. Veytsel, and V. E. Farber, "A megawatt power millimeter-wave phased-array radar," *IEEE AES Systems Magazine*, pp. 25–31, July 2000.
[53] Communications & Power Industries.
[54] M. E. Russell, A. Crain, A. Curran, R. A. Campbell, C. A. Drubin, and W. F. Miccioli, "Millimeter-wave radar sensor for automotive intelligent cruise control (ICC)," *IEEE Transactions on Microwave Theory and Techniques*, vol. 45, no. 12, Dec. 1997, pp. 2444–2453.
[55] G. M. Brooker, S. Scheding, M. V. Bishop, and R. C. Hennessy, "Development and application of millimeter wave radar sensors for underground mining," *IEEE Sensors Journal*, vol. 5, no. 6, Dec. 2005, pp. 1270–1280.
[56] M. Schlechtweg and A. Tessmann, "COBRA 94 ultra broadband experimental radar for ISAR applications," *International Conference on Signal Processing Applications and Technology (ICSPAT 2000) Proceedings*, pp. 1515–1519. www.icspat.com/papers/630mfi.pdf
[57] H. Essen, A. Wahlen, R. Sommer, G. Konrad, M. Schlechtweg, and A. Tessmann, "A very high bandwidth millimetre wave radar," *Electronic Letters*, vol. 41, no. 22, 2005, pp. 1247–1248.
[58] L. Samoska, E. Bryerton, M. Morgan, D. Thacker, K. Saini, T. Boyd, D. Pukala, A. Peralta, M. Hu, and A. Schmitz, "Medium power amplifiers covering 90–130 GHz for the ALMA telescope local oscillators," *IEEE MTT-S International Microwave Symposium Digest*, vol. 3, 2005, pp. 1583–1586.
[59] HiTeC colloquium by Mark Rodwell, UCA, Santa Barbara on "30–700 GHz transistor and diode integrated circuits," 12 Sept. 2005, DIMES, TU-Delft. http://www.dimes.tudelft.nl/live/pagina.jsp?id=677bb180-c90b-4b89-96a5-bdfcc552805d&lang=en

[60] L. V. Kasatkin and V. E. Chayka, *Semiconductor Devices of Millimeter Range* [in Russian], Veber Publishing House, Sevastopol, Ukraine, 2006, 319 pp.

[61] V. V. Kulish, O. V. Lysenko, V. I. Savchenko, and I. G. Majornikov, "Source of femto-second wave packages on the basis of two-stream free electron lasers," *Proceedings International Workshop on Microwaves, Radar and Remote Sensing MRRS 2005*, Sept. 19–21, Kiev, Ukraine, pp. 304–309.

[62] T. Sehm, A. Lento, and A. V. Raisanen, "Planar 64 element millimetre wave antenna," *Electron. Lett.*, vol. 35, no. 4, 1999, pp. 253–255.

[63] M. Ando and J. Hirokawa, "High gain and high efficiency planar antennas for various wireless systems in millimeter wave bands," *XXVIIth General Assembly of the International Union of Radio Science* (URSI), CAF.O.2 (Oral), pp. 716–719 (Maastricht, the Netherlands, 17–24 Aug. 2002).

[64] A. Roederer and J. Mosig, "COST: Co-Operation Scientifique & Technique on antennas—Results and perspectives, *"International Symposium on Antennas*, 12–14 Nov. 2002, Nice, France.

[65] Z. Popovic and A. Mortazawi, "Quasi-optical transmit/receive front ends," *IEEE Transactions on Microwave Theory and Techniques*, vol. 46, no. 11, Nov. 1998, pp. 1964–1975.

[66] K. Solbach and R. Schneider, "Review of antenna technology for millimeter wave automotive sensors." http://hft.uni-duisburg-essen.de/forschung/paper/MF-TuB3_Neue_Version.pdf

[67] R. W. McMillan, C. W. Trussell, Jr., R. A. Bohlander, J. C. Butterworth, and R. E. Forsythe. "An experimental 225 GHz pulsed coherent radar," *IEEE Transactions on Microwave Theory and Techniques*, vol. 39, no. 3, March 1991, pp. 555–562.

[68] P. Russer, "Si and SiGe millimeter-wave integrated circuits," *IEEE Transactions on Microwave Theory and Techniques*, vol. 46, no. 5, May 1998, pp. 590–603.

[69] H. Mizutani, M. Funabashi, M. Kuzuhara, and Y. Takayama, "Compact DC-60-GHz HJFET MMIC switches using ohmic electrode-sharing technology," *IEEE Transactions on Microwave Theory and Techniques*, vol. 46, no. 11, pt. 1, Nov. 1998, pp. 1597–1603.

[70] G. M. Rebeiz, *RF MEMS Theory, Design and Technology*, John Wiley, New York, 2003.

[71] Federal Intelligent Transportation Systems (ITS) program. www.itsoverview.its.dot.gov/

[72] ERTICO—Intelligent Transport Systems Europe. www.ertico.com

[73] N. A. Stanton and M. Pinto, "Will radar-based vision enhancement make driving safer? An experimental study of a hypothetical system on a driving simulator," *Proceedings of the I MECH E Part D Journal of Automobile Engineering*, vol. 215, no. 9, 20 Sept. 2001, pp. 959–967.

[74] C. A. MacCarley, "Advanced imaging techniques for traffic surveillance and hazard detection," *Intellimotion*, vol. 6, no. 2, 1997, pp. 6–8.

[75] R. Mende and H. Rohling, "New automotive applications for smart radar systems," http://www.smartmicro.de/New_Automotive_Applications_for_Smart_Radar_Systems_V5.pdf

[76] "A review of automotive radar systems devices and regulatory frameworks," Document: SP 4/01, April 2001, Australian Communications Authority, 9 pp. http://www.acma.gov.au/acmainterwr/radcomm/frequency_planning/spps/0104spp.pdf

[77] R. H. Rasshofer and K. Gresser, "Automotive radar and lidar systems for next generation driver assistance functions," *Advances in Radio Science*, vol. 3, 2005, pp. 205–209.

[78] K. M. Strohm, H.-L. Bloecher, R. Schneider, and J. Wenger, "Development of future short range radar technology," *European Microwave Week 2005, Conference Proceedings*, 3–7 Oct. 2005, Paris, pp. 165–168.

[79] E. G. Hoare, "Automotive millimetre-wave radar. Current applications and future developments." www.iee.org/Events/Hoare.pdf

[80] Intelligent Transportation Systems—Traffic Surveillance. www.calccit.org/itsdecision/serv_and_tech/Traffic_Surveillance/survoverview.htm

[81] "A summary of vehicle detection and surveillance technologies used in intelligent transportation systems," Prepared by L. E. Y. Mimbela and L. A. Klein, 30 Nov. 2000. www.nmsu.edu/ traffic/

[82] S. Nishikawa and H. Endo, "Applications of millimeter-wave sensors in ITS," *Furukawa Review*, no. 18, 1999, pp. 1–5. www.furukawa.co.jp/review/fr018/fr18_01.pdf

[83] R. Bishop, *Intelligent Vehicle Technology and Trends*, Artech House, Norwood, MA, 2005.

[84] Smart Microwave Sensors GmbH. www.smartmicro.de/home.html

[85] R. Mende, M. Behrens, and S. Milch, "A 24 GHz ACC radar sensor," Proc. International Radar Symposium IRS 2005, Berlin, 2005, pp. 91–95.

[86] J. Takezaki, N. Ueki, T. Minowa, and H. Kondoh, "Support system for safe driving. A step toward ITS autonomous driving," *Hitachi Review*, vol. 49, no. 3, 2000, pp. 107–114.

[87] R. Schneider and J. Wenger, "High resolution radar for automobile applications," *Advances in Radio Science*, vol. 1, 2003, pp. 105–111.

[88] I. Gresham, N. Jain, T. Budka, A. Alexanian, N. Kinayman, B. Ziegner, S. Brown, and P. Staecker, "A 76–77GHz pulsed-doppler radar module for autonomous cruise control applications," 2000, IEEE. www.anokiwave.com/pdffiles/radarpaper.pdf

[89] R. Mende, "The UMRR 24GHz radar sensor family for short and medium range applications," 8 April 2004.

[90] H. Rohling, A. Höβ, U. Lübbert, and M. Schiementz, "Multistatic radar principles for automotive RadarNet applications," *IRS 2002 International Radar Symposium*, Bonn, Germany, 2002, pp. 405–410.

[91] H. Rohling, and F. Fölster, "Radar network based on 77GHz FMCW sensors," *Proceedings 2nd International Workshop on Intelligent Transportation (WIT 2005)*, Hamburg, Germany, 2005, pp. 113–118.

[92] H. Rohling, "FMCW sensors for an automotive radar network," *Proceedings International Workshop on Microwaves, Radar and Remote Sensing MRRS 2005*, Sept. 19–21, Kiev, Ukraine, pp. 24–32.

[93] B. Rickett and R. Manor, "A vision of future applications for an automotive radar network," *WIT 2004 International Workshop on Intelligent*

Transportation, Hamburg, Germany, 2004. www.radarnet.org/publications/zip/wit_paper_2004.pdf

[94] University of Michigan Transportation Research Institute, "Road departure crash warning field operational test system architecture," May 2004. www.umtri.umich.edu/erd/RDCWWebsitesummary.ppt

[95] "The OSU autonomous vehicle." www.ece.osu.edu/citr/Demo97/osu-av.html

[96] Valeo Raytheon, "Valeo Raytheon's lane change assistance system." www.valeoraytheon.com

[97] F. Kruse, F. Fölster, M. Ahrholdt, M.-M. Meinecke, and H. Rohling, "Object classification with automotive radar," *Proc. International Radar Symposium IRS 2003*, Dresden, Germany, 2003, pp. 137–142.

[98] V. Winkler, J. Detlefsen, U. Siart, K. Böhm, and M. Wagner, "Automotive radar sensor extended by a communication capability," *Proc. International Radar Symposium IRS 2005*, Berlin, 2005, pp. 607–612.

[99] A. Zlocki and N. Zambou, "Application of WLAN vehicle-to-vehicle communication for automatic guidance of a vehicle driven in platoon," *Proceedings 2nd International Workshop on Intelligent Transportation (WIT 2005)*, Hamburg, Germany, 2005, pp. 125–129.

[100] G. Kramer, "Envisioning a radar-based automatic road transportation system," *IEEE Intelligent Systems*, May–June 2001, pp. 75–77 (also in *IEEE Intelligent Transportation Systems Council Newsletter*, vol. 3, no. 3, July 2001, pp. 3–6).

[101] AutoDrive System. www.eventhelix.com/ThoughtProjects/AutoDrive/

[102] O. O. Bezvesilniy and D. M. Vavriv, "Novel possibilities of Ka- and W-band dual-wavelength Doppler measurements for retrieving cloud and rain parameters," *Proceedings of the International Workshop on Microwaves, Radar and Remote Sensing MRRS-2005*, Kiev, Ukraine, 2005, pp. 71–75.

[103] R. J. Doviak and D. S. Zrnich, *Doppler Radar and Weather Observations*, Academic Press, New York, 1984. (2nd ed., 1993).

[104] J. B. Mead, A. L. Pazmany, S. M. Sekelsky, and R. E. McIntosh, "Millimeter-wave radars for remotely sensing clouds and precipitation," *Proceedings of the IEEE*, vol. 82, no. 12, Dec. 1994, pp. 1891–1906.

[105] F. J. Yanovsky, H. W. J. Russchenberg, and C. M. H. Unal, "Retrieval of information about turbulence in rain by using doppler-polarimetric radar," *IEEE Transactions on Microwave Theory and Techniques*, vol. 53, no. 2, 2005, pp. 444–450.

[106] H. B. Bluestein and A. L. Pazmany, "Observations of Tornadoes and other convective phenomena with a mobile, 3-mm wavelength, Doppler radar: The spring 1999 field experiment," *Bulletin of the American Meteorological Society*, vol. 81, no. 12, 2000, pp. 2939–2951.

[107] C. Weiss, H. B. Bluestein, A. L. Pazmany, and B. Geerts, "Fine-scale radar observations of a dryline during the International H2O Project (IHOP)," *The 84th AMS Annual Meeting*, Seattle, WA, 2004. Submitted to the *Sanders Symposium Monograph*, 25 Feb. 2005, 53 pp. http://ams.confex.com/ams/84Annual/techprogram/program_194.htm

[108] J. B. Mead, I. PopStefanija, P. Kollias, B. Albrecht, and R. Bluth, "Compact airborne solid-state 95 GHz FMCW radar system." http://artemis.rsmas.miami.edu/pubs/ppt/ams03_cfmcw_extended_abstract.pdf

[109] R. P. Bambha, J. R. Carswell, J. B. Mead, and R. E. McIntosh, "A compact millimeter wave radar for airborne studies of clouds and precipitation," *IEEE IGARSS Symposium*, 1998. http://abyss.ecs.umass.edu/uav-web/igarss98.pdf

[110] Wyoming Cloud Radar (also data communicated privately by Bart Geerts). http://www-das.uwyo.edu/wcr/

[111] NASA: The Science Mission Directorate. http://science.hq.nasa.gov/missions/satellite_24.htm

[112] S. M. Sekelsky and J. Carswell, "High power electronic scanning millimeter-wave radar system design," *IEEE Aerospace Conference*, 4–11 March 2006, pp. 1–6.

[113] 4D-Clouds Project, http://www.meteo.uni-bonn.de/Deutsch/4d-clouds/tools/miccyrad/

[114] CloudNet Project. http://cree.rdg.ac.uk/radar/cloudnet/

[115] CloudNet Publications. http://www.cloud-net.org/publications/publications.html

[116] D. S. Zrnic and A. V. Ryzhkov, "Recognition of nonmeteorological echoes with a dual-polarization radar," *Preprints 28th International Conference on Radar Meteorology*, 7–12 Sept. 1997, Austin, Texas, Amer. Meteor. Soc., pp. 5–6.

[117] A. Khandwalla, S. M. Sekelsky, and M. Quante, "Algorithms for filtering insect echoes from cloud radar measurements," *13th ARM Science Team Meeting Proceedings*, Broomfield, CO, March 31–April 4, 2003. www.arm.gov/publications/proceedings/conf13/extended_abs/khandwalla-a.pdf

[118] D. L. Glackin, "Earth remote sensing: An overview," http://www.aero.org/publications/crosslink/summer2004/01.html

[119] H. Schimpf, H. Essen, S. Boehmsdorff, and T. Brehm, "MEMPHIS—a fully polarimetric experimental radar," *Proc. IEEE IGARSS 2002*, vol. III, Toronto, 2002, pp. 1714–1716.

[120] M. Hägelen, H. Essen, T. Brehm, M. Ruegg, and E. Meier, "Traffic monitoring with millimeterwave monopulse SAR." Unpublished paper. Corresponding author: Dr. Helmut Essen, FGAN-Research Institute for High Frequency Physics and Radar.

[121] H. M. Chen, S. Lee, R. M. Rao, M. A. Slamani, and P. K. Varshney, "Imaging for concealed weapon detection," *IEEE Signal Processing Magazine*, March 2005, pp. 52–61.

[122] A. G. Denisov, V. N. Radzikhovsky, V. P. Gorishniak, S. E. Kuzmin, and B. M. Shevchuk, "Radiometric imaging system for concealed weapon detection," *12th International Crimean Conference Microwave & Telecommunication Technology*, Sept. 9–13, 2002, Sevastopol, Ukraine, pp. 106–107.

[123] A. Pergande and L. Anderson, "Video rate millimeter-wave camera for concealed weapons detection," *Proc. SPIE*, vol. 4373, 2001, pp. 35–39.

[124] C. A. Martin, S. E. Clark, J. A. Lovberg, and J. A. Galliano, "Real-time wide-field-of-view passive millimeter-wave imaging," *Proc. SPIE*, vol. 4719, 2002, pp. 341–349.

[125] H. Essen, H.-H. Fuchs, D. Nötel, F. Klöppel, P. Pergande, and S. Stanko, "Passive millimeter-wave imaging at short and medium range," Technologies for Optical Countermeasures II; Femtosecond Phenomena II and Passive Millimetre-Wave and Terahertz Imaging II, Proc. SPIE, vol. 5989, 2005, pp. 347–353.

[126] N. C. Currie, F. J. Demma, D. D. Ferris Jr., B. R. Kwasowsky, R. W. McMillan, and M. C. Wicks, "Infrared and millimeter-wave sensors for military special operations and law enforcement applications," *International Journal of Infrared and Millimeter Waves*, Springer, Netherlands, vol.17, no. 7, July 1996, pp. 1117–1138.

[127] Y.-W. Chang and M. Johnson, "Portable concealed weapon detection using millimeter wave FMCW radar imaging," Final report 189918 / Dr. NCJRS, 2001. www.ncjrs.gov/pdffiles1/nij/grants/189918.pdf

[128] A. Soule, "Norden veterans back on the radar with explosives detector," *Fairfield County Business Journal.* http://www.fairfieldcbj.com/archive/080706/0807060024.php

[129] The University of Texas at Arlington, IRIS—Concealed weapon detection. http://www.iris.uta.edu/projects2.html

[130] R. S. Blum, "Image fusion," Lehigh University. www.lehigh.edu/optics/Documents/2004OpenHouse/2004ARLWorkshop/13-5-blumImageFusion.pdf

[131] U.S. Department of Justice, Office of Justice Programs, National Institute of Justice. http://www.ojp.usdoj.gov/nij

[132] V. Gorishnyak, A. Denisov, S. Kuzmin, V. Radzikhovsky, and B. Shevchuk, "8 mm passive imaging system with 32 sensors," *European Radar Conference*, Amsterdam, 2004, pp. 333–336.

[133] QinetiQ Tarsier T1100 airport debris detection. www.QinetiQ.com/Tarsier

[134] P. D. L. Beasley, G. Binns, R. D. Hodges, and R. J. Badley, "Tarsier: A millimetre wave radar for airport runway debris detection," *1st European Radar Conference, Amsterdam*, Oct. 2004, pp. 261–264.

[135] W. Kreitmair-Steck, A. P. Wolframm, and A. Schuster, "Heliradar: The pilot's eye for flights in adverse weather conditions," Proc. SPIE, vol. 2736, Enhanced and Synthetic Vision 1996, J. G. Verly, Ed., May 1996, pp. 35–41.

[136] Bae Systems. "Autonomous approach landing capability improves situational awareness for air mobility platforms. http://www.na.baessystems.com/releasesDetail.cfm?a=316

[137] A. A. Vertiy and S. P. Gavrilov, "Millimeter wave tomograph for imaging of subsurface structures," *Proc. Mediterranean Microwave Symposium (MMS'2006)*, Genoa, Italy, 2006, pp. 416–418.

Chapter 11

Optical Generation and Transmission of Millimeter-Wave Signals

Ming-Tuo Zhou, Michael Sauer, Andrey Kobyakov, and John Mitchell

Contents

11.1 Introduction

Millimeter waves have found potential application in the fields of communications, radar, radiometry, spectroscopy, and radio astronomy, etc. Electronic generation of millimeter waves using oscillator and frequency multiplexers has been well investigated; however, such millimeter-wave sources are usually bulky and heavy. In addition, due to the relatively large air propagation loss, the free space transmission distance of millimeter-wave signals is generally relatively short. These facts limit the use of millimeter waves in many modern systems.

Optical generation and transmission of millimeter-wave signals are capable of alleviating these limitations effectively. A photodiode is by nature a square-law device; via photodetection, a radio-frequency (RF) signal from microwave frequency to terahertz can be generated by "photo-mixing" optical signals with the desired frequency difference. In order to generate pure-spectrum millimeter-wave signals, the beating optical waves are required to be coherent. As optical harmonic spectral components generated in optical modulations are naturally coherent in phase, harmonic generations are generally employed to generate pure millimeter waves. If individual lasers are used, phase-locking techniques must be used.

Optical fiber has long been recognized to be a suitable medium for transmission of millimeter waves. The very low loss of optical fiber (< 0.2 dB/km typical for single-mode fiber, < 2 dB/km for multimode fiber) allows for transmission of high-frequency signals over relatively long distances. An intrinsic (so called) radio-over-fiber (RoF) link is transparent to RF signal modulation formats and protocols; hence it enables antenna remoting and allows central operation of RF signals. In addition to single optical wavelength transmission, dense-wavelength-division multiplexing (DWDM) can be employed; hence, the capacity of a single fiber can be greatly increased and the optical networks for distributing millimeter-wave signals can be simplified in structure. However, effects other than fiber loss can affect signal transmission at higher frequencies, especially in the millimeter-wave range. These effects largely come from dispersion and fiber nonlinearity. Effects to be considered are chromatic dispersion (CD), polarization-mode dispersion (PMD), and modal dispersion leading to differential mode delay (DMD) in multimode fibers (MMFs). Nonlinear effects can usually be neglected for relatively short distance (a few kilometers) transmission due to typically low signal power, but it has been shown that especially self-phase modulation (SPM) and stimulated Brillouin scattering (SBS) can interact with the signal and may need to be considered.

The next section of this chapter first summarizes the roadmap of optical generation of millimeter waves and then introduces two representative generation technologies, one based on the Mach-Zehnder modulator (MZM) and the other using injection-phase locking. The third section first characterizes RoF links with respect to concepts, advantages, RF properties, and dispersion limitations and then introduces some enabling technologies such as external modulation, single-sideband modulation, DWDM techniques, and cost-effective link design methods. The fourth section presents properties and issues of multimode fiber RoF links.

11.2 Optical Generation of Millimeter-Wave Signals

For the optical generation and transmission of millimeter-wave signals, heterodyne techniques (mixing at the photodiode of two optical signals separated by the required frequency) have a number of significant advantages. The use of only two optical tones provides tolerance to the dispersion-induced fading that would occur with a directly modulated (three-tone) signal and has high efficiency because all the optical power contributes to the resultant RF tone. In this section we will examine a selection of the techniques available.

The available techniques can be divided into two broad classes:

- Single-stage techniques, where the generation of the millimeter wave and the modulation imposition are performed in the same stage
- Dual-stage techniques, where the generation of the millimeter wave and the modulation imposition are performed in two separate, cascaded stages

The relationship between these broad classes is illustrated in the roadmap of generation methods shown in Figure 11.1. Of these a subclass are the three-term techniques corresponding to the conventional intensity modulation of the optical signal at the millimeter-wave frequency. This generates an optical carrier centered between two modulation sidebands. Three-term approaches are based on the conventional intensity modulation of an optical signal using an optical intensity modulator such as an MZM or an electroabsorption modulator (EAM) [1] that can operate at the millimeter-wave frequencies. However, as the optical signal propagates along optical fiber, chromatic dispersion causes a relative phase change between the three components, leading to a cyclic variation of the generated power with fiber distance or frequency. At millimeter-wave frequencies, this effect limits the usefulness of three-term techniques (without dispersion compensation)

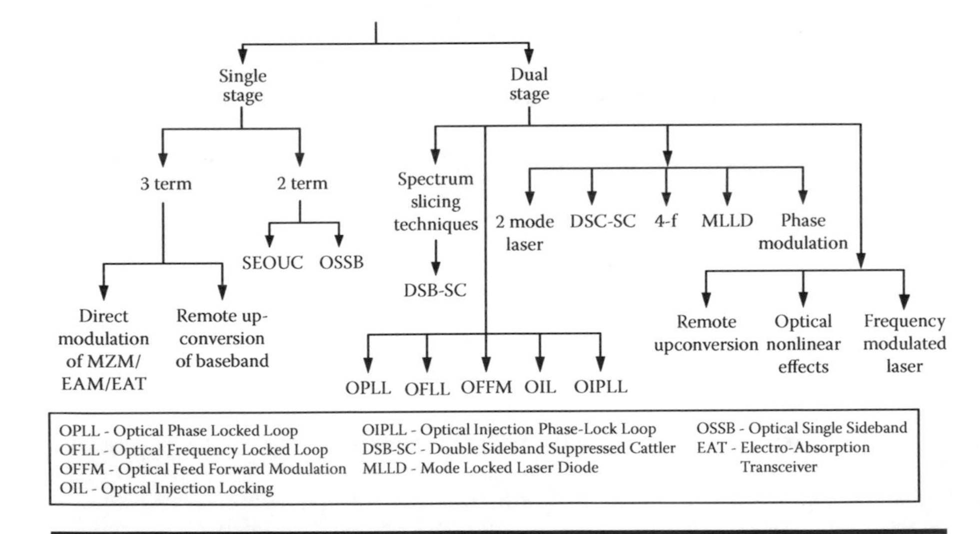

Figure 11.1 Roadmap of millimeter-wave generation and modulation imposition techniques.

to fiber reaches of only a few kilometers, due to link-length–dependent nulls in the detected signal power. The first null occurs at a fiber length of Z_1 meters, given by

$$Z_1 = \frac{c}{2D\lambda^2 {f_m}^2}, \tag{11.1}$$

where c is light speed in a vacuum, D is the fiber dispersion parameter, λ is the optical wavelength, and f_m is the millimeter-wave frequency.

In contrast, in two-term techniques, two optical components mix (heterodyne) in the photodetector, where, due to the E-field nonlinearity of the photodetector, an electrical signal at a frequency equal to the frequency separation of the two optical components is generated. Now the phase shift between the two terms will cause a constant phase offset only in the received signal, which in most cases is inconsequential. This will result in a received RF tone with no length-dependent fading and allows considerably longer fiber lengths to be used.

In order to generate a narrow-linewidth RF, the phase noise of the two optical terms must be correlated. Various methods have been proposed for obtaining correlated terms, such as obtaining the two terms from a single optical source [2,3], canceling the phase noise [4], or implementing phase-tracking feedback loops [5,6]. It is also possible to use a combination of optical phase modulation and filtering [7,8] or fiber-laser techniques [9]. Two specific techniques will now be discussed: the first uses an MZM to generate optical tones with a suppressed optical carrier, and the second locks two optical sources using a wideband optical injection phase lock loop.

11.2.1 Mach-Zehnder Modulator-Based Techniques

The basic operation of an MZM relies on a linear electro-optic effect. An applied voltage changes the refractive index of the electro-optic material, producing a phase shift for the optical signal propagating in the material. An integrated optical MZM structure transforms the induced optical phase shift to a change in intensity, with the device exhibiting a raised cosine intensity-voltage characteristic. With a dc bias of half the switching voltage, $(V_\pi/2)$, an approximate linear intensity-voltage response is achieved for small modulation indices. The intrinsic bandwidth of the electro-optic effect in $LiNbO_3$ is very high; the practical difficulty for the construction of millimeter-wave modulators is to achieve velocity matching between the propagating optical and electrical waves over the electrode interaction length. Modulation frequencies up to 75 GHz have been demonstrated [10], but the problem of velocity matching results in high optical insertion loss, high drive power requirements, and unit cost increasing with the device bandwidth. Operating the laser in continuous mode decreases the relative intensity noise (RIN) [11]; therefore, the dominant noise in this configuration is shot noise. However, MZM nonlinearity leads to intermodulation products and any induced chirp will cause a parasitic phase modulation.

A number of techniques for optical generation of millimeter waves based on the use of an MZM have been proposed. For example, a single-stage two-term technique, generating an optical single-sideband plus carrier (OSSB+C), has been demonstrated [12]. The modulated millimeter-wave signal is applied to an arm of the dual-electrode MZM, while a replica of the same signal is delayed by $\pi/2$ and applied to the other arm. This suppresses one of the optical sidebands, leaving just the optical carrier and one sideband. Although this method requires the dual-electrode MZM to be driven at the millimeter-wave frequency, the conversion efficiency and spectral purity demonstrated is good. A main advantage of the OSSB technique is that the millimeter-wave generation and the modulation imposition are carried out in a single stage, requiring a single MZM with dual arms, although the linearity of the MZM has been shown to be a limiting factor on the performance of the link [13] and an exact 90° phase shift is difficult to achieve with broadband modulation on the millimeter-wave carrier.

As proposed by O'Reilly et al. [14], double-sideband suppressed carrier (DSB-SC) signal generation makes use of a single laser source modulated by an MZM biased at the point of minimum transmission (V_π). If we consider a balanced MZM driven with a sinusoidal input, the following E-field response is obtained:

$$E_{out}(t) = \cos\left\{\frac{\pi}{2}[(1+\zeta) + \mu\cos(\omega_m t)]\right\} e^{j(\omega_o t)} \tag{11.2}$$

where ζ and μ are the normalized bias point (assuming $\zeta = 0$ is the minimum transmission point) and normalized drive voltage amplitude, V_{rf}/V_π, respectively, and ω_o is the optical carrier frequency. The first two terms of the Bessel function expansion of this expression are

$$\begin{aligned} E_{out}(t) = \frac{1}{2}J_0\left(\mu\frac{\pi}{2}\right)\cos\left[\frac{\pi}{2}(1+\zeta)\right] e^{j(\omega_o t)} \\ - J_1\left(\mu\frac{\pi}{2}\right)\sin\left[\frac{\pi}{2}(1+\zeta)\right] e^{j(\omega_o t \pm \omega_m t)} \\ + J_2\left(\mu\frac{\pi}{2}\right)\cos\left[\frac{\pi}{2}(1+\zeta)\right] e^{j(\omega_o t \pm 2\omega_m t)}. \end{aligned} \tag{11.3}$$

Equation (11.3) clearly shows a center optical term with two components separated by $2\omega_m$, which can be up to twice the bandwidth of the MZM. We see that if the modulator is biased at V_π ($\zeta \approx 0$), then the center term will be suppressed. In practice the maximum suppression achievable will

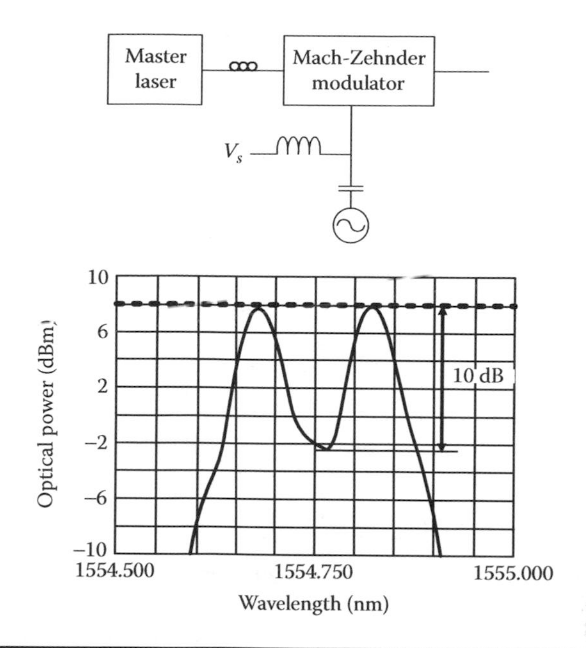

Figure 11.2 DSB-SC generation technique, showing the resultant two-term optical spectrum.

depend on the inherent extinction ratio of the device and the accuracy of the polarization alignment. The technique and resulting optical spectrum is illustrated in Figure 11.2. Driving the modulator with a sinusoid around this point generates the DSB-SC signal consisting of two optical components separated by twice the frequency of the drive signal. Since the two terms are harmonics of the same optical source, their phase noise is correlated, and as long as this correlation is maintained, the electrical signal generated by the heterodyning of these two components will have a narrow linewidth. An advantage of this method is that the MZM is driven at a frequency that is half that of the required millimeter-wave frequency, allowing the use of a lower bandwidth device; however, this results in a reduction of the conversion of MZM drive power to generated millimeter-wave power. The technique has been demonstrated in [14,15] and achieves millimeter-wave signals with high spectral purity.

The above technique can be modified to generate tones separated by four times the drive frequency [16]. The MZM is biased at the point of maximum transmission and driven by an RF signal with a frequency at a quarter of the required millimeter-wave frequency. The resulting optical spectrum consists of two components separated by four times the frequency of the drive source driving the MZM; however, this enhancement of the even-order harmonic terms does not lead to suppression of the optical carrier. To increase the conversion efficiency, this carrier term may be suppressed by an optical filter. Although this method allows the use of lower bandwidth devices, it exhibits a poor conversion of the electrical drive power to millimeter-wave power.

One advantage of these techniques is that it is possible to simultaneously up-convert a number of wavelengths using a single device, thus reducing the component count to a single high-frequency modulator. In this approach, baseband radio subcarriers directly modulate DWDM optical sources at an intermediate frequency. The modulated optical signals are then multiplexed and the composite signal is up-converted to millimeter-wave frequency using an MZM to perform DSB-SC modulation [17,18]. However, it must be noted that because the bias point of the MZM is wavelength dependent, the wavelength range is limited to reduce strong carrier regrowth and harmonic terms in the outer optical carriers.

11.2.2 Optical Locking Techniques

Distributed-feedback (DFB) semiconductor lasers are single mode and widely tunable, and it is reasonably straightforward to control the optical difference frequency between two lasers [19,20]. A problem with this approach is laser phase noise; since the two optical components separated by the millimeter-wave frequency originate in different lasers, the linewidth of the generated millimeter-wave electrical signal is equal to the sum of the individual laser linewidths. This is due to the phases of the two lasers not being correlated and the phase of each laser following a Wiener-Lévy random walk. However, techniques to reduce laser phase noise generally reduce either the robustness or the tunability compared to a monolithic laser.

A technique used to improve the electrical signal purity of the millimeter-wave signal is the optical phase locked loop (OPLL) [21,22]. This approach does not reduce the laser phase noise but employs a portion of the detected beat signal, which is mixed with a local oscillator, filtered, and fed back to a slave laser to track phase fluctuations of the master laser. Since the phase fluctuations are correlated, the detected beat signal exhibits good spectral purity. The fundamental problem with its implementation is that the loop-gain bandwidth, and hence the phase noise reduction, is limited by the practically achievable loop delay.

As an improvement, optical injection locking can be used [23]. In this technique, multiple sidebands are first induced in a seed laser, by either direct or external modulation. This signal is then fed to two frequency-controlled lasers whose free running optical frequencies are matched to sidebands separated by the desired frequency. With sufficient seed power, the phase noise of the two slave lasers then becomes correlated. Due to the correlation of the slave lasers locked to the sidebands of the master laser, a beat signal with very narrow linewidth is produced [24], but there are major stability problems that result in a small locking range.

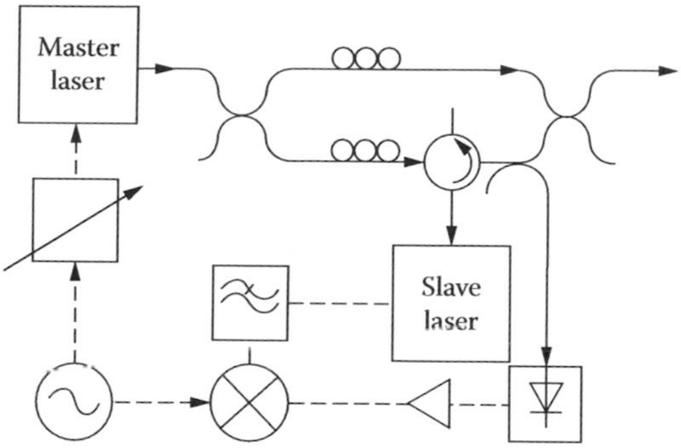

Figure 11.3 Schematic of the fiber-based optical injection phase lock loop system.

A combination of these two techniques, termed an optical injection phase-lock loop (OIPLL), has been demonstrated that not only achieves a millimeter-wave signal of high spectral purity but also has increased stability due to its wider locking range and its easy construction with fiber devices [6]. The system arrangement, shown in Figure 11.3, uses a master laser modulated at a third of the required millimeter-wave frequency, a portion of which is injected into a slave laser tuned to lock to the third harmonic. The phase-lock loop is formed by receiving the beating term between the master laser reflected from the slave laser facet and the slave laser term. It is the use of this reflected term that produces an inherently path-matched response. Using a subharmonically pumped double-balanced mixer, an error signal is produced by comparison with the reference signal that is used to drive the master laser. Experimental results have shown a 36-GHz carrier with a phase error variance of <0.005 rad^2 (in 100-MHz noise bandwidth) with a locking range >30 GHz [6].

11.3 Optical Fiber Transmission of Millimeter-Wave Signals

Optical fiber with ultra-low loss is the ideal medium for long-distance millimeter-wave signal transmission. However, the effect of fiber CD and PMD may cause signal degradation by introducing signal fading. In order to overcome the CD effect, optical single-sideband modulation can be used. An RoF link is transparent to RF signals, and RoF links can be widely used in many modern wireless communication systems. In addition to single-wavelength transmission, DWDM can greatly increase the system capacity. With different network topology — star, bus, and ring — different multiplexing and demultiplexing techniques need to be considered in DWDM RoF transmission. Finally, due to the relatively expensive components operating at millimeter-wave frequencies, cost-effective design should be considered for a millimeter-wave RoF system, for both downlink and uplink.

11.3.1 Millimeter-Wave Radio-over-Fiber Links

11.3.1.1 Concepts, Advantages, and Applications of RoF

RoF uses optical fiber links to distribute RF signals from a central office to a remote radio access point (RAP) or, conversely, to collect RF signals from a RAP. As shown in Figure 11.4, a basic full-duplex RoF link consists of a downlink and an uplink, each with a pair of fiber-connected optical transmitters (OTX) and optical receivers (ORX). At an OTX of downlink, RF signals to be delivered modulate an optical carrier; the modulated optical signal is coupled into an optical fiber and is delivered to a RAP. At the RAP, the subcarrier-bearing optical signal is photodetected and the recovered RF signals are radiated out to the end users after being amplified and filtered. In uplink, the recovered RF signals are fed into processing/demodulation circuits. Depending on the applications, the RF signals can be a single data carrier or a bunch of parallel carriers of a subcarrier multiplexing (SCM) system or an orthogonal frequency division multiplexing (OFDM) system; they can be narrow-bandwidth continuous waves or ultrawideband RF pulses; they can even be local oscillators or radar signals bearing no data.

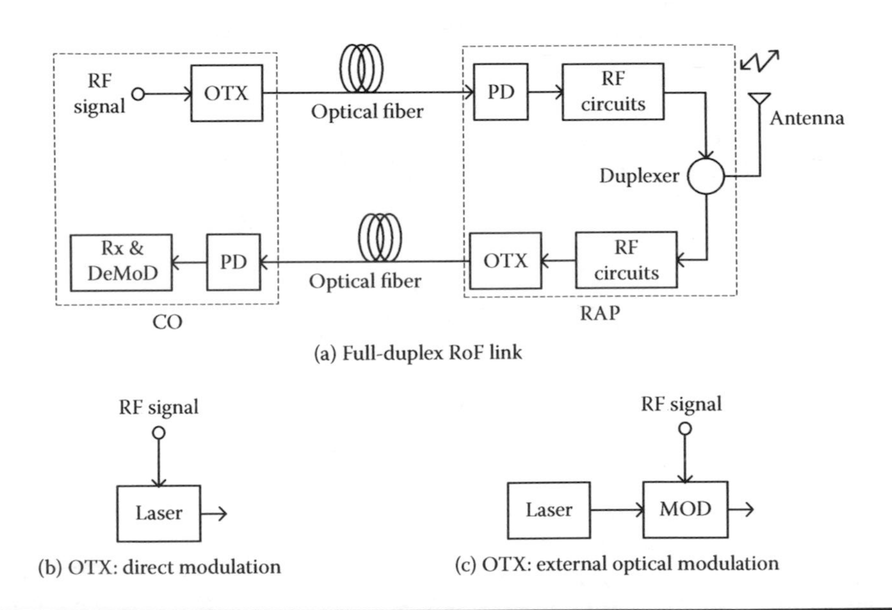

Figure 11.4 Basic full-duplex RoF link and optical transmitter. (OTX: optical transmitter; MOD: optical modulator; PD: photodetector; Rx & DeMoD: receiver and demodulator.)

A straightforward method to impose RF signals onto an optical carrier is optical direct modulation. However, this method is feasible only when the operating RF frequencies are below the modulation cut-off frequency of the laser diode used. So far, the highest cut-off frequency of lasers is reported to be about 40 GHz in laboratory [25]. Most of the commercially available lasers have modulation frequencies of about 10 GHz or less. For fiber transmission of higher frequencies such as millimeter-wave signals, optical external modulation is necessary. Moreover, external modulation is capable of providing better performance with respect to modulation chirp, achievable RF gain, noise figure, and so on [26].

The type of laser and fiber, optical wavelength, and connection distance of an RoF system depend on applications. For short-distance indoor and outdoor delivery of microwave signals below 10 GHz, such as IEEE 802.11 WLAN signals, low-cost vertical-cavity surface-emitting lasers (VCSELs) with an output at 850 nm plus multimode fiber can be used [27,28]. For relatively longer distance and higher frequencies, such as millimeter-wave signals, a single-mode fiber with much wider available bandwidth is necessary, as well as a DFB laser. In addition, the operation wavelength can be at 1300 nm or 1550 nm, where optical fiber has the lowest attenuation loss of 0.5 dB/km and 0.2 dB/km, respectively.

The main advantages of RoF are summarized as follows:

- *Transparent to data protocol and modulation formats of RF signals.* RoF provides optical delivery of RF signals but is not involved in the physical layer handling electrical signals. This makes RoF suitable to distribute a wide range of RF signals.
- *Very low transmission loss and huge available bandwidth.* Optical fiber has ultralow loss at wavelengths around 850 nm, 1310 nm, and 1550 nm. This property enables long-distance fiber transmission of RF signals with high air propagation loss (e.g., millimeter-wave signals); for single-mode fiber, there is more than 50 GHz bandwidth available in the above three wavelength windows taken together.
- *Capable of simplifying remote radio access points.* With RoF technology, a RAP can be required only to implement optical-to-electrical (O/E) and electrical-to-optical (E/O) conversion, RF signal radiation, and reception. Many electrical functions, such as frequency stabilization and conversion, local oscillator generation, data modulation and demodulation, and signal switching, can be centralized at a central office. It can lead to very simple RAPs and a reduction in system costs on installation, operation, and maintenance.

RoF technology can be used in microwave cellular networks, such as 2G and 3G systems, wireless LANs operating at both microwave and

Table 11.1 Main Features of a 60-GHz RoF System [29]

Wireless application	High-speed wireless LAN
Data rate	156 Mb/s
Radio frequencies	59.5 GHz (downlink), 61.5 GHz (uplink)
RF modulation	BPSK (binary phase shift keying)
Laser source	DFB laser
Optical modulation	External intensity modulation using EAM and OSSB filtering is implemented
Optical wavelength	1560.61 nm (downlink), 1552.52 nm (uplink 1), 1554.13 nm (uplink 2)
Fiber type	Single-mode fiber
Bit error rate (BER) achievable	$\leq 10^{-9}$

millimeter-wave frequencies, wireless personal communications such as IEEE 802.15 systems, broadband wireless access networks for fixed and mobile applications such as IEEE 802.16 systems, local multipoint distribution service (LMDS) for high-speed data and multimedia information transportation, intelligent transport systems (ITSs), satellite systems, as well as radar systems, and so forth. Features of a practically developed and demonstrated millimeter-wave RoF system for wireless LAN use can be found in Table 11.1.

11.3.1.2 Optical External Modulation for Millimeter-Wave RoF Links

Optical external modulation is necessary for millimeter-wave RoF transmission because the modulation cut-off frequencies of commercial lasers are below millimeter-wave frequencies. Mainly two types of external optoelectronic modulators are used in various millimeter-wave RoF systems: the Mach-Zehnder modulator and the electro absorption modulator.

A typical MZM consists of two planar optical waveguide arms on a $LiNbO_3$ substrate. The refractive index of the waveguide arms can be changed by applying electric field. When a laser output is coupled into an MZM, it is split into two paths; the light in the two paths propagates along the two arms in parallel and is combined at the end of the arms. Depending on the optical phase difference introduced by the applied DC voltages and the modulating electrical signals, the combined optical fields interfere constructively or destructively. If a voltage of V_π is applied along one arm, an optical phase shift of π is achieved. When DC bias $V_{dc1,2}$ and modulating signals $V_{ac1,2} \cdot s_{1,2}(t)$ are applied on the two arms, respectively, the modulator output can be expressed as

$$E_{mzm} = \frac{E_{in}}{\sqrt{2}} \left(e^{j\left(\frac{V_{dc1}}{V_\pi}\pi + \frac{V_{ac1}}{V_\pi}\pi \cdot s_1(t)\right)} + e^{j\left(\frac{V_{dc2}}{V_\pi}\pi + \frac{V_{ac2}}{V_\pi}\pi \cdot s_2(t)\right)} \right). \tag{11.4}$$

Let $\gamma_{1,2} = V_{dc1,2}/V_\pi$ and $\mu_{1,2} = V_{ac1,2}/V_\pi$; the MZM output becomes

$$E_{mzm} = \sqrt{2}E_{in}\cos\left[\frac{\gamma_1 - \gamma_2}{2}\cdot\pi + \frac{\mu_1 s_1(t) - \mu_2 s_2(t)}{2}\cdot\pi\right]e^{j\left(\frac{\gamma_1+\gamma_2}{2}\cdot\pi + \frac{\mu_1 s_1(t)+\mu_2 s_2(t)}{2}\cdot\pi\right)}. \tag{11.5}$$

Assuming $s_{1,2}(t)$ are sinusoidal waves with the same frequencies but have a phase difference, it is easy to find that various optical modulations can be achieved by biasing the two arms at different voltages:

- OSSB modulation: $\gamma_1 = 1/2$, $\gamma_2 = 0$, $\mu_1 = \mu_2 = \mu$, $s_1(t) = cos(\omega_m t)$, $s_2(t) = cos(\omega_m t \pm \pi/2)$ (small μ, i.e., small-signal modulation is assumed)
- Optical DSB-SC: $\gamma_1 = 1$, $\gamma_2 = 0$, $\mu_1 = \mu_2 = \mu$, $s_1(t) = cos(\omega_m t)$, $s_2(t) = cos(\omega_m t \pm \pi/2)$ (small-signal modulation assumed)

An EAM is either a bulk semiconductor-based or quantum-well-based device. The physics behind EAM is electroabsorption effect; i.e., the absorption coefficient of the material changes when an external electric field is applied [30]. When optical loss is ignored, the modulated output of an EAM can be expressed as

$$E_{eam} = E_{in}[1 + m\cdot s\ (t)]^{(1+j\alpha)/2}, \tag{11.6}$$

where E_{in} is the optical input of the EAM, $s(t)$ is a normalized RF modulating signal, m is the modulation index, and α is the chirp factor. Under the condition of small-signal modulation, the above equation can be written as

$$E_{eam} = E_{in}\left[1 + \frac{1+j\alpha}{2}m\cdot s\ (t)\right]. \tag{11.7}$$

If $s(t) = cos(\omega_m t)$, then it can be seen that the modulated output of EAM has a double-sideband spectrum. As introduced in the next subsection, for transmissions of millimeter-wave signals over standard dispersive fiber, the above optical double-sideband modulation introduces RF signal fading, and methods overcoming dispersion effect are required.

11.3.1.3 RF Properties of RoF Links

From the viewpoint of RF signal transmission, an RoF link is a simple transducer and can be characterized by transducer power gain (TPG), noise figure (NF), and spurious-free dynamic range (SFDR). For a better appreciation of the interaction between devices and link RF parameters, amplifierless intrinsic optical link with negligible effects of dispersion, and nonlinearity are generally discussed.

Transducer power gain is the ratio of a generator's available power to power delivered to a load by a transducer. For an intrinsic RoF link, the TPG is equal to the square of the product of the slope efficiency of the modulation device and the responsivity of the detection device [31]. In millimeter-wave RoF links with external modulation, the modulator slope efficiency is a derived parameter and can be improved by, for example, increasing the laser's output power and reducing the modulator's switching voltage if an MZM is used. In direct-modulation RoF links with p-i-n photodiode, the TPG is independent of optical power and is less than or equal to one; positive gain can be achieved if cascade laser is used.

The noise figure of an intrinsic RoF link is a measure of the degradation of the signal-to-noise ratio (SNR). If expressed in decibels, it is given by [31]

$$NF = 10 \cdot \log\left(\frac{n_{out}}{kT_0 B_N \cdot G_i}\right), \tag{11.8}$$

where k is Boltzmann's constant, $T_0 = 290$ K, B_N is the noise bandwidth of the electronic receiver, G_i is the intrinsic TPG, $n_{out} = G_i n_{in} + n_{link}$ is the total output noise, and n_{in} is the link input thermal noise. The link noise n_{link} includes thermal noise from the modulation device and photodetection circuits, relative intensity noise (RIN) from the optical source, and shot noise from photodetection. Typically, the latter two noise sources dominate the output noise and their contributions to NF are both inversely proportional to the intrinsic gain G_i because NF is the ratio between the output noise and the amplified (or attenuated) input thermal noise. Thus, increasing the intrinsic gain can reduce the NF.

An intrinsic RoF link has nonlinear components and will generate nonlinear distortion. Spurious-free dynamic range is a parameter characterizing the link nonlinearity and is defined as the output SNR when the power of the intermodulation products equals the power of the link output noise. The most important SFDR is the third order, which measures the intermodulation products that fall within the system bandwidth. The SFDR can be improved by linearization techniques, such as predistortion in a direct modulation link or concatenation of modulators in external links [31,32].

11.3.1.4 The Design of RoF Links

The design of a fiber-wireless system involves many factors. Two signal-to-noise ratios (SNRs) need to be considered: the optical SNR (OSNR) and the electrical SNR, which form the cumulative SNR in the concatenated fiber and wireless channel. If other effects (such as multipath spreading) can be ignored, the quality of the data transmission mainly depends on the SNR at the RF terminal. This SNR depends on the OSNR, the optical receiver amplifier gain, and the wireless channel loss. The relationship is given as follows [33]:

$$SNR = OSNR \left[\frac{1}{1 + (L_{wl}/G_{op})^2} \right], \quad (11.9)$$

where L_{wl} is the wireless channel loss and G_{op} is the optical receiver amplifier gain.

Hence, when the minimum SNR at an RF terminal (to achieve certain transmission quality) and the maximum wireless channel loss (for certain connection distance) are given, the optical gain and the OSNR need to be carefully designed. Certainly, the available optical gain in the design limits the OSNR.

In the fiber section, the OSNR is a function of the modulation index, E/O and O/E conversion losses, the fiber link loss (including optical loss from connectors, splitters, and optical fiber), and the optical noise. Only limited RF power can be modulated on optical sidebands. This is worse with OSSB modulation (to overcome fiber dispersion in millimeter-wave RoF links). The nonlinear distortion due to the optoelectronic modulators will also limit the millimeter-wave power that could be launched. The E/O and O/E conversion loss can be as high as 40 dB [34] due to impedance mismatch between the RF system and the laser diode at the E/O converter and the photodiode and the RF output at the O/E converter. Reactive matching techniques may decrease these losses; however, the link bandwith may be reduced [34].

11.3.2 Effect of Fiber Dispersion and Nonlinearity

Dispersion in an optical fiber is a phenomenon in which the propagation velocities of the various wave components (i.e., modes, frequencies) are different. A single-mode fiber is free from the relatively large intermodal dispersion found in multimode fibers because it is capable of supporting only the fundamental mode in normal operations. The dominant dispersion in a single-mode fiber is chromatic dispersion due to the slightly different propagation group velocities at different optical frequencies. In addition, originated from the nonideal symmetry of fiber, the orthogonally polarized components of the fundamental fiber mode have different mode indices, resulting in polarization-mode dispersion. Both CD and PMD may degrade the performance of a millimeter-wave signal transmitted along a single-mode fiber [35].

11.3.2.1 Signal Fading Induced by CD

If an optical signal with an angular frequency ω_o is slightly modulated by a pure sine wave at a frequency of f_m using a high-speed EAM, then (by putting $E_{in} = \sqrt{P_o}e^{j\omega_o t}$ into Equation (11.7) and assuming small-signal modulation) the modulated output can be rewritten as

$$E(0,t) = \sqrt{P_o}e^{j\omega_o t} + \frac{m\sqrt{(1+\alpha^2)P_o}}{4}\{e^{j[(\omega_o+\omega_m)t+\tan^{-1}\alpha]} + e^{j[(\omega_o-\omega_m)t+\tan^{-1}\alpha]}\}, \tag{11.10}$$

where P_o is the input light intensity and ω_m is the angular frequency of the modulating millimeter-wave signal. In Equation (11.10) the three terms in sequence are the optical carrier, the upper sideband, and the lower sideband, respectively. After being transmitted along a standard single-mode fiber of length z, the electric field seen by a photodetector (PD) is given by

$$E(z,t) = \sqrt{\frac{P_o}{L}}e^{j[\omega_o t-\beta(\omega_o)z]} + \frac{m\sqrt{(1+\alpha^2)P_o/L}}{4}\{e^{j[(\omega_o+\omega_m)t+\tan^{-1}\alpha-\beta(\omega_o+\omega_m)z]}$$

$$+e^{j[(\omega_o-\omega_m)t+\tan^{-1}\alpha-\beta(\omega_o-\omega_m)z]}\}, \tag{11.11}$$

where β is the fiber propagation constant and L is the fiber loss. As shown in Equation (11.11), at the three optical spectral frequencies, different phase delays are introduced because of the fiber CD. By square-law photodetection, the two optical sidebands beat with the optical carrier, and two millimeter-wave components at f_m with the same intensity are generated. At the PD, depending on the phase deviations between the optical spectral components introduced by CD, the two millimeter-wave components interfere constructively or destructively and signal fading occurs along the fiber length. The regenerated millimeter-wave signal and its normalized power at the PD output are given by

$$i_m(z,t) = \frac{m\sqrt{(1+\alpha^2)}RP_o}{L}\cos\left[\omega_m\left(t-\frac{z}{\upsilon_g}\right)\right]\cos\left(\frac{D\pi\lambda^2 f_m^2 z}{c}+\tan^{-1}\alpha\right), \tag{11.12}$$

and

$$P(f_m,z) = \left[\frac{\sqrt{(1+\alpha^2)}}{L}\cos\left(\frac{D\pi\lambda^2 f_m^2 z}{c}+\tan^{-1}\alpha\right)\right]^2, \tag{11.13}$$

respectively, where R is the responsivity of the PD, c is the light speed in a vacuum, υ_g is the light group velocity at ω_o, D is the fiber dispersion parameter, and λ is the light wavelength. The millimeter-wave signal vanishes at fiber length

$$Z_N = \frac{c\left[(2N-1)-\frac{\pi}{2}\tan^{-1}\alpha\right]}{2D\lambda^2 f_m^2}, \quad N = 1,2,3\ldots. \tag{11.14}$$

From the above equation, with parameters $f_m = 60$ GHz, $\lambda = 1550$ nm, and $D = 17$ ps/nm/km, the first point along the fiber length where the signal completely vanishes is limited to 1 km when the optical modulation is free of chirp (i.e., $\alpha = 0$). A positive chirp parameter shortens the fading distance, while a negative chirp parameter extends the distance. Extension of the fading distance by using a Mach-Zehnder modulator with a negative chirp has been experimentally demonstrated by Smith et al. [12]. However, the maximum transmission distance due to CD is still limited, and other methods such as optical single-sideband modulation/filtering shown in the next subsection are required.

11.3.2.2 Effect of Fiber CD on Self-Heterodyne Millimeter-Wave RoF Links

A millimeter-wave signal can be transmitted over fiber via optical self-heterodyne detection [35]. With this technique, a dual-frequency light source at a transmitter provides two phase-correlated optical signals with a center-frequency difference of a desired millimeter-wave frequency f_m. After fiber transmission, by square-law photodetection, a millimeter-wave signal at a frequency of f_m is generated by beating between the two optical signals.

Various concepts of dual-frequency light source have been proposed and investigated; e.g., dual-mode lasers [36], optical frequency shifting [37], and optical injection phase-locking [24]. With all of the above concepts, the two phase-correlated optical signals have fully correlated phase noise at the transmitter. Due to fiber CD, the two optical signals after fiber transmission experience a differential propagation delay given by $\Delta\tau_{CD} = D\lambda^2 f_m z/c$; thus, the phase noise of the two optical signals becomes partially correlated, or completely uncorrelated, and then a phase noise $\vartheta(t) - \vartheta(t - \Delta\tau_{CD})$ remains on the regenerated millimeter-wave signal. If the two optical signals have Lorentzian-shaped power spectra, the single-sideband power spectrum density of the regenerated millimeter-wave signal is given by [38]

$$S(f) = \delta(f)e^{-2\pi\Delta\upsilon_o\Delta\tau} + \frac{\Delta\upsilon_o}{\pi(\Delta\upsilon_o^2 + f^2)} \times \left\{1 - e^{-2\pi\Delta\upsilon_o\Delta\tau} \cdot \left[\cos(2\pi f\Delta\tau) + \frac{\Delta\upsilon_o}{f}\sin(2\pi f\Delta\tau)\right]\right\}, \quad (11.15)$$

where f is the offset from the millimeter-wave carrier and $\Delta\upsilon_o$ is the full-width half-maximum (FWHM) linewidth of the power spectra of the optical signals. The first term of Equation (11.15) indicates that the optical phase decorrelation due to fiber CD results in a decrease of the millimeter-wave signal power and consequently leads to a carrier-to-noise (CNR) penalty, given by

$$P_{cnr} = 10\log_{10}(e^{-2\pi\Delta\upsilon_o\Delta\tau_{CD}})[dB]. \quad (11.16)$$

The second term of Equation (11.15) characterizes the increase of the phase noise on the regenerated millimeter-wave signal. If the phase of the millimeter-wave signal bears information data, bit error may be introduced in demodulation due to the extra phase noise from the optical domain. In order to analyze the effect of the CD delay-induced phase noise on data transmission, the extra phase noise is modeled as a zero-mean Gaussian random variable with a phase variance given by

$$(\sigma_{\vartheta})^2 = \int_0^{B_n} \frac{2\Delta \upsilon_o}{\pi f^2} \cdot [1 - \cos(2\pi f \Delta \tau_{CD})] df, \tag{11.17}$$

where B_n is the millimeter-wave receiver noise bandwidth. Analyzed results show that, with a bit error rate (BER) of 10^{-9} and a CNR penalty due to delay-induced phase noise of 1 dB, with parameters of $f_m = 60$ GHz, $B_n = 100$ MHz, $\Delta \upsilon_o = 10$ MHz, $\lambda = 1550$ nm, and $D = 17$ ps/nm/km, the fiber distance for BPSK data transmission is limited under 20 km, while for 16 PSK, the distance becomes 2 km [38].

Finally, it should be noted that dispersion-shifted fibers with very low dispersion coefficients are available and potentially can be applied for millimeter-wave signal transmission where chromatic dispersion would be a limiting factor. Further, several dispersion compensation techniques are readily available. Examples of commonly used techniques are dispersion compensating fiber (DCF) and chirped fiber Bragg gratings [39].

11.3.2.3 Effect of PMD on Millimeter-Wave RoF Transmission

In the self-heterodyne millimeter-wave RoF link discussed above, the power of a regenerated millimeter-wave signal is related to polarization states of the two beating optical signals with respect to each other [35]. If at the fiber input the two optical signals are in arbitrary polarization states, then the normalized electrical fields can be represented by [40]

$$\vec{E} = \begin{bmatrix} \cos \psi \\ \sin \psi \cdot e^{j\phi} \end{bmatrix}, \tag{11.18}$$

and

$$\vec{E} = \begin{bmatrix} \cos(\psi + \Delta\psi) \\ \sin(\psi + \Delta\psi) \cdot e^{j(\phi + \Delta\phi)} \end{bmatrix} \cdot e^{j2\pi f_m}, \tag{11.19}$$

where ψ $(\psi + \Delta\psi)$ is the angle between polarization and the x-axis of the orthogonal principal axes of the optical fiber, ϕ $(\phi + \Delta\phi)$ is the polarization phase delay, $\Delta\phi$ stands for deviation of polarization phase delays, and $\Delta\psi$ represents mismatch of polarization between the two optical signals. When $\Delta\psi = 0$ and $\Delta\phi = 0$, the two optical signals have perfectly matched polarization states at the fiber input. Due to fiber PMD, a delay deviation $\Delta\tau_{PMD} = \Delta n/c \cdot z$ is introduced between the two optical signals after fiber transmission, where Δn is the difference of the polarization mode indices. Similar to the effect of fiber CD on millimeter-wave signal transmission, signal fading occurs because of PMD-induced delay. The normalized power of a regenerated millimeter-wave signal is given by

$$\begin{aligned} P_m = & [\cos\psi\cos(\psi+\Delta\psi)]^2 + [\sin\psi\sin(\psi+\Delta\psi)]^2 \\ & + 2[\cos\psi\cos(\psi+\Delta\psi)]\cdot[\sin\psi\sin(\psi+\Delta\psi)] \\ & \cdot\cos(2\pi f_m \Delta\tau_{PMD} + \Delta\phi). \end{aligned} \tag{11.20}$$

A polarization-preserving fiber has a relatively large and constant PMD with time. With such fiber, millimeter-wave power degradation due to PMD is determinable. Assume the two beating optical signals having perfectly matched polarization at fiber input; i.e., $\Delta\psi = 0$ and $\Delta\phi = 0$. It can be seen that when $\psi = 0$ there is no PMD-induced millimeter-wave power degradation since only one polarization component is excited along the fiber; however, when $\psi = \pi/4$, the millimeter-wave signal experiences the most serious signal fading and it vanishes totally at a certain fiber length [40]. When two optical signals have different polarization states with respect to each other at the fiber input end, the degradation of millimeter-wave power can be alleviated or aggravated, depending on the details of the polarization mismatch.

In a standard long-haul single-mode fiber, the PMD is stochastic and varies with time because of changes of environmental factors such as temperature, humidity, and mechanical pressure. The statistical variation of the actual PMD value over unit length is a Maxwellian distribution function determined by mean value $\Delta\tau_M$ over unit length. Thus, along a single-mode fiber cable, the millimeter-wave signal power degradation due to PMD-induced delay is random and the description of the polarization effect is statistical. For system design, the cumulative probability of millimeter-wave power degradation less than certain system margin can be calculated by integrating the Maxwellian distribution function over the target space [35,40].

It should be noted that modern optical fibers with a low PMD coefficient (<0.05 ps/$\sqrt{\text{km}}$) will not lead to signal degradation even after long transmission distances of $\gtrsim 100$ km.

11.3.2.4 Effect of Fiber Nonlinearity

Nonlinear fiber transmission effects need to be considered for long transmission distances and/or high launch power levels. Self-phase modulation (SPM) and stimulated Brillouin scattering (SBS) can distort the signal if the transmission system is not properly designed. Self-phase modulation results in phase modulation of the signal due to high intensity via the nonlinear gain coefficient of the fiber [41]. The effect of SPM on RF signal transmission has been studied [42]. SBS largely affects signals with strong carrier tones and can lead to a reduced received power by generating a backward propagating wave. If SBS can degrade the millimeter-wave signal transmission, measures should be taken to avoid it. An elegant solution in many cases may be the use of a fiber with increased SBS threshold [43]. Another possibility is to introduce an additional low-frequency phase modulation to the carrier signal in order to broaden the signal spectrum beyond the SBS gain bandwidth. Other than SPM and SBS, in RoF systems incorporating DWDM, cross-phase modulation (XPM) may introduce nonlinear crosstalk due to conversion of XPM-generated phase noise to amplitude noise through fiber CD [44,45].

In addition to fiber nonlinearity, nonlinear distortion due to optoelectronic modulators may need to be considered in RoF links [46,47].

11.3.3 Optical Single-Sideband Modulation

As discussed above, the fiber length of a millimeter-wave RoF link incorporating optical double-sideband (ODSB) modulation is seriously limited by CD. For relatively long-distance transmission, it is necessary to overcome this limitation. Several schemes alleviating ODSB limitation have been discussed, including OSSB modulation [12], OSSB filtering [29,48], midway optical phase conjugation [49], and soliton sampling [50]. Another technique to mitigate the dispersion effect would be an up-conversion technique in which an intermediate frequency (IF) signal is modulated onto an optical carrier by means of an external modulator or a directly modulated laser. A second modulator would up-convert the IF signal to millimeter-wave range. If this second modulator is an MZM and biased at minimum, a dual-sideband signal with suppressed carrier is generated. This technique has been widely studied [51–53] and is shown to lead to reasonably large transmission distances of tens of kilometers for millimeter-wave signals.

The idea of incorporating OSSB to combat CD-induced signal fading is direct and simple: if only one optical sideband is transmitted along the fiber with an optical carrier, then only one millimeter-wave component at f_m is generated in photodetection and there is no more signal fading. An OSSB power spectrum can be obtained by OSSB modulation or by tailoring an ODSB spectrum using an optical filter or nonlinear optical filtering [54].

Optical single-sideband modulation is generally based on a phase-shift method that has been devised to generate electrical single-sideband modulation [55]. In this technique, two carriers in quadrature are conventionally modulated by two modulating signals also in quadrature. The two generated DSB signals are combined and one of the sidebands is canceled due to destructive interference while the other pair is retained by constructive interference.

11.3.3.1 OSSB Modulation Using Dual-Electrode Mach-Zehnder Modulator

A dual-electrode MZM can be modeled as two integrated phase modulators in parallel. Its property of RF modulation is given by Equation (11.5). For OSSB modulation, one of the DC electrodes is grounded, while the other one is biased at $V_\pi/2$, where V_π is the switching voltage of the modulator. A modulating signal at f_m is equally split and applied to the two RF electrodes, with one of the millimeter-wave signals $\pm\pi/2$ phase shifted. Figure 11.5 schematically shows generation of OSSB signal using a dual-electrode MZM. If a continuous-wave optical signal with center frequency of f_o is coupled in, then the modulated output is given as

$$E_m = \frac{\sqrt{2P_o}}{2}\left\{\cos\left[\omega_o t + \mu\pi\cos(\omega_m t) + \frac{\pi}{2}\right]\right.$$
$$\left. + \cos\left[\omega_o t + \mu\pi\cos\left(\omega_m t \pm \frac{\pi}{2}\right)\right]\right\}, \qquad (11.21)$$

where $\mu = V_{ac}/V_\pi$, V_{ac} is the peak amplitude of the millimeter-wave modulating signals. The above equation can be rewritten as

$$E_m = \sqrt{2P_o}\sum_{n=-\infty}^{\infty} J_n(\mu\pi)\cos\left[(\omega_o + n\omega_m)t + \frac{(1\pm n)\pi}{4} + \frac{n\cdot\pi}{2}\right]$$
$$\times\cos\left[\frac{(-1\pm n)\pi}{4}\right] \cong \sqrt{P_o}J_0(\mu\pi)\cos\left(\omega_o t + \frac{\pi}{4}\right)$$
$$-\sqrt{2P_o}J_1(\mu\pi)\cos\left[(\omega_o \pm \omega_m)t\right], \, m \ll 1. \qquad (11.22)$$

In the above, $J_n(\cdot)$ is the nth-order Bessel function of the first kind.

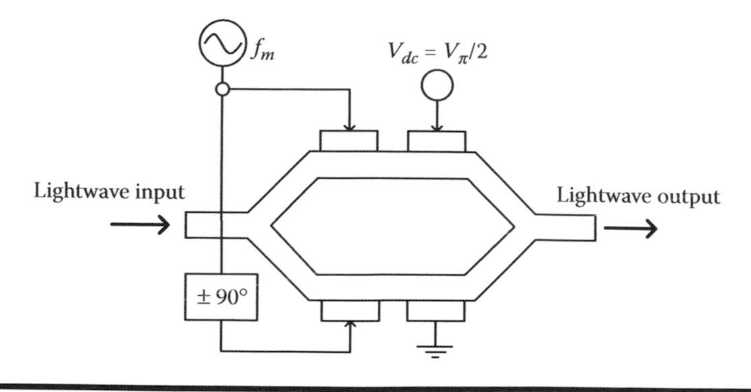

Figure 11.5 Optical single-sideband modulation using dual-electrode MZM with phase shift technique.

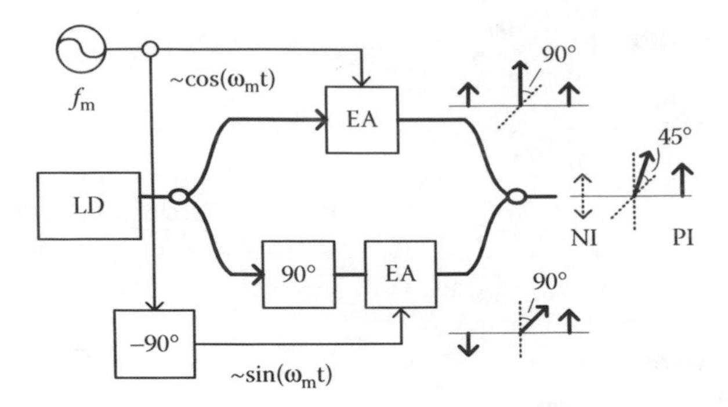

Figure 11.6 Schematic diagram of optical single-sideband modulation incorporating EAMs with phase shift technique [56].

11.3.3.2 OSSB Modulation Incorporating EAMs

An OSSB modulation scheme incorporating two EAMs in parallel with phase-shift technique is schematically shown in Figure 11.6 [56]. A laser output is split 50:50 and coupled separately into two EAMs. One path of the lightwave is 90° phase shifted compared to the other. The two paths of lightwave are modulated by two millimeter-wave signals in quadrature, respectively. When small-signal modulations are implemented and modulation chirp is ignored, the outputs of the two EAMs can be given by

$$E_1 = \sqrt{\frac{P_o}{2}}\left(\frac{m}{4}e^{-j\omega_m t} + 1 + \frac{m}{4}e^{j\omega_m t}\right)e^{j\omega_o t}, \tag{11.23}$$

and

$$E_2 = \sqrt{\frac{P_o}{2}}\left(-\frac{m}{4}e^{-j\omega_m t} + j + \frac{m}{4}e^{j\omega_m t}\right)e^{j\omega_o t}, \tag{11.24}$$

respectively. As indicated above, the phase difference between the two optical left sidebands, the optical carriers, and the two optical right sidebands are π, $\pi/2$, and 0, respectively. When the two paths of modulated lightwave are well matched in polarization, at the output of the second coupler, the two optical left sidebands interfere destructively and are eliminated, while the optical carriers and the optical right sidebands are retained. Assuming the coupling loss is ignored, the combined optical output is given by

$$E_c = \sqrt{P_o}e^{j(\omega_o t+\pi/4)} + \frac{m\sqrt{P_o}}{2\sqrt{2}}e^{j(\omega_o+\omega_m)t}. \tag{11.25}$$

In OSSB modulation using EAMs, rudimentary optical sidebands may be introduced due to nonideal polarization match and O/E phase shift of $\pi/2$ [56].

Other methods based on the phase-shift technique for generation of the OSSB spectrum include a scheme using a bidirectional single-electrode MZM [57] and a scheme of hybrid amplitude-phase modulation [58].

11.3.4 Millimeter-Wave RoF Transmission Incorporating DWDM

Dense-wavelength-division multiplexing is used to transmit optical wavelengths closely spaced along a single fiber to increase the capacity of an optical network. It is a promising technology for future fiber-supported millimeter-wave macro-/pico-cellular networks to simplify the structure, maintenance, management, and upgrade of the system. A schematic diagram of a DWDM RoF system distributing wireless signals is shown in Figure 11.7.

11.3.4.1 Simultaneous Electro-Optical Up-Conversion

For millimeter-wave DWDM RoF transmission, a conventional scheme is to modulate each optical wavelength with corresponding millimeter-wave carriers separately before wavelength multiplexing; thus, if N optical wavelengths are involved, one needs N optoelectronic modulators operating at millimeter-wave frequencies. A more cost-effective method is simultaneous electro-optical up-conversion [59]. In this technique, the optical wavelengths at λ_1 to λ_N are directly modulated by corresponding IF subcarriers first. The IF-modulated optical signals are multiplexed and coupled into a

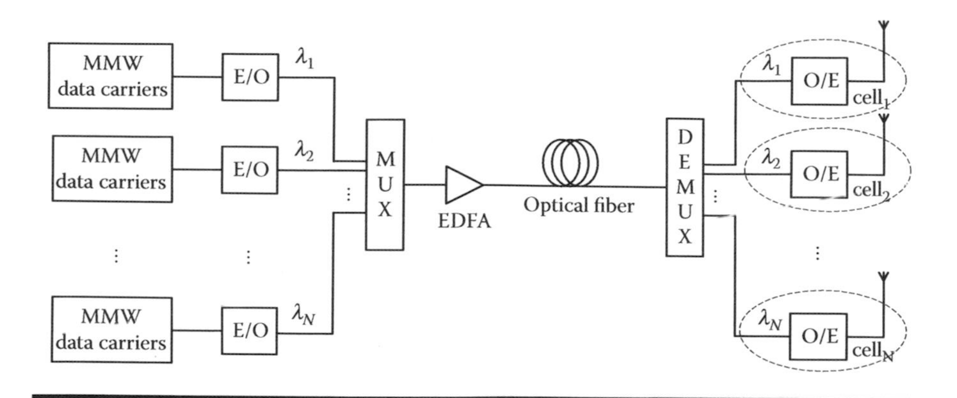

Figure 11.7 Schematic illustration of DWDM millimeter-wave RoF transmission. (EDFA: erbium-doped fiber amplifier; MUX: multiplexer; DEMUX: demultiplexer).

millimeter-wave optoelectronic external modulator, where all IF-modulated optical signals are up-converted by a millimeter-wave local oscillator (LO) signal at f_m. Compared to the conventional scheme, this technique requires only a single millimeter-wave electro-optical modulator (EOM). Since each wavelength is demultiplexed to a particular cell, a frequency plan for cellular management can be easily implemented by assigning different IF bands to different optical wavelengths. Furthermore, this technique generates a millimeter-wave LO at f_m at a remote access point, allowing down-conversion of the received millimeter-wave uplink signals to IF bands for IF-uplink fiber transmission, with advantages in cost reduction and fiber distance extension.

11.3.4.2 Wavelength Interleaving in Millimeter-Wave DWDM RoF Links

A challenge in system design of a millimeter-wave DWDM RoF link is that a single millimeter-wave carrier-modulated optical signal has a wide spectral width close to or more than 100 GHz. Consider N millimeter-wave carriers are modulated onto N optical wavelengths individually. The total spectral width can be $N \cdot 2f_m + (N-1) \cdot \Delta f$ when ODSB modulations and conventional wavelength multiplexing are implemented, where Δf is the DWDM channel space. When N and f_m are relatively large, the total spectral width may exceed the amplification window of an Erbium-doped fiber amplifier operating at 1.5 μm. With OSSB, the spectral width becomes smaller, $N \cdot f_m + (N-1) \cdot \Delta f$; however, the spectral efficiency is still relatively low. Wavelength interleaving (WI), as shown in Figure 11.8, is proposed and demonstrated for millimeter-wave DWDM RoF transmission. With this method, the optical spectra for millimeter-wave signal transmission overlap and lead to high spectral efficiency [60]. Assume OSSB is implemented, the optical carriers and the optical sidebands are equally spaced with a width of Δf, and $f_m = 3\Delta f$; the total spectral width is now a much smaller $(2N+1)/3 \cdot f_m$. With $N = 64$, $f_m = 60$ GHz, and $\Delta f = 20$ GHz, compared

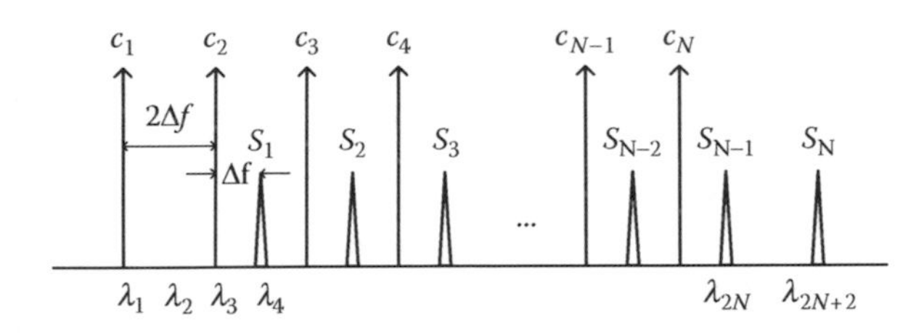

Figure 11.8 Optical wavelength interleaving for millimeter-wave DWDM RoF transmission. OSSB modulations are assumed; c_i and s_i stand for the ith optical carrier and its sideband. The channel space is Δf and the RF, $f_m = 3\,\Delta f$, is assumed.

to the foregoing two schemes (ODSB and OSSB without WI), by wavelength interleaving, the spectral efficiency is tripled and doubled, respectively.

11.3.4.3 Topology for Millimeter-Wave DWDM RoF Networks

As illustrated in Figure 11.9, there are three basic topologies for a millimeter-wave DWDM RoF network: star, bus, and ring. Among the three topologies, the DWDM transmissions between a central office (CO) and a remote node

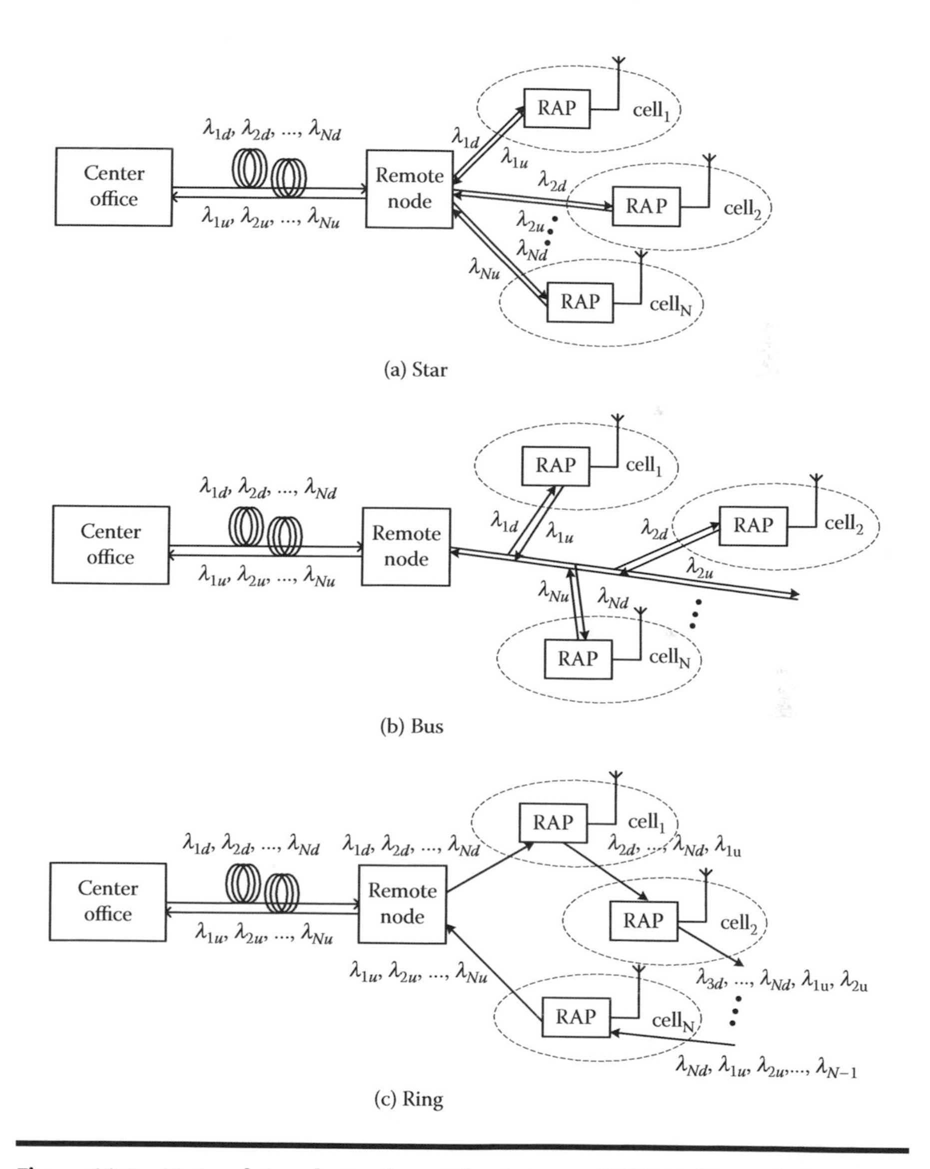

Figure 11.9 Network topologies for millimeter-wave DWDM fiber -wireless transmission.

(RN) are the same: all modulated downlink optical signals are multiplexed at a CO and distributed to a RN along a single fiber and all multiplexed uplink optical wavelengths are transmitted back to a CO from a RN. The difference lies in the section between a RN and RPAs. In a star structure, the N downlink wavelengths are demultiplexed at a RN and distributed to N cells separately. The uplink wavelength of each cell is transmitted back to the RN and multiplexed. In a bus structure, a fiber downlink bus and an uplink bus reach all of the RAPs and all of the cells download and upload their downlink and uplink wavelengths, respectively. With a ring topology, a single optical fiber connects all of the cells to a RN in a series manner; downlink and uplink wavelengths are dropped down and added in at each RAP.

11.3.4.4 Multiplexing, Demultiplexing, and Optical Add/Drop Multiplexing for DWDM RoF Links

Simultaneous up-conversion [59] can be used for multiplexing downlink wavelengths in a CO where the modulations are centered. For multiplexing individually modulated wavelengths, such as uplink wavelengths from cells at an RN with bus topology, other techniques must be used. A scheme using a $(2N+2) \times (2N+2)$ array waveguide grating (AWG) to multiplex N optical carriers and N single sidebands interleaved has been demonstrated [61]. It is based on the cyclic input–output character of a $(2N+2) \times (2N+2)$ AWG, where all of the sidebands S_i at input port IN_{2i-1} are directly multiplexed at the output port of OUT_1, and all the carriers C_i are combined at the same output port via loopbacks twice. A more direct and comprehensible method multiplexing N optical carriers and N sidebands by using an $N \times 2$ AWG has been proposed [62]. In that method, a pair of optical carrier C_i and sideband S_i is coupled into the ith input port of an $N \times 2$ AWG; all of the optical carriers are combined at the output port 1, while all sidebands are multiplexed at the output port 2, and then the carriers and sidebands are combined with the help of an optical circulator (OC) and a Fabry-Perot etalon (FPE) with a free spectral range (FSR) of the spacing between two neighboring carriers. One more multiplexing scheme is shown in Figure 11.10, where a $(2N-1) \times (2N-1)$ AWG, an OC, and a FPE are used to multiplex N optical carriers and N sidebands. The wavelength arrangement of the optical carriers and sidebands is the one shown in Figure 11.8, where wavelength interleaving is applied and $f_m = 3\Delta f$ is assumed; Δf is the channel space of the AWG. The input–output property of the $(2N-1) \times (2N-1)$ AWG used is shown in Table 11.2. By coupling optical carrier C_i (λ_{2i-1}) and its corresponding sideband S_i (λ_{2i+2}) into input port IN_{2i-1}, where $i = 1, 2, \ldots, N$, all optical carriers are coupled out at OUT_1 and all sidebands are at OUT_4. The optical carriers and sidebands are multiplexed by using an OC and an FPE with an $FSR = 2\Delta f$.

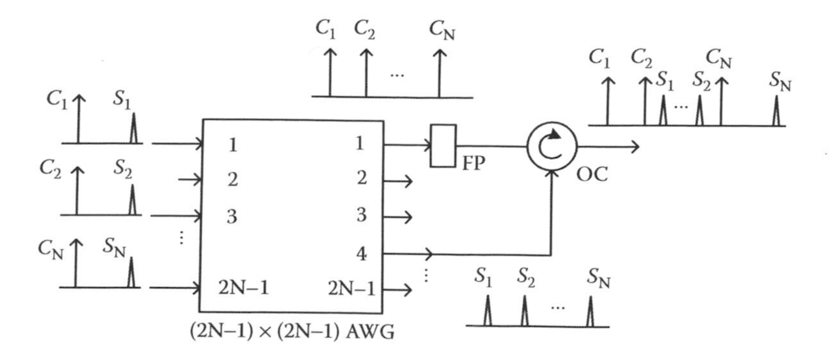

(a) Multiplexing scheme

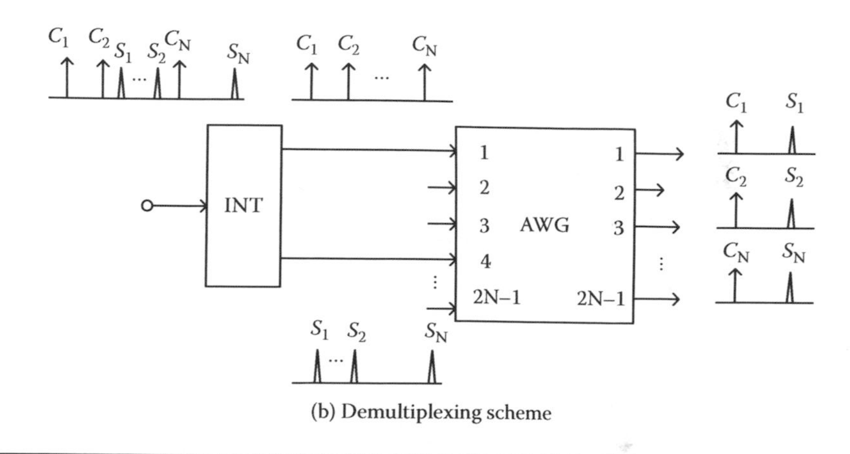

(b) Demultiplexing scheme

Figure 11.10 Multiplexing and demultiplexing schemes: (a) multiplexing N optical carriers and N sidebands using a $(2N - 1) \times (2N - 1)$ AWG; (b) demultiplexing N optical carriers and N sidebands using an optical interleaver and a $(2N - 1) \times (2N - 1)$ AWG. (INT: interleaver.)

Demultiplexing of the multiplexed uplink optical wavelengths is required at a CO in all three basic network topologies. Also, it is a must at a RN for distributing downlink wavelengths in a network with star structure. An $N \times 2$ AWG [62] for multiplexing can be used inversely for demultiplexing N optical carriers and N sidebands. Another scheme is to use an optical interleaver and a $(2N - 1) \times (2N - 1)$ AWG with the input–output characters listed in Table 11.2. The scheme is shown in Figure 11.10, where wavelengths interleaving with $f_m = 3\Delta f$ shown in Figure 11.8 are assumed. The multiplexed wavelengths are coupled into the input port of an optical interleaver with an FSR of $2\Delta f$. Due to the complementary spectral responses of the interleaver's output ports, the optical carriers and the

Table 11.2 Cyclic Input–Output Property of a (2*N* − 1) × (2*N* − 1) AWG

	IN_1	IN_2	IN_3	...	IN_{2N-3}	IN_{2N-2}	IN_{2N-1}
λ_1	OUT_1	OUT_2	OUT_3	...	OUT_{2N-3}	OUT_{2N-2}	OUT_{2N-1}
λ_2	OUT_2	OUT_3	OUT_4	...	OUT_{2N-2}	OUT_{2N-1}	OUT_1
λ_3	OUT_3	OUT_4	OUT_5	...	OUT_{2N-1}	OUT_1	OUT_2
...	...	...	...	...	...	...	...
λ_{2N-2}	OUT_{2N-2}	OUT_{2N-1}	OUT_1	...	OUT_{2N-5}	OUT_{2N-4}	OUT_{2N-3}
λ_{2N-1}	OUT_{2N-1}	OUT_1	OUT_2	...	OUT_{2N-4}	OUT_{2N-3}	OUT_{2N-2}

sidebands are separately presented at the interleaver's two output ports. The optical carriers and the sidebands are then respectively coupled into the first and the fourth input port of a cyclic $(2N-1) \times (2N-1)$ AWG. Due to the rotary input–output character of the AWG, at each odd output port OUT_{2i-1}, a pair of C_i and S_i can be obtained.

Unlike with star structure, a RAP in a fiber network with bus or ring topology drops its downlink wavelength and adds its uplink wavelength individually via an optical add/drop multiplexer (OADM). As shown in Figure 11.11a, when the multiplexed downlink wavelengths go through the *i*th RAP in a bus network, the RAP's downlink optical carriers at λ_{id} and sideband at λ'_{id} are dropped by using an OC and two fiber Bragg gratings with reflection wavelengths at λ_{id} and λ'_{id}, respectively; the uplink

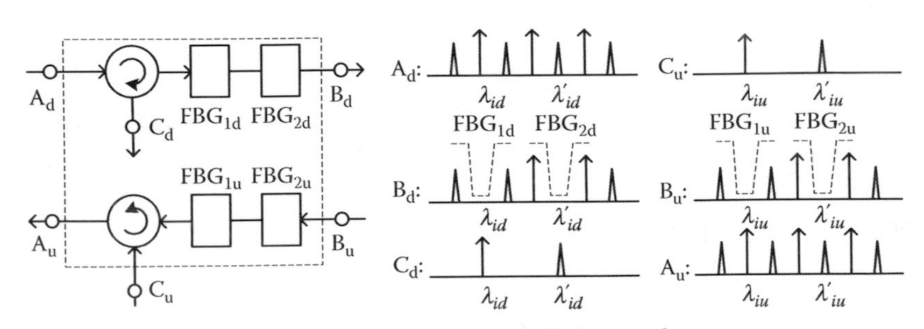

(a) Optical drop/add multiplexer for bus topology

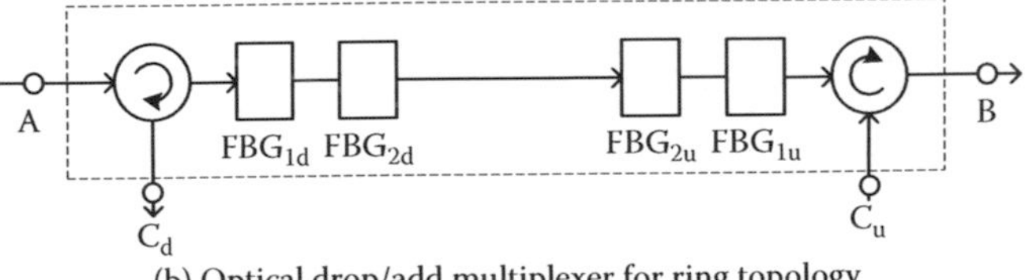

(b) Optical drop/add multiplexer for ring topology

Figure 11.11 OADM for (a) bus topology; and (b) ring topology.

optical carrier at λ_{iu} and sideband at λ'_{iu} are added into the upstream in a reverse manner. The OADM dropping and adding operations at a RAP in a ring network are similar to a bus network, except that the two functions are implemented in series (see Figure 11.11b), because in a ring network a single fiber is used to connect all RAPs of a wavelength group.

11.3.4.5 Unequally Spaced Channel Technology Minimizing Four-Wave Mixing Crosstalk in Millimeter-Wave DWDM RoF Links

Due to fiber nonlinearity, in a DWDM fiber transmission link, four-wave mixing (FWM) may occur and introduce nonlinear crosstalk. The generation efficiency of FWM increases when the fiber chromatic dispersion and channel space decrease. In conventional digital DWDM fiber links with ITU-T standard channel space, FWM crosstalk is generally minimized by fiber dispersion. However, when wavelength interleaving is applied for millimeter-wave DWDM RoF transmission, channel space can be less than 20 GHz. In this case, an FWM efficiency of more than 20 percent can be guaranteed even in a single-mode fiber with a chromatic dispersion of 15 ps/nm/km; hence, considerable nonlinear crosstalk can be introduced. An effective approach to minimizing FWM crosstalk is to unequally space the optical channels so that the FWM waves are generated at frequencies other than the signal channels [63]. Based on a frequency slot concept, a fast frequency allocation method of unequally spaced channels (USCs) can be used to minimize FWM crosstalk in millimeter-wave DWDM RoF links [64]. An example is illustrated in Table 11.3, where eight pairs of optical carriers and sidebands are assigned for eight 60-GHz millimeter-wave signal DWDM RoF transmissions. As shown in Figure 11.12, with USC allocation, the effect of FWM crosstalk on the bit error rate (BER) can be reduced [64].

Table 11.3 USC Allocation for Eight 60-GHz Millimeter-Wave Signal DWDM RoF Transmissions [64]

No.	*Without Wavelength Interleaving (carrier, sideband) (nm)*	*With Wavelength Interleaving (carrier, sideband) (nm)*
1	1550.2634, 1549.7829	1549.3027, 1548.8228
2	1549.6228, 1549.1247	1549.1427, 1548.6629
3	1548.9747, 1548.4950	1548.4950, 1548.0156
4	1548.3192, 1547.8399	1548.3192, 1547.8399
5	1547.6562, 1547.1773	1547.6562, 1547.1773
6	1546.9858, 1546.5073	1547.4646, 1546.9858
7	1546.3081, 1545.8300	1546.7864, 1546.3081
8	1545.6229, 1545.1453	1546.5791, 1546.1009

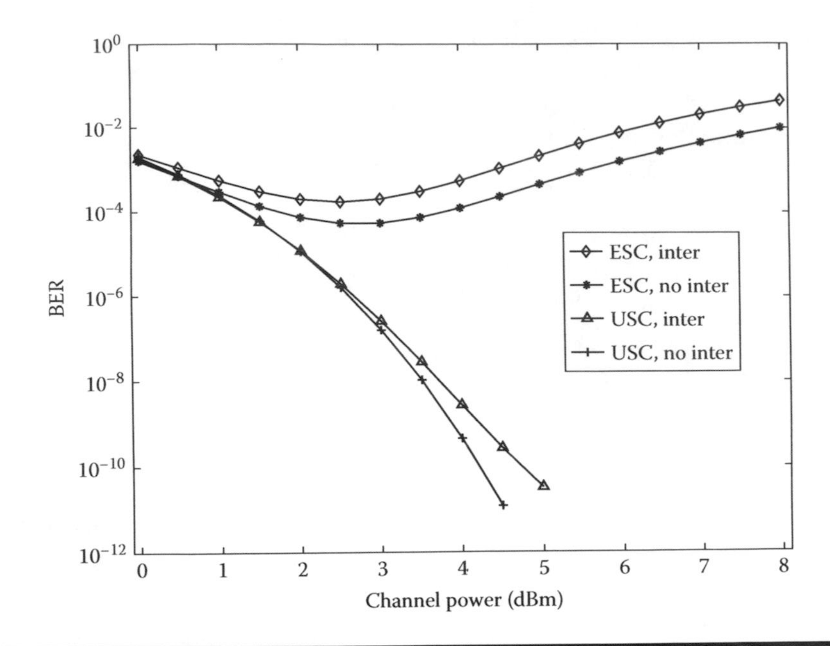

Figure 11.12 Simulated BER as a function of channel transmitter power for 622 Mb/s BPSK data transmission over 40-km dispersion-shifted fiber; $f_m = 60$ GHz, $N = 8$ [64].

11.3.5 Cost-Effective Design of Full-Duplex Millimeter-Wave RoF Links

With a normal radiation power, the free-space propagation distance of millimeter-wave signals can be limited to under hundreds or tens of meters due to high air propagation loss. Thus, numerous RAPs may be required for fullradio coverage in a millimeter-wave micro-/pico-cellular system, and then the cost of a RAP is critical to the development of such a millimeter-wave communication system. Cost-effective design of a RAP is closely related to the schemes of fiber downlink and fiber uplink transmission because different schemes require different components and structures of a RAP. In a full-duplex RoF link, a simplified design of RAPs needs to consider both fiber downlink and uplink.

11.3.5.1 Millimeter-Wave RoF Downlink with Remote Local Oscillator Delivery

To implement a fiber uplink in a full-duplex RoF system, IF over fiber is a relatively cost-effective option compared to direct millimeter-wave RoF transmission because the former requires no optoelectronic modulators operating in millimeter-wave bands at a RAP. In order to down-convert a received millimeter-wave uplink carrier to the IF band at a RAP, a millimeter-wave local-oscillator (LO) is required. It is more cost effective to provide a local oscillator remotely from a CO rather than at each of the very large number of RAPs. Basically, as shown in studies for remote millimeter-wave LO RoF delivery, optical heterodyne is employed, where two optical spectral components at f_o and $f_o + f_{LO}$ are transmitted and photodetected. Among the schemes remotely providing millimeter-wave LO, the one incorporating two dual-electrode MZMs has advantages in spectral efficiency, frequency stabilization, and simple configurations [65].

Phase noise is an important parameter to characterize the quality of an LO. Especially when a millimeter-wave data carrier is phase modulated, LO phase noise can introduce phase errors in demodulation and may limit the BER performance. In remote fiber delivery of millimeter-wave LO, additional phase noise converted from laser can be introduced due to fiber CD-induced propagation delay deviation between the two optical beating components at f_o and $f_o + f_{LO}$. Since the phase noise from optical domain and the original LO phase noise are statistically independent, the effect of the additional phase noise can be investigated solely. When the laser source used has a Lorentzian power spectral shape, the additional phase noise of the regenerated LO in dBc/Hz can be given by [66]

$$L_{\vartheta}(f)_{dBcHz} = \left(\frac{2\Delta\upsilon}{\pi f^2}[1 - \cos(2\pi f \Delta\tau_{LO})]\right)_{dBr/Hz} - 3dB, \qquad (11.26)$$

where $\Delta\upsilon$ is a 3-dB linewidth of the laser, f is the frequency deviation from the center frequency of the LO, and $\Delta\tau_{LO}$ is the delay deviation between the two optical signals introduced by fiber chromatic dispersion. With parameters of $f_{LO} = 37$ GHz, $z = 50$ km, $D = 17$ ps/nm/km, and $\Delta\upsilon = 50$ MHz, the remotely delivered LO has a phase noise degradation of 0.6 dB at 1 MHz deviation; for 1 Gbit/s BPSK data transmission, this amount degradation of LO phase noise may introduce an SNR penalty of 0.5 dB [66].

11.3.5.2 Optical Wavelength Reuse in Full-Duplex Millimeter-Wave RoF Links

In a full-duplex RoF system, a lightwave carrier is a must for fiber uplink transmission. A laser-free RAP can be possible if the required optical carrier is a portion of the downlink optical signal or is remotely provided from a CO via wavelength-division multiplexing. The former scheme is to reuse the downlink optical signal, and its advantage over the latter is that fewer wavelengths are required.

By reusing the downlink optical carrier for uplink transmission, a passive RAP becomes possible [67]. A passive pico-cell was demonstrated by using an electro absorption waveguide device, which simultaneously acts as a photodetector of fiber downlink and a modulator for fiber uplink [67]. The unbiased electroabsorption device has a small fundamental absorption; thus, the downstream lightwave is not completely converted to electrical current and the remaining part is modulated by applying the uplink RF carrier and is transmitted back to a CO. Similar photodetection and modulation can be realized using an asymmetric Fabry-Perot modulator/detector (AFPMD) [68]; a number of advantages have been demonstrated in polarization insensitivity and low optical insertion loss compared to an electroabsorption modulator/detector (EAMD). The AFPMD has an asymmetric Fabry-Perot cavity with a highly reflective (91 percent) bottom, a partially reflective (35 percent) top, and a multiple quantum well (MQW) structure sandwiched. A portion of downlink optical signal is detected and the reflected is modulated by an uplink RF signal in the MQW region. In a wireless system with FDD or TDD mode, full-duplex operation can be achieved by EAMD and AFPMD.

Recovering a portion of the downlink optical carrier for uplink reuse can avoid the use of the specially designed EAMD or AFPMD. Two configurations incorporating a narrow-band fiber Bragg grating and an optical circulator have been demonstrated [69] for recovering a downlink optical carrier. A simpler scheme using an ultra-low-cost all-fiber optical interleaver (AFOI) was proposed and demonstrated [70]. As shown in Figure 11.13a, at a CO a laser output is weakly modulated by a millimeter-wave downlink carrier at f_m. The modulated optical downlink signal is transmitted to an RAP and coupled into a 2×2 AFOI. The AFOI makes use of cascaded fiber couplers linked with differential delay lines in a forward lattice structure [71]. The FSR of the AFOI is designed to be $2f_m$. At the output port 2 of the AFOI, when a peak of the spectral response at the port is aligned to the optical carrier at f_o, the optical carrier passes while the optical sidebands at $f_o \pm f_m$ are suppressed; thus, the downlink optical carrier is recovered at the output port 2 and can be used for fiber uplink transmission.

In the above scheme, due to the power conservation property of the passive interleaver used, the spectral response at the output port 1 is complementary to the output port 2. Thus, at the output port 1, the optical carrier at f_o is attenuated while the sidebands pass; i.e., the carrier-to-sideband ratio (CSR) is suppressed and the optical modulation depth is improved. This is helpful to improve the link performance, since generally weak modulation is employed in a millimeter-wave RoF link because a high-frequency optoelectronic modulator has limited linear operation region. Figure 11.13b shows the measured optical spectra of a downlink optical signal, a recovered optical carrier, and a CSR-suppressed optical signal, with $f_m = 30$ GHz and ODSB modulation, by using a low-cost and simple AFOI. In RoF links with lower radio frequencies, such as cellular radios (900 MHz), personal communication systems (1.8 GHz), and IEEE 802.11g

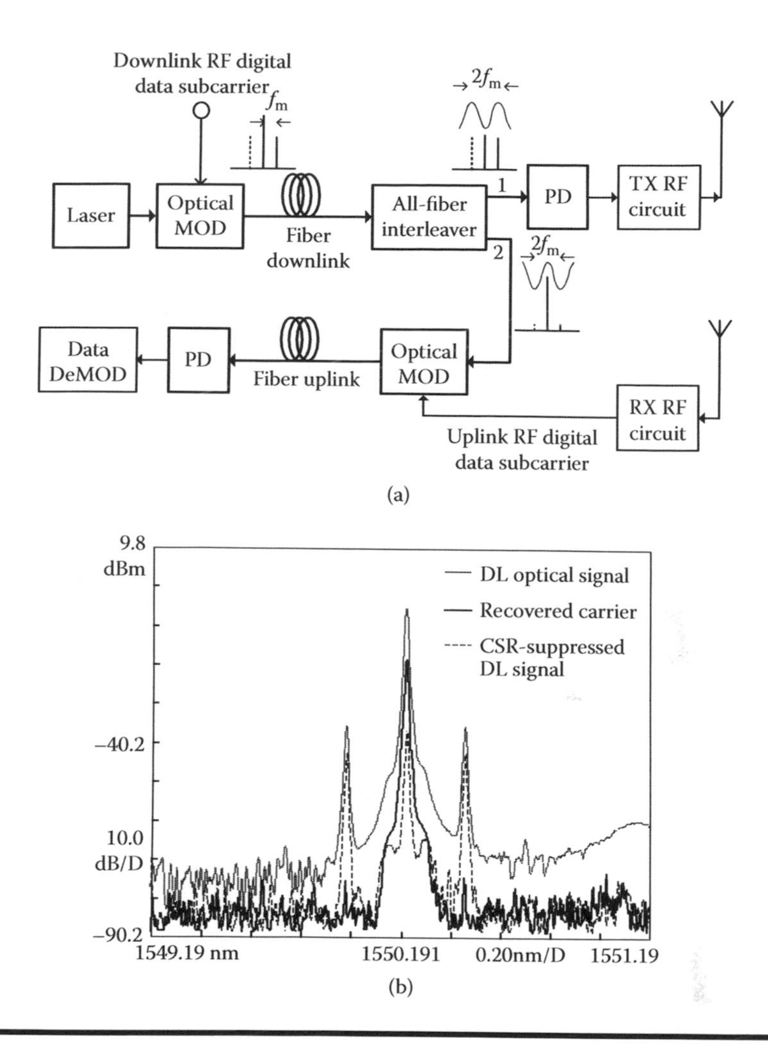

Figure 11.13 Optical carrier recovery using an all-fiber interleaver for full-duplex fiber-wireless transmission: (a) schematic diagram of configuration; (b) measured optical spectra [70].

(2.4 GHz), separating the optical carrier and sidebands with an AFOI is a rather challenging task. Instead, an ultra-narrow optical bandpass filter can be used for CSR suppression [72].

11.4 Multimode Radio-over-Fiber Links

Fiber-optic RF signal transmission is typically considered over single-mode fiber. In recent years, multimode fiber has increasingly been investigated for RF signal transmission as an alternative to SMF links. Its main attraction comes from wide availability in buildings and on campuses, potentially low overall system cost, reuse of existing infrastructure, and higher tolerances due to large fiber core sizes. It is interesting, therefore, to understand the capabilities of multimode fiber (MMF) for transmission of high-frequency signals.

The main limitation of MMF comes from DMD, which describes the relative propagation delay of the mode groups of the multimode fiber [73]. Depending on the fiber design, large variations of this parameter can be observed in practice. There are two main types of MMF: old legacy fibers with a 62.5 μm core size and modern high-bandwidth fibers with 50 μm core size. They are not easily compatible due to the mismatch of the core sizes when interconnected, which potentially leads to high losses. The 62.5-μm fibers were developed for low-data-rate systems using light-emitting diodes (LEDs) and have a specified distance-bandwidth product of 200 MHz×km at 850 nm or 500 MHz×km at 1300 nm. This fiber type is known as OM1 fiber. The distance-bandwidth product of this fiber type is only guaranteed if an LED light source launch into the fiber is used, which is called overfilled mode launch since all mode groups are fully excited. If used with laser sources (as would have to be the case for RF signal transmission), the actual bandwidth may decrease. However, selected fibers can show very high bandwidths that can be exploited for RF signal transmission, but this is not ensured and there are many samples that do not show high bandwidth.

The actual bandwidth depends on the specific fiber core variations, from production processes, which are unknown in practice. Since the only purpose for using this fiber type would be in the reuse of existing infrastructure based on legacy 62.5-μm fiber, the very low guaranteed bandwidth is a serious problem for RF transmission. It typically does not help in practice for a reliable and robust system design that selected fiber samples show higher bandwidths. In most applications the system needs to run over all existing fibers and a worst-case system design is required. For RF signal transmission, especially millimeter-wave signal transmission, 50-μm MMF is a much better choice due to its higher bandwidth. Two types are commonly considered: OM2 fiber, with a guaranteed distance-bandwidth product of 500 MHz×km both at 850 nm and 1300 nm, and OM3 fiber, with a bandwidth larger than 2000 MHz×km at 850 nm. There exist commercial products with minimum bandwidths of even 4500 MHz×km. Such fibers are much more suitable for RF system design. Several representative fiber frequency responses for 62.5-μm and 50-μm MMF, as measured with a network analyzer and an 850-nm VCSEL designed for 10 Gb/s operation, are shown in Figure 11.14. It is evident that 62.5-μm fiber can occasionally have very large bandwidth, but low bandwidth fibers are frequently encountered.

The actual DMD to consider for signal transmission strongly depends on the launch condition of the signal into the fiber; i.e., the mode excitation profile at launch and mode-coupling effects in the fiber at discontinuities

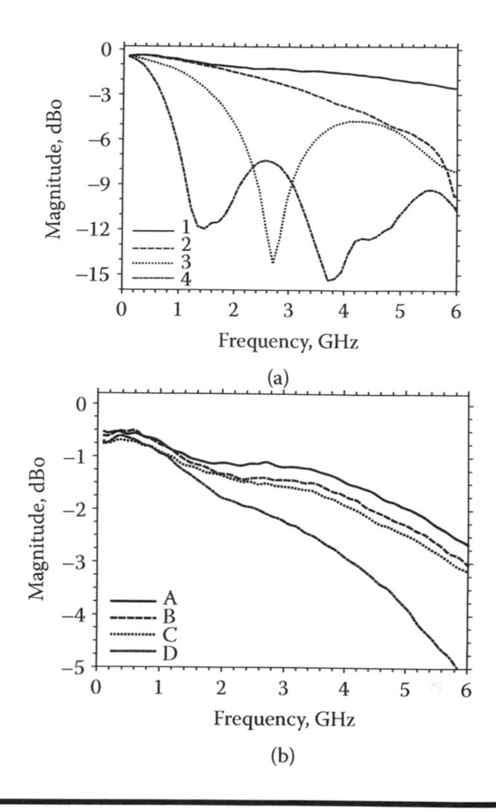

Figure 11.14 Frequency response of several representative 300-m MMFs: (a) 62.5-μm fibers, (b) 50-μm fibers.

(e.g., connectors). While the launch condition can be controlled or manipulated by specially designed launch techniques, the mode coupling along the fiber path is almost impossible to control. Several effects play a role: tolerance variations at connectors, pressure points along the fiber cable (e.g., tight bends of the deployed fiber), temperature variations, mechanical vibrations caused (e.g., by fans), and human interaction from fiber handling, such as touching fibers in large patch panels. Such effects lead to uncontrolled mode coupling and will affect the frequency response of the specific MMF link. Only the minimum bandwidth of the specific fiber type is a reliable parameter and can be used for system designs that need to show robust performance under practical deployment scenarios.

As explained above, multimode fiber is typically described by its distance-bandwidth product, of which the exact definition depends on the intended application and fiber type. Corning Inc. [74] provides typically used definitions for all common MMF types. For fibers for RF signal transmission, the so-called laser launch is relevant and typically only a few mode groups are excited and used for signal transmission. As described, mode-selective launch can increase the usable bandwidth of MMFs enormously but potentially has stability and repeatability issues if device tolerances are taken into account.

It is further known that MMF has bandpass characteristics beyond the –3 dB bandwidth of the fiber link. It has been proposed and demonstrated [75] that this passband frequency region above the –3 dB bandwidth point can be used for signal transmission; therefore, significantly increasing the fiber transmission capability at higher frequencies. However, the exact frequency response of an MMF is impossible to predict, and in many cases significant transmission nulls are observed (see Figure 11.14a). If the signal frequency falls on these low-power regions, the transmission loss would be very high. These bandpass regions and transmission nulls in the frequency response of the MMF are caused by constructive or destructive interference of a few modes that originally were excited by mode-selective laser launch. Thus, uncontrolled mode coupling at discontinuities, as discussed above, will lead to variations of this performance. This bandpass region, therefore, is not stable in its response, and care needs to be taken if it is to be exploited for RF signal transmission. While several experiments have shown that signals can be transmitted beyond the –3 dB bandwidth, these studies were typically performed under laboratory conditions, which are more stable and controlled than actual field deployments with unknown specifics of the particular installed fibers. Although digital transmission systems use wide bandwidth signals and frequency-selective transmission nulls can be overcome by signal processing, the inherently band-limited nature of RF signal transmission makes it impossible to rely on signal processing techniques for recovering a strongly attenuated signal. If the frequency of the RF signal falls into a frequency fading dip of the fiber frequency response, the signal will suffer from the high loss. This high attenuation may last only for a certain time period since variations in mode coupling eventually will lead to constructive interference and lower transmission loss, but these processes are relatively slow in nature. It therefore is suggested and demonstrated in the experiments described below to rely on the low-pass region of the MMF link for RF signal transmission.

An important question regarding the usage of MMF for RF transmission is the reliability of the channel; i.e, a guaranteed bandwidth for an intended transmission distance. Since MMF performance (distance-bandwidth product) typically varies widely, it is important to understand the minimum guaranteed bandwidth. A question that often is discussed is the scaling of the distance-bandwidth product. Typically, investigations are performed at a certain transmission length, but the results need to be scaled to another length. The scaling of MMF has been investigated for multiple fiber samples [76,77], and although there is some significant variability in the results, the

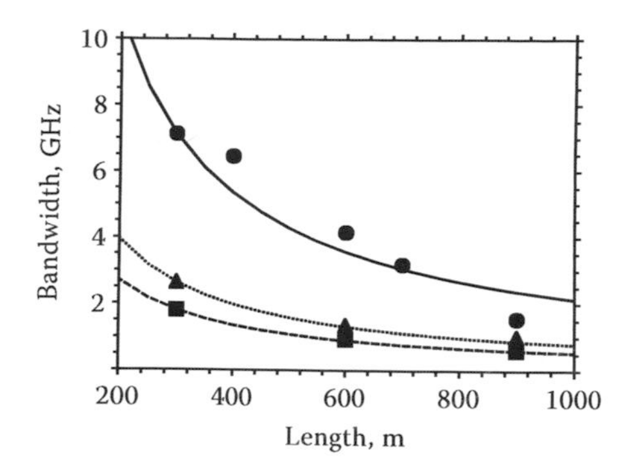

Figure 11.15 Bandwidth scaling with fiber length.

overall conclusion is that Z^{-1} scaling applies. Figure 11.15 shows some representative measurement results. Therefore, the simple formula that relates bandwidth BW with the MMF's length Z

$$BW(Z) = BW_m \frac{Z_m}{Z} \tag{11.27}$$

applies, where BW_m is the bandwidth measured for an MMF of the length Z_m. Equation (11.27) can also be used for calculating an expected maximum transmission distance for a certain frequency and a given fiber type. The deviation from the Z^{-1} dependence in practical realizations is caused by slight variations in the signal launch for the different samples.

Another investigation centered on the bandwidth stability for launch from either side of the fiber [77]. Here, the frequency response of an MMF was measured with a network analyzer and for signal launch from either side into the fiber. The same transmission laser and receiver were used in order to ensure equal launch conditions. The –3 dB frequency is reported. This experiment proves that the fiber core design is uniform over the fiber length of a few hundred meters (up to 900 m in the case under investigation), which are typical transmission distances for MMF systems. The results of this investigation are shown in Figure 11.16.

Further, a study on the reliability of the RF link on connector offsets was conducted [27]. For this experiment, a fully modulated WLAN signal according to the IEEE 802.11g standard at 2.4 GHz was transmitted over MMF. After detection, the error vector magnitude (EVM) was measured with a vector signal analyzer (VSA). For ideal alignment of two MMF cores (representing center alignment of connectors), an EVM of 2.4 percent rms was measured. Then the two fiber cores were misaligned relative to each other in a controlled way by using a splicer (Figure 11.17a) and the EVM was reported for all offsets measured. The measured performance is depicted

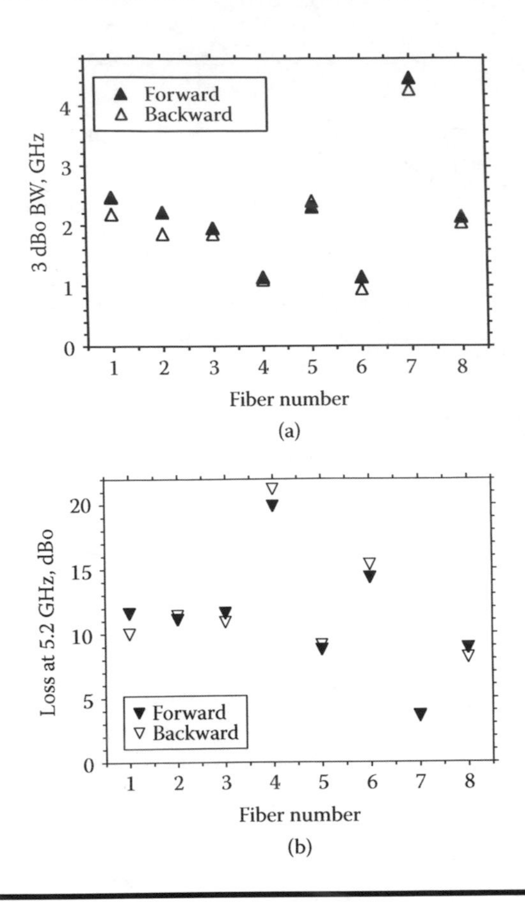

Figure 11.16 MMF bandwidth (a) and RF loss (b) for signal launch from both ends of the fiber.

in Figure 11.17b, and it can be seen that ±3 μm offset can be tolerated with minor EVM degradation but that larger offset values rapidly lead to significantly degraded performance.

Also, the stability of the RF signal quality after MMF transmission was investigated [27,77] by measuring the EVM of the received signal for WLAN after transmission over several MMFs with different bandwidths. These experiments were conducted both at 2.4 GHz and 5.2 GHz with WLAN signals according to the IEEE 802.11g standard. Twenty fiber samples of both types were used: 62.5 μm and 50 μm. These samples were selected to show a wide range of bandwidths and all had length of 300 m, which is a typical maximum in-building distance. Figure 11.18 shows the EVM measured after transmission for the investigation at 5.2 GHz for all samples versus the –3 dB bandwidth (Figure 11.18a) and RF loss at 5.2 GHz (Figure 11.18b). It is clear that all fibers that have a sufficiently high distance-bandwidth product leading to a link bandwidth of >5 GHz show only minor degradation relative to

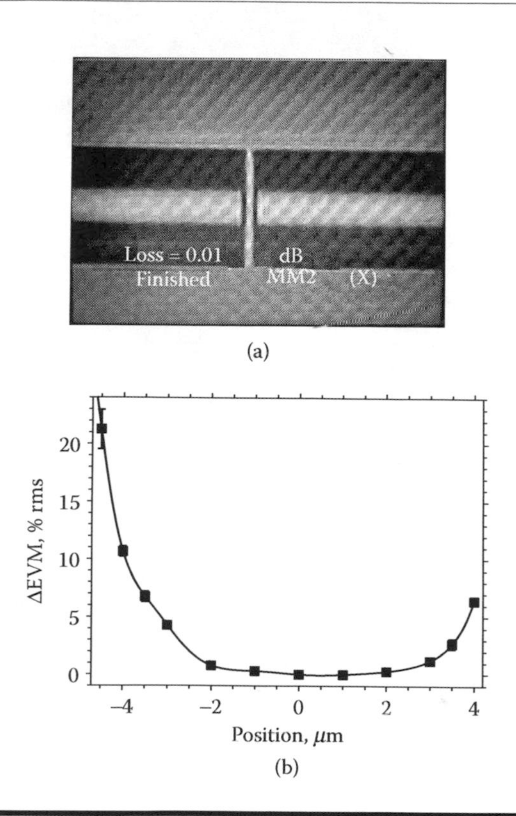

Figure 11.17 (a) Offset launch micrograph. (b) The measured increase in the EVM vs. launch position.

the back-to-back measurement. Lower bandwidth links sometimes lead to low EVM, which would represent cases where the bandpass characteristics of MMF could be successfully exploited, but many samples show unacceptably poor performance. Therefore, it is concluded that the use of MMF links with bandwidths lower than the RF carrier frequency leads to unpredictable performance and should not be considered for field applications.

The capability for RF signal transmission at a longer wavelength and specifically at 1300 nm was also investigated [78]. WLAN radio transmission at 2.4 GHz and in a 5-GHz band over a record distance of 1.1 km multimode and > 30 km over standard single-mode fibers was demonstrated with low EVM, < 4.9 percent rms, using a 1310-nm high-speed single-mode AlGaInAs/InP VCSEL. This experiment demonstrates that with newly emerging low-cost, long-wavelength laser technologies, RF signal transmission over meaningful distances can be achieved also at this wavelength and MMF could be an interesting alternative to single-mode fiber for short-distance links.

To achieve millimeter-wave frequency transmission, direct modulation of lasers typically cannot be considered. An often-applied solution is the

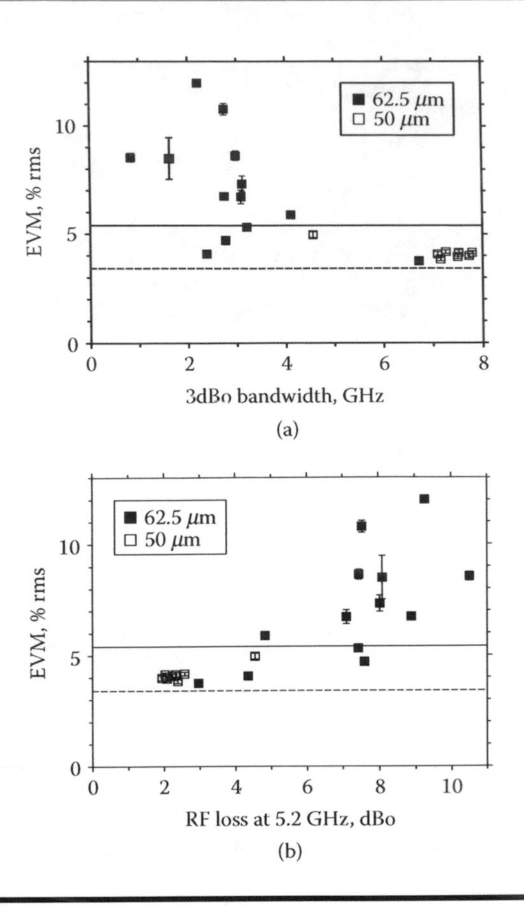

Figure 11.18 EVM measurements of 21 MMF samples at 300 m. EVM vs. (a) measured 3-dB bandwidth and (b) RF loss at 5.2 GHz.

use of external modulators. This technique can also be used for MMF. At 1300 nm, modulator technologies based on $LiNbO_3$ can be readily applied. However, all other techniques for millimeter-wave signal generation can be used for MMF transmission. High-bandwidth photodetectors for MMF also are feasible and are available on the market. If the generation technology requires single-mode fiber (as would be the case for $LiNbO_3$ modulators), the signal can be readily launched into MMF with no additional loss, and even controlled selective mode launch can then easily be applied. An interesting approach for high-frequency signal generation and MMF transmission has been demonstrated [79]. There, a frequency multiplying technique was developed that relies on single-mode fiber technology for signal generation. For the transmission part, however, MMF could be readily used. As an example, if one considers modern MMF with >4500 MHz×km, a 38-GHz millimeter-wave signal could be reliably transported over more than 100 m, which would be enough distance for many in-building or local system applications.

In summary, the distance-bandwidth product governs the MMF transmission distance that can be achieved. Today, MMF with a specified distance-bandwidth product of >4500 MHz×km is commercially available. Selected fiber samples can achieve even higher values, or mode-selective launch may be used to improve this value. With such fiber, even millimeter-wave signal transmission becomes feasible over short distances in the range of hundreds of meters. Depending on the application, MMF may be a cost-effective and robust alternative to single-mode fiber. As for millimeter-wave signal generation, all external modulator techniques or other optical frequency generation techniques can be used and single-mode fiber launch into MMF can be advantageously deployed for mode-selective launch.

11.5 Conclusion

This chapter presented concepts, limitations, and key technologies of optical generation and transmission of millimeter-wave signals. Both harmonic generation using the Mach-Zehnder modulator and optical phase-locking techniques were introduced. Signal degradations due to fiber CD and PMD in single-mode fiber millimeter-wave links as well as nonlinearity were presented. As a solution to fiber CD effect, OSSB modulation was covered in detail. DWDM technologies to increase the system capacity and simplify the system structure were also introduced. Finally, MMF links for transmission of both RF signals and millimeter waves were discussed.

References

[1] Westbrook L., and Moodie D.G., "Simultaneous bi-directional analogue fiber-optic transmission using an electroabsorption modulator," *Electronics Letters*, Vol. 32, No. 19, pp. 1806–1807, Sept. 1996.

[2] Griffin R.A., Lane P.M., and O'Reilly J.J., "Radio over fiber distribution using an optical millimeter wave/DWDM overlay," *OFC/IOOC'99 Technical Digest*, San Diego, CA, paper WD6, 1999.

[3] O'Reilly J.J., and Lane P.M., Fiber supported optical generation and delivery of 60 GHz signals, *Electronics Letters*, Vol. 30, No. 16, pp. 1329–1330, Aug. 1994.

[4] Griffin R.A., and Kitayama K., "Optical millimeter-wave generation with high spectral purity using feed forward optical field modulation," *Electronics Letters*, Vol. 34, No. 8, pp. 795–796, April 1998.

[5] Braun R., Grobkopf G., and Rhode D., "Optical millimeter-wave generation and transmission technologies for mobile communications, an overview," *IEEE Microwave Systems Conference (NTC'95)*, 1995, pp. 239–242, May 1995.

[6] Johansson L.A., and Seeds A.J., "Millimeter-wave modulated optical signal generation with high spectral purity and wide-locking bandwidth using a fiber integrated optical injection phase-lock loop," *IEEE Photonics Technology Letters*, Vol. 12, No. 6, pp. 690–692, June 2000.

[7] Qi G., Yao J., Seregelyi J., Paquet S., and Belisle C., "Optical generation and distribution of continuously tunable millimeter-wave signals using an optical phase modulator," *OSA/IEEE Journal of Lightwave Technology*, Vol. 23, No. 9, pp. 2687–2695, Sept. 2005.

[8] Yu J., Jia Z., Xu L., Chen C., Wang T., and Chang G.-K., "DWDM optical millimeter-wave generation for radio-over-fiber using an optical phase modulator and an optical interleaver," *IEEE Photonics Technology Letters*, Vol. 18, No. 13, pp. 1418–1420, July 2006.

[9] Lai Y., Zhang W., Williams J.A.R., and Bennion I., "An optical millimeter wave fiber laser," *Optical Fiber Communications Conference 2003, OFC 2003*, pp. 238–239, 23–28 March 2003.

[10] Agrawal G.P., *Fiber-Optic Communication Systems*, 2nd edition, John Wiley & Sons, New York p. 246, 1997.

[11] Agrawal G.P., *Fiber-Optic Communication Systems*, 3rd edition, Wiley, New York, 2002.

[12] Smith G.H., Novak D., and Ahmed Z., "Overcoming chromatic dispersion effects in fiber-wireless systems incorporating external modulators," *IEEE Transactions on Microwave Theory and Techniques*, Vol. 45, No. 8, pp. 1410–1415, Aug. 1997.

[13] Kurniawan T., Nirmalathas A., Lim C., Novak D., and Waterhouse R., "Performance analysis of optimized millimeter-wave fiber radio links," *IEEE Transactions on Microwave Theory and Techniques*, Vol. 54, No. 2, Part 2, pp. 921–928, Feb. 2006.

[14] O'Reilly J.J., Lane P.M., Heidemann R., and Hofstetter R., "Optical generation of very narrow linewidth millimeter wave signals," *Electronics Letters*, Vol. 28, No. 25, pp. 2309–2311, Dec. 1992.

[15] Schmuck H., Heidemann R., and Hofstetter R., "Distribution of 60 GHz signals to more than 1000 base stations," *Electronics Letters*, Vol. 30, No. 1, pp. 59–60, Jan. 1994.

[16] O'Reilly J.J., et al., "RACE R2005: Microwave optical duplex antenna link," *IEEE Proceedings Journal of Optoelectronics*, Vol. 140, No. 6, pp. 385–391, Dec. 1993.

[17] Vergnol E., Cadiou J.F., Carenco A., and Kazmierski C., "New modulation scheme for integrated single side band lightwave source allowing fiber transport up to 256 QAM over 38 GHz carrier," *Proceedings OFC*, Baltimore, MD, Paper FH3-1, 2000.

[18] Sauer M., Kojucharow K., Kaluzni H., Sommer D., and Nowak W., "Simultaneous electro-optical upconversion to 60 GHz of uncoded OFDM signals," *International Topical Meeting on Microwave Photonics 1998*, Princeton, NJ, pp. 219–222, 1998.

[19] Wake D., Lima C. R., and Davies P. A. "Transmission of 60 GHz signals over 100Km of optical fiber using a dual mode semiconductor laser source," *IEEE Photonics Technology Letters*, Vol. 8, No. 4, pp. 578–580, April 1996.

[20] Tun T.S., et al., "Calibration of optical receivers and modulators using an optical heterodyne technique," *Proceedings of IEEE MTT-S*, pp. 1067–1070, 1988.

[21] Kawanishi S., et al., "Wideband frequency measurement of optical receivers using optical heterodyne detection," *IEEE Journal of Lightwave Technology*, Vol. 7, No. 1, pp. 1242–1243, Jan. 1989.

[22] Braun R.P., Grosskopf G., Rohde D., and Schmidt F., "Optical millimeter-wave generation and transmission experiments for mobile 60 GHz band communications," *Electronics Letters*, Vol. 32, No. 7, pp. 626–628, March 1996.

[23] Noel L., Marcenac D., and Wake D., "20 Mbit/s QPSK radio fiber transmission over 100 Km of standard fiber at 60 GHz using a master slave injection locked DFB laser source," *Electronics Letters*, Vol. 32, No. 20, pp. 1895–1897, Sept. 1996.

[24] Ramos R.-P., Grosskopf G., Rodhe D., and Schmidt F., "Low-phase-noise millimeter-wave generation at 64 GHz and data transmission using optical sideband injection locking," *IEEE Photonics Technology Letters*, Vol. 10, No. 5, pp. 728–730, May 1998.

[25] Weisser S., et al., "Dry-etched short-cavity ridge-waveguide MQW lasers suitable for monolithic integration with direct modulation bandwidth up to 33 GHz and low drive currents," *Proceedings of ECOC 1994*, pp. 973–976, 1994.

[26] Cox C. H., III, *Analog Optical Links — Theory and Practice*, Cambridge University Press, Cambridge, UK, 2004.

[27] Sauer M., Kobyakov A., Herley J.E., and George J., "Experimental study of radio frequency transmission over standard and high-bandwidth multimode optical fibers," *Proceeding of International Topical Meeting on Microwave Photonics 2005 (MWP2005)*, Seoul, Korea, pp. 99–102, 2005.

[28] Yee M.L., Chung H.L., Tang P.K., Ong L.C., Luo B., Zhou M.-T., Shao Z., and Fujise M., "850 nm radio-over-fiber EVM measurements for IEEE 802.11g WLAN and cellular signal distribution," *European Microwave Conference*, Sept. 10–14, 2006, Manchester, UK, pp. 882–885.

[29] Choi S.T., Yang K.S., Shimizu S., Tokuda K., and Kim Y.H., "A 60-GHz point-to-multipoint millimeter-wave fiber-radio communication system," *IEEE Transactions on Microwave Theory and Techniques*, Vol. 54, No. 5, pp. 1953–1960, May 2006.

[30] Cartledge J. C., and Christensen B., "Optimum operating points for electroabsorption modulators in 10 Gb/s transmission systems using nondispersion shifted fiber," *Journal of Lightwave Technology*, Vol. 16, No. 3, pp. 349–357, March 1998.

[31] Cox C. H., III, Ackerman E. I., Betts G. E., and Prince J. L., "Limits on the performance of RF-over-fiber links and their impact on device design," *IEEE Transactions on Microwave Theory and Techniques*, Vol. 54, No. 2, pp. 906–920, Feb. 2006.

[32] Fernando X. N., and Sesay A. B., "Adaptive asymmetric linearization of microwave fiber optic links for wireless access,"*IEEE Transactions on Vehicular Technology*, Vol. 51, No. 6, pp. 1576–1596, Nov. 2002.

[33] Fernando X. N., and Anpalagan A., "On the design of optical fiber based wireless access systems," *Proceedings of the International Conference on Communications, ICC2004*, pp. 3550–3555, Paris, June, 2004.

[34] Microwave Fiber Optics Group, "A system designer's guide to RF and microwave fiber optics," *Technical Report*, Ortel Corporation, 1999.

[35] Hofstetter R., Schmuck H., and Heidemann R., "Dispersion effects in optical millimeter-wave system using self-heterodyne method for transport and generation," *IEEE Transactions on Microwave Theory and Techniques*, Vol. 43, No. 9, pp. 2263–2269, Sept. 1995.

[36] Wake D., Lima C. R., and Davies P. A., "Optical generation of millimeter-wave signals for fiber-radio systems using a dual-mode DFB semiconductor laser," *IEEE Transactions on Microwave Theory and Techniques*, Vol. 43, No. 9, pp. 2270–2278, Sept. 1995.

[37] Hori T., Park K.-H., Kawanishi T., and Izutsu M., "Generation of CW millimeter wave signals in a lithium niobate nonlinear optical waveguide using modulated optical input," *Japanese Journal of Applied Physics*, Vol. 39, pp. L667–L669, 2000.

[38] Gliese U., Nørskov S., and Nielsen T.N., "Chromatic dispersion in fiber-optic microwave and millimeter-wave links," *IEEE Transactions on Microwave and Techniques*, Vol. 44, No. 10, pp. 1716–1724, Oct. 1996.

[39] Kitayama K.I., "Ultimate performance of optical DSB signal-based millimeter-wave fiber-radio system: Effect of laser phase noise," *Journal of Lightwave Technology*, vol. 17, no. 10, 1774–1781, Oct. 1999.

[40] Zhou M.-T., Zhang Y., Xiao S.-Q., and Fujise M., "Signal power degradation due to polarization mismatch in dual-wavelength millimeter-wave fiber-radio systems," *Proceeding of the 6th International Conference on ITS Telecommunications (ITST2006)*, Chengdu, pp. 478–481, 2006.

[41] Agrawal G.P., *Nonlinear Fiber Optics*, 3rd edition, Academic Press, New York, 2001.

[42] Ramos F., Marti J., Polo V., and Fuster J. M., "On the use of fiber-induced self-phase modulation to reduce chromatic dispersion effects in microwave/millimeter-wave optical systems," *IEEE Photonics Technology Letters*, Vol. 10, No. 10, pp. 1483–1485, 1998.

[43] Kobyakov A., Kumar S., Chowdhury D.Q., Ruffin A.B., Sauer M., Bickham S.R., and Mishra R., "Design concept for optical fibers with enhanced SBS threshold," *Optics Express*, Vol. 13, pp. 5338–5346, 2005.

[44] Chen W.H. and Way W. I., "Multichannel single-sideband SCM/DWDM transmission systems," *Journal of Lightwave Technology*, Vol. 22, No. 7, pp. 1679–1693, 2004.

[45] Cheng L., Aditya S., Li Z., and Nirmalathas A., "Nonlinear distortion due to cross-phase modulation in microwave fiber-optic links with optical SSB or electro-optic upconversion," *Transactions on Microwave Theory and Techniques*, IEEE Vol. 55, No. 1, pp. 176–184, Jan 2007.

[46] Cheng L., Aditya S., Li Z., and Nirmalathas A., "Generalized analysis of subcarrier multiplexing in dispersive fiber-optic links using mach-zehnder external modulator," *Journal of Lightwave Technology*, Vol. 24, No. 6, pp. 2296–2304, 2006.

[47] Wu C. and Zhang X., "Impact of nonlinear distortion in radio over fiber systems with single-sideband and tandem single-sideband subcarrier modulations," *Journal of Lightwave Technology*, Vol. 24, No. 5, pp. 2076–2090, 2006.

[48] Kuri T., Kitayama K., Stöhr A., and Ogawa Y., "Fiber-optic millimeter-wave downlink system using 60 GHz-Band external modulation," *Journal of Lightwave Technology*, Vol. 17, No. 5, pp. 799–806, May 1999.

[49] Sotobayashi H., and Kitayama K., "Cancellation of the signal fading for 60 GHz subcarrier multiplexed optical DSB signal transmission in nondispersion shifted fiber using midway optical phase conjugation," *Journal of Lightwave Technology*, Vol. 17, No. 12, pp. 2488–2497, Dec. 1999.

[50] Khosravani R., Lee S., Hayee M.I., and Willner A.E., "Soliton sampling of subcarrier multiplexed signals to suppress dispersion-induced RF power fading," *IEEE Photonics Technology Letter*, Vol. 12, No. 9, pp. 1275–1277, Sept. 2000.

[51] Griffin R.A., Lane P.M., and O'Reilly J.J., "Dispersion-tolerant subcarrier data modulation of optical millimeter-wave signals," *Electronics Letters* Vol. 32, No. 24, pp. 2258–2260, 1996.

[52] Fuster J.M., Marti J., and Corral J. L., "Chromatic dispersion effects in electro-optical upconverted millimeter-wave signals," *Electronics Letters*. Vol. 33, No. 23, pp. 1969–1970, 1997.

[53] Sauer M., and Nowak W., "Simultaneous upconversion of several channels in millimeter-wave subcarrier transmission systems for wireless LANs at 60 GHz," *Microwave Photonics Conference 1997*, paper FR1-1, 1997.

[54] Shen Y., Zhang X., and Chen K., "Optical single sideband modulation of 11-GHz RoF system using stimulated Brillouin scattering," *IEEE Photonics Letters*, Vol. 17, No. 6, pp. 1277–1279, June 2005.

[55] Carlson A.B., *Communication Systems*, 3rd edition, McGraw-Hill, New York, 1986.

[56] Zhou M.-T., Sharma A.B., Shao Z.-H., and Fujise M., "Optical single-sideband modulation at 60 GHz using electro-absorption modulators," *IEEE International Topical Meeting on Microwave Photonics 2005 (MWP2005)*, Seoul, Korea, pp. 121–124, 2005.

[57] Loayssa A., Lim C., Nirmalathas A., and Benito D., "Design and performance of the bidirectional optical single-sideband modulator," *Journal of Lightwave Technology*, Vol. 21, No. 4, pp. 1071–1082, April 2003.

[58] Davies B., and Conradi J., "Hybrid modulator structures for subcarrier and harmonic subcarrier optical single sideband," *IEEE Photonics Technology Letters*, Vol. 10, No. 4, pp. 600–602, April 1998.

[59] Griffin R.A., "DWDM aspects of radio-over-fiber," *IEEE 13th Annual Meeting Lasers and Electro-Optics Society 2000 Annual Meeting (LEOS 2000)*, Vol. 1, pp. 76–77, 13–16 Nov. 2000.

[60] Toda H., Yamashita T., Kitayama K., and Kuri T., "A DWDM millimeter-wave fiber-radio system by optical frequency interleaving for high spectral efficiency," *International Topical Meeting on Microwave Photonics (MWP2001)*, Tu-3.3, Long Beach, CA, pp. 85–88, 2001.

[61] Bakaul M., Nirmalathas A., Lim C., Novak D., and Waterhouse R., "Efficient multiplexing scheme for wavelength-interleaved DWDM millimeter-wave fiber-radio systems," *IEEE Photonics Technology Letters*, Vol. 17, No. 12, pp. 2718–2720, Dec. 2005.

[62] Toda H., Yamashita T., Kuri T., and Kitayama K., "Demultiplexing using an arrayed-waveguide grating for frequency-interleaved DWDM millimeter-wave radio-on-fiber systems," *Journal of Lightwave Technology*, Vol. 21, No. 8, pp. 1735–1741, Aug. 2003.

[63] Forghieri F., Tkach R.W., and Chraplyvy A.R., "WDM systems with unequally spaced channels," *Journal of Lightwave Technology*, Vol. 13, No. 5, pp. 889–897, May 1995.

[64] Zhou M.-T., Sharma A.B., Zhang J.-G., Shao Z.-H., and Fujise M., "Extended method of fast allocation of USC's for millimeter-wave DWDM-ROF transmission over ZDSF," *International Journal of Infrared and Millimeter Waves*, Vol. 26, No. 5, pp. 617–635, May 2005.

[65] Zhou M.-T., Sharma A.B., Zhang J.-G., and Parvez F., "An improved configuration for radio over fiber transmission with remote local-oscillator delivery by using two dual-Mach-Zehnder modulators in parallel," *IEICE Transactions on Fundamentals*, Vol. E86-A, No. 6, pp. 1374–1381, June 2003.

[66] Zhou M.-T., Zhang J.-G., Sharma A.B., Zhang Y., and Fujise M., "Design of millimeter-wave fiber-wireless downlink with remote local-oscillator delivery by using dual-electrode Mach-Zehnder modulators configured for optical single-sideband modulations," *Optics Communications*, Vol. 269, No. 1, pp. 69–75, Jan. 2007.

[67] Wake D., Johansson D., and Moodie D.G., "Passive pico-cell: A new concept in wireless network infrastructure," *Electronics Letters*, Vol. 33, No. 5, pp. 404–406, Feb. 1997.

[68] Liu C.P., Chuang C.H., Karlsson S., Kjebon O., Schatz R., Yu Y., Tsegaye T., Krysa A.B., Roberts J.S., Seeds A.J., "High-speed 1.56-μm multiple quantum well asymmetric Fabry-Perot modulator/detector (AFPMD) for radio-over-fiber applications," *International Topical Meeting on Microwave Photonics 2005 (MWP2005)*, Seoul, Korea, pp. 137–140, Oct. 2005.

[69] Nirmalathas A., Novak D., Lim C., and Waterhouse R. B., "Wavelength reuse in the WDM optical interface of a millimeter-wave fiber-wireless antenna base station," *IEEE Transaction on Microwave Theory and Techniques*, Vol. 49, No. 10, pp. 2006–2012, Oct. 2001.

[70] Zhou M.-T., Wang Q.J., Zhang Y., Zhang Y., Soh Y.C., and Fujise M., "Cost-effective optical modulation depth enhancement and optical carrier recovery in millimeter-wave fiber-wireless links using an all-fiber optical interleaver," *IEEE International Topical Meeting on Microwave Photonics (MWP2006)*, Grenoble, France, P35, Oct. 2006.

[71] Wang Q.J., Zhang Y., and Soh Y.C., "Efficient structure for optical interleavers using superimposed chirped fiber Bragg gratings," *IEEE Photonics Technology Letters*, Vol. 17, No. 2, pp. 387–389, Feb. 2005.

[72] Gu X.J., He Y., Kosek H., and Fernando X., "Transmission efficiency improvement in microwave fiber-optic link using sub-picometer optic

bandpass filter in," *Photonic Applications in Nonlinear Optics, Nanophotonics, and Microwave Photonics*, edited by R. A. Morandotti, H.E. Ruda, and J.P. Yao, SPIE Press, Bellingham, WA, 2005.

[73] Snyder A.W., and Love J.D., *Optical Waveguide Theory*, Chapman and Hall, London, 1983.

[74] Corning Inc., White Paper CO4249, www.corning.com/opticalfiber

[75] Wake D., Dupont S., Lethien C., Vilcot J.-P., and Decoster D., "Radio frequency transmission of 32-QAM signals over multimode fiber for distributed antenna system applications," *Electronics Letters*, Vol. 37, No. 17, pp. 1087–1089, 2001.

[76] Johnson C.P., "Characterizing bandwidth/length uniformity in high-speed data communication multimode optical fiber," *14th Annual Meeting IEEE LEOS*, Vol. 2, pp. 893–894, 2001.

[77] Sauer M., Kobyakov A., Fields L., Annunziata F., Hurley J.E., and George J., "Experimental investigation of multimode fiber bandwidth requirements for 5.2 GHz WLAN signal transmission," *Optical Fiber Communication Conference 2006*, paper JThB27, 2006.

[78] Kobyakov A., Sauer M., Nishiyama N., Chamarti A., Annunziata F., Hurley J.E., Caneau C., George J., and Zah C.-E., "802.11a/g WLAN radio transmission at 1.3 μm over 1.1 km multimode and >30 km standard single-mode fiber using InP VCSEL," *European Conference on Optical Communications 2006*.

[79] Ng'oma A., Koonen A.M.J., Tafur-Monroy I., Boom H.P.A.vd., Smulders P.F.M., and Khoe G.D., "Optical frequency up-conversion in multimode and single-mode fiber radio systems," *Proceedings of SPIE*, Vol. 5466, 169–177, 2004.

Index

M

N

O

P

R

S

T

U

V

W